Reductions in Organic Chemistry

Reductions
in
Organic Chemistry

Second Edition

Miloš Hudlický
Virginia Polytechnic Institute and State University

ACS Monograph 188

American Chemical Society
Washington, DC

Library of Congress Cataloging-in-Publication Data

Hudlicky, Milos, 1919–
     Reductions in organic chemistry / Miloš Hudlický.—2nd ed.
         p. cm.—(ACS monograph : 188)
     Includes bibliographical references (p.   –   ) and indexes.
     ISBN 0–8412–3344–6
     1. Reduction (Chemistry) 2. Organic compounds—Synthesis.
     I. Title. II. Series.

     QD281.R4H83   1996
     547´.23—dc20   96–26758
                                                                         CIP

This book was produced from camera-ready copy provided by the author.

PRINTED IN THE UNITED STATES OF AMERICA

Sic ait, et dicto citius tumida aequora placat
collectasque fugat nubes solemque REDUCIT.

Publius Vergilius Maro

All springs REDUCE their currents to mine eyes.
William Shakespeare, *The Tragedy of Richard the Third*

Foreword

ACS MONOGRAPH SERIES was started by arrangement with the interallied Conference of Pure and Applied Chemistry, which met in London and Brussels in July 1919, when the American Chemical Society undertook the production and publication of Scientific and Technological Monographs on chemical subjects. At the same time it was agreed that the National Research Council, in cooperation with the American Chemical Society and the American Physical Society, should undertake the production and publication of Critical Tables of Chemical and Physical Constants. The American Chemical Society and the National Research Council mutually agreed to care for these two fields of chemical progress.

The Council of the American Chemical Society, acting through its Committee on National Policy, appointed editors and associates to select authors of competent authority in their respective fields and to consider critically the manuscripts submitted. The first Monograph appeared in 1921. Since 1944, the Scientific and Technological Monographs have been combined in the series.

These Monographs are intended to serve two principal purposes: first, to make available to chemists a thorough treatment of a selected area in a form usable by persons working in more or less unrelated fields so that they may correlate their own work with a larger area of physical science; and second, to stimulate further research in the specific field treated. To implement this purpose, the authors of Monographs give extended references to the literature.

About the Author

MILOŠ HUDLICKÝ, who is a native of Czechoslovakia, obtained his Ph.D. from the Technical University in Prague, Czechoslovakia. After spending the year 1948 at the Ohio State University as a UNESCO postdoctoral fellow, he taught as an assistant professor and, later, as an associate professor at the Technical University in Prague until 1958. He then worked as a research associate in the Research Institute of Pharmacy and Biochemistry in Prague. After the Russian occupation of Czechoslovakia in 1968, he moved to the United States, where he was offered a professorship at Virginia Polytechnic Institute and State University.

His field of interest is organofluorine chemistry. He has written 68 research papers, 19 review papers, 29 patents, and 16 books (8 in English). His chefs-d'oeuvre are *Chemistry of Organic Fluorine Compounds, Reductions in Organic Chemistry,* and *Oxidations in Organic Chemistry,* published in 1976, 1984, and 1990, respectively.

Contents

Categories of Reductions

Reduction of Specific Types of Organic Compounds

Procedures

xiii

Preface to the First Edition

Thirty years after my book *Reduction and Oxidation* was first published in Czech (1953) I am presenting this volume as an entirely new reference work.

Myriads of reductions reported in the literature have been surveyed in countless review articles and a respectable row of monographs, most of which are listed in the bibliography at the end of this book. With the exception of two volumes on reduction in Houben-Weyl's *Methoden der Organischen Chemie*, the majority of the monographs deal mainly with catalytic hydrogenation, reductions with hydrides, and reductions with metals.

This book encompasses indiscriminately all the types of reductions and superimposes them over a matrix of types of compounds to be reduced. The manner of arrangement of the compounds to be reduced is a somewhat modified Beilstein system and is explained in the introduction. Numerous tables summarize reducing agents and correlate them with the starting compounds and products of the reductions. Reaction conditions and yields of reductions are mentioned briefly in the text and demonstrated in 175 examples of reductions of simple types of compounds and in 50 experimental procedures.

The material for the book has been collected from original papers (till the end of 1982) after screening *Organic Syntheses*, Theilheimer's *Synthetic Methods*, Harrison and Harrison's *Compendium of Synthetic Methods*, Fieser and Fieser's *Reagents for Organic Synthesis*, and my own records of more or less systematic scanning of main organic chemistry journals. The book is far from being exhaustive, but it represents a critical selection of methods which, in my view and according to my experience, give the best results based on yields. In including a reaction preference was given to simple rather than complicated compounds, to most completely described reaction conditions, to methodical studies rather than isolated examples, to synthetically rather than mechanistically oriented papers, and to generally accessible rather than rare journals. Patent literature was essentially omitted. Consequently many excellent chemical contributions could not be included, for which I apologize to their authors. Unfortunately this was the only way in which to turn in not an encyclopedia but rather a "Pocket Dictionary of Reductions".

I would like to thank Ms. Tammy L. Henderson for secretarial help and Ms. Melba Amos for the superb drawings of 175 schemes and equations. To my colleague Dr. H. M. Bell I owe thanks for reading some of the procedures, and to my son Dr. Tomáš Hudlický I am grateful for reading my manuscript and for his critical remarks. My special gratitude is due to my wife Alena for her patience in "holding the fort" while I was spending evenings and weekends immersed in work, for her efficient help in proofreading and indexing, and for her releasing me from my commitment never to write another book.

Finally, I would like to express my appreciation of the excellent editorial and graphic work involved in the publishing of my book. For them, my thanks are extended to Ellis Horwood and his publishing team.

Miloš Hudlický
Blacksburg, Virginia

March 16, 1983

Preface to the Second Edition

There is very little to be added to the Preface to the First Edition. Because of favorable reviews of the book, the scope and organization of the book remain the same. The parts involving diastereoselective and enantioselective reductions have been expanded because the number of chiral catalysts in homogeneous catalytic hydrogenation and of chiral hydrides has increased considerably over the past 10 years (the literature is included through the end of 1993). Special sections on these two topics and on biochemical reductions have been added, and many examples throughout the book document the importance of these methods in the modern organic synthesis.

I acknowledge the assistance of my wife, Alena, in helping me with proofreading and in maintaining a milieu favorable for my writing. I also appreciate critical comments of Professor Stanley M. Roberts of the University of Exeter, England, and of my son, Professor Tomáš Hudlický of the University of Florida in Gainesville. I am grateful to Professor Emeritus Miloslav Ferles of the Institute of Chemistry and Technology in Prague, Czech Republic, for his thorough review of my manuscript.

My thanks are due to Angie Miller and Wanda Ritter for their meticulous typing of the text, tables, and especially formulae, and to the editors of the American Chemical Society Books Department, Janet S. Dodd and Paula M. Bérard, for a thorough editorial job and the production of the book.

Miloš Hudlický
Blacksburg, VA 24061–0212

November 17, 1995

Introduction

In the development of reductions in organic chemistry, zinc, iron, and hydrogen sulfide are among the oldest reducing agents, having been used since the 1840s. Two discoveries mark the most important milestones: catalytic hydrogenation (1897) and reduction with metal hydrides (1947). Each accounts for about one-fourth of all reductions, the remaining half of the reductions being due to electroreductions and reductions with metals, metal salts, and inorganic as well as organic compounds.

The astronomical number of reductions of organic compounds described in the literature makes an exhaustive survey impossible. The most complete treatment is published in Volumes 4/lc and 4/ld of the Houben-Weyl compendium *Methoden der Organischen Chemie* and in Volume 8 of Barry M. Trost and Ian Fleming's *Comprehensive Organic Synthesis*. Other monographs dealing mainly with sections of this topic are listed in the bibliography.

Most of the reviews on reductions deal with a particular method, such as catalytic hydrogenation, or with particular reducing agents, such as complex hydrides and metals. This book gives a kind of a cross section. Reductions are discussed according to what bond or functional group is reduced by different reagents. Special attention is paid to selective reductions that are suitable for the reduction of one particular type of bond or function without affecting another present in the same molecule. Where appropriate, stereochemistry of the reactions is mentioned.

The criteria considered for selecting a reaction are the quality of the description of the reduction and the yields of the products. Where indicated in the original paper, isolated yields are quoted. Gas-liquid chromatography and nuclear magnetic resonance (NMR) spectroscopy contributed to better identification of the products and even to correction of some results. On the other hand, yields based only on these two methods may be misleading, as they may differ from the yields of the products after their isolation.

Types of compounds are arranged according to the following system: hydrocarbons and basic heterocycles; hydroxy compounds and their ethers; mercapto compounds, sulfides, disulfides, sulfoxides and sulfones, sulfenic, sulfinic, and sulfonic acids and their derivatives; amines, hydroxylamines, hydrazines, and hydrazo and azo compounds; carbonyl compounds and their functional derivatives; carboxylic acids and their functional derivatives; and organometallic compounds. In each chapter, halogen, nitroso, nitro, diazo, and azido compounds follow the parent compounds as their substitution derivatives. More details are indicated in the table of contents. In polyfunctional derivatives, reduction of a particular function is most of the time mentioned in the place of the highest functionality. Reduction of acrylic acid, for example, is described in the chapter on acids rather than functionalized ethylene, and reduction of ethyl acetoacetate is discussed in the chapter on esters rather than in the chapter on ketones.

Systematic description of reductions of bonds and functions is preceded by discussion of methods, mechanisms, stereochemistry, and scopes of reducing agents. Correlation tables show what reagents are suitable for conversion of individual types of compounds to their reduction products. More detailed reaction conditions are indicated in schemes and equations, and selected laboratory procedures demonstrate the main reduction techniques.

Safety First

In addition to general rules and regulations for the work in the laboratory, very thoroughly described in the publication *Prudent Practices Book*, some reductions require special attention.

In *catalytic hydrogenation* using elemental hydrogen, special precautions must be taken to prevent potential explosions and/or fires.

Before any work with hydrogen tanks or lecture bottles all connections, metal or glass, must be checked for leakage. If enough hydrogen (more than 4%) is present in the laboratory, an explosion could occur on switching on any regular electrical switch or if scintillating motors are running in the laboratory.

In a *low-pressure hydrogenation* carried out in glass equipment, after the apparatus shown on pp 4 and 293 has been filled with hydrogen and tested for leakage, the hydrogenation flask A is disconnected. Before adding a catalyst, the compound to be hydrogenated, and a solvent, the flask must be flushed with an inert gas such as nitrogen or argon to remove any residual hydrogen present in the flask. If the catalyst were added immediately after the disconnection of the flask with some hydrogen still in it, a small explosion could take place, because noble metal and Raney nickel catalysts cause igniting of the hydrogen–air mixture.

The catalyst should be placed in the flushed flask first, followed by the compound to be hydrogenated or its solution, best by using a pipet or hypodermic syringe. If for any reason the catalyst is to be added after the solution has been in the flask already, for example, if additional catalyst has to be used, the catalyst must be added under the blanket of an inert gas. Otherwise the vapors of volatile solvents and the air could catch fire at the surface of the catalyst. The filled hydrogenation flask is then attached to the apparatus, and the hydrogenation is carried out.

After the reaction is finished, the heterogeneous catalyst must be filtered off. The contents of the disconnected hydrogenation flask are poured over a fluted filter paper for gravity filtration. Suction filtration is somewhat risky and can be carried out only with small amounts of catalyst (less than 1 g) or under a cover of an inert gas. The metals after the reduction are often pyrophoric and can cause fire if the water aspirator that is used for the suction filtration is not disconnected before all the solvent covering the catalyst has passed through the filter. Like many other reactions, low-pressure catalytic hydrogenation and the workup of the products are best carried out in a hood.

In the *high-pressure catalytic hydrogenation* (pp 5 and 297), especially one carried out at pressures of 250–350 atm and high temperatures (250–300 °C), a thorough calculation must precede loading of the autoclave. The initial pressure of hydrogen, charged into the cold autoclave, will rise on heating according to the gas law. Also, total pressure will be increased by pressure caused by evaporation of volatile solvents. As the hydrogenation proceeds, the pressure will decrease, corresponding to the consumption of hydrogen.

Under the high pressure of hydrogen, many solvents may not evaporate, but the heat of hydrogenation (about 30 kcal per double bond) may raise the temperature considerably, especially if large amounts of the compounds are used. At high enough temperatures, the critical temperature of the compound, of the solvent, and of the product could be exceeded; in that event, the liquids gasify and the pressure increases even further. The critical temperature of methanol is 240 °C at 79 atm, and that of ethanol is 243 °C at 63 atm.

Although exact calculations of the pressure during the hydrogenation are difficult, at least an approximate estimation can be made. Furthermore, the nominal volume of the autoclave is diminished by the volume of the compounds in it.

It is advisable to not fill the autoclave to more than one-third of its volume and to use no more material than would generate one-half of the nominal pressure. Maybe these precautions are overdone for most of the hydrogenations, but sometimes explosions happen for entirely unknown reasons, as in the high-pressure hydrogenation of amides using dioxane as the solvent.

Work with hydrides and complex hydrides also requires strict precautions. Even hydrides that were considered safe, such as sodium borohydride, caused fire and explosion (p 28). Hydrides and complex hydrides should not be stored in ground-glass bottles. Twisting of the glass stopper in a bottle of lithium aluminum hydride caused fire and burning out of some 25 g of the material (p 28). All commercially available hydrides and complex hydrides are now accompanied by Material Safety Data Sheets, which describe the conditions under which they can be handled safely.

The work with pyrophoric hydrides, for example, boranes, alanes, and some complex hydrides, must be done under inert atmosphere of nitrogen or argon, and addition of their solutions is best carried out by hypodermic syringes through rubber septa attached to the openings of the reaction flasks. Such reagents are now delivered in special bottles (septa bottles) fitted with closures allowing for withdrawal with hypodermic syringe after injecting the same volume of an inert gas into the bottle to maintain pressure in the bottle the same as it was before the withdrawal.

In *reductions with alkali metals*, two steps require attention. Cutting lithium, best after hammering it into a metal sheet, or cutting sodium can be carried out in the laboratory atmosphere, using a dry knife and dry support material (paper). Eye protection must be used, and hands should be protected by rubber or plastic gloves. Cutting potassium cannot be carried out in this way because it catches fire in contact with most air. It must be cut under hexane, benzene, or other inert solvents.

The second step is handling of residues of these metals, either after their cutting prior to the reduction, or after the reduction, if there are unreacted metals left in the reaction vessels. Discarding these metal residues is best carried out by dissolution in ethanol. Decomposition with water would cause explosions and fires. A hood with the front window pulled almost completely down is the best place for such disposals.

Residues after reductions with zinc dust usually contain unreacted metal because an excess of metal is often used in the reductions. Such residues, especially remaining on a filter paper, must not be discarded into waste paper baskets in the laboratory because the paper could catch fire by heat generated by oxidation of zinc powder.

Reductions by gaseous compounds such as hydrogen sulfide or sulfur dioxide should be done in a hood because these gases, especially hydrogen sulfide, are extremely toxic. A good precaution is to assume that most other gases and volatile liquids are toxic unless proven otherwise.

Abbreviations and Proprietary Names

Alpine-Borane	*B*-isopinocampheyl-9-borabicyclo[3.3.1]nonane
Alpine-Hydride	lithium *B*-isopinocampheyl-9-borabicylo[3.3.1]nonyl hydride
9-BBN	9-borabicyclo[3.3.1]nonane
Binal-H	lithium 1,1'-binaphthyl-2.2'-dihydroxy(ethoxy)-aluminum hydride
Binap	2,2'-bis(diphenylphosphino)-1,1'-binaphthyl
Carbitol	2-(2-ethoxyethoxy)ethanol or diethylene glycol monoethyl ether
Celite	silicon dioxide, diatomaceous earth, infusorial earth, or kieselguhr
Cellosolve	2-ethoxyethanol
Chiraphos	2,3-bis(diphenylphosphino)butane
Dec	decomposition
Dibal-H	diisobutylaluminum hydride
Diglyme	2-methoxyethyl ether, or diethylene glycol dimethyl ether
Disiamylborane	di-*sec*-amylborane, bis(3-methyl-2-butyl)borane
DMF	dimethylformamide
DMSO	dimethyl sulfoxide
Florisil	activated magnesium silicate
HMPA	hexamethylphosphoramide, or hexamethylphosphoric triamide
K-Glucoride	potassium 9-*O*-(1,2:5,6-diisopropylidene-α-*D*-glucofuranosyl)-9-borabicyclo[3.3.1]nonane
K-Selectride	potassium tris(*sec*-butyl)borohydride
L-Selectride	lithium tris(*sec*-butyl)borohydride
LDA	lithium diisopropylamide
Monoglyme	ethylene glycol dimethyl ether or 1,2-dimethoxyethane
NAD$^+$, NADH	nicotinamide adenine dinucleotide, reduced
NADP$^+$, NADPH	nicotinamide adenine dinucleotide phosphate, reduced
NB-Enanthrane, NB-Enanthride	2-(2'-benzyloxy)ethyl-3-(9-borabicyclo[3.3.1]nonyl)-6,6-dimethylbicyclo[3.1.1]heptane
Red-Al, Vitride	sodium bis(2-methoxyethoxy)aluminum hydride
RT	room temperature
Super-hydride	lithium triethylborohydride
Thexylborane	*tert*-hexylborane, 2,3-dimethyl-2-butylborane
THF	tetrahydrofuran
Tos	tosyl, or *p*-toluenesulfonyl
Vitride	Red-Al

Categories of Reductions

CHAPTER 1

Catalytic Hydrogenation

The first catalytic hydrogenation recorded in the literature is reduction of acetylene and ethylene to ethane in the presence of platinum black (*von Wilde,* 1874) [*1*]. However, the widespread use of catalytic hydrogenation did not start until 1897 when *Sabatier* and his co-workers developed the reaction between hydrogen and organic compounds to a universal reduction method (Nobel Prize, 1912) [*2*]. In the original work, hydrogen and vapors of organic compounds were passed at 100–300 °C over copper or nickel catalysts. This method of carrying out the hydrogenation has now been almost completely abandoned, and the only instance of hydrogenation still carried out by passing hydrogen through a solution of a compound to be reduced is the Rosenmund reduction (p 203).

After proceeding through several stages of development, catalytic hydrogenation is now carried out essentially in two ways: low-temperature, low-pressure hydrogenation in glass apparatus at temperatures up to about 100 °C and pressures of 1–4 atm, and high-pressure processes at temperatures of 20–400 °C and pressures of a few to a few hundred (350) atm. The first high-pressure catalytic hydrogenations were carried out over iron and nickel oxide by *Ipatieff* [*3*]. An apparatus for low-pressure hydrogenation usually consists of a glass flask attached to a liquid-filled graduated container connected to a source of hydrogen and to a reservoir filled with a liquid (water or mercury) (Figure 1). The progress of the hydrogenation is followed by measuring the volume of hydrogen used in the reaction (Procedure 1, p 293).

A special self-contained glass apparatus was designed for catalytic hydrogenation using hydrogen evolved by decomposition of sodium borohydride (Figure 2). It consists of two Erlenmeyer flasks connected in tandem. Hydrogen generated in the first flask by decomposition of an aqueous solution of sodium borohydride with acid is introduced into the second flask containing a soluble salt of a catalyst and a compound to be reduced in ethanolic solution. Hydrogen first reduces the salt to the elemental metal that catalyzes the hydrogenation. The hydrogenation can also be accomplished in situ in just one (the first) Erlenmeyer flask, which, in this case, contains ethanolic solution of the catalytic salt to which an ethanolic solution of sodium borohydride is added followed by concentrated hydrochloric or acetic acid and the reactant [*4*] (Procedure 2, p 295).

High-pressure hydrogenation requires rather sophisticated (and expensive) hydrogenation autoclaves that withstand pressures up to about 350 atm and can be heated and rocked to ensure mixing (Figure 3). For medium-pressure, small-scale hydrogenations (in the laboratory, really high-pressure hydrogenations are fairly rare nowadays), a simple apparatus may be assembled from stainless steel cylinders (available in sizes of 30, 75, and 500 mL) attached to a hydrogen tank and a pressure gauge by means of copper or stainless steel tubing and swage-lock valves and unions (Figure 4).

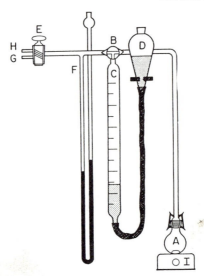

Figure 1. Apparatus for hydrogenation at low pressure. Key: A, ground-glass flask with a magnetic stirring bar; B, three-way stopcock; C, graduated tube; D, reservoir with a liquid; E, three-way stopcock; F, manometer; G, outlet to aspirator; H, outlet to hydrogen source; and I, magnetic stirrer.

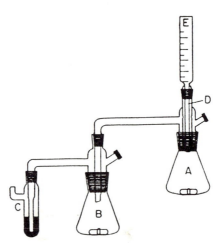

Figure 2. Apparatus for hydrogenation with hydrogen generated from sodium borohydride. Key: A, Erlenmeyer flask with a magnetic stirring bar and an injection port (generator of hydrogen); B, Erlenmeyer flask with a magnetic stirring bar and an injection port (reactor); C, mercury-filled back-suction prevention valve; D, automatic dispenser; and E, delivery buret.

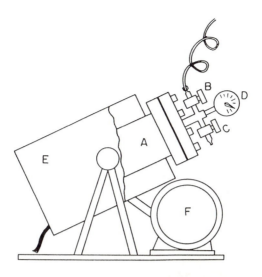

Figure 3. *Apparatus for high-pressure hydrogenation. Key: A, autoclave; B, hydrogen inlet valve; C, pressure release valve; D, pressure gauge; E, heating mantle; and F, rocking device.*

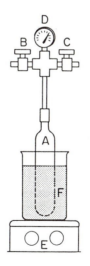

Figure 4. *Apparatus for hydrogenation at medium pressure. Key: A, stainless steel bomb; B, hydrogen inlet valve; C, pressure release valve; D, pressure gauge; E, hot plate–stirrer; and F, oil bath.*

Stirring is provided by magnetic stirring bars, and heating is accomplished by means of oil baths. Such apparatus can withstand pressures up to about 125 atm. The progress of the reaction is controlled by measuring the decrease in the pressure of hydrogen (Procedure 3, p 296).

MECHANISM AND STEREOCHEMISTRY OF CATALYTIC HYDROGENATION

With a negligible number of exceptions, such as reduction of organolithium compounds [5], elemental hydrogen does not react with organic compounds at temperatures below about 480 °C. The reaction between gaseous hydrogen and an organic compound takes place only at the surface of a catalyst that adsorbs both the hydrogen and the organic molecule and thus facilitates their contact. Even under these circumstances, the reaction has an activation energy of some 6.5–16 kcal/mol, as has been measured for all nine transition elements of Group VIII for the reaction between hydrogen and propene [6]. For this particular reaction, the catalytic activities of the nine metals decrease in the sequence shown as follows and do not exactly match the activation energies (E_{act} in kilocalories per mole):

1	Element	Rh	>	Ir	>	Ru	>	Pt	>	Pd	>	Ni	>	Fe	>	Co	>	Os	[6]
	E_{act} (kcal/mol)	13.0		15.0		6.5		16.0		11.0		13.0		10.0		8.1		7.4	

Although all of these elements catalyze hydrogenation, only platinum, palladium, rhodium, ruthenium, and nickel are currently used. In addition, some other elements and compounds were found useful for catalytic hydrogenation: copper (to a very limited extent); oxides of copper and zinc combined with chromium oxide; rhenium heptoxide, heptasulfide, and heptaselenide; and sulfides of cobalt, molybdenum, and tungsten.

At the surface of a catalyst, the organic compounds react with individual atoms of hydrogen that become attached successively through a half-hydrogenated intermediate [7]. Because the attachment takes place at a definite time interval, reactions such as hydrogen–hydrogen (or hydrogen–deuterium) exchange, *cis–trans* isomerism, and even allylic bond shift occur. Nevertheless, the time interval between the attachment of the two hydrogen atoms must be extremely short because catalytic hydrogenation is often stereospecific and usually gives predominantly *cis* products (where feasible) resulting from the approach of hydrogen from the less hindered side of the molecule. However, different and sometimes even contradictory results were obtained over different catalysts and under different conditions [8, 9, 10, 11, 12].

Catalytic hydrogenation is subject to steric effects to the degree that it is faster with less-crowded molecules. The rate of hydrogenation of alkenes decreases with increasing branching at the double bond [13], and so do the heats of hydrogenation.

Electronic effects do not affect catalytic hydrogenation too strongly, although they may play some role: hydrogenation of styrene is about 3 times as fast as that of 1-hexene [14].

2	$CH_2=CHC_3H_7$	$CH_2=CC_2H_5$ $\quad\quad\quad\|$ $\quad\quad\quad CH_3$	$CH_3C=CHCH_3$ $\quad\|$ $\quad CH_3$	$CH_3C=CCH_3$ $\quad\|\quad\quad\|$ $\quad CH_3\ CH_3$	[13]
Relative Rate	1.00	0.62	0.06	0.02	
(kcal/mol)	30.1	28.5	26.9	26.6	

Catalytic hydrogenation giving optically active compounds was accomplished over palladium on poly(β-*S*-aspartate) and on poly(γ-*S*-glutamate) [15]. Good to excellent stereoselectivity was obtained by *homogeneous catalytic hydrogenation* over chiral catalysts, usually of rhodium–phosphine type [16, 17, 18, 19, 20, 21, 22, 23] (p 15).

More general statements about catalytic hydrogenation are difficult to make because the results are affected by many factors, such as the catalyst, its supports, its activators or inhibitors, solvents [24], pH of the medium [24] (*Auwers–Skita rule*: acidic medium favors *cis* products; neutral or alkaline medium favors *trans* products [25]), and to a certain extent temperature and pressure [8].

CATALYSTS

Many catalysts, certainly those most widely used (such as platinum, palladium, rhodium, ruthenium, nickel, and Raney nickel) and catalysts for homogeneous hydrogenation (such as tris(triphenylphosphine)rhodium chloride) are now commercially available. Procedures for the preparation of catalysts are therefore described in detail only in the cases of the less common ones (p 298). Guidelines for use and dosage of catalysts are given in Table 1. More comprehensive correlation of catalytic hydrogenation of functional groups and types of catalysts is published in the Engelhard Chemicals Catalog (p 15).

Platinum

The original forms of platinum referred to in older literature (colloidal platinum or platinum sponge) are hardly ever used. Instead, catalysts of more reproducible activities gained ground. *Platinum oxide* (PtO_2) *(Adams' catalyst)* [26, 27], made by fusing chloroplatinic acid and sodium nitrate, is considered among the most powerful catalysts. It is a brown, stable, nonpyrophoric powder that, in the process of hydrogenation, is first converted to black, very active platinum. Each molecule of this catalyst requires two molecules of hydrogen, and this amount must be subtracted from the final volume of hydrogen used in exact hydrogenations. In most hydrogenations the difference is minute because the amount of platinum oxide necessary for hydrogenation is very small (1–3%). Platinum oxide is suitable for almost all hydrogenations. It is subject to activation by some metal salts, especially stannous chloride and ferric chloride [28] (p 11), and to deactivation by sulfur and other catalytic poisons [29] (p 11). It withstands strong organic and mineral acids and is therefore suited for hydrogenation of aromatic rings and heterocycles.

**Table 1. Recommended Reaction Conditions for Catalytic Hydrogenation
of Selected Types of Compounds**

Starting Compound	Product	Catalyst	Catalyst/Compound Ratio (wt %)	Temp. (°C)	Pressure (atm)
Alkene	alkane	5% Pd (C)	5–10%	25	1–3
		PtO$_2$	0.5–3%	25	1–3
		Raney Ni	30–200%	25	1
			10%	25	50
Alkyne	alkene	4% Pd (C)	10% + quinoline	25	1
		5% Pd(BaSO$_4$)	2% + 2% quinoline	20	1
				25	1
		Lindlar catalyst	10% + 4% quinoline		
	alkane	PtO$_2$	3%	25	1
		Raney Ni	20%	25	1–4
Carbo- cyclic aromatic	hydro- aromatic	PtO$_2$	6–20%, AcOH	25	1–3
		5% Rh(Al$_2$O$_3$)	40–60%	25	1–3
		Raney Ni	10%	75– 100	70–100
Hetero- cyclic aromatic (pyridines)	hydro- aromatic	PtO$_2$	4–7%, AcOH or HCl/MeOH	25	1–4
		5% Rh(C)	20%, HCl/MeOH		
		Raney Ni	2%	65– 200	130
Aldehyde, ketone	alcohol	PtO$_2$	2–4%	25	1
		5% Pd (C)	3–5%	25	1–4
		Raney Ni	30–100%	25	1
Halide	hydro- carbon	5% Pd (C) or 5% Pd (BaSO$_4$)	1–15%, KOH	25	1
			30–100%, KOH	25	1
		Raney Ni	10–20%, KOH	25	1
Nitro compd., azide	amine	PtO$_2$	1–5%	25	1
		Pd (C)	4–8%	25	1
		Raney Ni	10–80%	25	1–3
Oxime, nitrile	amine	PtO$_2$	1–10%, AcOH or HCl/MeOH	25	1–3
		5% Pd (C)	5–15%, AcOH	25	1–3
		Raney Ni	3–30%	25	35–70

Note: References to the original papers are to be found on p 323.

Another form of very active elemental platinum is obtained by reduction of chloroplatinic acid by sodium borohydride in ethanolic solution. Such platinum hydrogenated 1-octene 50% faster than did platinum oxide [30].

To increase the contact of a catalyst with hydrogen and the compounds to be hydrogenated, platinum or other metals are precipitated on materials having large surface areas such as activated charcoal, silica gel, alumina, calcium carbonate, barium sulfate, and others. Such *supported catalysts* are prepared by hydrogenation of solutions of the metal salts, for example, chloroplatinic acid, in aqueous suspensions of activated charcoal or other solid substrates [31].

Supported catalysts, which usually contain 5, 10, or 30 wt % platinum, are very active and frequently pyrophoric.

The support exerts a certain effect on the hydrogenation, especially its rate, but sometimes even its selectivity. It is therefore advisable to use the same type of catalyst when duplicating experiments.

Platinum catalysts are universally used for hydrogenation of almost any type of compound at room temperature and atmospheric or slightly elevated pressures (1–4 atm). They are usually not used for hydrogenolysis of benzyl-type residues, and they are completely ineffective in reductions of acids or esters to alcohols. At elevated pressures (70–210 atm) platinum oxide converts aromatics to perhydroaromatics at room temperature very rapidly [8].

Palladium

Palladium catalysts closely resemble platinum catalysts. *Palladium oxide* (PdO) is prepared from palladium chloride and sodium nitrate by fusion at 575–600 °C [32, 33]. Elemental palladium is obtained by reduction of palladium chloride with sodium borohydride [30, 34]. *Supported palladium catalysts* are prepared with 5–10% palladium on charcoal, calcium carbonate, and barium sulfate [35]. Sometimes a special support can increase the selectivity of palladium. Palladium on strontium carbonate (2%) was successfully used for reduction of just the γ, δ-double bond in an α, β, γ, δ-unsaturated ketone [36].

Palladium catalysts are more often modified for special selectivities than platinum catalysts. Palladium prepared by reduction of palladium chloride with sodium borohydride (Procedure 4, p 298) is suitable for the reduction of unsaturated aldehydes to saturated aldehydes [34]. Palladium on barium sulfate deactivated with sulfur compounds, most frequently the so-called quinoline-S obtained by boiling quinoline with sulfur [37], is suitable for the Rosenmund reduction [38] (p 203). Palladium on calcium carbonate deactivated by lead acetate *(Lindlar's catalyst)* is used for partial hydrogenation of acetylenes to *cis*-alkenes [39] (p 58) (Procedure 5, p 298).

Palladium catalysts can be applied in strongly acidic media and in basic media, and are especially suited for hydrogenolyses such as cleavage of benzyl-type bonds (p 211). They do not reduce carboxylic groups.

Rhodium

A very active elemental *rhodium* is obtained by reduction of rhodium chloride with sodium borohydride [30]. *Supported rhodium catalysts,* usually 5% on carbon or alumina, are especially suited for hydrogenation of aromatic systems [40]. A mixture of rhodium oxide and platinum oxide is also used for this purpose and proved better than platinum oxide alone [41, 42]. Unsaturated halides containing vinylic halogens are reduced at the double bond without hydrogenolysis of the halogen [43]. Compounds of rhodium complexed with phosphines are used in homogeneous hydrogenations (p 15).

Nickel

Nickel catalysts are universal and are widely used not only in the laboratory but also in industry. The supported form, *nickel* on *kieselguhr* or *infusorial earth*, is prepared by precipitation of nickel carbonate from a solution of nickel nitrate by sodium carbonate in the presence of infusorial earth and by reduction of the precipitate with hydrogen at 450 °C after drying at 110–120 °C. Such catalysts work at temperatures of 100–200 °C and pressures of hydrogen of 100–250 atm [44].

Nickel catalysts of very high activity are obtained by the *Raney process*. An alloy containing 50% nickel and 50% aluminum is heated with 25–50% aqueous sodium hydroxide at 50–100 °C. Aluminum is dissolved and leaves nickel in the form of very fine particles. It is then washed with large amounts of distilled water and finally with ethanol. Depending on the temperature of dissolution, on the content of aluminum, and on the method of washing, Raney nickel of varied activity can be produced. It is categorized by the symbols W1–W8 [45]. The most active is Raney nickel W6, which contains 10–11% aluminum, has been washed under hydrogen, and has been stored under ethanol in a refrigerator. Its activity in hydrogenation is comparable to that of the noble metals and may decrease with time. Some hydrogenations can be carried out at room temperature and atmospheric pressure [46]. Many Raney nickel preparations are pyrophoric in the dry state. Some, like Raney nickel W6, may react extremely violently, especially when dioxane is used as a solvent, the temperature is higher than about 125 °C, and large quantities of the catalyst are used. If the ratio of the Raney nickel to the compound to be hydrogenated does not exceed 5%, the hydrogenation at temperatures above 100 °C is considered safe [46]. The temperature in the autoclave may rise considerably because of heat of hydrogenation. *Utmost precautions are in order during high-temperature–high-pressure hydrogenations with Raney nickel* (*see* p xxiii).

Raney nickel can be used for reduction of practically any functions. Many hydrogenations can be carried out at room temperature and atmospheric or slightly elevated pressure (1–3 atm) [46] (Procedure 6, p 298). At high temperatures and pressures, even the acids and esters that are difficult to reduce are converted to alcohols [45] (p 214). Raney nickel is not poisoned by sulfur and is used for desulfurization of sulfur-containing compounds [47] (p 119, 145) (Procedure 7, p 299). Its disadvantages are difficulty in calculating dosage (it is usually measured as a suspension rather than weighed) and ferromagnetic properties that preclude the use of magnetic stirring.

Nickel of activity comparable to Raney nickel is obtained by reduction of nickel salts, for example, nickel acetate, with 2 mol of sodium borohydride in an aqueous solution and by washing the precipitate with ethanol [13, 48] (Procedure 8, p 299). Such preparations are designated *P-1* or *P-2* and can be conveniently prepared in situ in a special apparatus [4] (Procedure 2, p 295). They contain a high percentage of nickel boride, are nonmagnetic and nonpyrophoric, and can be used for hydrogenations at room temperature and atmospheric pressure. Nickel *P-2* is especially suitable for semihydrogenation of acetylenes to *cis*-alkenes [48] (p 58).

Nickel precipitated from aqueous solutions of nickel chloride by aluminum or zinc dust is referred to as *Urushibara catalyst* and resembles Raney nickel in its activity [49].

Another highly active, nonpyrophoric nickel catalyst is prepared by reduction of nickel acetate in tetrahydrofuran by sodium hydride at 45 °C in the presence of *tert*-amyl alcohol (which acts as an activator). Such catalysts, referred to as *Nic catalysts,* compare with *P* nickel boride and are suitable for hydrogenations at room temperature and atmospheric pressure and for partial reduction of acetylenes to *cis*-alkenes [50].

The Raney nickel process applied to alloys of aluminum with other metals produces *Raney iron, Raney cobalt, Raney copper,* and *Raney zinc.* These catalysts are used very rarely and only for special purposes.

Other Catalysts

Catalysts suitable specifically for reduction of carbon–oxygen bonds are based on oxides of copper, zinc, and chromium *(Adkins' catalysts).* The so-called copper chromite (which is not necessarily a stoichiometric compound) is prepared by thermal decomposition of ammonium chromate and copper nitrate [51]. Its activity and stability are improved if barium nitrate is added before the thermal decomposition [52]. Similarly prepared zinc chromite is suitable for reductions of unsaturated acids and esters to unsaturated alcohols [53]. These catalysts are used specifically for reduction of carbonyl- and carboxyl-containing compounds to alcohols. Aldehydes and ketones are reduced at 150–200 °C and 100–150 atm, whereas esters and acids require temperatures up to 300 °C and pressures up to 350 atm. Because such conditions require special equipment and because all reductions achievable with copper chromite catalysts can be accomplished by hydrides and complex hydrides, the use of Adkins' catalyst in the laboratory is now very limited.

Ruthenium catalysts [54] resemble the other noble metal catalysts but are used less frequently. Ruthenium complexes with phosphines are applied in homogeneous hydrogenation (p 15). *Rhenium catalysts,* that is, rhenium heptoxide [55], rhenium heptasulfide [56], and rhenium heptaselenide [57], all require temperatures of 100–300 °C and pressures of 100–300 atm. Rhenium heptasulfide is not sensitive to sulfur and is more active than *molybdenum* and *cobalt sulfides* in hydrogenating oxygen-containing functions [56, 58].

ACTIVATORS AND DEACTIVATORS OF CATALYSTS

The efficiency of catalysts is affected by the presence of some compounds. Even small amounts of alien admixtures, especially with noble metal catalysts, can increase or decrease the rate of hydrogenation and, in some cases, even inhibit the hydrogenation completely. Moreover, some compounds can influence the selectivity of a catalyst.

Minute quantities of zinc acetate or ferrous sulfate enhance hydrogenation of the carbonyl group in unsaturated aldehydes and cause preferential hydrogenation to unsaturated alcohols [59]. As little as 0.2% palladium present in a platinum-on-carbon catalyst deactivates platinum for hydrogenolysis of benzyl groups and halogens [31]. Admixture of stannous chloride (7% of the weight of platinum dioxide) increases the rate of hydrogenation of valeraldehyde 10 times, and admixture of 6.5% of ferric chloride, eight times [28].

On the other hand, some compounds slow down the uptake of hydrogen and may even stop it at a certain stage of hydrogenation. Addition of lead acetate to palladium on calcium carbonate makes the catalyst suitable for selective hydrogenation of triple to double bonds (the Lindlar catalyst) [39].

The strongest inhibitors of noble metal catalysts are sulfur and most sulfur compounds. With the exception of modifying a palladium-on-barium sulfate catalyst [38] or platinum oxide [60] for the Rosenmund reduction, the presence of sulfur compounds in materials to be hydrogenated over platinum, palladium, and rhodium catalysts is highly undesirable. Except for hexacovalent sulfur compounds such as sulfuric acid and sulfonic acids, most sulfur-containing compounds are *catalytic poisons* and may inhibit hydrogenation very strongly. If such compounds are present as impurities, they can be removed by contact with Raney nickel, which combines with almost any form of sulfur to form nickel sulfide. Shaking or stirring of a compound contaminated with sulfur-containing contaminants with Raney nickel makes possible subsequent hydrogenation over noble metal catalysts.

Many nucleophiles act as inhibitors of platinum, palladium, and rhodium catalysts. The strongest are mercaptans, sulfides, cyanides, and iodides; weaker are ammonia, azides, acetates, and alkalis [29, 31].

Acidity or alkalinity of the medium plays a very important role. Hydrogenation of aromatic rings over platinum catalysts requires an acid medium. Best results are obtained when acetic acid is used as the solvent. Addition of sulfuric or perchloric acid is an advantage, and in the hydrogenation of pyridine compounds, it is necessary. Strongly basic piperidine produced by reduction of weakly basic pyridine acts as a deactivator of the catalyst.

On the other hand, addition of tertiary amines accelerates hydrogenation of some compounds over Raney nickel [61]. In hydrogenation of halogen derivatives over palladium [62] or Raney nickel [63], the presence of at least one equivalent of sodium or potassium hydroxide was found necessary.

EFFECTS OF THE AMOUNT OF CATALYST, SOLVENT, TEMPERATURE, AND PRESSURE

The *ratio of the catalyst to the compound* to be hydrogenated has, within certain limits, a strong effect on the rate of hydrogenation. Platinum group catalysts are used in amounts of 1–3% of the weight of the metal, whereas Raney nickel requires much larger quantities. Doubling the amount of a nickel catalyst in the hydrogenation of cottonseed oil from 0.075% to 0.15% doubled the reaction rate [64]. With 10% as much Raney nickel as ester, practically no hydrogenation of the ester took place, even at 100 °C. With 70%, hydrogenation at 50 °C was more rapid than hydrogenation at 100 °C with 20–30% catalyst [61]. Some hydrogenations over Raney nickel and copper chromite went well only if 1.5 times as much catalyst as reactant was used [65].

The best *solvents* for hydrogen (pentane and hexane) are not always good solvents for the reactants. Methanol and ethanol, which dissolve only about one-third the amount of hydrogen as pentane and hexane dissolve, are used most frequently. Other solvents for hydrogenations are ether (rarely), thiophene-free benzene, cyclohexane, dioxane, and acetic acid, the last one being especially useful in the catalytic hydrogenations of aromatics over platinum metal catalysts. Too volatile solvents are not desirable. In hydrogenations at atmospheric pressures they make exact reading of the volume of consumed hydrogen tedious and inaccurate, and in high-temperature and high-pressure hydrogenations they contribute to the pressure in the autoclave.

Water does not dissolve many organic compounds, but it can be used as a solvent, especially in hydrogenations of acids and their salts. It may have some deleterious effects; for example, it enhanced hydrogenolysis of vinylic halogens [66].

The pH of the solution plays an important role in the steric outcome of the reaction. Acidic conditions favor *syn*-addition, and basic conditions favor *anti*-addition of hydrogen [25].

Temperature affects the rate of hydrogenation but not as much as is usual with other chemical reactions. The rise in temperature from 50 to 100 °C and from 25 to 100 °C caused, respectively, fourfold and 12-fold increases in the rates of hydrogenation of esters over Raney nickel W6 [61]. The increased rate of hydrogenation may reduce the selectivity of the particular catalyst.

The *pressure of hydrogen,* as expected, increases the rate of hydrogenation considerably, but not to the same extent with all compounds. In the hydrogenation of esters over Raney nickel W6, the rate doubles with an increase from 280 to 350 atm [61]. In hydrogenation of cottonseed oil in benzene over nickel at 120 °C, complete hydrogenation is achieved at 30 atm in 3 h, at 170 atm in 2 h, and at 325 atm in 1.5 h [64]. High pressure favors *syn* hydrogenation where applicable [10] but decreases the stereoselectivity [8].

The seemingly peripheral effect of *mixing* must not be underestimated. Hydrogenation, especially heterogeneous hydrogenation, is a reaction of three phases; therefore, good contact must take place not only between the gas and the liquid but also between hydrogen and the solid, the catalyst. The stirring must provide for frequent, or better still, permanent contact between the catalyst and the gas. Shaking and fast magnetic stirring are therefore preferred to slow rocking [67].

CARRYING OUT CATALYTIC HYDROGENATION

After a decision has been made as to what type of hydrogenation, what catalyst, and what solvent are to be used, careful calculation should precede carrying out the reaction.

In atmospheric or low-pressure hydrogenation, the volume of hydrogen needed for a partial or total reduction should be calculated. This calculation is imperative for partial hydrogenations when the reduction has to be interrupted after the required volume of hydrogen has been absorbed. In exact calculations, the vapor pressure of the solvent used must be considered because it contributes to the total pressure in the apparatus. If oxide-type catalysts are used, the amount of hydrogen needed for reduction of the oxides to the metals must be included in the calculation.

In high-pressure hydrogenations, the calculations are even more important. It is necessary to take into account, in addition to the pressure resulting from heating of the reaction mixture to a certain temperature, an additional pressure increase caused by the temperature rise owing to considerable heat of hydrogenation (approximately 30 kcal/mol per double bond). In hydrogenations of large amounts of compounds in low-boiling solvents, the reaction heat may raise the temperature inside the autoclave above the critical temperature of the components of the mixture. In such a case, when one or more components of the mixture gasify, the pressure inside the container rises considerably and may even exceed the pressure limits of the vessel. *A sufficient leeway for such potential or accidental pressure increases should be ensured* [67] (*see* p xxiii).

The calculated amounts of the catalyst, reactant and solvent (if needed) are then placed in the hydrogenation vessel. *Utmost care must be exercised in loading the hydrogenation container with catalysts that are pyrophoric, especially when highly volatile and flammable solvents like ether, methanol, ethanol, cyclohexane, or benzene are used* (*see* p xxiii). The solution should be added to the catalyst in the container. If the catalyst must be added to the solution, this step should be done under a blanket of an inert gas to prevent ignition.

The hydrogenation vessel is attached to the source of hydrogen, evacuated, flushed with hydrogen once or twice to displace any remaining air (oxygen may inhibit some catalysts), pressurized with hydrogen, and shut off from the hydrogen source. Then stirring and heating (if required) are started.

After the hydrogenation is over and the apparatus is cold, the excessive pressure is bled off. In high-temperature, high-pressure hydrogenations carried out in robust autoclaves, it takes considerable time for the assembly to cool down. If correct reading of the final pressure is necessary, it is advisable to let the autoclave cool overnight.

Isolation of the products is usually carried out by filtration. Suction filtration is faster and preferable to gravity filtration. *When pyrophoric noble metal catalysts and Raney nickel are filtered with suction, the suction must be stopped before the catalyst on the filter paper becomes dry. Otherwise it can ignite and cause fire.* Where feasible, centrifugation and decantation should be used for the separation of the catalyst. Sometimes the filtrate contains colloidal catalyst that has passed through the filter paper. Stirring of such filtrate with activated charcoal followed by another filtration usually solves this problem. Evaporation of the filtrate and crystallization or distillation of the residue completes the isolation.

Heterogeneous catalytic hydrogenation has also been applied to the **preparation of chiral compounds**, but only if a chiral auxiliary group had been incorporated prior to the hydrogenation [68].

3   [68]

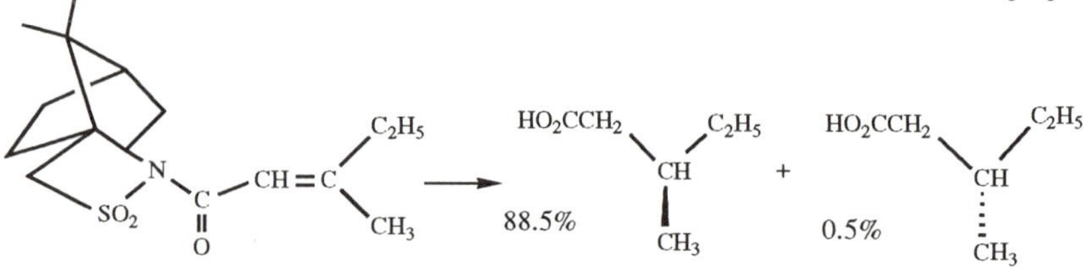

1. H$_2$/Pd(C), EtOH, RT, 7 atm
2. LiOH, aq. THF, RT 18 h

CATALYTIC TRANSFER OF HYDROGEN

A special kind of catalytic hydrogenation is catalytic hydrogen transfer achieved by heating of a compound to be hydrogenated in a solvent with a catalyst and a hydrogen donor—a compound that gives up its hydrogen. The hydrogen donors are hydrazine [*69, 70, 71*], formic acid [*72, 73, 74*], ammonium formate [*75, 76, 77, 78*], triethylammonium formate [*79, 80, 81*], cyclohexene [*82, 83*], cyclohexadiene [*84, 85*], tetralin [*86*], pyrrolidine [*87*], indoline [*87*], 1,2,3,4-tetrahydroquinoline [*87*], triethylsilane [*88*], 1-benzyl-1,4-dihydronicotine amide [*89*], and others; the catalysts are platinum, palladium, rhodium, or Raney nickel [*90*] (Procedure 9, p 299).

Catalytic hydrogen transfer results usually in *cis (syn)* addition of hydrogen and is sometimes more selective than catalytic hydrogenation with hydrogen gas.

HOMOGENEOUS CATALYTIC HYDROGENATION

In the past 25 years, **homogeneous hydrogenation** has been developed. It is catalyzed by compounds soluble in organic solvents. Of a host of complexes of noble metals, *tris(triphenylphosphine)rhodium chloride (the Wilkinson catalyst* [*467*] (structure **4**) is now used most frequently and is commercially available [*91, 92, 93*] (Procedure 10, p 300).

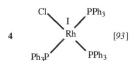

4

[*93*]

Tris(triphenylphosphino-
rhodium) chloride

Homogeneous catalytic hydrogenation is often carried out at room temperature and atmospheric pressure and hardly ever at temperatures exceeding 100 °C and pressures higher than 100 atm. It is less effective and more selective than heterogeneous hydrogenation and is therefore more suitable for selective reductions of polyfunctional compounds, for example, hydrogenation of isolated double bonds in preference to the conjugated ones [*94*], conversion of α,β-unsaturated aldehydes to saturated aldehydes [*93*] (Procedure 10, p 300), and replacement of chlorine in imidoyl chlorides without saturation of the double bond [*95*]. It also causes fewer rearrangements and less isotope exchange and is therefore convenient for deuteration. The disadvantages of homogeneous hydrogenation are lower commercial availability of homogeneous catalysts and more complicated isolation of products.

The great advantage of homogeneous catalytic hydrogenation over heterogeneous hydrogenation lies in the ease of achieving stereoselectivity in addition of hydrogen to prochiral substrates [*21, 22*]. Although stereoselective reductions have been accomplished by heterogeneous catalytic hydrogenation of compounds into which chiral auxiliary groups such as camphorsultam have been incorporated [68] (equation 3), homogeneous diastereo- and enantioselective catalytic hydrogenation are much more widespread.

In these cases the chirality is introduced by a chiral catalyst, consisting of a noble metal, most often rhodium [20, 96, 97, 98, 99, 100, 101, 102] or ruthenium [103, 104, 105], complexed by bidentate ligands, usually chiral diphosphines and dienes such as norbornadiene, trans-1,5-cyclooctadiene, or cyclopentadienes in the form of metallocenes [22, 23, 106] (structure 5). The most promising bidentate ligands seem to be the Noyori's atropoisomeric 2,2'-bis(diphenylphosphino)-1,1'-binaphthyl [Binap] [105] (structure 6) and bis(diphenylphosphino)butane [Chiraphos] [107] (structure 7).

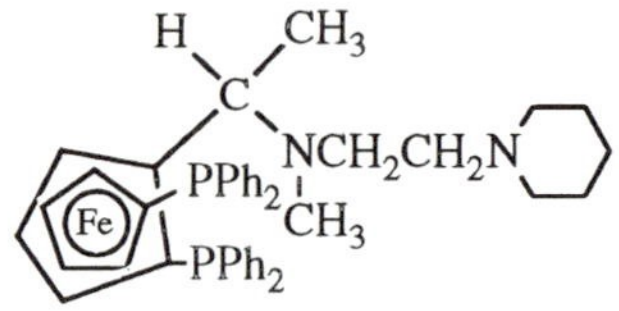

1-[N-methyl-N-2-piperidinoethyl)]-(S)-
1',2-bis(diphenylferrocenyl)aminoethane

(R)-Binap

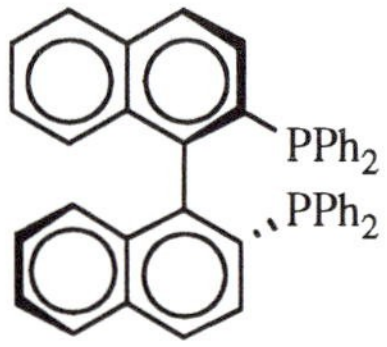

(R)-2,2'-Bis(diphenylphosphino)-
1,1'-binaphthyl

(S,S)-Chiraphos

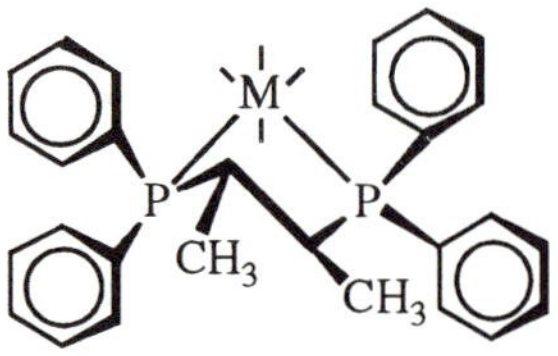

2S,3S-bis(diphenyl-
phosphino)butane

The mechanism is rather complex and consists of several consecutive steps in which the diene component is first replaced by a solvent, usually methanol, ethanol, or ethyl acetate, and this in turn is replaced by a substrate, containing two π bonds, that can form five-, six- or seven-membered complexes with the metal. Hydrogen is then bound to the metal, thus increasing its coordination number to six (and the valence from I to III) and is finally transferred to the substrate, giving predominantly one stereoisomer [21, 22] (equation 8).

So far such chiral homogeneous hydrogenations have been performed with high stereoselectivity in reductions of α-ketoesters to α-hydroxy esters [108], of β-keto esters to β-hydroxy esters [103, 105], and especially of α,β-unsaturated α-acylamido acids or esters, which are reduced to optically active amino acid derivatives [23, 96, 97, 100, 101, 102].

For industrial applications, water-soluble chiral catalysts are desirable. These are complexes of rhodium with phosphines containing quaternary ammonium groups [97] or sulfonyl groups [99].

[21,22]

8

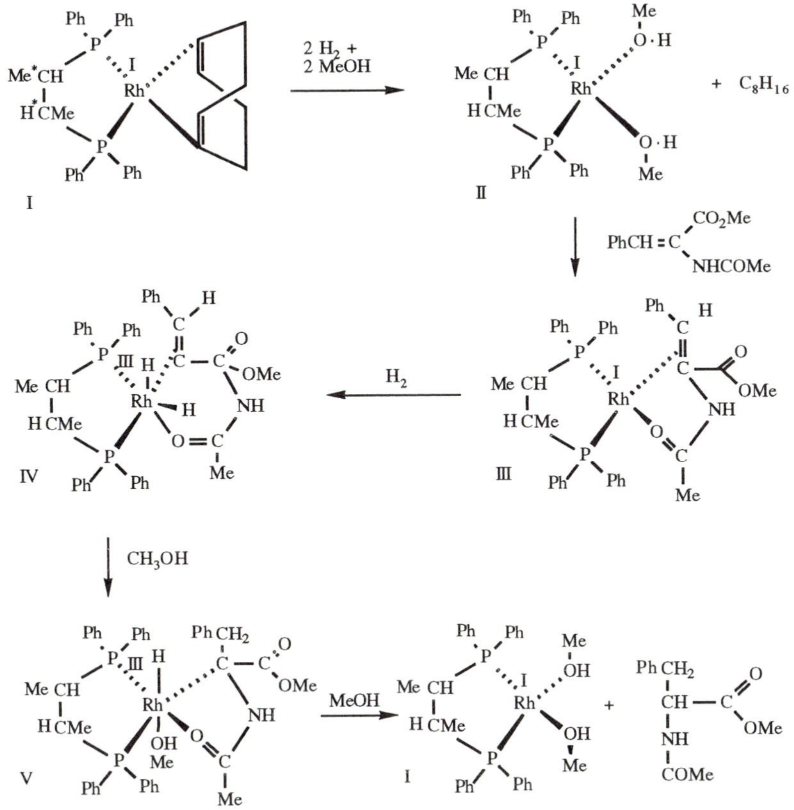

I

2 H$_2$ + 2 MeOH

II

+ C$_8$H$_{16}$

PhCH=C(CO$_2$Me)(NHCOMe)

IV

H$_2$

III

CH$_3$OH

V

MeOH

I

+ PhCH$_2$CH(NHCOMe)C(=O)OMe

SUMMARY

Catalytic hydrogenation is the most universal reduction method. Most unsaturated systems can be reduced catalytically. On the other hand, high regioselectivity in hydrogenating multifunctional systems is not always good enough with the most common catalysts. However, the variety of special catalysts, now available commercially, makes selective reductions feasible, as can be found in specific examples. Heterogeneous catalytic hydrogenation is stereoselective in reductions leading to *cis–trans* isomerism, but is not enantioselective. Only after chiral catalysts for homogeneous hydrogenation were developed could catalytic hydrogenation compete with stereo- and enantioselectivity of chiral complex hydrides. Still, most stereoselective reductions are achieved by chiral complex hydrides and by biochemical reductions (p 48, 50).

The great advantage of heterogeneous catalytic hydrogenation is a very easy isolation of the products. This aspect is especially important for large-scale industrial reductions for which, as a rule, complex hydride reductions are not suitable. Provided the necessary safety measures are applied (p xxiii), catalytic hydrogenation is a safe reaction.

CHAPTER 2

Reduction With Hydrides and Complex Hydrides

Fifty years after the introduction of catalytic hydrogenation into the methodology of organic chemistry, another discovery of comparable importance was published: synthesis [109] and applications [110] of lithium aluminum hydride and sodium borohydride [111].

Treatment of lithium hydride with aluminum chloride gives *lithium aluminum hydride*, which, with additional *aluminum chloride*, affords *aluminum hydride, alane* [109] (equations **9** and **10**), (Procedure 11, p 301).

9 $4 \text{ LiH } + \text{ AlCl}_3 \xrightarrow{\text{Et}_2\text{O}} \text{ LiAlH}_4 + 3 \text{ LiCl}$ [109]
M.W. 37.95, m.p. 125°C (dec.)

10 $3 \text{ LiAlH}_4 + \text{ AlCl}_3 \xrightarrow{\text{Et}_2\text{O}} 4 \text{ AlH}_3 + 3 \text{ LiCl}$ [109]
M.W. 27.01

A boron analog, *sodium borohydride*, was prepared by reaction of sodium hydride with trimethyl borate [111] (equation **11**) or with sodium fluoroborate and hydrogen [112] (equation **12**), and gives, on treatment with boron trifluoride or aluminum chloride, *borane (dibora*ne) [113] (equations **13** and **14**).

11 $4 \text{ NaH } + \text{ B(OMe)}_3 \longrightarrow \text{ NaBH}_4 + 3 \text{ MeONa}$ [111]
M.W. 37.83

12 $4 \text{ NaH } + \text{ NaBF}_4 \xrightarrow[360°\text{C, 50 atm}]{\text{H}_2} \text{ NaBH}_4 (35\%) + 4 \text{ NaF}$ [112]

13 $3 \text{ NaBH}_4 + 4 \text{ BF}_3 \longrightarrow 4 \text{ BH}_3 + 3 \text{ NaBF}_4$ [113]
M.W. 13.83

14 $2 \text{ BH}_3 \longrightarrow \text{ B}_2\text{H}_6$
M.W. 13.83   M.W. 27.66

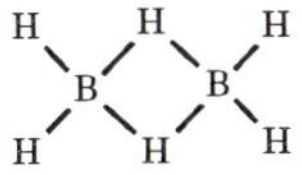

Borane is a strong Lewis acid and forms complexes with many Lewis bases. Some of them, such as complexes with tetrahydrofuran, dimethyl sulfide, and trimethylamine (structure groups **15** and **16**), are sufficiently stable to have been made commercially available.

15 $\text{BH}_3 \cdot \text{THF}$ $\text{BH}_3 \cdot \text{SMe}_2$ $\text{BH}_3 \cdot \text{NMe}_3$ BH_2Cl
M.W. 85.94   M.W. 75.97   M.W. 72.95   M.W. 48.28

16 $\text{BH}_3 \cdot \text{NH}_3$ $\text{BH}_3 \cdot \text{Me}_3\text{CNH}_2$ $\text{BH}_3 \cdot \text{C}_5\text{H}_5\text{N}$
M.W. 30.86   M.W. 86.97   M.W. 92.93

Some others should be handled with caution (*see* p xxiii). *A spontaneous explosion of a 1M solution of borane in tetrahydrofuran stored at less than 15 °C out of direct sunlight has been reported [114].*

Complex aluminum and boron hydrides can contain other cations. The following compounds are prepared by metathetical reactions of lithium aluminum hydride or sodium borohydride with the appropriate compounds of other metals: *sodium aluminum hydride [115, 116], potassium aluminum hydride [116], magnesium aluminum hydride [117], lithium borohydride [118], potassium borohydride [119], calcium borohydride [120], zinc borohydride [121],* and *tetrabutylammonium borohydride [122]* (structure groups **17–19**).

17	$NaAlH_4$ [*115*]		$KAlH_4$ [*116*]		$Mg(AlH_4)_2$ [*117*]
	M.W. 54.00		M.W. 70.11		M.W. 86.33
18	$LiBH_4$ [*118*]		KBH_4 [*119*]		$Ca(BH_4)_2$ [*120*]
	M.W. 21.78 m.p. 275-280°C		M.W. 53.94		M.W. 69.76
19	$Zn(BH_4)_2$ [*121*]		Bu_4NBH_4 [*122*]		
	M.W. 95.06		M.W. 257.30 m.p. 103-104°C		

Lithium aluminum hydride and sodium borohydride react with alcohols and form alkoxyaluminohydrides and alkoxyborohydrides: most widely used are *lithium trimethoxy- [123]* and *triethoxyaluminum hydride [124], lithium diethoxyaluminum hydride [124], lithium tris(tert-butoxy)aluminum hydride [125]* (Procedure 12, p 301), *lithium tris(triethylmethoxy)aluminum hydride [126], sodium bis(2-methoxyethoxy)aluminum hydride* (Vitride, Red-Al) [*127, 128*], and *sodium trimethoxyborohydride [129].* *Sodium borohydride and alkylamines form sodium dimethylaminoborohydride [130], sodium-*tert-*butylamino-borohydride [130], sodium diethylpiperidinoaluminum hydride [131], and tetrabutylammonium triacetoxyborohydride [132]* (structure groups **20–23**).

20	$LiAlH(OMe)_3$ [*123*] M.W. 128.03		$LiAlH(OEt)_3$ [*124*] M.W. 170.11		$LiAlH_2(OEt)_2$ [*124*] M.W. 126.06
21	$LiAlH(OCMe_3)_3$ [*125*] M.W. 254.27 m.p. 319°C (dec.)		$LiAlH(OCEt_3)_3$ [*126*] M.W. 380.50		$NaAlH_2(OCH_2CH_2OMe)_2$ Red-Al, Vitride [*127,128*] M.W. 202.16
22	$NaBH(OMe)_3$ [*129*] M.W. 127.91		$NaBH_3NMe_2$ [*130*] M.W. 81.90		$NaBH_3NHBu$-*t* [*130*] M.W. 109.95
23	$NaAlHEt_2NC_5H_{10}$ [*131*] M.W. 193.24		$Bu_4NBH(OAc)_3$ [*132*] M.W. 431.41		

Replacement of hydrogen by alkyl groups gives compounds such as *lithium triethylborohydride* (Super-Hydride) [*133*], *lithium tris(sec-butyl)borohydride* [*134*] (L-Selectride), *potassium tris(sec-butyl)borohydride* (K-Selectride) [*135*], *potassium triphenylborohydride* [*136*], and *potassium 9-s-amyl-9-borabicyclo-[3.3.1]nonane* [*137*]. Replacement of hydrogen by a cyano group yields *sodium cyanoborohydride* [*138*], a compound stable even at low pH (down to ~3), and *tetrabutylammonium cyanoborohydride* [*139*] (structure groups **24–26**).

24	LiBHEt₃	[*133*]	LiBH(CHMeEt)₃	[*134*]	KBH(CHMeEt)₃	[*135*]	

$$24 \quad LiBHEt_3 \quad [133] \qquad LiBH(CHMeEt)_3 \quad [134] \qquad KBH(CHMeEt)_3 \quad [135]$$

Superhydride   L-Selectride   K-Selectride
M.W. 105.94   M.W. 190.11   M.W. 222.27

$$25 \quad KBHPh_3 \quad [136]$$

M.W. 278.37

$$K\left[Me_2CHCHMeBH \right] \quad [137]$$

M.W. 232.25

$$26 \quad NaBH_3CN \quad [138] \qquad Bu_4NBH_3CN \quad [122]$$

M.W. 62.84   M.W. 282.30
m.p. >242°C (dec.)

Addition of alane and borane to alkenes affords a host of alkylated alanes and boranes with various reducing properties (and sometimes bizarre names): *diisobutylalane* (Dibal-H) [*140*], *disiamylborane* [*141, 142*], tert-*hexylborane* (*thexylborane*) [*143*] (structure group **27**), *its B-chloro-* [*144*] *and B-bromo-derivatives* [*145*] (structure group **28**), *and 9-borabicyclo[3.3.1]nonane* (9-BBN), prepared from borane and 1,5-cyclooctadiene [*146*] (equation **29**).

$$27 \quad AlH(CH_2CHMe_2)_2 \quad [140] \qquad BH(CHMeCHMe_2)_2 \quad [142] \qquad BH_2(CMe_2CHMe_2) \quad [143]$$

Dibal-H   Disiamylborane   Thexylborane
M.W. 142.22, b.p. 140°C/4 mm   M.W. 286.31   M.W. 97.99

$$28 \quad BHCl(CMe_2CHMe_2) \quad [144] \qquad BHBr(CMe_2CHMe_2) \quad [145]$$

M.W. 132.43   M.W. 176.89

29   ⬡ + BH₃ ⟶ (BH) ≡ HB ≡ HB   [*146*]

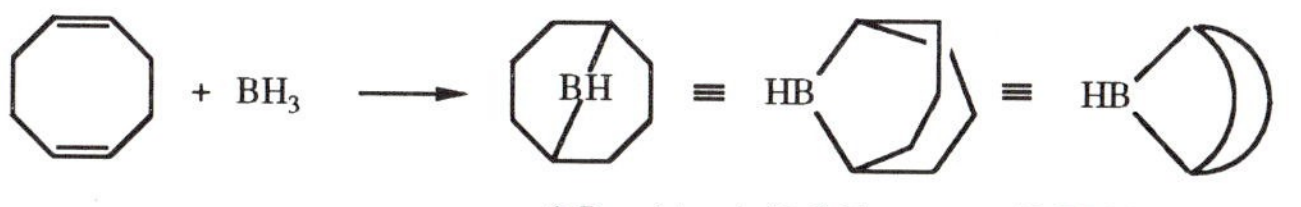

9-Borabicyclo[3.3.1]nonane (9-BBN)
M.W. 122.02 (dimer), m.p. 150-152°C

Because of the desirability of preparing enantiomerically pure compounds, several chiral hydrides have been developed: *isopinocampheylborane* [*143, 147*] *and diisopinocampheylborane (B-di-3-pinanylborane)* [*148*], both prepared from borane and optically active α-pinene; *diisopinocampheylchloroborane* [*149*] (equation **30**); *B-isopinocampheyl-9-borabicyclo[3.3.1]nonane* (also called B-*3-pinanyl-9-borabicyclo[3.3.1]nonane, 3-pinanyl-9-BBN*, and Alpine-Borane), prepared from 9-borabicyclo[3.3.1]nonane and α-pinene) [*150*] (equation **31**); NB-Enanthrane, prepared from 9-borabicyclo[3.3.1]nonane and nopol benzyl ether) [151] (equation 32); *lithium B-isopinocampheyl-9-borabicyclo[3.3.1]nonyl hydride* (Alpine-Hydride) [152]; *oxazaphosphalidineborane* [153]; *diphenyloxazaboralidine* [154] (structure group 33); *potassium 9-O-(1,2:5,6-di-O-isopropylidene- α-D- glucofuranosyl) - 9- borabicyclo [3.3.1] -nonane* [155] (K-Glucoride); and especially Noyori's *lithium 1,1'-binaphthyl-2,2'-dihydroxy(ethoxy)aluminum hydride* (Binal-H) [156] (structure group 34). A review paper surveys asymmetric reductions with chiral boron compounds [157]. Many chiral complex hydrides are now available in both enantiomeric forms.

30   + BH₃ ⟶   [148] $\xrightarrow{\text{HCl}}$ [149]

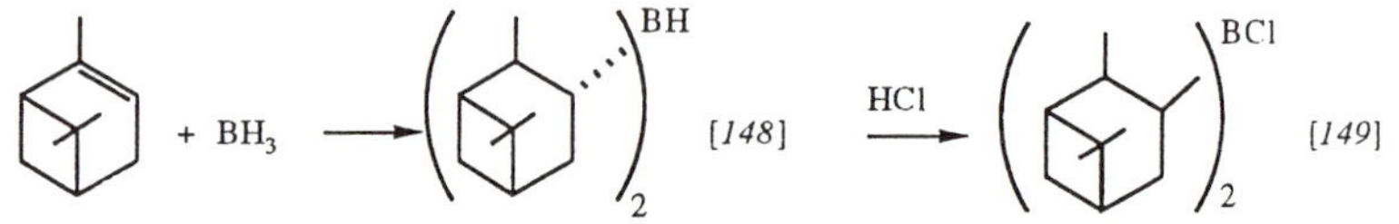

Diisopinocampheylborane
B-di-3-pinanylborane
M.W. 286.29

Diisopinocampheylchloroborane
M.W. 320.73

31   + HB⟶   [150]

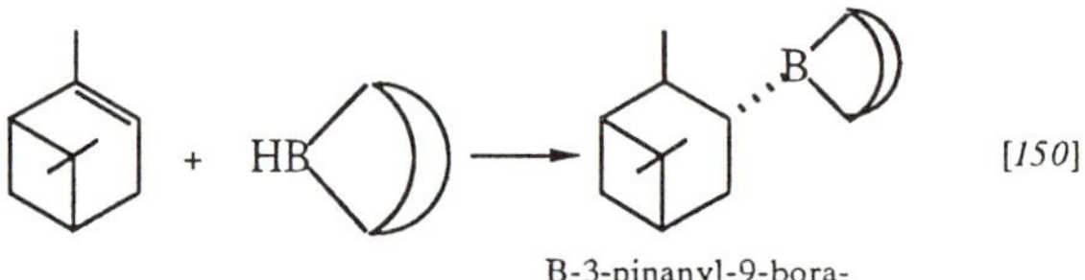

B-3-pinanyl-9-bora-
bicyclo[3.3.1]nonane
(Alpine-Borane)
M.W. 197.24

32   OCH₂Ph   + HB⟶   OCH₂Ph   NB-Enanthrane
NB Enantride   [151]

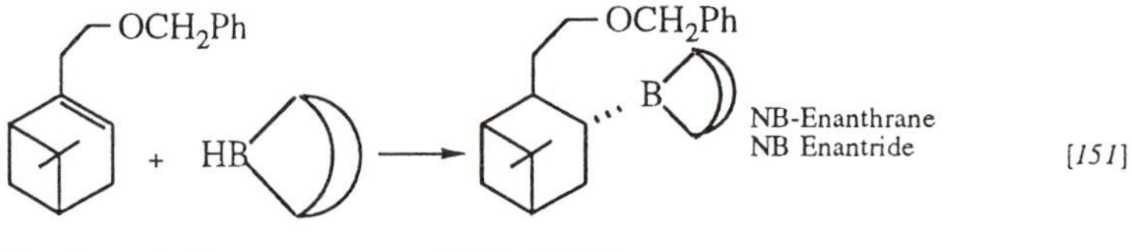

Nopol benzyl ether   M.W. 317.38

33   Li   BH₃   [152]   P   [153]   N   C₆H₅   C₆H₅   O   [154]

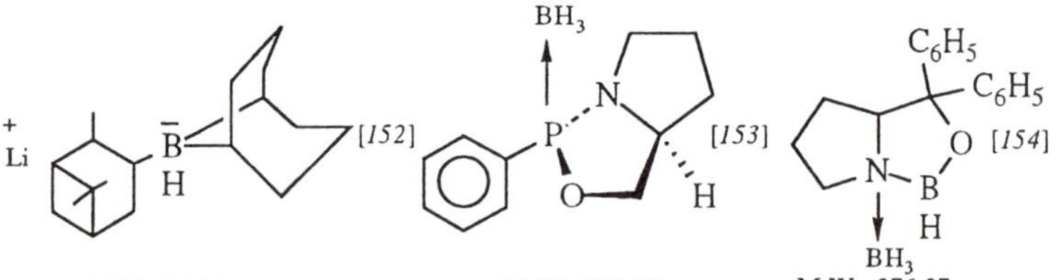

M.W. 266.21   M.W. 221.07   M.W. 276.97

R and S-Alpine-
Hydride

Hydrides and complex hydrides of aluminum and boron reduce practically all functionalities. However, isolated carbon–carbon double bonds are reduced only exceptionally with lithium aluminum hydride activated by chlorides of iron, cobalt, and nickel [158].

Lithium aluminum hydride and alanes are frequently used for the preparation of hydrides of other metals. Diethylmagnesium is converted to

34 K$^+$ [155]

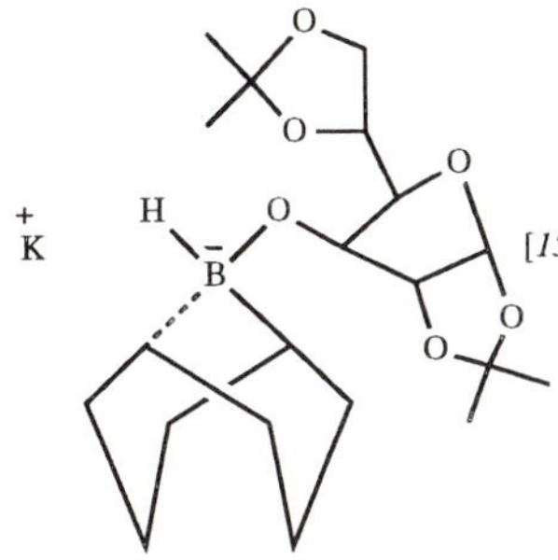

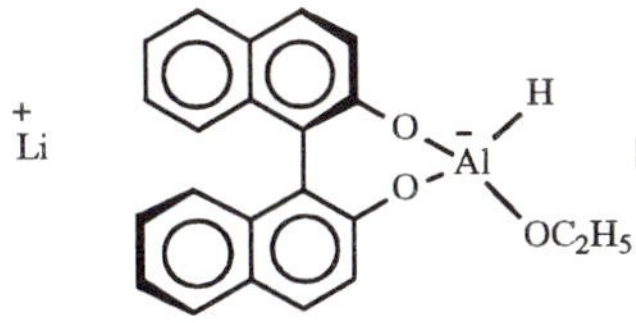

K-Glucoride
M.W. 420.38

Li$^+$ [156]

Binal-H
M.W. 364.28

magnesium hydride [159]. Alkylchlorosilanes are transformed to *alkylsilanes* [160] (equation 35). *Triethylsilane* reduces alkyl halides [161] and desulfurizes thiol esters to aldehydes [162]. *Phenylsilane* [163] deoxygenates phosphine oxides [164], and *dimethylphenylsilane* [165] is used for stereoselective reduction of ketones [166]. *Trichlorosilane* with trialkylamines reduces polyhalogen compounds [167] and converts carboxyl groups into methyls [168] (structure group 36).

35 4 Et$_3$SiCl + LiAlH$_4$ $\longrightarrow$ 4 Et$_3$SiH + LiCl + AlCl$_3$ [160]
M.W. 116.28
b.p. 109°C, d 0.728

36 PhSiH$_3$ [163]   Me$_2$PhSiH [165]   Cl$_3$SiH [167]
M.W. 108.21   M.W. 136.27   M.W. 135.46

Alkyl and aryltin chlorides are reduced to mono-, di-, and trisubstituted stannanes of which *dibutyl-* and *diphenylstannanes* and *tributyl-* and *triphenylstannanes* are used most frequently [169, 170, 171] (equations 37–39). Their specialty is replacement of halogens in all types of organic halides, but they can also selectively reduce α,β-conjugated double bonds [171] and aldehydes to alcohols [172]. A specially modified stannane, (o-*dimethylaminomethyl)phenyldimethylstannane*, hydrogenolyzes aliphatic and aromatic halogens and reduces acetylenic ketones to acetylenic alcohols [173, 174] (equation 40).

Reduction of cuprous chloride with sodium borohydride in the presence of trivalent phosphorus compounds gives complex *copper hydride*, which is a highly selective agent for the preparation of aldehydes from acyl chlorides [175]. Another copper hydride, a hexamer of [Ph$_3$PCuH]$_6$ (M.W. 2560.92), reduces double bonds conjugated with carbonyl groups, either in the presence of chlorotrimethylsilane or hydrogen [176]. *Triphenylphosphinocopper hydride hexamer* is used for conjugate reduction of α,β-unsaturated ketones [176] (equation 41).

37 4 Bu$_3$SnCl + LiAlH$_4$ $\longrightarrow$ 4 Bu$_3$SnH + LiCl + AlCl$_3$ [170]
M.W. 291.05
b.p. 80°C/0.4 mm, d 1.082

38   4 Ph$_3$SnCl + LiAlH$_4$ $\longrightarrow$ 4 Ph$_3$SnH + LiCl + AlCl$_3$ [170]
                                                    M.W. 351.02

39   2 Ph$_2$SnCl$_2$ + LiAlH$_4$ $\longrightarrow$ 2 Ph$_2$SnH$_2$ + LiCl + AlCl$_3$ [169]
                                                    M.W. 274.92

40   [benzene ring with CH$_2$NMe$_2$ and SnMe$_2$Br substituents] + LiAlH$_4$ $\longrightarrow$ [benzene ring with CH$_2$NMe$_2$ and SnMe$_2$H substituents]   [173,174]
                                                    M.W. 283.96

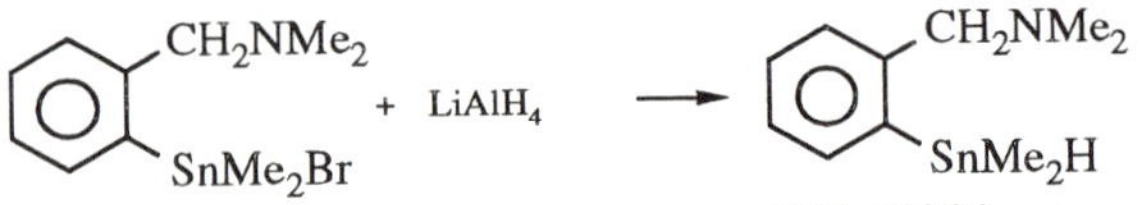

41   CuCl + NaBH$_4$ + 2 Ph$_3$P $\longrightarrow$ (Ph$_3$P)$_2$Cu[B with H H / H H] + NaCl   [175]

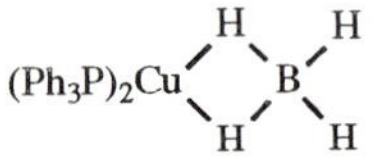

In more recent publications a new, more systematic nomenclature for hydrides and complex hydrides has been adopted. Examples of both nomenclatures are as follows:

Old	New
Lithium aluminum hydride	Lithium tetrahydroaluminate
	Lithium tetrahydridoaluminate
Sodium borohydride	Sodium tetrahydroborate
	Sodium tetrahydridoborate
Lithium trialkoxyaluminum hydride	Lithium trialkoxyhydridoaluminate
Sodium bis(2-methoxyethoxy)aluminum hydride	Sodium bis(2-methoxyethoxy)dihydroaluminate
Diisobutylaluminum hydride	Diisobutylalane
Tributyltin hydride	Tributylstannane

Because the majority of publications quoted here use the older terminology, it is still used predominantly herein.

MECHANISM, STOICHIOMETRY, AND STEREOCHEMISTRY OF REDUCTIONS WITH HYDRIDES

The reaction of complex hydrides with carbonyl compounds can be exemplified by the reduction of an aldehyde with *lithium aluminum hydride*. The reduction is assumed to involve a **hydride transfer from a nucleophile**, tetrahydroaluminate ion, onto the carbonyl carbon as a place of the lowest electron density. The alkoxide ion thus generated complexes the remaining aluminum hydride and forms an alkoxytrihydroaluminate ion. This intermediate reacts with a second molecule of the aldehyde and forms a dialkoxy-dihydroaluminate ion that reacts with the third molecule of the aldehyde and forms a trialkoxyhydroaluminate ion. Finally, the fourth molecule of the aldehyde converts the aluminate to the ultimate stage of tetraalkoxyaluminate ion that on contact with water liberates four molecules of an alcohol, aluminum hydroxide, and lithium hydroxide. Four molecules of water are needed to hydrolyze the tetraalkoxyaluminate. The individual intermediates really exist and can also be prepared by a reaction of lithium aluminum hydride with alcohols. In fact, they themselves are, as long as they contain at least one unreplaced hydrogen atom, reducing agents (equation **42**).

$$
\begin{array}{c}
\textbf{42} \quad \underset{\overset{\curvearrowleft}{H^- \underline{Al}H_3}}{RCH{=}O} \quad \overset{+}{Li} \quad \longrightarrow \quad RCH_2 - \underset{\underset{\underline{Al}H_3}{|}}{O} \quad \overset{+}{Li} \quad \overset{RCH{=}O}{\longrightarrow} \quad RCH_2 - \underset{\underset{\overset{|}{\underset{OCH_2R}{}}}{\underset{A\bar{l}H_2}{|}}}{O} \quad \overset{+}{Li}
\end{array}
$$

$$
\downarrow RCH{=}O
$$

$$
\begin{array}{c}
\underset{Al(OH)_3}{\overset{LiOH}{+}} \; 4\,RCH_2OH \quad \overset{4\,H_2O}{\longleftarrow} \quad RCH_2 - \underset{\underset{OCH_2R}{|}}{O} \quad \overset{+}{Li} \\
RCH_2O{-}\,A\bar{l}{-}OCH_2R \quad \overset{RCH{=}O}{\longleftarrow} \quad RCH_2 - \underset{\underset{OCH_2R}{|}}{O} \quad \overset{+}{Li} \\
A\bar{l}H{-}OCH_2R
\end{array}
$$

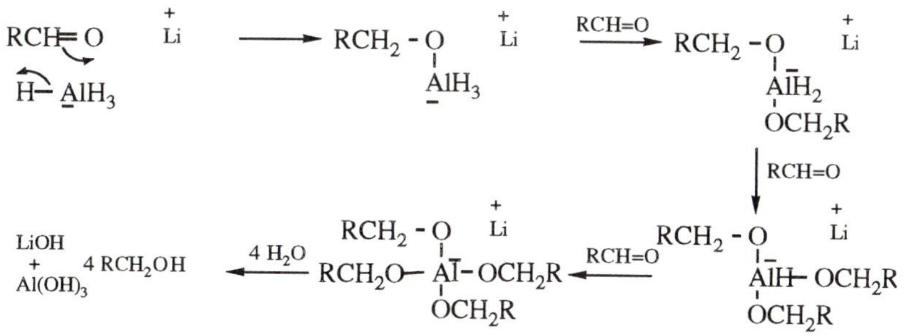

Equation **42** shows that 1 mol of lithium aluminum hydride can reduce as many as four molecules of a carbonyl compound, aldehyde, or ketone. *The stoichiometric equivalent of lithium aluminum hydride* is therefore one-fourth of its molecule, that is, 9.5 g/mol, as much as 2 g or 22.4 L of hydrogen. Decomposition of 1 mol of lithium aluminum hydride with water generates four molecules of hydrogen, four hydrogens from the hydride, and four from water.

The stoichiometry determines the ratios of lithium aluminum hydride to other compounds to be reduced. Esters or tertiary amides treated with one hydride equivalent (one-fourth of a molecule) of lithium aluminum hydride are reduced to the stage of aldehydes or their nitrogen analogs. To reduce an ester to the corresponding alcohol, two hydride equivalents, that is, 0.5 mol of lithium aluminum hydride, are needed because, after the reduction of the carbonyl, hydrogenolysis requires one more hydride equivalent (equation **43**).

$$
\textbf{43} \quad RC\underset{\underset{OR'}{\diagdown}}{\overset{\overset{O}{\diagup\!\!\diagup}}{}} \quad \overset{0.25\ LiAlH_4}{\longrightarrow} \quad RCH\underset{\underset{OR'}{\diagdown}}{\overset{\overset{OH}{\diagup}}{}} \quad \rightleftharpoons \quad RCHO + R'OH
$$

$$
\downarrow 0.25\ LiAlH_4
$$

$$
RCH_2OH \; + \; R'OH
$$

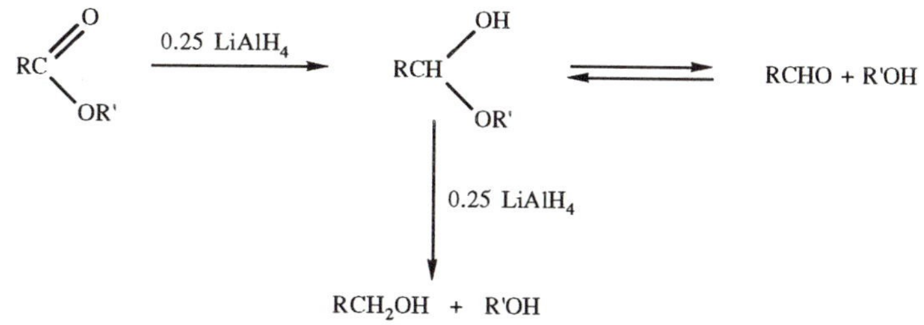

Free acids require still an additional hydride equivalent because their acidic hydrogens combine with one hydride ion of lithium aluminum hydride forming acyloxytrihydroaluminate ion. Complete reduction of free carboxylic acids to alcohols requires 0.75 mol of lithium aluminum hydride (equation **44**).

44 $4\ RC\overset{O}{\underset{OH}{\diagup\kern-1em\diagdown}}\ +\ LiAlH_4\ \longrightarrow\ 4\left[RC\overset{O}{\underset{O}{\diagup\kern-1em\diagdown}}\right]^{-}\overset{3+}{Al}\overset{+}{Li}\ +\ 4\,H_2$

$4\left[RC\overset{O}{\underset{O}{\diagup\kern-1em\diagdown}}\right]^{-}\overset{3+}{Al}\overset{+}{Li}\ +\ 2\,LiAlH_4\ \longrightarrow\ 4\left[RCH_2O\right]^{-}\overset{3+}{Al}\overset{+}{Li}\ +\ 2\,LiAlO_2$

$2\downarrow H_2O$

$4\ RCH_2OH\ +\ LiAlO_2$

$4\ RCO_2H\ +\ 3\,LiAlH_4\ +\ 2\,H_2O\ =\ 4\,RCH_2OH\ +\ 3\,LiAlO_2\ +\ 4\,H_2$

The same amount, 0.75 mol, is needed for reduction of monosubstituted amides to secondary amines. Unsubstituted amides require one full mole of lithium aluminum hydride because one half reacts with two acidic hydrogens and the second half achieves the reduction.

The stoichiometry of lithium aluminum hydride reductions with other compounds such as nitriles, epoxides, and sulfur- and nitrogen-containing compounds is discussed in original papers [123, 177, 178, 179] and surveyed in Table 2.

In practice often an excess of lithium aluminum hydride over the stoichiometric amount is used to compensate for the impurities in the assay. In truly precise work, when it is necessary to use an exact amount, either because any overreduction is to be avoided or because kinetic and stoichiometric measurements are to be carried out, it is necessary to determine the contents of the pure hydride in the solution analytically (Procedure 13, p 301).

Theoretical equivalents of various hydrides are listed in Table 2.

A similar mechanism and stoichiometry underlie reactions of organic compounds with *lithium* and *sodium borohydrides*. With modified complex hydrides, the stoichiometry depends on the number of hydrogen atoms present in the molecule.

The mechanism of reduction by *boranes* and *alanes* differs somewhat from that of complex hydrides. The main difference is in the entirely different chemical nature of the two types. Whereas complex hydride anions are strong **nucleophiles** that attack the places of the lowest electron density, boranes and alanes are *electrophiles* and combine with that part of the organic molecule that has a free electron pair [180]. Alkoxyboranes or alkoxyalanes are formed by a hydride transfer until all three hydrogen atoms from the boron or aluminum have been transferred and the borane or alane is converted to alkoxyboranes (trialkylborates) or alkoxyalanes (equation **45**).

In contrast to the ionic (nucleophilic or electrophilic) mechanism of reductions with complex hydrides and hydrides of boron and aluminum, respectively, the mechanism of the reactions with stannanes is predominantly *free radical* in nature. Reductions are enhanced by initiators such as azobisiso-butyronitrile (AIBN) and others and are usually carried out in aprotic solvents.The polarity of the solvent may influence the mechanism of reduction and even the results [174] (equation **46**).

Table 2. Stoichiometry of Hydrides and Complex Hydrides

Starting Compound	Product	Stoichiometric Hydride Equivalent[a]
RCl, RBr, RI	RH	1
RNO_2	RNH_2	6
$2 RNO_2$	$RN=NR$	8
ROH	RH	2
$R_2\overset{O}{\overset{/\backslash}{C}}-CR'_2$	$R_2C(OH)CHR'_2$	1
RSSR	RSH	2
RSO_2Cl	RSH	6
RCHO, RR'CO	RCH_2OH, RR'CHOH	1
RCO_2H	RCH_2OH	3
RCOCl	RCHO	1
	RCH_2OH	2
$(RCO)_2O$	$2 RCH_2OH$	4
RCO_2R'	RCHO + R'OH	1
	RCH_2OH + R'OH	2
$RCONR_2'$	RCHO + HNR'_2	1
	$RCH_2NR'_2$	2
RCONHR'	RCH_2NHR	3
$RCONH_2$	RCH_2NH_2	4
RCN	RCHO	1
	RCH_2OH	2

Source: Abstracted from A. Hajos, *Complex Hydrides and Related Reducing Agents in Organic Synthesis*, Elsevier, New York, 1979.

[a]One hydride equivalent is equal to 0.25 equivalent of $LiAlH_4$ or $NaBH_4$, 0.33 equivalent of BH_3 or AlH_3, 0.5 equivalent of $NaAlH_2$ (OCH_2CH_2OMe), and 1 equivalent of $LiAlH(OR)_3$ or R_3SnH.

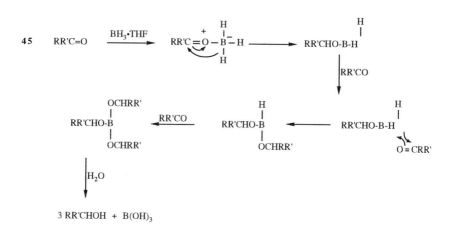

46   [174]

$$
C_6H_5CCH_2Br \ \ + \ \ \text{[aryl]} \ SnMe_2H \xrightarrow[\substack{MeOH \\ 20°C}]{} \ \ C_6H_5CHCH_2Br \ \ \ 87\%
$$

with the reaction branching:

	OH	
MeOH 20°C	C₆H₅CHCH₂Br	87%
THF 20°C	O‖ C₆H₅CCH₃	86%

A similar dichotomy applies to the mechanisms of reductions with silanes, which can be affected by free radical initiators [161], or they can work in strongly polar solvents such as trifluoroacetic acid [181].

Hydrides and complex hydrides tend to approach the molecule of a compound to be reduced from the less hindered side *(steric approach control).* If a relatively uninhibited function is reduced, the final stereochemistry is determined by the stability of the product *(product development control).* In addition, torsional strain in the transition state affects the steric outcomes of the reduction.

In many cases, the reduction agent itself also influences the result of the reduction, especially if it is bulky and the environment of the function to be reduced is crowded. A more detailed discussion of stereochemistry of reduction with hydrides is found in the section on ketones (p 149, 154, 155, 155 and 162).

HANDLING OF HYDRIDES AND COMPLEX HYDRIDES

Many of the hydrides and complex hydrides are now commercially available. Some of them require certain precautions in handling. Lithium aluminum hydride is usually packed in polyethylene bags enclosed in metal cans. It should not be stored in ground-glass bottles.

Opening of a bottle in which some particles of lithium aluminum hydride were squeezed between the neck and the stopper caused a fire [67]. Lithium aluminum hydride must not be crushed in a porcelain mortar with a pestle. Fire and even explosion may result from contact of lithium aluminum hydride with small amounts of water or moisture. Sodium bis(2-methoxyethoxy)aluminum hydride (Vitride or Red-Al) delivered in benzene or toluene solutions also may ignite in contact with water. Borane (diborane) ignites in contact with air and is therefore kept in solutions in tetrahydrofuran or in complexes with amines and sulfides. Powdered lithium borohydride may ignite in moist air. A severe explosion resulting in serious injury occurred when a screw-cap glass bottle containing some 25 g of sodium borohydride was opened [182]. Sodium cyanoborohydride, on the other hand, is considered safe.

Some of the hydrides and complex hydrides are moisture-sensitive, and therefore hypodermic syringe techniques and work under an inert atmosphere are advisable and sometimes necessary. Because of high toxicity of some of the hydrides (borane) and complex hydrides (sodium cyanoborohydride), and for other safety reasons, all work with these compounds should be done in hoods. Specifics about handling individual hydrides and complex hydrides are to be found in the particular papers or reviews [179].

Reductions with hydrides and complex hydrides are usually carried out by mixing solutions. Only sodium borohydride and some others are sometimes added portionwise as solids. Because some of the complex hydrides such as lithium aluminum hydride are not always completely pure and soluble without residues, it is wise to place the solutions of the hydrides in the reaction flask and add the reactants or their solutions from separatory funnels or by means of hypodermic syringes.

If, however, an excess of the reducing agent is to be avoided and a solution of the hydride is to be added to the compound to be reduced (by means of the so-called "inverse technique"), special care must be taken to prevent undissolved particles from clogging the stopcock of a separatory funnel.

When sparingly soluble compounds have to be reduced, they may be transported into the reaction flask for reduction by extraction in a Soxhlet apparatus surmounting the flask [*110, 183*].

In many reactions, especially with compounds that are only partly dissolved in the solvent used, efficient stirring is essential. A paddle-type "trubore" stirrer is preferable to magnetic stirring, particularly when larger amounts of reactants are handled.

Solubilities of the most frequently used hydrides and complex hydrides in most often used solvents are listed in Table 3. Not only the solubility, but also the boiling points of the reactants must be considered when choosing the solvent in case the reduction requires heating.

Isolation of the product varies from one reducing agent and from one reduced compound to another. The general feature is the decomposition of the reaction mixture after the reaction with ice or water followed by acid or alkali. The reaction of the products of *lithium aluminum hydride* reduction with water is very exothermic, especially for reductions in which an excess of lithium aluminum hydride has been used. In such cases it is advisable to decompose the unreacted hydride by addition of ethyl acetate (provided its reduction product, ethanol, does not interfere with the isolation of the products). Then normal decomposition with water is carried out followed by acids [*110*] or bases [*184*].

Table 3. Solubilities of Selected Complex Hydrides in Common Solvents

Hydride Solvent	H_2O	MeOH	EtOH	Et_2O	THF	Dioxane	$(CH_2OMe)_2$	Diglyme
$LiAlH_4$				35–40	13	0.1	7 (12)[a]	5 (8)[a]
$LiBH_4$		~2.5 (dec.)	~2.5 (dec.)	2.5	28	0.06		
$NaBH_4$	55 (88.5)[b]	16.4 (dec.)	20 (dec.)		0.1		0.8	5.5
$NaBH_3CN$	212	very sol.	slightly sol.	0	37.2			17.6
$NaBH(OMe)_3$						1.6 (4.5)[a]		
$LiAlH(OCMe_3)_3$				2	36		4	41

Note: All values are given in grams per 100 g of solvent at 25 ± 5 °C.
Source: Abstracted from A. Hajos, *Complex Hydrides and Related Reducing Agents in Organic Synthesis*, Elsevier, New York, 1979.
[a] At 75 °C.
[b] At 60 °C.

If the reduction has been carried out in ether, the ether layer is separated after the acidification with dilute hydrochloric or sulfuric acid. Sometimes, especially when not very pure lithium aluminum hydride has been used, a gray voluminous emulsion is formed between the organic and aqueous layers. Suction filtration of this emulsion over a fairly large Büchner funnel is often helpful. In other instances, especially in the reductions of amides and nitriles when amines are the products, decomposition with alkalis is in order. With certain amounts of sodium hydroxide of proper concentration a granular byproduct, sodium aluminate, may be separated without problems [*184*].

Isolation of products from the reductions with *sodium borohydride* is in the majority of cases much simpler. The reaction is carried out in aqueous or aqueous–alcoholic solutions, and so extraction with ether is usually sufficient. Acidification of the reaction mixture with dilute mineral acids may precede the extraction.

Decomposition of the reaction mixtures with water followed by dilute acids applies also to the reductions with *boranes* and *alanes*. Modifications are occasionally needed, for example hydrolysis of esters of boric acid and the alcohols formed in the reduction. Heating of the mixture with dilute mineral acid or dilute alkali is sometimes necessary.

The domain of hydrides and complex hydrides is reduction of carbonyl functions (in aldehydes, ketones, acids, and acid derivatives). With the exception of boranes, which add across carbon–carbon multiple bonds and afford, after hydrolysis, hydrogenated products, isolated carbon–carbon double bonds resist reduction with hydrides and complex hydrides. However, a conjugated double bond may be reduced by some hydrides, as well as a triple bond to the double bond. Many other functions are reduced by the hydride reagents or their modifications. Examples of applications of hydrides are shown in Procedures 14–28 (pp 302-308).

SUMMARY

In contrast to catalytic hydrogenation, complex hydrides do not reduce isolated multiple bonds. Being polar compounds, they have higher affinity to polar bonds: carbon–oxygen and carbon–nitrogen multiple bonds and multiple bonds conjugated with these functions. Within the systems of such polar bonds, very high selectity is achieved by using variously modified hydrides and complex hydrides. Multifunctional systems of polar bonds are selectively reduced better than they are by catalytic hydrogenation.

The most important feature of reductions with hydrides and complex hydrides is stereo- and enantioselectivity using chiral reagents. Enantiomeric excesses in some cases approach those obtained by biochemical reduction, and the yields are usually higher.

Isolation of products of reduction with hydrides and complex hydrides is often more lengthy than that of heterogeneous catalytic hydrogenation, because the workup in two-phase systems of water and organic solvents requires more time, especially if water-miscible solvents have been used and emulsions are formed. Work with some hydrides and complex hydrides requires strict safety precautions, as described on p xxiii.

CHAPTER 3

Electroreduction and Reductions with Metals

Dissolving metal reductions were among the first reductions of organic compounds discovered some 130 years ago. Although overshadowed by more universal catalytic hydrogenation and metal hydride reductions, metals are still used for reductions of polar compounds and selective reductions of specific types of bonds and functions. Almost the same results are obtained by electrolytic reduction.

MECHANISM AND STEREOCHEMISTRY

Reduction is defined as acceptance of electrons. Electrons can be supplied by an electrode (cathode) or else by dissolving metals. If a metal goes into solution it forms a cation and gives away electrons. A compound to be reduced, for example, a ketone, accepts one electron and changes to a radical anion A. In the absence of protons the radical anion may accept another electron and form a dianion B. Such a process is not easy because it requires an encounter of two negative species, an electron and a radical anion, and the two negative sites are close together. It takes place only with compounds that can stabilize the radical anion and the dianion by resonance.

Rather than accepting another electron, the radical anion A may combine with another radical anion and form a dianion of a dimeric nature C. This intermediate is formed more readily in the aromatic series.

In the presence of protons, the initial radical anion A is protonated to a radical D, which has two options: either to couple with another radical to form a pinacol E, or to accept another electron to form an alcohol F. The pinacol E and the alcohol F may also result from double protonation of the doubly charged intermediates C and B, respectively (equation **47**).

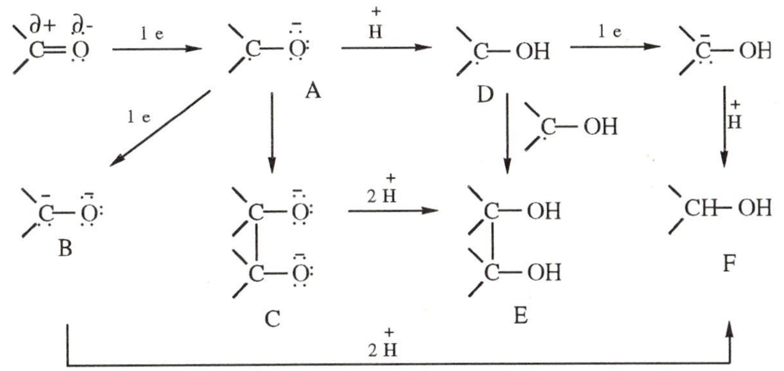

A similar process may take place in the reduction of polar compounds with single bonds. By accepting an electron, a halogen, hydroxy, sulfhydryl, or amino derivative dissociates into a radical and an anion. In aprotic solvents the two radicals combine. In halogen derivatives, the result is Wurtz synthesis. In the presence of protons the anion is protonated and the radical accepts another electron to form an anion that, after protonation, gives a hydrocarbon or a product in which the substituent has been replaced by hydrogen (equation **48**).

$$\textbf{48} \quad R-X \xrightarrow{\text{1 e}} R\cdot + \bar{X}$$

R-R ⟵ R· / \ 1 e ⟶ $\bar{R}$: $\xrightarrow{\overset{+}{H}}$ RH

X = Hal, OH, SH, NH$_2$

The reductions just described are quite frequent with compounds containing polar multiple bonds such as carbon–oxygen, carbon–nitrogen, and nitrogen–oxygen [*185*]. Even carbon–carbon bond systems can be reduced in this way, but only if the double bonds are conjugated with a polar group or at least with another double bond or aromatic ring. Dissolving metal reduction of an isolated carbon–carbon double bond is extremely rare and has no practical significance. On the other hand, reduction of conjugated dienes and aromatic rings is feasible and becomes easier with increasing resonance of the charged intermediates (anthracene and phenanthrene are reduced much more readily than naphthalene and that in turn more readily than benzene). Reductions are especially successful when the source of protons (compounds with acidic hydrogens, alcohols, water, or acids) is present in the reducing medium. In such cases dimerization reactions are suppressed and good yields of fully reduced products are obtained (e.g., the *Birch reduction* with alkali metals, and reduction with zinc and other metals).

Because of the stepwise nature of the reductions by acceptance of electrons, the net result of reductions with metals is usually *anti* or *trans* addition of hydrogen. Thus disubstituted acetylenes give predominantly *trans*-alkenes, *cis*-alkenes give racemic mixtures and *threo* forms, and *trans*-alkenes give *meso* or *erythro* forms. Reduction of 4-oxocyclohexanecarboxylic acid with sodium amalgam gave *trans*-4-hydroxycyclohexanecarboxylic acid (with both substituents equatorial) [*186*], and reduction of 2-isopropylcyclohexanone with sodium in moist ether gave 90–95% *trans*-2-isopropylcyclohexanol [*187*]. In other cases conformational effects play an important role [*188*].

ELECTROLYTIC REDUCTION

The most important factor in electrolytic reduction (electroreduction) is the nature of the metal used as a *cathode*. Metals of low overvoltage such as platinum (0.005–0.09 V), palladium, nickel, and iron give results of reduction similar to those obtained with catalytic hydrogenation [*189*]. Cathodes made of metals of high overvoltage such as copper (0.23 V), cadmium (0.48 V), lead (0.64 V), zinc (0.70 V), or mercury (0.78 V) produce results similar to those of dissolving metal reductions.

Another important factor in electroreduction is the *electrolyte*. Most electrolytic reductions are carried out in more or less dilute sulfuric acid, but some are done in alkaline electrolytes such alkali hydroxides, alkoxides, or solutions of salts like tetramethylammonium chloride in methanol [*190*] or lithium chloride in alkylamines [*191, 192*].

Results may also differ with electrolytic cells that are divided (by a diaphragm separating the cathode and anode spaces) or undivided [*191*].

The apparatus for electrolytic reductions (Figure 5) may be an open vessel for work with nonvolatile compounds, or a closed container fitted with a reflux condenser for work with compounds of high vapor pressure. The electro-reduction in exothermic, and cooling must sometimes be applied [*193, 194, 195*].

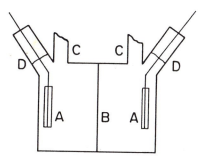

Figure 5. Electrolytic cell. Key: A, electrodes; B, diaphragm; C, reflux condensers; and D, glass seals.

The size of electrodes and the current passed through the electrolytic cell determine the current density, which in most reductions is in the range 0.05–0.2 A/cm^2. The current in turn is determined by the imposed voltage and conductivity of the electrolyte. The amount of electricity used for electroreduction of 1 g-equiv of a compound is 96,500 C (26.8 A-h). In practice about twice the amount is used (current yield of 50%) (Procedure 29, p 308).

Examples of electroreduction are partial reduction of aromatic rings [*194, 196*], reduction of halo epoxides to epoxides [*197*], desulfurization of sulfoxides [*198*], and reduction of α, β-unsaturated ketones [*199*].

ALKALI METALS

The reducing powers of metals parallel their relative electrode potentials, that is, potentials developed when the metal is in contact with a normal solution of its salts. The potential óf hydrogen being equal to 0, the potentials or electrochemical series of some elements are as shown in Table 4.

Metals with large negative potentials, like alkali metals and calcium, are able to reduce almost anything, even carbon–carbon double bonds. Metals with low potentials (e.g., iron and tin) can reduce only strongly polar bonds (such as nitro groups but, generally, not carbonyl groups).

Sodium, lithium, and *potassium* dissolve in liquid ammonia (m.p. –77.7 °C, b.p. –33.4 °C). They form blue solutions that are stable unless traces of metals such as iron or its compounds are present. These trace metals catalyze the reaction between the metal and ammonia to form alkali amide and hydrogen.

**Table 4. Physical Constants and Relative
Electrode Potentials of Some Metals**

Metal	Li	K	Na	Al	Zn	Fe	Sn
Relative electrode potential (volts)	–2.9	–2.9	–2.7	–1.34	–0.76	–0.44	–0.14
Atomic weight	6.94	39.10	22.99	26.98	65.37	55.85	118.69
Melting point (°C)	179	63.7	97.8	660	419.4	1535	231.9
Density	0.53	0.86	0.97	2.70	7.14	7.86	7.28

Solubilities of lithium, sodium, and potassium in liquid ammonia at its boiling point are (in grams per 100 g) 10.4, 24.5, and 47.8, which correspond to mole ratios of 0.25, 0.18, and 0.21, respectively [200]. In practice usually more dilute solutions are used. The majority of dissolving metal reductions are carried out in the presence of proton donors, most frequently methanol, ethanol, and *tert*-butyl alcohol (this is the *Birch reduction*, Procedure 30, p 309). The function of these donors is to protonate the intermediate anion radicals and thus to cut down on undesirable side reactions such as dimerization of the radical anion and polymerization. Other proton sources such as *N*-ethylaniline and ammonium chloride are used less often. Cosolvents such as tetrahydrofuran help increase mutual miscibility of the reaction components.

The reductions are usually carried out in three-necked flasks fitted with efficient cold finger reflux condensers filled with dry ice and acetone, mechanical or magnetic stirrers, and inlets for ammonia and the metal. Ammonia may be introduced as a liquid from the bottom of an ammonia cylinder or as a gas that condenses in the apparatus. If ammonia is introduced as a gas, the reaction flasks should be cooled with dry ice–acetone baths to accelerate the condensation. When ammonia without traces of water is necessary, it can be first introduced into a flask containing pieces of sodium and then evaporated into the proper reaction flask where it condenses. The opening for the introduction of ammonia is also used for adding the pieces of metal to the flask. After the reduction has been completed, the reflux condenser is removed, and ammonia is allowed to evaporate. Unreacted metal is decomposed by addition of ammonium chloride or, better still, finely powdered sodium benzoate. Evaporation of ammonia is slow. It is best done overnight. It can be accelerated by gentle heating of the reaction flask. The final workup depends on the chemical properties of the products. A similar technique can be used for reductions with calcium.

Reductions in liquid ammonia run at atmospheric pressure at a temperature of –33 °C. If higher temperatures are necessary for the reduction, other solvents of alkali metals are used: methylamine (b.p. –6.3 °C) [201], ethylamine (b.p. 16.6 °C) [202, 203], and ethylenediamine (b.p. 116–117 °C) [204].

Many reductions with sodium are carried out in boiling alcohols: in methanol (b.p. 64 °C), ethanol (b.p. 78 °C), 1-butanol (b.p. 117–118 °C), and isoamyl alcohol (b.p. 132 °C). More intensive reductions are achieved at higher temperatures. For example, reduction of naphthalene with sodium in ethanol gives 1,4-dihydronaphthalene, whereas in boiling isoamyl alcohol tetralin is formed.

Reductions carried out by adding sodium to a compound in boiling alcohols require large excesses of sodium and alcohols. A better procedure is to carry out such reductions by adding a stoichiometric quantity of an alcohol and the compound in toluene or xylene to a mixture of toluene or xylene and a calculated amount of a dispersion of molten sodium [*205*] (p 213). Reductions with sodium can also be accomplished in moist ether [*187*] and other solvents (Procedures 31 and 32, pp 309 and 310).

A form of sodium suitable for reductions in aqueous media is *sodium amalgam*. It is easily prepared in concentrations of 2–4% by dissolving sodium in mercury (Procedure 33, p 310).

Reductions with sodium amalgam are very simple. The amalgam is added in small pieces to an aqueous or aqueous alcoholic solution of the compound to be reduced contained in a heavy-walled bottle. The contents of the stoppered bottle are shaken energetically until the solid pieces of the sodium amalgam change to liquid mercury. After all the sodium amalgam has reacted, the mercury is separated and the aqueous phase is worked up. Sodium amalgam generates sodium hydroxide. If the alkalinity of the medium is undesirable, an acid or a buffer such as boric acid [*206*] must be added to keep the pH of the solution in the desirable range.

Reductions with sodium amalgam are fairly mild. Only easily reducible groups and conjugated double bonds are affected. With the availability of sodium borohydride, the use of sodium amalgam is dwindling even in the field of saccharides, where sodium amalgam has been widely used for reduction of aldonic acids to aldoses.

MAGNESIUM AND CALCIUM

Applications of *magnesium* are rather limited. Halogen derivatives, especially bromides, iodides, and imidoyl chlorides, react with magnesium to form Grignard reagents that, on decomposition with water or dilute acids, give compounds in which the halogen has been replaced by hydrogen [*207, 208*]. The reduction can also be carried out by the simultaneous action of isopropyl alcohol [*209*] and magnesium activated by iodine. Another specialty of magnesium is reduction of ketones to pinacols carried out by magnesium amalgam. The amalgam is prepared in situ by adding a solution of mercuric chloride in the ketone to magnesium turnings submerged in benzene [*210*].

Calcium in liquid ammonia hydrogenolyzes benzyl ethers [*211*]; in ethylenediamine or its mixture with methylamine, calcium reduces alkynes to alkenes or alkanes [*212*]; in butylamine and ethylenediamine in the presence of *tert*-butyl alcohol, calcium achieves partial reduction of aromatic rings [*213*]. Both magnesium and calcium reduce azides to amines [*214*].

ALUMINUM

Reductions with aluminum are carried out almost exclusively with aluminum amalgam. This is prepared by immersing strips of a thin aluminum foil in a 2% aqueous solution of mercuric chloride for 15–60 s, decanting the solution, rinsing the strips with absolute ethanol, then rinsing with ether, and cutting them with scissors into pieces of approximately 1 cm^2 [215, 216]. In aqueous and nonpolar solvents, aluminum amalgam reduces cumulative double bonds [217], ketones to pinacols [218], halogen compounds [219], nitro compounds [220, 221], azo compounds [222], azides [223], oximes [224], and quinones [225], and cleaves sulfones [215, 226, 227] and phenylhydrazones [228] (Procedure 34, p 311). Aluminum activated by iodine in liquid ammonia reduces disulfides to mercaptans [229].

Several reductions have been achieved by dissolving a *nickel–aluminum alloy* containing usually 50% of aluminum in 20–50% aqueous sodium hydroxide in the presence of a reducible compound. Under such conditions elemental hydrogen is generated by dissolution of aluminum, and Raney nickel is formed in this process, and therefore such reductions have to be considered catalytic hydrogenations rather than dissolving metal reductions. Saturation of pyridine rings certainly points to catalytic hydrogenation [230] (Procedure 35, p 311).

ZINC

Zinc has been, and still is, after sodium the most abundantly used metal reductant. It is available mainly in the form of zinc dust and in a granular form referred to as mossy zinc.

Zinc dust is frequently covered with a thin layer of zinc oxide that deactivates its surface and causes induction periods in reactions with compounds. This disadvantage can be removed by proper activation of zinc dust immediately prior to use. Such an activation can be achieved by a 3–4-min contact with very dilute (0.5–2%) hydrochloric acid followed by washing with water, ethanol, acetone, and ether [231]. Similar activation is carried out in situ by a small amount of anhydrous zinc chloride [232] or zinc bromide [233] in alcohol, ether, or tetrahydrofuran. Another way of activating zinc dust is by its conversion to a zinc–copper couple by stirring it (180 g) with a solution of 1 g of copper sulfate pentahydrate in 35 mL of water [234].

Mossy zinc is activated by conversion to zinc amalgam by brief immersion in a dilute aqueous solution of mercuric chloride and decantation of the solution before the reaction proper (40 g of mossy zinc, 4 g of mercuric chloride, 4 mL of concentrated hydrochloric acid, and 40 mL of water [235]). This type of activation is especially used in the Clemmensen reduction, which converts carbonyl groups to methylene groups [236] (Procedure 36, p 312).

In the Clemmensen reduction (which does not proceed through a stage of an alcohol) the indispensable strong acid protonates the oxygen and ultimately causes its elimination as water after a transient formation of an unstable organozinc species. A carbene-like intermediate is formed and gives an alkene [237]. When acetophenone is reduced with zinc in deuterium oxide, it gives α,α-dideuteroethylbenzene and styrene. If the reduction is carried out with ω,ω,ω-trideuteroacetophenone, α,β,β-trideuterostyrene is obtained [238].

Reductions with zinc are carried out in aqueous [236] as well as anhydrous solvents [239] and at different pHs of the medium. The choice of the reaction conditions is very important because entirely different results may be obtained under different conditions. Reduction of aromatic nitro groups in alkali hydroxides or aqueous ammonia gives hydrazo compounds, but reduction in aqueous ammonium chloride gives hydroxylamines, and reduction in acidic medium gives amines (p 96). Of organic solvents the most efficient seem to be *N,N*-dimethylformamide [240] and acetic anhydride [231]. However, alcohols have been used much more frequently. The majority of reductions with zinc are carried out in acids: hydrochloric, sulfuric, formic, and especially acetic. Strongly acidic medium is essential for some reductions such as the Clemmensen reduction (Procedures 36–38, p 312).

Reductions with zinc are usually performed in hot or refluxing solvents, and some take many hours to complete. After the reaction is over, the residual zinc is filtered off, and the products are isolated in ways depending on their chemical nature. If zinc dust has been used, care must be taken in disposing of it because it may be pyrophoric in the dry state.

Zinc is used to a limited extent for reductions of double bonds conjugated with strongly polar groups, for semireduction of alkynes to *trans* alkenes and partial reduction of some aromatic compounds. It is especially suited for reductions of carbonyl groups in aldehydes and ketones to methyl or methylene groups, respectively. It is also able to reduce carbonyl compounds to alcohols and pinacols, is excellent for replacement of halogens by hydrogen, and is, with the exception of catalytic hydrogenation and reductions with complex hydrides, the most versatile reducing agent for aromatic nitro compounds and other nitrogen derivatives. Its reducing properties may be strongly enhanced by addition of *nickel dichloride*. Zinc thus activated reduces even isolated double bonds [241] and saturates pyridine rings [241].

IRON

Iron was one of the first metals employed for the reduction of organic compounds more than 140 years ago. It is used in the form of filings. Best results are obtained with 80-mesh grain [242]. Although some reductions are carried out in dilute or concentrated acetic acid, the majority are performed in water in the presence of small amounts of hydrochloric acid, acetic acid, or salts such as ferric chloride, sodium chloride (as little as 1.5–3%) [242], and ferrous sulfate [243]. Under these conditions iron is converted to iron oxide, Fe_3O_4. Methanol or ethanol is used to increase the solubility of the organic material in the aqueous medium [243] (equation **49**) (Procedure 39, p 312).

$$\textbf{49} \quad 4\,RNO_2 + 9\,Fe + 4\,H_2O \longrightarrow 4\,RNH_2 + 3\,Fe_3O_4$$

Iron is occasionally used for the reduction of aldehydes [244], replacement of very reactive halogens [245, 246], deoxygenation of amine oxides [247], and desulfurizations [248, 249], but its principal domain is the reduction of aromatic nitro compounds (p 96). The advantages over other reducing agents are (1) the possibility of using almost neutral or only a slightly acidic medium and (2) the selectivity of iron, which reduces few functions beside nitro groups.

TIN

Once a favorite reducing agent for the conversion of aromatic nitro compounds to amines, tin is used nowadays only occasionally. The reasons it is not used are its relative unavailability; the necessity for strongly acidic media; and the fact that most reductions achieved by tin can also be accomplished by stannous chloride, which is soluble in aqueous acids and some organic solvents. Reductions with tin require heating with hydrochloric acid, sometimes in the presence of acetic acid or ethanol.

Tin was used for reduction of acyloins to ketones [250], of enediones to diketones [251], of quinones [252], and especially of aromatic nitro compounds [253] (Procedure 40, p 313).

CHAPTER 4

Reductions with Metal Compounds

Compared with tin, *stannous chloride* is only one-half as efficient because it furnishes only two electrons per mole (tin donates four electrons). This disadvantage is more than offset by the solubility of stannous chloride in water, hydrochloric acid, acetic acid, alcohols, and other organic solvents. Thus reductions with stannous chloride can be carried out in homogeneous media (Procedure 41, p 313).

Stannous chloride is used most frequently for the reduction of nitro compounds [254, 255, 256] and quinones [257, 258]. It is also suitable for conversion of imidoyl chlorides [259] and nitriles [260] to aldehydes, for transformations of diazonium salts to hydrazines [261], for reduction of oximes [262], and for deoxygenation of sulfoxides to sulfides [263].

Divalent chromium salts show very strong reducing properties. They are prepared by reduction of chromium(III) compounds with zinc [264] or a zinc–copper couple and form dark blue solutions that are extremely sensitive to air. The most frequently used salts are chromous chloride [265], chromous sulfate [266], and less often chromous acetate [267]. Reductions of organic compounds, for example alkyl halides [267], are carried out in homogeneous solutions in aqueous methanol [268], acetone [269], acetic acid [270], *N,N*-dimethyl-formamide [271], or tetrahydrofuran [272] (Procedure 42, p 313).

Divalent chromium reduces triple bonds to double bonds (*trans* where applicable) [273, 274]; enediones to diones [275]; epoxides to alkenes [270]; and aromatic nitroso, nitro, and azoxy compounds to amines [268]; divalent chromium also deoxygenates amine oxides [269] and replaces halogens with hydrogen [276, 277].

Titanous chloride (*titanium trichloride*) is applied in aqueous solutions, sometimes in the presence of solvents that increase the miscibility of organic compounds with the aqueous phase [278, 279]. Its applications are reduction of nitro compounds [280] and nitroso compounds [281] and cleavage of nitrogen–nitrogen bonds [282], but it is also an excellent reagent for deoxygenation of sulfoxides [283] and amine oxides [278] (Procedure 43, p 314).

Solutions of *low-valence titanium chloride (titanium dichloride)* are prepared in situ by reduction of solutions of titanium trichloride in tetrahydrofuran or 1,2-dimethoxyethane with lithium aluminum hydride [284, 285], with lithium or potassium [286], with magnesium [287, 288], or with a zinc–copper couple [289, 290]. Such solutions effect hydrogenolysis of halogens [288], deoxygenation of epoxides [284], and reduction of aldehydes and ketones to alkenes [285, 287, 289] (pp 113, 142, 151, 156 and 164). Applications of titanium chlorides in reductions have been reviewed [291] (Procedure 44, p 314).

Cerium(III) iodide prepared in situ from an aqueous solution of ceric sulfate by reduction with sulfur dioxide followed by sodium iodide reduces α-bromoketones to ketones in 82–96% yields [292].

Aqueous solutions of *vanadous chloride* (*vanadium dichloride*) are prepared by reduction of vanadium pentoxide with amalgamated zinc in hydrochloric acid

[*293*]. Reductions are carried out in solution in tetrahydrofuran at room temperature or under reflux. Vanadium dichloride reduces α-halo ketones to ketones [*294*], α-diketones to acyloins [*295*], quinones to hydroquinones [*295*], sulfoxides to sulfides [*296*], and azides to amines [*297*] (Procedure 45, p 314).

Molybdenum trichloride is generated in situ from molybdenum pentachloride and zinc dust in aqueous tetrahydrofuran. It is used for conversion of sulfoxides to sulfides [*296*].

Ferrous sulfate is a gentle reducing agent. It is applied in aqueous solutions mainly for reduction of aromatic nitro compounds containing other reducible functions such as aldehyde groups that remain intact [*298*].

Nickel dichloride and zinc reduce alkenes to alkanes, arylalkenes to arylalkanes, nitro compounds to amines, quinolines to 1,2,3,4-tetrahydroquinolines, carbonyl compounds to hydroxy compounds, and nitriles to amines and alcohols [*241*].

Sodium stannite reduces nitroalkenes and is one of the reagents suitable for replacement of a diazonium group in aromatic compounds by hydrogen [*299*].

Two special uses of *sodium arsenite* (applied in aqueous alkaline solutions) are partial reduction of trigeminal halides to geminal halides [*300*] and reduction of aromatic nitro compounds to azoxy compounds [*301*].

Sodium iodide in aqueous tetrahydrofuran or dioxane reduces α-chloro-, α-bromo-, and α-iodoketones to ketones [*302*]. α-Chloro- and α-bromoketones can also be reduced by treatment with sodium iodide and chlorotrimethylsilane in acetonitrile at room temperature [*303*].

Samarium diiodide [*1500*], prepared from samarium and 1,2-diiodoethane [*304, 305*], reduces alkynes to *cis*-alkenes [*306*], halogen derivatives [*307*], and α,β-conjugated double bonds [*308*]; deoxygenates epoxides and sulfoxides [*304*] and phosphine oxides [*309*]; reduces nitro groups [*310*], carbonyl groups, and oximes [*304*]; and cleaves sulfones [*311*], ethers, and esters [*312*] (Procedure 46, p 315).

CHAPTER 5

Reductions with Nonmetal Compounds

HYDROGEN IODIDE

Hydrogen iodide (M.W. 127.93, m.p. –55 °C, b.p. –35.4 °C, density 5.56),* one of the oldest reducing agents, is available as a gas but is used mainly in the form of its aqueous solution, hydriodic acid. Azeotropic hydriodic acid boils at 127 °C, has a density of 1.70, and contains 57% of hydrogen iodide. So-called fuming hydriodic acid is saturated at 0 °C and has a density of 1.93–1.99.

Hydrogen iodide dissociates at higher temperatures to iodine and hydrogen, which effects hydrogenations. The reaction is reversible. Its equilibrium is shifted to the decomposition of hydrogen iodide by the reaction of hydrogen with organic compounds to be reduced, but it can also be affected by the removal of iodine; this can be accomplished by allowing iodine to react with phosphorus to form phosphorus triiodide, which decomposes in the presence of water to phosphorous acid and hydrogen iodide. In this way, by adding phosphorus to the reaction mixture, hydrogen iodide is recycled and the reducing efficiency of hydriodic acid is enhanced [313] (Procedure 47, p 316).

Reductions with hydriodic acid are usually accomplished by refluxing organic compounds with the azeotropic (57%) acid. Acetic acid is added to increase miscibility of the acid with the organic compound. If more energetic conditions are needed, the heating has to be done in sealed tubes. This technique is necessary when fuming hydriodic acid is employed. At present very few such reactions are carried out.

Milder reductions with hydriodic acid can be accomplished by using more dilute hydriodic acid, or solutions of hydrogen iodide prepared from *alkaline iodides* and hydrochloric, acetic, or trifluoroacetic acids in organic solvents [314, 315].

Hydriodic acid is a reagent of choice for reduction of alcohols [316], some phenols [317], some ketones [318, 319], quinones [313], halogen derivatives [314, 320], sulfonyl chlorides [321], diazo ketones [322], azides [323], and even some carbon–carbon double bonds [324]. Under very drastic conditions at high temperatures even polynuclear aromatic compounds and carboxylic acids can be reduced to saturated hydrocarbons, but such reactions are hardly ever used nowadays.

In very reactive halogen derivatives such as α-bromo ketones [325] and α-bromothiophene [326], the halogens are replaced by hydrogen with *hydrogen bromide* in acetic acid provided phenol is added to react with the evolved bromine and to affect favorably the equilibrium of the reversible reaction. Hydrogen bromide in acetic acid also reduces azides to amines [323].

---

*Density as given in this book refers to the ratio of specific gravity of the compound to specific gravity of water, usually designated D or d in the literature.

SULFUR COMPOUNDS

*Hydrogen sulfide** (M.W. 34.08) is probably the oldest reducing agent applied in organic chemistry, having been used by Wöhler in 1838. It is a very toxic gas with a foul odor (m.p. −82.9 °C, b.p. −60.1 °C, density 1.19); it is sparingly soluble in water (0.5% at 10 °C, 0.4% at 20 °C, and 0.3% at 30 °C) and more soluble in anhydrous ethanol (1 g in 94.3 mL at 20 °C) and ether (1 g in 48.5 mL at 20 °C). It must be handled in well-ventilated hoods. It is usually applied in basic solutions in pyridine [327], piperidine [328], and most frequently aqueous ammonia [329, 330] where it acts as *ammonium sulfide* or *ammonium hydrogen sulfide* (Procedure 48, p 316).

Similar reducing effects are obtained from *alkali sulfides* [331, 332], *hydrosulfides*, and *polysulfides* [333]. A peculiar reaction believed to be due to sodium polysulfide formed in situ by refluxing sulfur in aqueous–ethanolic sodium hydroxide is a conversion of *p*-nitrotoluene to *p*-aminobenzaldehyde [334]. Oxidation of the methyl group by the polysulfide generates hydrogen sulfide, which then reduces the nitro group to the amino group.

The reducing properties of organic compounds of sulfur, such as *methyl mercaptan* [335] (M.W. 48.11, m.p. −123 °C, b.p. 6.2 °C, density 0.87), and *ethyl mercaptan* [336] (M.W. 62.13, m.p. −144 °C, b.p. 35 °C, density 0.84), apply to polyhalides, and *thiophenol* (M.W. 110.18, m.p. −14.8 °C, b.p. 169 °C, density 1.08) is used for conversion of dithioacetals to thio ethers [337]. *Dimethyl sulfide* reduces hydroperoxides to alcohols and ozonides to aldehydes while being converted to dimethyl sulfoxide [338].

The main field of applications of hydrogen and alkali sulfides is reduction of nitrogen functions in nitro compounds [327, 330, 333], nitroso compounds [339], azo compounds [340], and azides [341] (all are reduced to amines). Less frequent are reductions of carbonyl groups [328, 342] and sulfonyl chlorides (to sulfinic acids) [343].

The use of *sulfur dioxide* as a reductant is very limited, mainly to the reduction of quinones to hydroquinones [344]. Sulfur dioxide is a gas of pungent odor, soluble in water (17.7% at 0 °C, 11.9% at 15 °C, 8.5% at 25 °C, and 6.4% at 35 °C), ethanol (25%), methanol (32%), ether, and chloroform. It is used for deoxygenation of amine oxides and reduction of quinones (Procedure 49, p 316).

More abundant are reductions with *sodium sulfite*, which is applied in aqueous solutions (solubility 24%). Its specialties are reduction of peroxides to alcohols [345], sulfonyl chlorides to sulfinic acids [346], aromatic diazonium compounds to hydrazines [347], and partial reduction of geminal polyhalides [348] (Procedure 50, p 317).

A very strong reducing agent is *sodium hydrosulfite (hyposulfite* or *dithionite)* $Na_2S_2O_4$ (M.W. 174.13). Like sulfites, it is applied in aqueous solutions. It is widely used in the dyestuff industry for reducing vat dyes. In the laboratory, the main applications are reductions of reactive halides [349], double bonds conjugated with carbonyl groups [350, 351], nitroso compounds [352], nitro compounds [353, 354, 355], azo compounds [356], and azides [357];

---

*Reduction with hydrogen sulfide and its salts is referred to as the *Zinin* reduction (*Org. React.* (N.Y.), **1973**, *20*, 455).

cleavage of sulfones [*358*]; and conversion of aromatic arsonic and stibonic acids to arseno compounds [*359*] and stibino compounds [*360*], respectively. On rare occasions, even carbonyl groups have been reduced by boiling solutions of sodium hydrosulfite in aqueous dioxane or *N*,*N*-dimethylformamide [*361*], and a pyrazine ring was converted to a 1,4-dihydropyrazine [*362*] (Procedures 51 and 52, p 317).

Sodium formaldehyde sulfoxylate (rongalite), $HOCH_2OSONa$ (M.W. 118.08), is used for selective reduction of halo ketones [*363*].

DIIMIDE AND HYDRAZINE

Diimide (diimine or *diazene)*, N_2H_2 or $HN=NH$, is an ephemeral species that results from decomposition with acids of potassium azodicarboxylate [*364, 365*], from the reaction of hydroxylamine with potassium hydroxide and ethyl acetate in *N*,*N*-dimethylformamide [*366*], and from thermal decomposition of anthracene-9,10-diimine [*367, 368*] and hydrazine [*369, 370*] and its derivatives [*371*]. Although this species has not been isolated, its transient existence has been proven by mass spectroscopy and by its reactions in which it hydrogenates organic compounds with concomitant evolution of nitrogen [*372*] (equation **50**).

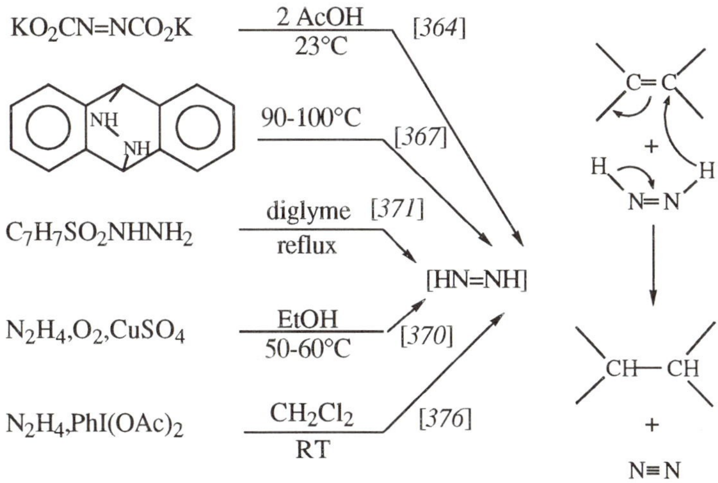

The reaction of diimide with alkenes is a concerted addition, proceeding through a six-membered transition state. It results in predominant *cis (syn)* addition of hydrogen and is therefore very useful for stereospecific introduction of hydrogen onto double bonds. If N_2D_2 is used, deuterium is added to the double bond. No hydrogen–deuterium exchange takes place, as it sometimes does in catalytic hydrogenations [*372, 373*]. Applications of diimide are very rare and are mainly limited to hydrogenation of carbon–carbon multiple bonds [*370, 374, 375, 376, 377, 378*] and azo compounds to hydrazo compounds [*364*].

Hydrazine (M.W. 32.05, b.p. 113.5 °C, density 1.011) is usually available and used in the form of its hydrate (M.W. 50.06, m.p. –51.7 °C, b.p. 120.1 °C, density 1.032). Its reducing properties are due to its thermal decomposition to hydrogen and nitrogen. Similar decomposition takes place when hydrazine derivatives such as hydrazones or hydrazides are heated in the presence of alkali. Simple refluxing of hydrazine hydrate with oleic acid causes hydrogenation to stearic acid [*379*], and heating with aromatic hydrocarbons at 160–280 °C effects complete hydrogenation of the aromatic rings [*380*]. Most likely, such hydrogenations are due to the transient formation of diimide, which is formed from hydrazine in the presence of air. However, the hydrogenation of oleic acid took place even when the air was excluded [*379*].

Hydrazine is a very effective hydrogen donor in the *catalytic transfer of hydrogen* achieved by refluxing compounds to be reduced with hydrazine in the presence of palladium or Raney nickel [*71, 381, 382, 383*].

Most reactions with hydrazine are carried out with aldehydes and ketones in the presence of alkali. The reduction proper is preceded by formation of hydrazones that decompose in alkaline medium at elevated temperatures to nitrogen and compounds in which the carbonyl oxygen has been replaced by two hydrogens. The same results are obtained by alkaline–thermal decomposition of ready-made hydrazones of the carbonyl compounds. Both reactions are referred to as the *Wolff–Kizhner* reduction [*384*].

The original reduction carried out in ethanol required heating in sealed tubes. The use of solvents with higher boiling points (ethylene glycol, di- and triethylene glycols) allows the reaction to be carried out in open vessels [*385, 386, 387*] (*Huang–Minlon reduction*) (Procedure 53, p 317). Conversion of carbonyl groups to methylene groups can also be achieved by decomposition of semicarbazones. Another important reduction in which hydrazine is implicitly involved is the *McFadyen–Stevens reduction* of acids to aldehydes. The acid is converted to its benzenesulfonyl or *p*-toluenesulfonyl hydrazide, and this is decomposed by sodium carbonate at 160 °C [*388, 389*].

Hydrazine and its methyl and dimethyl derivatives reduce disulfides to mercaptans [*390*].

PHOSPHORUS COMPOUNDS

Both inorganic and organic phosphorus compounds are used relatively infrequently, but their reducing properties are sometimes very specific.

Phosphonium iodide, PH_4I (M.W. 161.93, b.p. 62.5 °C, density 2.86), is prepared somewhat tediously from white phosphorus and iodine [*391*] and was used, together with fuming hydriodic acid, for reduction of a pyrrole ring to a 2,5-dihydropyrrole ring [*392*]. Its applications are very rare.

Diphosphorus tetraiodide, P_2I_4 (M.W. 507.62, m.p. 125–128 °C) , which is commercially available, deoxygenates benzylic alcohols [*393*].

Hypophosphorous acid, H_3PO_2 (M.W. 66.00), usually available as its 50% aqueous solution (density 1.274), is a specific reagent for replacement of aromatic diazonium groups by hydrogen under very mild conditions (at 0–25 °C) and in fair to good yields [*394, 395*] (Procedure 54, p 318). Its monosodium salt, *sodium hypophosphite*, NaH_2PO_2 (M.W. 87.98), is used for partial and selective reduction of trigeminal trihalides [*396*] (p 84). *Ammonium hypophosphite*, $H_2PO_2NH_4 \cdot H_2O$ (M.W. 101.0), prepared from hypophosphorous acid and ammonia, hydrogenates carbon–carbon double bonds and acetylenes to alkenes in the presence of palladium on carbon [*397*] (Procedure 55, p 318).

Organic derivatives of phosphorus excel by the strong tendency of trivalent phosphorus to combine with oxygen or sulfur to form phosphine oxides and phosphates or their sulfur analogs.

Lithium diphenyl phosphide (Ph_2PLi, M.W. 192.11), prepared in situ from lithium and diphenylphosphinous chloride, Ph_2PCl, deoxygenates acyloins to ketones [*398*].

Triphenylphosphine, Ph_3P (M.W. 262.29, m.p. 79–81 °C, b.p. 377 °C), reacts spontaneously with all kinds of organic peroxides in ether, petroleum ether, or benzene and extracts one atom of oxygen per peroxide bond [*399*]. It also extrudes sulfur and converts thiiranes to alkenes [*400*] and performs other reductions [*401*].

Triphenylphosphonium iodide, Ph_3PHI (M.W. 390.18), hydrogenolyzes α-haloketones to ketones [*402*].

Of *trialkyl phosphites* the most frequently used is triethyl phosphite, $(EtO)_3P$ (M.W. 166.16, b.p. 156 °C, density 0.969), which combines with sulfur in thiiranes [*400, 403*] and gives alkenes in respectable yields. In addition, it can extrude sulfur from sulfides [*404*], convert α-diketones to acyloins [*405*], convert α-keto acids to α-hydroxy acids [*406*], reduce nitroso compounds to hydroxylamines [*407*], and deoxygenate amine oxides [*408*] (Procedure 56, p 319).

Strongly reducing properties are also characteristic of *diethyl phosphite*, $(EtO)_2PHO$ (M.W. 138.10, b.p. 50–51 °C/2 mm, density 1.072), which is suitable for partial reduction of geminal dihalides [*409*]; of *sodium O,O-diethylphosphorotellurate* $(EtO)_2PTeONa$, which replaces halogens by hydrogen in α-halo ketones [*410*] and transforms epoxides to alkenes [*411*]; and of *ethyl hypophosphite*, $EtOPH_2O$, which converts sulfinates to disulfides [*412*].

Strong affinity for oxygen and sulfur is also exhibited by phosphines such as *tris(dimethylamino)phosphine* (hexamethylphosphorous triamide) and *tris(diethylamino)phosphine* (hexaethylphosphorous triamide). The former converts diphenyl-*o,o'*-dicarboxaldehyde to phenanthrene-9,10-oxide [*413*], and both reduce disulfides to sulfides [*414, 415*]. *Hexamethylphosphoric triamide* (hexamethylphosphoramide, HMPA) is used for replacement of a diazo group by hydrogen [*780*].

BORON COMPOUNDS (EXCEPT BORANES)

Overwhelming applications of boron compounds are reductions with organoboranes and borohydrides (p 19). Only a few boron compounds that do not contain hydrogen bonded to boron show reducing properties: *Boron tribromide, dimethylboron bromide,* and *9-bromo-9-borabicyclo[3.3.1]nonane* deoxygenate sulfoxides to sulfides [*416*]. Diisopinocampheylchloroborane [*149*] is used for stereoselective reduction of ketones.

SILICON COMPOUNDS (EXCEPT SILANES)

Hexamethyldisilane, $Me_3SiSiMe_3$ (M.W. 146.37) functions as a source of hydrogen in catalytic hydrogenolysis of α-halo ketones and nitriles [*417*].

Derivatives of silanes such as *chlorotrimethylsilane*, $(CH_3)_3SiCl$ (M.W. 108.64) [*167*]; *iodotrimethylsilane*, $(CH_3)_3SiI$ (M.W. 200.10) [*418*]; and *methyltrichlorosilane*, CH_3SiCl_3 (M.W. 149.48), with *sodium iodide* are [*419*] used for carbon–oxygen bond cleavage in acyloins [*418*], in ethers and esters [*419*], for replacement of halogens in α-haloketones [*419*], and for deoxygenation of sulfoxides to sulfides [*419*] (Procedure 57, p 319).

OTHER ORGANIC COMPOUNDS

Reductive properties of *alcohols* (most frequently ethanol and isopropyl alcohol) derive from their relatively easy dehydrogenation to aldehydes or ketones. They therefore act as hydrogen donors in catalytic transfer of hydrogen in the presence of catalysts such as Raney nickel [*420*]. In the presence of cuprous oxide [*421*] or under irradiation with ultraviolet light [*422*], *ethanol* (M.W. 46.07, m.p. –117.3 °C, b.p. 78.3 °C, density 0.79) is used for replacement of aromatic diazonium groups by hydrogen (methanol is unsuitable for this purpose) [*395*]. Also an aliphatic diazo group is replaced by two hydrogens in this way [*423*]. *Isopropyl alcohol* (M.W. 60.11, m.p. –89.5 °C, b.p. 82.4 °C, density 0.786) under irradiation with ultraviolet or γ rays hydrogenolyzes a carbon–chlorine and even carbon–fluorine bond in polyhalogenated compounds and in α-positions to carbonyl groups [*424, 425, 426, 427, 428*].

Isopropyl alcohol is also used for reduction of aldehydes and ketones to alcohols. This can be done either under ultraviolet irradiation [*429*], or much more often in the presence of anhydrous basic catalysts such as aluminum isopropoxide (the *Meerwein–Ponndorf–Verley reduction*) [*430*]. The reduction is a base-catalyzed transfer of hydride ion from isopropoxide ion to the carbonyl function, thus yielding the alkoxide ion of the desired alcohol and acetone. By removal of acetone by distillation, the equilibrium of the reversible reaction is shifted in favor of the reduction of the starting carbonyl compound (p 153). The reduction with alcohols is very selective and particularly suitable for the preparation of unsaturated alcohols from unsaturated carbonyl compounds (Procedure 58, p 319). Halogens can occasionally be replaced by hydrogen [*431, 432*] .

A few examples of reduction of organic compounds by *aldehydes* are recorded. Heating of aqueous *formaldehyde* (M.W. 30.03, m.p. –92 °C, b.p. –21 °C, density 0.815) with ammonium chloride gives a mixture of all possible methylamines (the *Eschweiler reaction*) [*433*]. Aqueous formaldehyde is also used for replacement of a diazonium group by hydrogen [*395*]. *p*-Tolualdehyde is reduced by treatment with formaldehyde in alkaline medium at 60–70 °C (*crossed Cannizzaro reaction*) [*434*]. Salicylaldehyde in alkaline medium reduced aromatic ketones to alcohols [*435*]. Reducing sugars such as glucose were used for the reduction of nitro compounds to azo compounds [*436*] and azoxy compounds [*437*].

Formic acid, anhydrous (M.W. 46.03, m.p. 8.5 °C, b.p. 100.8 °C, density 1.22) or a 90% aqueous solution, is an excellent hydrogen donor in catalytic hydrogen transfer carried out by heating in the presence of copper [*72*], nickel [*72*], or palladium [*74*]. Also its salts, *ammonium formate* [*75, 77, 78, 438*] and *triethylammonium formate* [*79, 80, 439*] are used for the same purpose in the presence of palladium. Conjugated double bonds, triple bonds, aromatic rings, and nitro compounds are hydrogenated in this way. Also *formamide* is used as a reducing agent for a selective replacement of a diazonium group by hydrogen [*440*].

Reductions with formic acid alone or in the presence of amines are effected at temperatures of 130–190 °C. Under these conditions formic acid reduces conjugated double bonds [*441*], pyridine rings in quaternary pyridinium salts to 1,2,3,6-tetrahydro- and hexahydropyridines [*442*], quinolines to 1,2,3,4-tetrahydroquinolines [*443*], and enamines to amines [*444*]. Formic acid is a reducing agent in methylation of primary amines will formaldehyde [*445*]. The most important reduction involving formic acid is the *Leuckart reaction*, achieved by heating of an aldehyde or a ketone with ammonium formate or formamide, or formic acid in the presence of a primary or secondary amine. The final products are primary, secondary, or tertiary amines. The Leuckart reduction is an alternative to reductive aminations by hydrogen and gives good yields of amines [*446, 447*] (Procedure 59, p 320).

Organic compounds of sulfur and phosphorus are included in the section on reduction with sulfur and phosphorus compounds (pp 42 and 45).

Reducing effects of *organometallic compounds* have been known for a long time. Treatment of ketones and acid derivatives with some Grignard reagents, especially isopropyl- and *tert*-butylmagnesium halides, leads sometimes to reduction rather than to addition across the carbonyl function [*448, 449*]. The reduction takes place at the expense of the alkyl group in the Grignard reagent, which is converted to an alkene. Although not widely used, reductions with Grignard reagents often give good to excellent yields [*450, 451*].

Similar reductions were effected by heating of *triethylaluminum* *(triethylalane)* in ether with aldehydes and ketones. The alcohols are formed by hydrogen transfer from the alkyl groups, which are transformed to alkenes.

Analogous reductions are achieved by *trialkylboranes* [*452*]. These reactions, although different in nature from the reductions by hydrides and complex hydrides, were among the first applications of boranes and alanes for the reduction of organic compounds.

1,4-Dihydropyridine derivatives easily transfer two hydrogen atoms onto reducible compounds. Thus diethyl 1,4-dihydro-2,6-dimethyl-3,5-dicarboxylate (the Hantzsch ester) reduced β-nitrostyrene to 2-phenyl-1-nitroethane [*1470*] (Procedure 60, p 320).

Derivatives of *1,4-dihydronicotinamide*, for example, 1-benzyl-1,4-dihydronicotinamide, are used for replacement of halogen by hydrogen in heterocyclic compounds [*453*]. Nucleotides and nucleosides of 1,4-dihydronicotinamide are involved in enantioselective biochemical reductions.

SUMMARY OF ELECTROREDUCTION AND REDUCTIONS WITH METALS AND WITH METAL AND NONMETAL COMPOUNDS

Like *electrolytic reductions*, *reductions by metals* can be applied only to polar systems, that is, to multiple bonds between carbon and oxygen and carbon and nitrogen. The most electronegative metals, lithium, potassium, and sodium, reduce acetylenes to olefins. Isolated carbon–carbon double bonds are resistant, but carbon–carbon multiple bonds conjugated with the polar functions or other double bonds as in aromatic rings, are reduced. Less electronegative zinc still can reduce double bonds conjugated with polar functions, but aluminum, magnesium, tin, and iron can be applied only to polar multiple bonds between carbon and oxygen and carbon and nitrogen and to multiple bonds between two hetero atoms such as in nitro compounds.

Many reductions that were carried out with metals can now be achieved with complex hydrides. With the exception of sodium and especially potassium, which requires special safety measures (p xxiii), all other metal reductions do not pose safety problems.

Compared with metals, reductions with *metal compounds* have an advantage of taking place in homogeneous solutions. The host of classical metal compounds such as stannous chloride, titanium trichloride, sodium stannite, sodium arsenite, ferrous sulfate, and others has been enriched by some strong reducing agents: Divalent chromium compounds and samarium diiodide can reduce acetylenes to olefins, titanium dichloride couples ketones to alkenes, and all of them hydrogenolyze halogen compounds. Most of the metal compounds are used for reduction of halogen compounds and compounds of nitrogen. They are suitable for partial reductions and for selective reductions of multifunctional compounds, for example, reduction of halocarbonyl compounds to carbonyl compounds.

With the exception of hydrogen iodide and diimide, which effected rather rare and unimportant carbon–carbon double bond reductions, all other *nonmetal reducing agents* affect only polar double bonds between carbon and hetero atoms or between two hetero atoms. The nonmetal compounds are useful as selective reducing agents, capable of distinguishing between different bonds of unsaturated systems. Different sulfur compounds, for example, are suitable for selective reductions of polynitro compounds.

Some of the organic reducing agents require catalysis. Reductions with alcohol are catalyzed by aluminum alkoxide or promoted by active irradiation. Some unsaturated hydrocarbons and especially formic acid in the presence of catalysts achieve the same results as would heterogeneous catalytic hydrogenation. They function as hydrogen donors in the catalytic hydrogen transfer.

Some of the compounds, for example, hydrogen sulfide, sulfur dioxide, and diimide, are toxic and are best used in well-ventilated hoods (p xxiii).

BIOCHEMICAL REDUCTIONS

The earliest recorded microbial reductions are probably reductions with *baker's yeast (Saccharomyces cerevisiae)* that contains a chiral enzyme oxidoreductase. They are carried out under aerobic conditions without nutrients [454] or with saccharides such as glucose [454, 455, 456, 457, 458] or sucrose [459, 460, 461] as sources of carbon and hydrogen. Other sources of hydrogen are alcohols [462], especially ethanol, which also acts as a source of carbon [457, 460, 462, 463]. Reductions are usually performed in aqueous suspensions, often buffered to maintain the proper pH, but they can also be run in organic solvents [462].

To facilitate handling of the yeast, a suspension of baker's yeast is converted to tiny beads by treatment with sodium alginate in ratios 1:1 to 4:1 followed by immersion in 10% aqueous calcium chloride [*1496*]. In this form, the separation of the yeast after the reaction is much easier. The temperature is kept between 30 and 40 °C, and the reaction times vary from several hours to a few days. The progress is monitored analytically.

Isolation of the products after the removal of the yeast, usually by centrifugation or filtration, is done by extraction. Yields vary from 10 to 90%, and enantiomeric excesses are usually very high, often up to 99%. Stereoselectivity is affected by the structure of the compound to be reduced [*455*] and by the relative concentration of the substrate [*455*]. Variations in these factors are due to the presence of more than one dehydrogenase in some organisms and may even lead to reduction products of opposite configurations [*455*].

Mechanistically the reduction occurs by hydride transfer from the reduced coenzyme *nicotinamide adenine dinucleotide* (NADH) or *nicotinamide adenine dinucleotide phosphate* (NADPH) to the *Re* or *Si* surface of a trigonal system containing a π bond with the generation of NAD$^+$ or NADP$^+$. The regeneration of NADH or NADPH occurs by reduction with electrons from donors such as alcohols or saccharides [*455, 464*]. The reductions with whole cells do not need the coenzyme because it is contained in the organisms. If isolated and purified enzymes, *dehydrogenases*, are used the coenzyme must be added and recycled.

AD is alcohol dehydrogenase.

Baker's yeast is used for reductions of carbon–carbon and carbon–oxygen double bonds. Thus double bonds are stereoselectively reduced in allylic alcohols [*463*]; in α,β-unsaturated ketones, acids, and esters [*465, 466*]; and in α,β-unsaturated nitro compounds, while the other functions are left intact [*456*]. α,β-Unsaturated aldehydes are reduced to saturated alcohols [*457*]. Most reductions are applied to ketones [*455, 458, 459*] and keto acids and their esters [*460, 461, 462, 466*]. Occasionally, even aromatic nitro groups are reduced to amino groups [*1485*] and halogens are replaced by hydrogens [*465*].

Besides *Saccharomyces cerevisiae* many other microorganisms effect stereoselective reductions, especially of ketones [*464*]. Reductions are achieved by various enzymes and are carried out at mild temperatures (30–40 °C, and infrequently at 70 °C by thermostable organisms) and give good to high yields and high enantiomeric excesses. The enzymes responsible for such reductions were isolated and identified: *alcohol dehydrogenase* from *Saccharomyces cerevisiae*, *Lactobacillus kefir* [*468*], *Pseudomonas sp.* [*469*], *Thermoanaerobicum brockii* [*468*], and horse liver [*470*]; *lactate dehydrogenase* from *Bacillus stearothermophilus* [*471*] and different organs of animals [*472*]. Many of the enzymes are now commercially available [*472*].

A selective reduction of carbon–carbon double bonds is accomplished with *Corynebacterium equii* [*473*] and *Clostridium kluyveri* [*465*]. Microbial reductions are used for preparation of chiral alcohols from ketones [*468, 469,*

474, 475, 476], from diketones [*470, 477*], and especially from keto acids [*471, 472*] and their esters [*466, 478, 479*] (Procedures 61–63, pp 320-322).

The most frequently used microorganisms for reduction of ketones to alcohols are *Aspergillus niger* [*475*], *Bacillus stearothermophilus* [*471*], *Clostridium kluyveri* [*4 6 5*], *Curvularia falcata* [*4 7 4*], *Clostridium thermoaceticum* [*480*]. *Geotrichum candidum* [*475*], *Hansenula anomala* [*479*], *Lactobacillus kefir* [*468*], *Pseudomonas sp.* [*469*], *Rhodotorula rubra* [*476, 477*], *Sporobolomyces pararoseus* [*476*], *Streptomyces sp.* [*475*], *Thermo-anaerobicum brockii* [*468, 478*], *Xanthomonas oryzae* [*481*], and others [*464*] (Procedures 61–63, pp 320–322).

DIASTEREOSELECTIVE AND ENANTIOSELECTIVE REDUCTIONS

Two diastereomers or enantiomers may have different physiological effects, so the demand for the preparation of diastereomerically and enantiomerically pure pharmaceuticals resulted in a strong effort to prepare stereochemically pure isomers and not mixtures of stereoisomers that would require laborious separation or resolution.

In the past 30 years, purely chemical stereoselective reductions have been discovered. Only those that give enantiomeric excesses higher than about 90% are of practical importance.

Heterogeneous catalytic hydrogenation has only very limited potential in this respect. It has been used for a reduction of compounds into which optically active auxiliary groups had been incorporated [*68, 1499*].

Much more general is *homogeneous catalytic hydrogenation* over chiral catalysts including Binap [*105*], and Chiraphos [*107*] (p 16). It was applied to diastereoselective and enantioselective reductions of ketones (p 150, 154 and 162), diketones (p 180), keto acids (p 200), unsaturated esters (p 218), and keto esters (p 226) and gave high yields and high diastereomeric and enantiomeric excesses.

Similar results were obtained by using *chiral complex hydrides* as are described on p 21-23 and shown in equations **30–34**. High yields and variable diastereomeric and enantiomeric excesses were obtained in reductions of aliphatic ketones (p 149), aromatic ketones (p 154), cyclic ketones (p 162), acetylenic ketones (p 169), and keto esters (p 224).

The most important enantioselective reducing reagents have been critically reviewed [*1487, 1490*].

Biochemical reductions usually provide very high enantiomeric excesses but not always good yields. They were applied to aliphatic ketones (p 150), aromatic ketones (p 154), cyclic ketones (p 162), hydroxy ketones (p 174), diketones (p 178), and keto esters (p 226). *Prelog's rule* determines the configuration of chiral alcohols prepared by reduction of ketones: If the ketone is placed with the larger group on the observer's left, the hydroxyl group formed is closer to the observer (p 163).

Reduction of Specific Types of Organic Compounds

CHAPTER 6

Reduction of Hydrocarbons and Aromatic Heterocycles

CYCLOALKANES

On catalytic hydrogenation, some cycloalkanes give alkanes in quantitative yields: Cyclopropane and cyclobutane yield propane and butane over nickel at 120 and 200 °C [*482*], respectively, and cyclopentane is almost quantitatively cleaved to pentane over 12% platinum on carbon at 300 °C [*483*]. The relatively easy fission of the cyclopropane ring must be kept in mind when designing reductions of functions in compounds containing such rings. In 1,2-dimethyl-1,2-diphenylcyclopropane, the ring is cleaved between carbons 1 and 2 by hydrogen over platinum, palladium, or Raney nickel at 25 °C and 1 atm [*484*].

Disubstituted and especially trisubstituted cyclopropanes are hydrogenolyzed by treatment with trialkylsilanes and trifluoroacetic acid (ionic hydrogenation, applicable to compounds forming tertiary carbonium ions) [*485*]. 1,1,2-Trimethylcyclopropane affords 2,3-dimethylbutane in a 65% yield. The reduction is not entirely stereoselective. 1-Methylbicyclo[3.1.0]hexane gives 83% *cis*- and 7% *trans*-1,2-dimethylcyclopentane, whereas 1-methylbicyclo[4.1.0]heptane yields 10.7% *cis*- and 64.3% *trans*-1,2-dimethylcyclohexane [*485*].

ALKENES AND CYCLOALKENES

Isolated double bonds in alkenes and cycloalkenes are best reduced by catalytic hydrogenation. Addition of hydrogen is very easy and takes place at room temperature and atmospheric pressure over almost any noble metal catalyst, Raney nickel, or nickel catalysts prepared in special ways such as P-1 nickel [*13*] or complex catalysts referred to as Nic [*50*] (Procedures 1 and 2, pp 293–296).

The rates of hydrogenation approximately parallel the heats of hydrogenation [*486*] and are subject to steric effects. Monosubstituted alkenes are reduced faster than di-, tri-, and tetrasubstituted ones [*14*] to such an extent that they can be reduced preferentially, especially over a ruthenium catalyst [*487*]. Half-hydrogenation times (relative to 1-pentene) of the hydrogenation of alkenes over P-1 nickel boride are illustrative: 1-pentene, 1.0; 2-methyl-1-butene, 1.6; 2-methyl-2-butene, 16; and 2,3-dimethyl-2-butene, 45 [*13*]. The lengths of the chains attached to the double-bond carbon and their branching have little effect on the rate of hydrogenation [*13*]; however, if one of the substituents at the double bond is a benzene ring, the hydrogenation is even faster than that of the 1-alkene. The aromatic ring is hydrogenated much more slowly. Hydrogenation rates (relative to benzene) of some alkenes and cycloalkenes over nickel on alumina are as follows: styrene, 900; 1-hexene, 306; cyclopentene, 294; cyclohexene, 150; 4-methyl-1-cyclohexene, 134; 1-methyl-1-cyclohexene, 5; and benzene, 1 [*14*].

Isomerizations may take place over some catalysts more than over others. Regio- as well as stereoisomers may be formed [*13, 50*]. Such side reactions are undesirable in the reduction of prochiral alkenes, and especially in the hydrogenation of some cycloalkenes.

Reduction of cycloalkenes is as easy as that of alkenes, and again catalytic hydrogenation is the method of choice. The rates of hydrogenation are slightly lower than those of 1-alkenes and are affected by the ring size and by the catalyst used. Double bonds in strained cycloalkenes are hydrogenated even faster than in 1-alkenes [488]. Half-hydrogenation times (relative to 1-octene) in hydrogenations over W-2 Raney nickel and P-1 nickel boride (in parentheses) are, respectively, 1-octene, 1.3 (1.0); cyclopentene, 2.0 (1.3); cyclohexene, 3.5 (2.5); cyclooctene, 5.3 (2.0); and norbornene, about 90% of that of octene [13]. In a competitive experiment over a nickel boride catalyst, 92% of the hydrogen was taken up by norbornene and only 8% by cyclopentene [488].

The stereochemistry of the products is strongly affected by the catalyst and slightly affected by the pressure of hydrogen [489]. Hydrogenation of 1,2-dimethylcyclohexene over platinum oxide gives 81.8% and 95.4% cis-1,2-dimethylcyclohexane at 1 and 500 atm, respectively. When hydrogenated over palladium on charcoal or alumina at 1 atm, 1,2-dimethylcyclohexene yields 73% trans- and 27% cis-1,2-dimethylcyclohexane. An increase in the hydrogen pressure in the yields of cis-1,2-dimethylcyclohexane from 2,3-dimethyl-cyclohexene and 2-methyl-1-methylenecyclohexane results in a slight decrease of the ratio of cis to trans isomer [489] (equation 51).

[489]

51

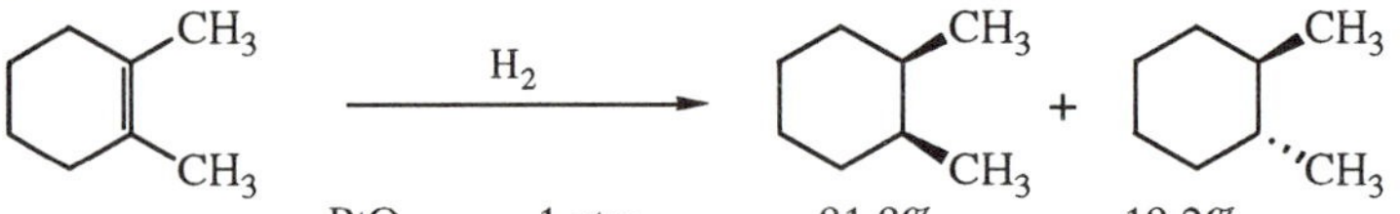

	PtO$_2$	1 atm	81.8%	18.2%
	AcOH	4 atm	83.3%	16.7%
	25 °C	500 atm	94.5%	5.5%
	Pd(C)	1 atm	27%	73%
	AcOH			
	25 °C			
	PtO$_2$	1 atm	76.6%	23.4%
	AcOH	4 atm	70.6%	29.4%
	25 °C	500 atm	69.7%	30.3%
	Pd(C)	1 atm	27%	73%
	AcOH			
	25 °C			
	PtO$_2$	1 atm	70.2%	29.8%
	AcOH	4 atm	65.7%	30.0%
	25 °C	500 atm	67.3%	32.7%
	Pd(C)	1 atm	39%	61%
	AcOH,			
	25 °C			

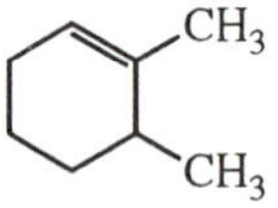

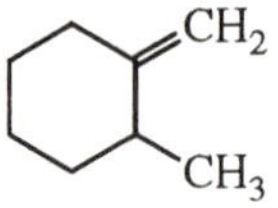

Hydrogenation over Raney nickel is even less stereoselective. 2-, 3-, and 4-Methyl-1-methylenecyclohexanes give different mixtures of *cis*- and *trans*-dimethylcyclohexanes, depending not only on the structure of the starting alkene but also on the method of preparation and on the freshness of the catalysts. The composition of the stereoisomers ranges from 27–72% *cis* to 28–73% *trans* [*490*].

Highly stereoselective reductions of carbon–carbon double bonds are achieved by *homogeneous catalytic hydrogenation* using chiral catalysts. Examples are shown in sections on reduction of unsaturated alcohols, ketones, and acids and their esters.

In addition to catalytic hydrogenation, a few other methods of reduction can be used for saturation of carbon–carbon double bonds. However, their practical applications are no match for catalytic hydrogenation.

Lithium aluminum hydride in the presence of *titanium trichloride, cobalt dichloride,* and *nickel dichloride* converts alkenes to alkanes in almost quantitative yields [*158*].

The *indirect reduction by boranes* consists of two steps: addition of a borane across the double bond and decomposition of the alkylborane by an organic acid, usually propionic acid [*491*] (equation **52**).

5 2   [*491*]

$$C_4H_9CH{=}CH_2 \xrightarrow[BF_3\cdot Et_2O]{NaBH_4} C_4H_9CH_2CH_2{-}BH_2 \longrightarrow C_4H_9CH_2CH_3 \quad 91\%$$

$$\text{(with } O{-}CC_2H_5 \text{)} \qquad + \qquad O{=}C{-}OBH_2$$
$$\qquad\qquad\qquad\qquad\qquad \overset{|}{C_2H_5}$$

So-called *ionic hydrogenation* using trialkylsilanes and trifluoroacetic acid is applicable mainly to trisubstituted alkenes and cycloalkenes, compounds that form tertiary carbonium ions (*181, 485, 492*]. 1-Methylcyclohexene is converted to methylcyclohexane in 67–72% yields on heating for 10 h at 50 °C with one equivalent of triethylsilane and two to four equivalents of trifluoroacetic acids [*181, 485*] (Procedure 27, p 308). 3-Methyl-5α-cholest-2-ene is reduced to 3β-methyl-5α-cholestane in a 66% yield with triethylsilane and trifluoroacetic acid in methylene chloride at room temperature [*181*]. The reduction with triethylsilane is not stereoselective. 1,2-Dimethylcyclopentene gives 60.5% *cis*- and 27.5% *trans*-1,2-dimethylcyclopentane, whereas 1,2-dimethylcyclohexene affords 12.4% *cis*- and 61.6% *trans*-1,2-dimethylcyclohexane [*485*]. Most mono- and disubstituted alkenes do not undergo this reduction. Thus the ionic hydrogenation might be a nice complement to catalytic hydrogenation, which reduces preferentially the nonsubstituted double bonds. Unfortunately not enough experimental material is yet available to prove the usefulness of the ionic method.

The reduction of an isolated carbon–carbon double bond by other methods is an exception. Occasionally an isolated double bond has been reduced *electrolytically* [*195*] and by *dissolving metals*. Sodium under special conditions with the use of *tert*-butyl alcohol and hexamethylphosphoramide as solvents reduces alkenes and cycloalkenes in 40–100% yields [*493*]. 3-Methylene-cholestane, for example, affords 3-methylcholestane in a 74% yield on heating for 2 h at 55–57 °C with lithium in ethylenediamine [*494*]. 1-Undecene is converted to undecane in a 78–88% yield by refluxing with *nickel dichloride and zinc* in methanol [*241*].

A compound that can accomplish addition of hydrogen across isolated double bonds is *diimide (diazene)*. A series of cycloalkenes is hydrogenated with more or less stereospecificity using disodium azodicarboxylate as a source of diimide [*495*]. This compound is most probably an intermediate generated in the reaction of alkenes with hydrazine in the presence of air, copper, and carbon tetrachloride. l-Hexene is thus reduced to hexane in a 38.7% yield [*369*].

Reductions of double bonds in dienes, unsaturated halides, alcohols, amines, carbonyl compounds, acids, and derivatives are discussed in the appropriate sections.

DIENES AND POLYENES

The behavior of dienes and polyenes, both open chain and cyclic, toward reduction depends especially on the respective positions of the double bonds. Carbon–carbon double bonds separated by at least one carbon atom behave as independent units and can be partly or completely reduced by *catalytic hydrogenation* or by *diimide*. *cis,trans,trans*-1,5,9-Cyclododecatriene treated with hydrazine and air in the presence of copper sulfate (diimide in situ) gives *cis*-cyclododecene in a 51–76% yield [*370, 496*]. It appears as if the *trans* double bonds were reduced preferentially.

If the double bonds are structurally different to such an extent that their rates of hydrogenation differ considerably, they can be reduced selectively. Monosubstituted double bonds are saturated in preference to di- and trisubstituted ones [*13, 497, 498*]. When 2-methyl-1,5-hexadiene is hydrogenated over nickel boride with 1 mol of hydrogen, only the unbranched double bond is reduced, giving a 95% yield of 2-methyl-1-hexene [*498*].

The monosubstituted double bond in 4-vinylcyclohexene is hydrogenated over P-1 nickel in preference to the disubstituted double bond in the ring, giving 98% 4-ethylcyclohex-1-ene [*13*]. Similarly in limonene the double bond in the side chain is reduced, and the double bond in the ring is left intact if the compound is treated with hydrogen over 5% platinum on carbon at 60 °C and 3.7 atm (97.6% yield) [*497*].

Under the conditions that cause migration of the double bonds, for example, treatment with sodium and especially potassium, the isolated double bonds can become conjugated, and thus they undergo reduction by metals. Some macrocyclic cycloalkadienes are reduced (predominently to *trans*-cycloalkenes) by treatment with potassium on alumina in a hydrogen atmosphere [*499*].

In systems of **conjugated double bonds**, catalytic hydrogenation usually gives a mixture of all possible products. Conjugated dienes and polyenes can be reduced by *metals: sodium, potassium,* or *lithium*. The reduction is accomplished by 1,4-addition, which results in the formation of a product with only one double bond and products of coupling and polymerization. Isoprene is reduced in a 60% yield to 2-methyl-2-butene by sodium in liquid ammonia [*500*]. Reduction of cyclooctatetraene with sodium in liquid ammonia gives a

61% yield of a mixture of 1,3,6- and 1,3,5-cyclooctatrienes, the 1,3,5-isomer being a result of isomerization of the initially formed 1,3,6-triene [*501*].

Side reactions can be partly decreased when the reduction by alkali metals is carried out in the presence of proton donors such as triphenylmethane, fluorene, or secondary amines (*N*-alkyanilines). The products of the initial 1,4-addition are protonated as formed before coupling and polymerization start [*502, 503, 504*]. 1,3-Butadiene is reduced with lithium in ether in the presence of *N*-ethylaniline to *cis*-2-butene in a 92.5% yield [*503*], and similarly *cis,cis*-1,3-cyclooctadiene in the presence of *N*-methylaniline is reduced to *cis*-cyclooctene in a 98% yield [*504*]. With sodium in liquid ammonia the yields of reduction of conjugated dienes are much lower [*502*]. Selective reduction of a less substituted double bond can be accomplished by diimide (diazine) [*377, 378*] (equation 53).

53

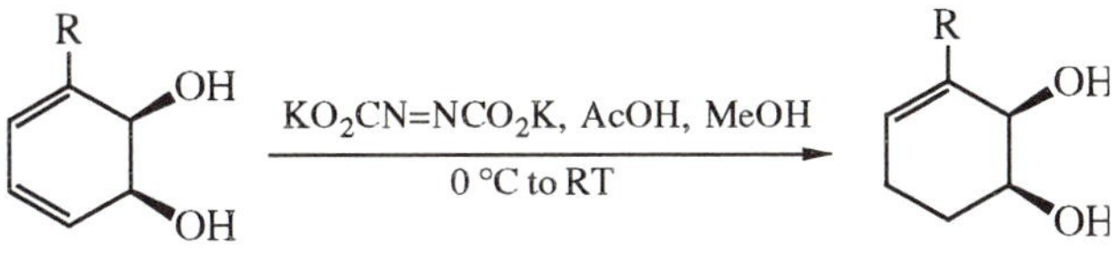

| R = CH$_3$ | 83% | [*377*] |
| R = CH$_2$CH$_2$Br | | [*378*] |

Double bonds conjugated with aromatic rings and with carbonyl, carboxyl, nitrile, and other functions are readily reduced by catalytic hydrogenation and by metals. These reductions are discussed in the appropriate sections.

ALKYNES AND CYCLOALKYNES

Partial reduction of acetylenes (alkynes) gives alkenes; total reduction gives alkanes. Total reduction is no problem because it can be easily accomplished by catalytic hydrogenation, and rarely accomplished also with diazene. Partial reduction is more desirable, especially with internal alkynes, which, depending on the reduction method, can be converted either to *cis*- or *trans*-alkenes. It thus provides a means for obtaining the particular olefinic stereoisomers. **Catalytic hydrogenation gives exclusively or at least predominantly *cis*-alkenes; reduction with metals and metal salts gives predominantly *trans*-alkenes.** Hydrides afford both stereoisomers depending on the nature of the hydride and reaction conditions. Reduction with hydrides and metals or metal salts stops with negligible exceptions at the stage of alkenes, but catalytic hydrogenation gives alkenes only under strictly controlled conditions.

The rates of hydrogenation of double and triple bonds do not differ appreciably. With some catalysts the double bond is hydrogenated even faster than the triple bond. For this reason, it is very difficult to prevent total hydrogenation by stopping the reaction after the consumption of just 1 mol of hydrogen.

Partial hydrogenation of the triple bond is achieved with catalysts over which the rate of hydrogenation decreases after the absorption of the first mole of hydrogen: *nickel* and *palladium.*

Hydrogenation using Raney nickel is carried out under mild conditions and gives *cis*-alkenes from internal alkynes in yields ranging from 50 to 100% [*505, 506, 507, 508, 509*]. Half-hydrogenation of alkynes is also achieved over nickel prepared by reduction of nickel acetate with sodium borohydride (*P-2* nickel or nickel boride) [*498, 510, 511*] or by reduction with sodium hydride in the presence of *tert*-amyl alcohol (catalyst referred to as Nic) [*50*], or by reduction of nickel bromide with potassium–graphite [*512*]. Other catalysts are palladium on charcoal [*513*], on barium sulfate [*514, 515*], on barium carbonate [*516*], or on calcium carbonate [*515*], especially if the noble metal is partly deactivated by quinoline [*517, 518*], lead acetate [*39*], or potassium hydroxide [*519*]. The most popular catalyst is palladium on calcium carbonate deactivated by lead acetate *(Lindlar's catalyst)* [*39*]. However, potassium hydroxide–deactivated palladium is claimed to be even better. Stereoselectivity as high as 97% of the *cis* (Z) product with only 2% of the *trans* (E) isomer and 1% of the saturated compound can be achieved in hydrogenation of diphenylacetylene over a quinoline-deactivated Pd catalyst, prepared by treatment of palladium acetate with sodium hydride and *tert*-amyl alcohol in tetrahydrofuran at 40 °C. The reaction gives best yields in ethanol or ethanol–tetrahydrofuran at 20–21 °C and 1 atm [*520*]. Palladium on carbon has also been used for semihydrogenation of acetylenes using ammonium hypophosphite [*397*] or trimethylammonium formate in lieu of hydrogen [*79*]. With terminal alkynes where no stereoisomeric products are formed, reduction with *sodium* in liquid ammonia gives alkenes of higher purity than catalytic hydrogenation [*509*].

Reduction of alkynes by *lithium aluminum hydride* in tetrahydrofuran or its mixture with diglyme gives high yields of *trans* isomers [*521*]. When toluene is used as the solvent, 2-pentyne yields 97.6% pentane at 125 °C. This rather unusual total reduction of alkyne to alkane is probably due to the presence of hydrogen resulting from the decomposition of the hydride at 125 °C in the autoclave. Lithium aluminum hydride was proven earlier to be a catalyst for hydrogenation of alkenes and alkynes with elemental hydrogen [*522*].

Lithium aluminum hydride and nickel dichloride in tetrahydrofuran reduce 1-octyne to 1-octene at room temperature in a 99% yield and 2-hexyne to *cis*-2-hexene in a 96% yield [*158*].

Lithium methyldiisobutylaluminum hydride prepared in situ from diisobutylalane and methyllithium reduces 3-hexyne to pure *trans*-3-hexene in an 88% yield [*523*]. On the other hand, *diisobutylalane* gives a 90% yield of *cis*-3-hexene [*523, 524*]. Deuterodiborane prepared in situ from lithium aluminum deuteride and boron trifluoride etherate is used for converting cyclodecyne to *cis*-1,2-dideuterocyclodecene [*525*] (43% yield, 81% purity). *Borane* and *disiamylborane* (di-*sec*-amylborane) give, respectively, 68–80% and 90% high-purity *cis*-3-hexene [*141*] (equation **54**).

Another hydride, *magnesium hydride,* prepared in situ from lithium aluminum hydride and diethylmagnesium, reduces terminal alkynes to 1-alkenes in 78–98% yields in the presence of cuprous iodide or cuprous *tert*-butoxide, and

54

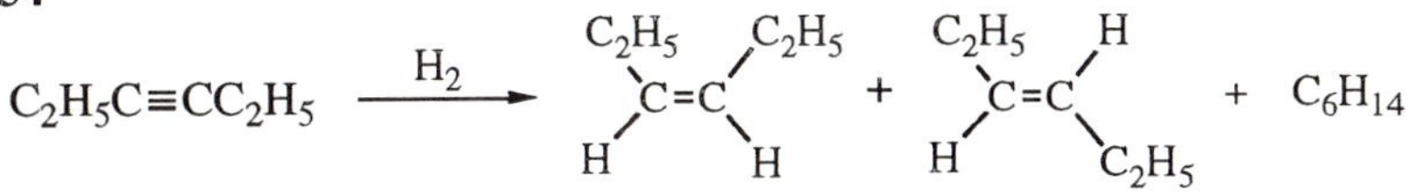

$$C_2H_5C{\equiv}CC_2H_5 \xrightarrow{H_2} \underset{H}{\overset{C_2H_5}{>}}C{=}C\underset{H}{\overset{C_2H_5}{<}} \;+\; \underset{H}{\overset{C_2H_5}{>}}C{=}C\underset{C_2H_5}{\overset{H}{<}} \;+\; C_6H_{14}$$

		cis	trans	C_6H_{14}
[50]	Nic, MeOH, 25 °C, 1 atm, 80% (isol.)[a]	97%		3%
[498]	NiB$_2$	98–99%		
[510]	P-2 Ni, EtOH, 97% (80% isol.)[a]	99–100%		
[50]	Nic, EtOH, 25 °C, 1 atm	98%		
[521]	LiAlH$_4$, THF, diglyme, 117–150 °C, 91–97%[a]	2–4%	96–98%	
[524]	*iso*-Bu$_2$AlH, <45 °C, 91%[a]	95%	3%	
[523]	LiAlHMe(*i*-Bu)$_2$ 88%[a]		100%	
[141]	NaBH$_4$·BF$_3$, diglyme, 25 °C, 68–80%[a]	98–99%		
[141]	*sec*-Am$_2$BH, diglyme, 90%[a]	98–99%		
[493]	Na, HMPA,[b] *tert*-BuOH		14%	70%

[a]Overall yield.
[b]Hexamethylphosporic triamide.

2-hexyne to pure *cis*-2-hexene in 80–81% yields [*159*]. Reduction of alkynes by lithium aluminum hydride in the presence of transition metals give alkenes with small amounts of alkanes. Internal acetylenes are reduced predominantly but not exclusively to *cis*-alkenes [*526, 527*].

Alkynes can be reduced *electrolytically*. Internal alkynes give 65–80% yields of *cis*-alkenes when electrolyzed in 10% sulfuric acid in ethanol at a spongy nickel cathode [*189*], or predominantly *trans*-alkenes if the electrolysis is carried out in a methylamine solution of lithium chloride. The yields of the alkenes and the ratios of *trans*- to *cis*-alkenes vary depending on whether the electrolysis is carried out in divided or undivided cells (24–80% yields and product composition of 89–99% *trans*-alkene) [528].

A universal method for conversion of internal alkynes to alkenes is reduction with *alkali metals* in liquid ammonia. Open-chain acetylenes are reduced in high yields to *trans*-alkenes with sodium [509] or lithium [529]. Cycloalkenes are obtained by sodium reduction of cycloalkynes in 14-membered rings [188, 530]. Dissolving-metal reduction of medium- and large-ring cycloalkynes is not stereospecific, the results depending on the metal used and experimental conditions, especially the presence of alcohols as proton donors in the liquid ammonia used as a solvent.

Cyclononyne is reduced by sodium in liquid ammonia to pure *trans*-cyclononene in a 63.5% yield [531]. Cyclodecyne is converted by sodium or potassium in liquid ammonia to a mixture containing more than 90% *cis*-cyclodecene [530]. Reduction of cyclodecyne with sodium in liquid ammonia gives 72–96% yields of a product containing 96–98% *cis*- and 2–4% *trans*-cyclodecene, and reduction of cyclododecyne affords 17.5–47% cyclododecene consisting of 81–92% *trans*-cyclododecene [530]. The composition of the product of reduction of 4,4,7,7-tetramethylcyclodecyne with lithium in liquid ammonia is 99% *trans*-4,4,7,7-tetramethylcyclodecene; with sodium in liquid ammonia, 35% *cis*- and 37% *trans* product, and in the presence of ethanol, 83% *cis*- and 17% *trans*-4,4,7,7-tetramethylcyclodecene [188]. The predominant reduction to the *cis* isomers is probably a result of reduction of the corresponding 1,2-cycloalkadiene [532] formed by isomerization of the cycloalkyne in the alkaline medium [533].

Simple acetylenes, that is, those in which the triple bond is not conjugated with an aromatic ring, carbonyl, or carboxyl function, are not reduced by sodium in amyl alcohol or by zinc. Conjugation, however, changes the reducibility of the triple bond dramatically *(vide infra)*.

Lithium in ethylamine reduces 5-decyne at –78 °C to *trans*-5-decene in a 55% yield and at 17 °C to decane in a 19% yield [203]. *Calcium* in ethylenediamine or its mixture with methylamine converts alkynes to *trans*-alkenes or alkanes in yields of 70–88%. The *trans*-alkenes are accompanied by their position isomers [212].

In **acetylenes containing double bonds,** the triple bond is selectively reduced by controlled treatment with hydrogen over special catalysts such as palladium deactivated with quinoline [517] or lead acetate [39], or with triethylammonium formate in the presence of palladium [79]. 1-Ethynylcyclohexene is hydrogenated to 1-vinylcyclohexene over a special nickel catalyst (Nic) in an 84% isolated yield [50].

Stereoselective reductions can be further accomplished by reagents that usually do not reduce double bonds (unless conjugated with carbonyl or carboxyl functions): *lithium aluminum hydride* [534] or *zinc* [535].

Diacetylenes having an internal and a terminal triple bond can be reduced selectively at the internal triple bond if they are first converted to sodium acetylides at the terminal bond by sodamide prepared in situ from sodium in liquid ammonia. Undeca-1,7-diyne is in this way converted to *trans*-7-undecene-1-yne in a 75% yield [536].

Complete reduction of alkynes to alkanes is easily accomplished by catalytic hydrogenation, especially using palladium [*537, 538*], platinum oxide, and active nickel catalysts [*508*]. Reduction of 3-hexyne to *trans*-3-hexene (14%) and hexane (70%) with sodium in *tert*-butyl alcohol and hexamethylphosphoric triamide is an exception [*493*].

If the triple bond is conjugated with an aromatic ring or with carbonyl groups, complete reduction can be achieved with reagents that are capable of reducing conjugated double bonds (p 65).

Reduction of triple bonds in compounds containing aromatic rings and functional groups is dealt with in the appropriate sections.

CARBOCYCLIC AROMATIC COMPOUNDS

Reduction of carbocyclic aromatic compounds is generally more difficult than that of alkenes, dienes, and alkynes. Any reduction destroys resonance-stabilized systems and consequently must overcome resonance energy. The heat of hydrogenation of the *cis*-disubstituted carbon–carbon double bond in *cis*-2-butene is 28.6 kcal. Because the resonance energy of benzene is 36 kcal, uptake of two hydrogen atoms is an endothermic reaction. In partial hydrogenation of naphthalene, whose resonance energy is 61 kcal, destruction of one aromatic ring means loss of resonance energy of only 25 kcal (61 – 36 kcal). Reduction of the 9,10-double bonds in anthracene and phenanthrene is accompanied by losses of resonance energy of only 12 and 20 kcal, respectively. In accordance with the energetics, the reduction of naphthalene is easier than that of benzene, and reduction of anthracene and phenanthrene in turn is easier than that of naphthalene. The trend is noticeable in the increasing ease of reduction of these four aromatic compounds, be it by catalytic hydrogenation, reduction with metals, or reduction with other compounds.

The *catalytic hydrogenation* of benzene and its homologs is more difficult than that of unsaturated aliphatic hydrocarbons, and more difficult than that of carbonyl functions, nitriles, halogen compounds, nitro compounds, and others. It is, therefore, possible to hydrogenate these functions preferentially. Rates of hydrogenation of benzene homologs decrease with increasing numbers of substituents [*14*]. If the relative rate of hydrogenation of benzene over nickel on alumina equals 1, the relative rates of hydrogenation of its homologs are as follows: for toluene, 0.4; xylene, 0.22; trimethylbenzene, 0.1; tetramethyl-benzene, 0.04; pentamethylbenzene, 0.005; and hexamethylbenzene, 0.001 [*14*].

Hydrogenation of benzene and its homologs at room temperature and atmospheric or slightly elevated pressure (2–3 atm) requires very active catalysts: platinum oxide, noble metals, and possibly very active Raney nickel. Excellent results are obtained by hydrogenation using *Adams' catalyst (platinum dioxide)* [*539, 540*]. Benzene and its homologs are completely hydrogenated in 2–26 h using 0.2 g of the catalyst per 0.2 mol of the aromatic compound, acetic acid as the solvent, temperatures of 25–30 °C, and hydrogen pressures of 2–3 atm [*539*]. Under high pressure (215 atm) such hydrogenations are completed within 12–30 min [*8*]. Even better results are obtained using *rhodium–platinum oxides* prepared by fusion of rhodium chloride and chloroplatinic acid (or ammonium chloroplatinate) in 3:1 ratio with sodium nitrate in the manner used for the preparation of platinum dioxide. Hydrogenations of aromatic compounds over this catalyst take place at 24–26 °C and at atmospheric pressure at times considerably shorter than with platinum dioxide alone [41] (equation 55).

5 5

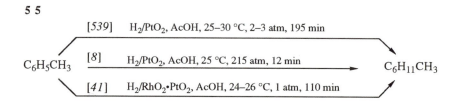

$$C_6H_5CH_3 \qquad\qquad\qquad\qquad C_6H_{11}CH_3$$

[539] H_2/PtO_2, AcOH, 25–30 °C, 2–3 atm, 195 min

[8] H_2/PtO_2, AcOH, 25 °C, 215 atm, 12 min

[41] H_2/RhO_2·PtO_2, AcOH, 24–26 °C, 1 atm, 110 min

Hydrogenation over *rhodium* (5% on alumina) in acetic acid at room temperature and 3–4 atm of hydrogen is successfully used for reduction of benzene rings in compounds containing functions that would be hydrogenolyzed over platinum or palladium catalyst [541]. Palladium at low temperature usually does not reduce benzene rings and is therefore suited for hydrogenolysis of benzyl derivatives (pp 210 and 211).

Specially prepared highly active *Raney nickel* (W-6) can also be used for hydrogenation of some benzene homologs under the same mild conditions that are used with noble metals [46]. In biphenyl, selective hydrogenation of one benzene ring is possible.

Hydrogenation of benzene and its homologs over other nickel catalysts requires higher temperatures (170–180 °C) and/or higher pressures [44, 49, 542].

Copper chromite (*Adkins' catalyst*) does not catalyze hydrogenation of benzene rings.

The stereochemistry of catalytic hydrogenation of benzene homologs is somewhat controversial. The old simplistic view that platinum, palladium, and osmium favor *cis* products whereas nickel favors *trans* products, and that *cis* products rearrange to *trans* products when heated with nickel at higher temperatures (175–180 °C) [542] has had to be modified as a result of later observations. Treatment for 1 h of *cis*-1,2-, *cis*-1,3-, and *cis*-1,4-dimethyl-cyclohexanes over platinum oxide at 85 °C and 84 atm gives only up to 3%, and over Raney nickel at 200 °C and 84 atm only up to 4% of the corresponding *trans* isomers [9]. Hydrogenation using platinum oxide over a 35–85 °C temperature range and up to 8.4 atm of pressure gives up to 15.6% *trans*-1,2-, 19% *trans*-1,3-, and 33.4% *trans*-1,4-dimethylcyclohexane from *o*-, *m*-, and *p*-xylenes, respectively. The corresponding contents of the *trans* isomers after treatment of the three xylenes over Raney nickel at 198–200 °C at 8.6 atm are 11%, 21%, and 5%, respectively (less than over platinum!) [9]. Other authors found that increase in hydrogen pressure from 0.5 to 300 atm increased the amounts of *cis* isomers from 88% to 93% for *cis*-1,2-, from 77% to 85% for *cis*-1,3-, and from 70% to 82% for *cis*-1,4-dimethylcyclohexane over platinum oxide at 25 °C [10]. Above all, the ratios of *cis* to *trans* product depend also on the groups linked to the benzene ring [9, 10].

Whereas catalytic hydrogenation always converts benzene and its homologs to cyclohexane derivatives, *electrolysis* using platinum electrodes and lithium chloride in anhydrous methylamine yields partially hydrogenated products.

When the electrolytic cell is undivided, the products are 1,4-cyclohexadiene and its homologs. When the cell is divided by an asbestos diaphragm, cyclohexene and its homologs are obtained in high purities and yields ranging from 44% to 85% [*191, 194*] (Procedure 29, p 308).

Benzene and its homologs can be converted to the corresponding cyclohexadienes and cyclohexenes, and even cyclohexanes, by treatment with *dissolving metals*: *lithium, sodium, potassium*, or *calcium* in liquid ammonia or amines. Conversions are not complete, and the ratio of cyclohexadienes to cyclohexenes depends on the metal used, the solvent, and the presence of hydrogen donors (alcohols) added to the ammonia or amine [*543, 544, 545*].

Addition of benzene in methanol to an excess (25–100%) of sodium in liquid ammonia affords 84–88% yields of 1,4-cyclohexadiene [*543*]. *o*-Xylene under the same conditions gives a 70–92% yield of 1,2-dimethyl-1,4-cyclohexadiene [*543, 546*]. Lithium in neat ammonia at 60 °C gives 91% 1,4-cyclohexadiene and 9% cyclohexene (58.4% conversion), and calcium under the same conditions yields 21% cyclohexadienes and 79% cyclohexene in conversions of 13–60% [*544*]. If benzene dissolved in ether is added to the compound $Ca(NH_3)_6$, preformed by dissolving calcium in liquid ammonia and evaporating the ammonia, a 75% conversion to pure cyclohexene is achieved [*545*].

Similar results are achieved when benzene is reduced with alkali metals in anhydrous methylamine at temperatures of 26–100 °C. Best yields of cyclohexene (up to 77.4%) are obtained with lithium at 85 °C [*201*]. Ethylamine [*202*] and especially ethylenediamine are even better solvents [*204*]. Benzene is reduced to cyclohexene and a small amount of cyclohexane [*202, 204*]; ethylbenzene treated with lithium in ethylamine at –78 °C gives 75% l-ethylcyclohexene, whereas at 17 °C a mixture of 45% l-ethylcyclohexene and 55% ethylcyclohexane is obtained [*202*]. Xylenes (*m*- and *p*-) yield nonconjugated 2,5-dihydro derivatives, 1,3-dimethyl-3,6-cyclohexadiene and 1,4-dimethyl-1,4-cyclohexadiene, respectively, on reduction with sodium in liquid ammonia in the presence of ethanol (in poor yields) [*547*]. Better yields are obtained with calcium in butylamine and ethylenediamine. *p*-Xylene is reduced to 1,4-dimethylcyclohexa-1,4-diene and 1,4-dimethylcyclohex-1-ene [*213*] (equation **56**).

Reduction of biphenyl with sodium or calcium in liquid ammonia at –70 °C affords mainly l-phenylcyclohexene [*548*], whereas with sodium in ammonia at 120–125 °C mainly phenylcyclohexane is formed [*544*].

Dissolving metal reductions of the benzene rings are especially important with functional derivatives of benzene such as phenols, phenol ethers, and carboxylic acids (pp 109, 112 and 113).

Benzene and its homologs can also be reduced to the corresponding hexahydro derivatives by heating with anhydrous *hydrazine*. Best yields are obtained with mesitylene (92%). However, this reaction is not very attractive because it requires heating in autoclaves at 250 °C [*380*].

Aromatic hydrocarbons with side chains containing double bonds can be easily reduced by catalytic hydrogenation regardless of whether the bonds are isolated or conjugated. Double bonds are saturated before the aromatic ring is reduced. Hydrogenation of styrene to ethylbenzene is one of the fastest catalytic hydrogenations [*14*].

5 6   [*213*]

CH$_3$—⟨O⟩—CH$_3$ ⟶ CH$_3$—⟨=⟩—CH$_3$ + CH$_3$—⟨=⟩—CH$_3$

                              62–70%   8–9%

Ca, BuNH$_2$, H$_2$NCH$_2$CH$_2$NH$_2$
t-BuOH, THF, 0 °C

Saturation of double bonds can also be achieved by *catalytic transfer of hydrogen* [*82, 84*]. Styrene is quantitatively reduced to ethylbenzene by 1,4-cyclohexadiene in the presence of iodine under ultraviolet irradiation [*84*], and *trans*-stilbene is reduced to 1,2-diphenylethane in a 90% yield by refluxing for 46 h with cyclohexene in the presence of palladium on charcoal [*82*].

Double bonds conjugated with benzene rings are reduced *electrolytically* [*195*]. Where applicable, stereochemistry can be influenced by using either catalytic hydrogenation or *dissolving metal* reduction [*549*]. Indene is converted to indane by sodium in liquid ammonia in an 85% yield [*550*] and acenaphthylene is converted to acenaphthene in an 85% yield by reduction with *lithium aluminum hydride* in 2-(2-ethoxyethoxy)ethanol (carbitol) at 100 °C [*551*]. The benzene ring is not inert toward alkali metals. Nuclear reduction may accompany reduction of the double bond. Styrene treated with *lithium in methylamine* affords 25% 1-ethylcyclohexene and 18% ethylcyclohexane [*552*].

Aluminum amalgam in wet tetrahydrofuran reduces the cumulene 1,1,4,4-tetraphenylbutatriene to 1,1,4,4-tetraphenylbuta-1,2-diene, which on hydrogenation over Raney nickel is converted to 1,1,4,4-tetraphenylbutane [*217*] (equation **57**).

57   [*217*]

$(C_6H_5)_2C=C=C=C(C_6H_5)_2$ $\xrightarrow[\substack{THF, H_2O \\ 25 °C}]{AlHg}$ $(C_6H_5)_2C=C=CH–CH(C_6H_5)_2$ 70%

                                                    H$_2$ | Raney Ni, 25 °C

                                        $(C_6H_5)_2CHCH_2CH_2CH(C_6H_5)_2$

Triple bonds in side chains of aromatic compounds can be reduced to double bonds or can be completely saturated. The outcome of such reductions depends on the structure of the acetylene and on the method of reduction. If the triple bond is not conjugated with the benzene ring, it can be handled in the same way as in aliphatic acetylenes. In addition, *electrochemical reduction* in a solution of lithium chloride in methylamine has been used for partial reduction to alkenes (*trans* isomers, where applicable) in 40–51% yields (with 2,5-dihydroaromatic alkenes as byproducts) [*528*]. Arylacetylenes with triple bonds conjugated with benzene rings can be *hydrogenated over Raney nickel* to *cis* olefins [*505, 508*], or to arylalkanes over palladium [*538*] or rhenium sulfide catalyst [*57*]. Electroreduction in methylamine containing lithium chloride gives 80% yields of arylalkanes [*528*].

Reduction of diphenylacetylene with *sodium* in methanol or with *zinc* yields *trans*-stilbene, and reduction with sodium in ethanol gives 1,2-diphenylethane. Unfortunately the yields were not reported. *Samarium diiodide* in the presence of cobalt dichloride–tris(triphenyl)phosphine and isopropyl alcohol catalyst converts diphenylacetylene to *cis*-stilbene [306]. Reduction with *diimide* generated from hydrazine by iodobenzene diacetate gives *cis*-stilbene [376], whereas diimide prepared from hydrazine by oxidation with oxygen and copper affords 1,2-diphenylethane [374]. An excellent reagent for partial *trans* reduction of acetylenes, *chromous sulfate*, converts phenylacetylene to styrene in an 89% yield [273]. Reduction of diphenylacetylene to *cis*-stilbene by *lithium aluminum hydride* in the presence of dichlorides of iron, cobalt, and nickel is achieved in 72–80% yields [158] (equation **58**), and with *ammonium hypophosphite* in an 85% yield (Procedure 55, p 318).

58

$$C_6H_5C\equiv CC_6H_5 \longrightarrow C_6H_5CH{=}CHC_6H_5 \ + \ C_6H_5CH_2CH_2C_6H_5$$

[508]	H_2, Raney Ni, MeOH, 4 atm	87% (*cis*)	
[538]	H_2, Pd, AcOH, 25 °C		99.3%
[158]	$LiAlH_4$, $FeCl_2$, THF, –78 °C to RT, 1 h	86% (*cis*)	
[158]	$LiAlH_4$, $CoCl_2$, THF, –78 °C to RT, 12 h	72% (*cis*) 12% (*trans*)	
[158]	$LiAlH_4$, $NiCl_2$, THF, –78 °C to RT, 12 h	75% (*cis*)	
[528]	electro, LiCl, $MeNH_2$	2% (*trans*)	78%
[306]	SmI_2, $CoCl_2 \cdot 4PPh_3$, Me_2CHOH, THF, RT	59%	
[376]	N_2H_4, $PhI(OAc)_2$, CH_2Cl_2, RT	59% (*cis*)	
[374]	N_2H_4, O_2 Cu		80%

Condensed aromatic hydrocarbons are reduced much more easily than those of the benzene series, both catalytically and by dissolving metals.

Depending on the catalysts, *catalytic hydrogenation* converts **naphthalene** to 1,2,3,4-tetrahydronaphthalene (tetralin), or *cis*- or *trans*-decahydronaphthalene (decalin) [8, 553, 554, 555]. Tetrahydronaphthalene is converted to *cis*-decalin by hydrogenation over platinum oxide [8] (equation **59**).

Dissolving metals reduce naphthalene to a host of products depending on the reaction conditions. Heating naphthalene with sodium to 140–145 °C [556] or treating naphthalene with sodium in liquid ammonia at temperatures below –60 °C gives 1,4-dihydronaphthalene, which slowly rearranges to 1,2-dihydronaphthalene [548]. 1,2-Dihydronaphthalene has the double bond conjugated with the remaining benzene ring, so it is reduced to 1,2,3,4-tetrahydronaphthalene [548]. This is also the final product when the reduction is carried out with enough sodium in liquid ammonia at –33 °C [548]. In the presence of alcohols, sodium in liquid ammonia reduces naphthalene [557, 558] and 1,4-dihydronaphthalene to 1,4,5,8-tetrahydronaphthalene [558].

5 9

[553]	$H_2/CuCr_2O_4$, 200 °C, 150–200 atm	80%		
[8]	H_2/PtO_2, AcOH, 25 °C, 130 atm		71%	21%
[554]	H_2/Pt		75%	25%
[554]	N_2/Ni, 160–162 °C		22%	78%

Lithium in ethylamine converts naphthalene to a mixture of 80% 1,2,3,4,5,6,7,8-octahydronaphthalene ($\Delta^{9,10}$-octalin) and 20% 1,2,3,4,5,6,7,10-octahydronaphthalene [559, 560]. *Potassium* in methylamine reduces tetralin to $\Delta^{9,10}$-octalin in 95% yield [201], whereas *sodium* in ammonia and methanol yields 98% 1,2,3,4,5,8-hexahydronaphthalene [561]. Lithium in ethylenediamine reduces tetralin to 68.7% $\Delta^{9,10}$-octalin and 18.5% *trans*-decalin [204]. *Calcium* in butylamine and ethylenediamine in the presence of *tert*-butyl alcohol transforms naphthalene to 1,4,5,8-tetrahydro- and 1,2,3,4,5,8-hexahydro-naphthalenes and transforms tetralin to 1,2,3,4,5,8-hexahydro-, 1,2,3,4,5,6,7,8-octahydro-, and 2,3,4,5,6,7,8,10-octahydronaphthalenes [213] (equation 60).

Reduction of naphthols, naphthyl ethers, naphthylamines, and naphthoic acids is discussed on pp 109, 113, 130 and 195, respectively. **Acenaphthylene** heated with *lithium aluminum hydride* in 2-(2-ethoxyethoxy)ethanol (carbitol) at 120 °C gives a 97% yield of acenaphthene [551]. **Anthracene** is reduced very easily to the 9,10-dihydro compound by *catalytic hydrogenation* [86] and by *sodium* [562]. Further reduction is achieved by catalytic hydrogenation using different catalysts and reaction conditions [86, 555] and by *lithium* in ethylenediamine [563]. *Calcium* in butylamine and ethylenediamine in the presence of *tert*-butyl alcohol affords 1,4,5,8,9,10-hexahydro- and 1,2,3,4,5,6,7,8,9,10-decahydro-anthracenes [213] (equation 61).

Phenanthrene is converted to partially and totally reduced phenanthrene by *catalytic hydrogenation* [564, 565, 566] (Procedure 3, p 296). Partial reduction to 1,2,3,4-tetrahydrophenanthrene can also be achieved by *sodium* [567] (equation 62).

HETEROCYCLIC AROMATIC COMPOUNDS

Reduction of **furan** and its homologs is readily achieved by *catalytic hydrogenation* using palladium [32] or Raney nickel [568] as catalysts. Highly energetic conditions are to be avoided because hydrogenated furan rings are easily hydrogenolyzed [569, 570] (equation 63).

Benzofuran is reduced over Raney nickel at 90 °C and 85 atm to 2,3-di-hydrobenzofuran and at 190 °C and 119 atm to octahydrobenzofuran, and further hydrogenation over copper chromite at 200–300 °C causes hydrogenolysis to *cis*- and *trans*-2-ethylcyclohexanol [569].

6 0

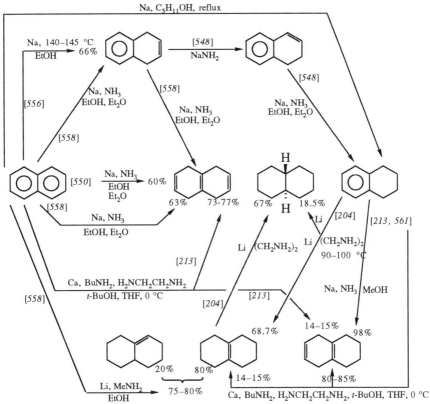

6 1

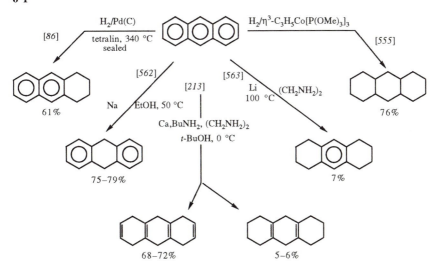

6 2   [564]

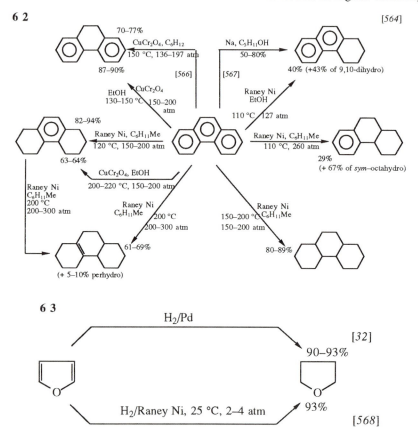

6 3

3,4-Dihydrobenzopyran **(chromane)** and its methyl homologs are reduced by *lithium* in ethylamine and dimethylamine to 3,4,5,6,7,8-hexahydrobenzopyrans in 84.5% yields [571].

Catalytic hydrogenation of **thiophene** poses a problem because noble metal catalysts are poisoned and Raney nickel causes desulfurization. The best catalysts proved to be cobalt polysulfide [572], dicobalt octacarbonyl [573], rhenium heptasulfide [56], and rhenium heptaselenide [57]. The last two require high temperatures (230–260 °C, 250 °C) and high pressures (140 °C, 322 atm) and give 70% and 100% tetrahydrothiophene (also called thiophane or thiolane), respectively.

Procedures similar to those used for the catalytic hydrogenation of furan and thiophene can usually be applied to the functional derivatives of these heterocycles.

Pyrrole and its homologs can be reduced completely or partially. Complete reduction is best achieved by *catalytic hydrogenation*. Pyrrole itself is hydrogenated with difficulty, and not in high yields, over platinum [574], nickel [44], Raney nickel, or copper chromite [575]. Over Raney nickel at 180 °C and 200–300 atm, only a 47% yield of pyrrolidine is obtained; 48% of the pyrrole remains unreacted.

On the other hand, pyrrole homologs and derivatives are readily hydrogenated and give better yields than pyrrole itself, especially if the substituents are attached to nitrogen [*575, 576, 577*]. 1-Methyl- and 1-butylpyrroles hydrogenated over platinum oxide give high yields of the corresponding *N*-alkylpyrrolidines [*574*]. 2,5-Dimethylpyrrole in acetic acid over rhodium at 60 °C and 3 atm affords 70% *cis*-2,5-dimethylpyrrolidine [*578*], and 2-butylpyrrole in acetic acid over platinum oxide at room temperature and 3 atm gives 94% 2-butylpyrrolidine [*576*]. 1-Phenylpyrrole is hydrogenated over Raney nickel at 135 °C and 200–300 atm to 63% 1-phenylpyrrolidine and 30% 1-cyclohexylpyrrolidine [*575*]. 2-Phenylpyrrole gives, over Raney nickel at 165 °C and 250–300 atm, 15% 2-phenylpyrrolidine and 27% 2-cyclohexyl-pyrrolidine, whereas over copper chromite at 200 °C it gives a 55% yield of 2-phenylpyrrolidine [*577*]. Some pyrrole derivatives may be hydrogenated over platinum oxide at room temperature [*579*], and others over Raney nickel at 50–180 °C and 200–300 atm in yields up to 98% [*575*]. Copper chromite as the catalyst is inferior to Raney nickel because it requires temperatures of 200–250 °C [*575*] (equation **64**).

6 4

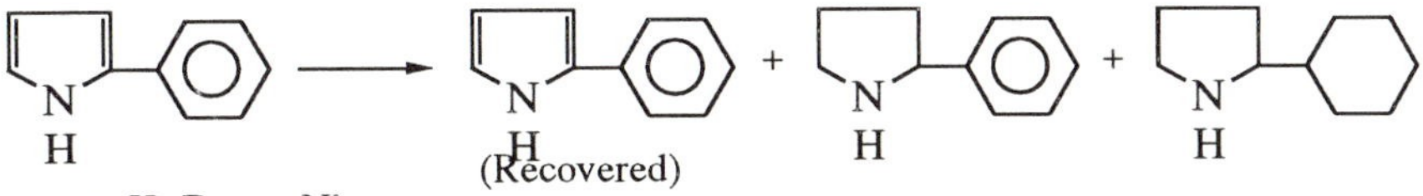

[*577*] H$_2$/Raney Ni
165 °C, 250–300 atm   40%   15%   27%

[*577*] 200 °C, 250–300 atm   69%
H$_2$/CuCr$_2$O$_4$
200 °C, 250–300 atm   20%   55%

Partial and total reduction of *N*-methylpyrrole is achieved with *zinc* [*580*], and of a pyrrole derivative to a dihydropyrrole derivative with *phosphonium iodide* [*392*]. The pyrrole ring is not reduced by sodium.

Pyrrole derivatives having double bonds in the side chains are first reduced at the double bonds and then in the pyrrole ring. 2-(2-Butenyl)pyrrole gives an 88% yield of 2-butylpyrrole over platinum oxide in ether; further hydrogenation in acetic acid gives a 94% yield of 2-butylpyrrolidine [*576*].

Reduction of the pyrrole ring in **benzopyrrole (indole)** is discussed on p 72 [*577*].

Pyridine and its homologs can be reduced completely to hexahydro derivatives, or partially to dihydro- and tetrahydropyridines. *Catalytic hydrogenation* is faster than with the corresponding benzene derivatives and gives only completely hydrogenated products. Partial reduction can be achieved by different methods.

Hydrogenation of the pyridine ring takes place under very mild conditions using palladium [*581*], platinum oxide [*581, 582*], or rhodium [*578, 581, 583*]. With these metals the reaction must be carried out in acidic media, best in acetic acid, because the hydrogenated products are strong bases that deactivate the catalysts. Hydrogenation over Raney nickel [*584*] and copper chromite [*51*] requires high temperatures and high pressures. Use of alcohols as solvents for these hydrogenations should be avoided because alkylation on nitrogen could occur. A catalyst that does not require an acidic medium and can be used in the presence of alcohols is ruthenium dioxide [585]. A thorough treatment of the catalytic hydrogenation of pyridine was published in *Advances in Catalysis* [586].

Complete saturation of the pyridine rings in pyridine and its homologs and derivatives can be achieved by *nickel–aluminum alloy* in 0.5 M potassium hydroxide at room temperature in yields of 51–90% [230] (equation **65**).

Lithium aluminum hydride dissolves in pyridine and forms lithium tetrakis-(*N*-dihydropyridyl)aluminate, which itself is a reducing agent for purely aromatic ketones [587, 588].

6 5

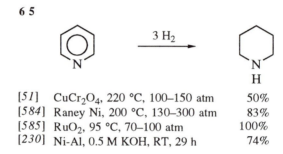

[51]	CuCr$_2$O$_4$, 220 °C, 100–150 atm	50%
[584]	Raney Ni, 200 °C, 130–300 atm	83%
[585]	RuO$_2$, 95 °C, 70–100 atm	100%
[230]	Ni-Al, 0.5 M KOH, RT, 29 h	74%

The total and partial reduction of pyridine and its homologs and derivatives is achieved by *alane* [196], by *electroreduction* [196], and by reduction with *sodium* in alcohols [196, 589]. All these three methods give mixtures of tetrahydropyridines and hexahydropyridines [589]. The heteropolar bonds are more receptive to electrons than the carbon–carbon bonds. Consequently the reduction initially gives 1,2-dihydro- and 1,4-dihydropyridines. These rearrange according to the scheme shown in equation **66**.

The 1,2-dihydro intermediate is further reduced to 1,2,3,6-tetrahydro- and 1,2,5,6-tetrahydro products (identical if the starting pyridine compound is symmetrical). The 1,4-dihydro intermediate rearranges to the 3,4-dihydro compound, which, on further reduction, gives a 1,2,3,4-tetrahydro derivative; this undergoes an internal enamine–imine shift and is ultimately reduced to the hexahydropyridine product. Both tetrahydro derivatives contain carbon–carbon double bonds isolated from the nitrogen and do not undergo further reduction to piperidines (equation **66**).

Reductions of this type were applied to pyridine [590], alkylpyridines [196], and many pyridine derivatives: alcohols [591], aldehydes [591], ketones [591], and acids [592]. In compounds containing both pyridine and benzene rings, sodium exclusively reduces the pyridine ring [593]. Pyridine and its homologs are quantitatively reduced to piperidine and its homologs by *sodium* in refluxing ethanol (the *Ladenburg reduction*) [594] but are not reduced by zinc and other similar metals.

The **pyridine ring** is easily reduced in the form of its **quaternary salts** to hexahydro derivatives by *catalytic hydrogenation* [595] and to tetrahydro and

6 6

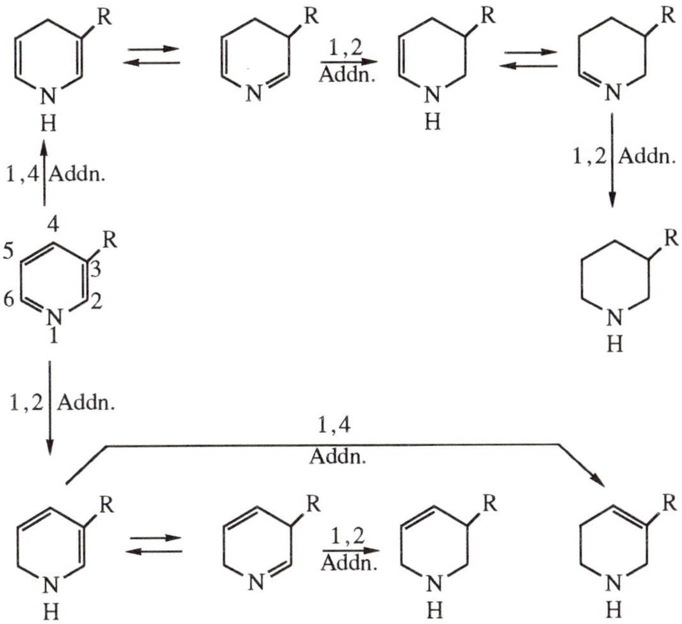

hexahydro derivatives by reduction with *alane (aluminum hydride)* [596], *lithium aluminum hydride* [596, 597], *sodium aluminum hydride* [598], *sodium bis(2-methoxyethoxy)aluminum hydride* [598], *sodium borohydride* [596, 599], *potassium borohydride* [600], *sodium* in ethanol [592, 601], and *formic acid* [442, 589, 602]. Reductions with hydrides give predominantly 1,2,5,6-tetrahydro derivatives, whereas *electroreduction* [597] and reduction with formic acid give more hexahydro derivatives [589] (equation **67**).

Double bonds in **alkenylpyridines** may be *hydrogenated* under mild conditions (Raney nickel at room temperature) to give alkylpyridines [601]. If

6 7

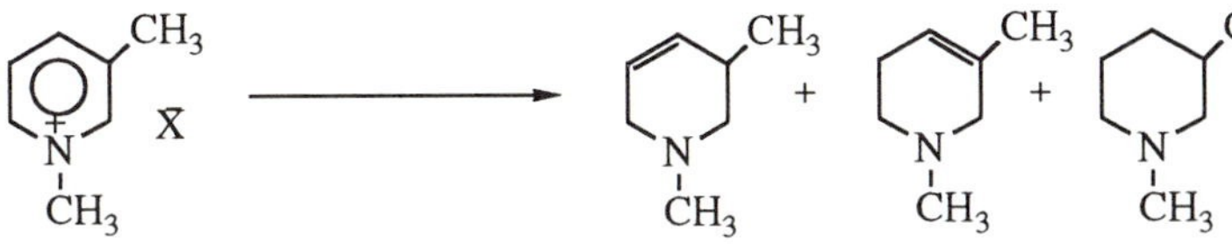

X		Yield	%	%	%
[596,599]	I, NaBH$_4$	79.1%	5.0	91.5	3.5
[596,597]	I, LiAlH$_4$	98%	15.5	84.5	0
[596]	I, AlH$_3$	88%	16.0	84.0	0
[600]	I, KBH$_4$	72%	7.5	82.5	10
[442]	HCO$_2$H	68%	11	45	55
[597]	MeOSO$_3$, electroredn.			17	72
[598]	I, NaAlH$_4$	44%	20	35	5
[598]	I, NaAlH$_2$ (OCH$_2$CH$_2$OMe)$_2$	51%	27	32	20

the double bond is conjugated with the pyridine ring, *sodium* in alcohol will reduce both the double bond and the pyridine ring in good yields [*601*].

In pyridylpyrrole derivatives, the pyridine ring is hydrogenated preferentially, giving piperidylpyrroles, which are further hydrogenated to piperidylpyrrolidines over platinum oxide in acetic acid at room temperature and 2 atm [*603*].

The *N*-methylpyrrole ring in β-nicotyrine (1-methyl-2-(3'-pyridyl)pyrrole) is reduced to nicotine both by catalytic hydrogenation [*604*] and by *zinc* [*605*] in 40% and 12% yields, respectively.

The double bond in **indole** and its homologs and derivatives is reduced easily and selectively by *catalytic hydrogenation* over platinum oxide in ethanol and fluoroboric acid [*606*], by *sodium borohydride* [*607*], *sodium cyanoborohydride* [*607*], *borane* [*608, 609*], *sodium in ammonia* [*610*], *lithium* [*611*], and *zinc* [*612*]. Reduction with *sodium borohydride* in acetic acid can result in alkylation on nitrogen giving *N*-ethylindoline [*607*] (equation **68**).

6 8

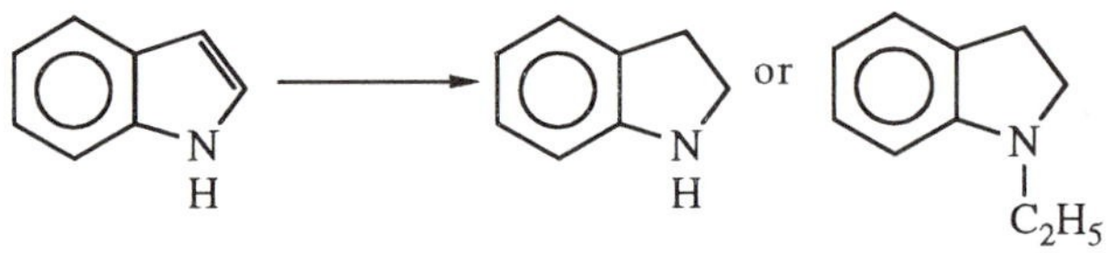

[*607*]	NaBH₃CN	88%	
[*607*]	NaBH₄, AcOH		86%
[*608*]	BH₃•NEt₃	80%	
[*609*]	BH₃•C₅H₅N, AcOH	86%	
[*612*]	Zn, H₃PO₄, 70-80°C	64%	

Catalytic hydrogenation may selectively reduce the double bond or reduce the aromatic ring as well, depending on the reaction conditions used [*577, 613, 614*]. Only rarely has the benzene ring been hydrogenated in preference to the pyrrole ring [*613, 615*] (equation **69**).

In indole's isomer **pyrrocoline** (1-azabicyclo[4.3.0]nonatetraene), *catalytic hydrogenation* over palladium in acidic medium reduces the pyrrole ring [*616*] and in neutral medium the pyridine ring [*617*] (equation **70**).

Carbazoles, too, can be reduced partially to dihydrocarbazoles by *sodium* in liquid ammonia [*610*], or to tetrahydrocarbazoles by sodium in liquid ammonia and ethanol [*610*] or by *sodium borohydride* [*607*]. Carbazole is converted by *catalytic hydrogenation* over Raney nickel or copper chromite to 1,2,3,4-tetrahydrocarbazole, 1,2,3,4,10,11-hexahydrocarbazole, and dodeca-hydrocarbazole in good yields [*577*].

In **quinoline** and its homologs and derivatives, the pyridine ring is usually reduced first. *Sodium* in liquid ammonia converts quinoline to 1,2-dihydro-quinoline [*618*]. 1,2,3,4-Tetrahydroquinoline is obtained by *catalytic hydrogenation* (56–100% yields) [*44, 619*] and by reduction with *borane* [*609*], *sodium cyanoborohydride* [*620*], *nickel–aluminum alloy* [*230*], and *zinc* and

6 9

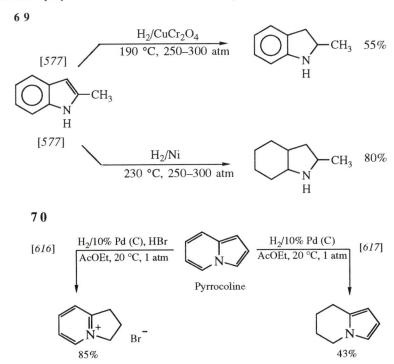

7 0

Pyrrocoline

85%   43%

nickel chloride [*241*]. 5,6,7,8-Tetrahydroquinoline results from hydrogenation over platinum oxide or 5% palladium or rhodium on carbon in trifluoroacetic acid (69–84% yields) [*621*]. Vigorous hydrogenation gave *cis*- and *trans*-decahydroquinoline [*44, 619*] (equation **71**).

7 1

[*618*]	Na, NH$_3$	75.5%		
[*618*]	H$_2$, Raney Ni, EtOH	56%		
[*44*]	H$_2$, Ni, 150 °C, 160 atm	96%		
[*619*]	H$_2$, Pt, 40 °C, 2–3 atm	100%		
[*51*]	H$_2$, CuCr$_2$O$_4$, 190 °C, 150–200 atm	100%		
[*619*]	H$_2$, Pt, H$_2$PtCl$_6$, AcOH, H$_2$O, 40 °C, 2–3 atm		20%	80%
[*619*]	H$_2$, Pt, H$_2$PtCl$_4$, AcOH, HCl, 40 °C, 2–3 atm		65%	35%
[*609*]	BH$_3$·C$_5$H$_5$N, AcOH	71%		
[*620*]	NaBH$_3$CN, AcOH	71%		
[*230*]	Ni–Al alloy, 0.5 M KOH, MeOH, RT	84%		
[*241*]	NiCl$_2$, Zn, MeOH, reflux 12 h	70%		

Quinoline homologs and derivatives, including those with double bonds in the side chains, are reduced selectively by *catalytic hydrogenation* over platinum oxide (side-chain double bonds), and to dihydro- and 1,2,3,4-tetrahydroquinolines by *sodium* in butanol [622], *zinc* [241], and *triethylammonium formate* [443, 622]. Catalytic hydrogenation of quinoline and its derivatives has been thoroughly reviewed [586].

Isoquinoline is converted to 1,2,3,4-tetrahydroisoquinoline in an 89% yield by reduction with *sodium* in liquid ammonia and ethanol [623] and in a 74% yield by treatment with *nickel–aluminum alloy* in potassium hydroxide [230], and to a mixture of 70–80% *cis-* and 10% *trans*-decahydroisoquinoline by *catalytic hydrogenation* over platinum oxide in acetic and sulfuric acid [624]. Without sulfuric acid the hydrogenation stops at the tetrahydro stage. Catalytic hydrogenation of isoquinoline and its derivatives is the topic of a review in *Advances in Catalysis* [586] (equation **72**).

7 2

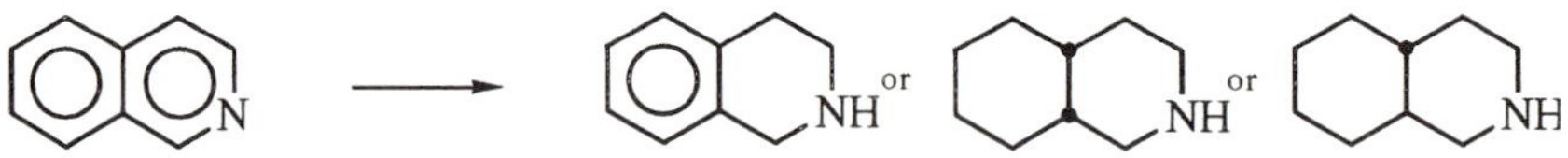

[623]	Na, NH$_3$, EtOH	89%		
[230]	Ni–Al alloy, 0.5 M KOH, RT, 43 h	74%		
[624]	H$_2$/PtO$_2$, AcOH, H$_2$SO$_4$, RT, 1 atm, 8 h		70–80%	10%

Dehydroquinolizinium salts containing quaternary nitrogen at the bridgehead of bicyclic systems are easily *hydrogenated over platinum* to quinolizidines [625] (equation **73**).

7 3   [625]

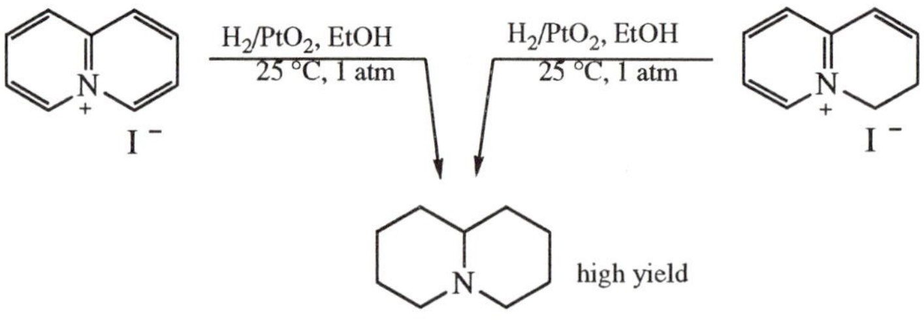

Quinolizidine

Reaction conditions used for reduction of **acridine** [577, 626], partly hydrogenated **phenanthridine** [627], and partly hydrogenated **benzo[f]quinoline** [627] are shown in equations **74–77**. *Hydrogenation over platinum oxide* in trifluoroacetic acid at 3.5 atm reduces only the carbocyclic rings in acridine [628] and mainly the carbocyclic rings in **benzo[h]quinoline** [628].

7 4

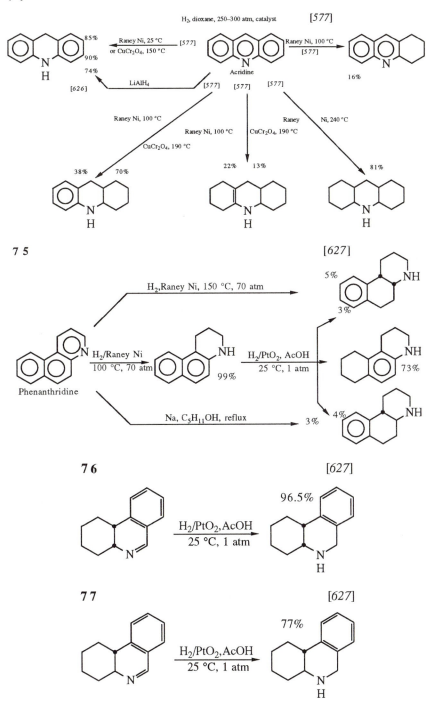

7 5

7 6

7 7

Aromatic heterocycles containing two nitrogen atoms are best reduced by catalytic hydrogenation. Other reduction methods may cleave some of the rings. Even catalytic hydrogenation causes occasional hydrogenolysis.

Pyrazole is *hydrogenated* over palladium on barium sulfate in acetic acid to 4,5-dihydropyrazole (Δ^2-pyrazoline), and 1-phenylpyrazole is hydrogenated at 70–80 °C to l-phenylpyrazolidine [629]. In benzopyrazole *(indazole)* and its homologs and derivatives, the six-membered ring is hydrogenated preferentially to give 4,5,6,7-tetrahydroindazoles over platinum, palladium, or rhodium in acetic, hydrochloric, sulfuric, and perchloric acid solutions in 45–96% yields [630].

Imidazole is converted by *hydrogenation* over platinum oxide in acetic anhydride to 1,3-diacetylimidazolidine in 80% yield, and **benzimidazole** similarly to 1,3-diacetylbenzimidazoline in 86% yield [631]. Benzimidazole is very resistant to hydrogenation over platinum at 100 °C and over nickel at 200 °C and under high pressure, but 2-alkyl- or 2-aryl-substituted benzimidazoles are reduced in the benzene ring rather easily. 2-Methylbenzimidazole is hydrogenated over platinum oxide in acetic acid at 80–90 °C to 2-methyl-4,5,6,7-tetrahydrobenzimidazole in an 87% yield [632]. *Lithium aluminum hydride* reduces benzimidazole to 2,3-dihydrobenzimidazole [626].

The behavior of six-membered rings with two nitrogen atoms depends on the position of the nitrogen atoms. **1,2-Diazines (pyridazines)** are very stable to catalytic hydrogenation [633]. **1,3-Diazine (pyrimidine)** and its 2-methyl, 4-methyl, and 5-methyl homologs are easily *hydrogenated* in aqueous hydrochloric acid over 10% palladium on charcoal to 1,4,5,6-tetrahydropyrimidines in yields of 97–98% [634].

1,4-Diazines (pyrazines) afford hexahydro derivatives by *catalytic hydrogenation* over palladium on charcoal at room temperature and 3–4 atm [635, 636] (yield of 2-butylpiperazine is 62% [635], and yield of 2,5-diphenylpiperazine is 80% [636]). *Electrolytic reduction* gives unstable 1,2-, 1,4-, and 1,6-dihydropyrazines [636].

In condensed systems containing rings with two nitrogen atoms, the heterocyclic ring is reduced preferentially most of the time. **4-Phenylcinnoline** is *hydrogenated* in acetic acid over palladium almost quantitatively to 4-phenyl-1,4-dihydrocinnoline [637], and this over platinum to 4-phenyl-1,2,3,4-tetrahydrocinnoline [638]. **4-Methylcinnoline** gives either 4-methyl-1,4-dihydrocinnoline or 3-methylindole on *electrolytic reduction* under various conditions [639] (equation **78**).

Quinazoline was *reduced by hydrogen* over platinum oxide to 3,4-dihydroquinazoline [640], and by *sodium borohydride* in trifluoroacetic acid to 1,2-dihydroquinazoline [641] (equation **79**).

Reduction of **quinoxalines** is carried out by *catalytic hydrogenation* [642, 643, 644, 645], with *sodium borohydride* [641], or with *lithium aluminum hydride* [626] to different stages of saturation (equation **80**).

Phthalazine and its homologs and derivatives are easily hydrogenolyzed. *Electroreduction* in alkaline medium gives 1,2-dihydrophthalazines [646]. In acidic media [646] and on *catalytic hydrogenation* [647], the ring is cleaved to yield o-xylene-α,α'-diamine (equation **81**).

7 8 [*639*]

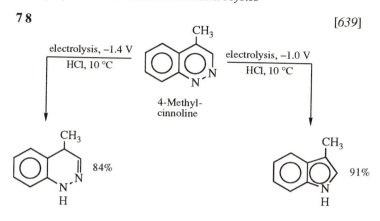

7 9

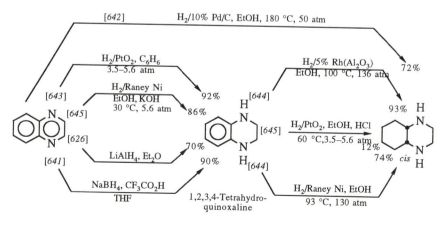

8 1

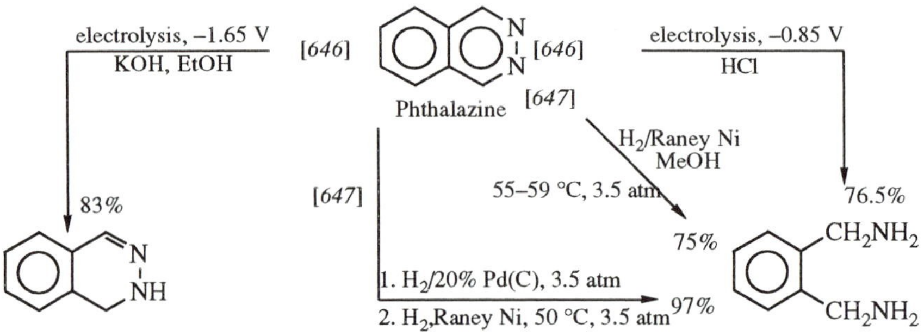

Phthalazine [647]

83%

[647]

Phenazine (9,10-diazaanthracene) is partially reduced by *lithium aluminum hydride* to 9,10-dihydrophenazine (90% yield) [626] and totally reduced *by catalytic hydrogenation* over 10% palladium on carbon in ethanol at 180 °C and 50 atm to tetradecahydrophenazine (80% yield) [642]. Catalytic hydrogenation of **1,10-phenanthroline** affords 1,2,3,4-tetrahydro- and *sym*-octahydrophenanthroline [648] (equation **82**).

8 2   [648]

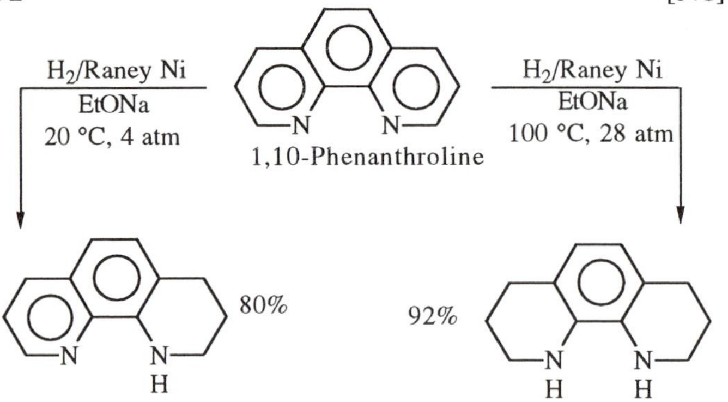

Because of easily controlled reaction conditions, *catalytic hydrogenation* is used frequently for partial reductions of complex natural products and their derivatives [649, 650]. A random example is reduction of yobyrine [649] (equation **83**).

8 3 [*649*]

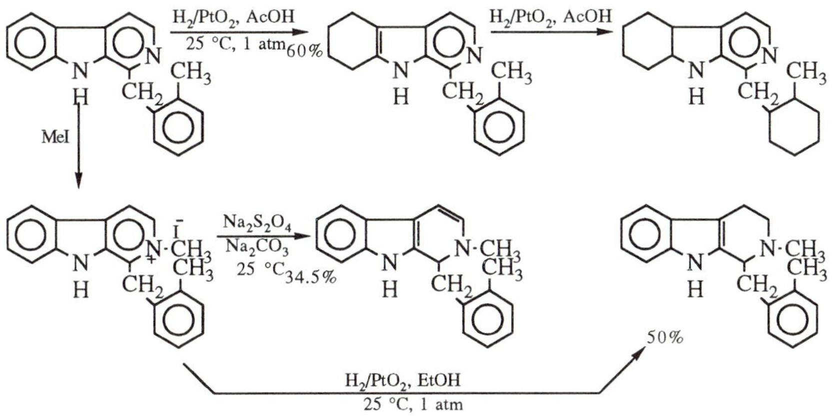

SUMMARY

A universal method for reduction of all the systems described in this chapter is catalytic hydrogenation. Reduction of an isolated double bond by hydrides, electroreduction, and alkali metals does not have practical significance. Conjugated dienes and acetylenes can be reduced to alkenes by some hydrides and metals and even some metal compounds, if the triple bond is conjugated with aromatic rings. Total reduction of carbocyclic aromatic compounds is achieved only by catalytic hydrogenation. Partial reduction is accomplished by alkali metals. Partial and complete reduction of aromatic heterocycles is carried out by catalytic hydrogenation, and partial reduction is also achieved with complex hydrides and alkali metals.

CHAPTER 7

Reduction of Halogen Derivatives of Hydrocarbons and Basic Heterocycles

For the replacement of halogen by hydrogen—hydrogenolysis of the carbon–halogen bond—many reduction methods are available. Breaking of carbon–halogen bonds is involved, so the ease of hydrogenolysis decreases in the series iodide, bromide, chloride, and fluoride and increases in the series aliphatic < aromatic < vinylic < allylic < benzylic halides. In catalytic hydrogenation the rate of reduction increases from primary to tertiary halides, whereas in reductions with complex hydrides, where the reagent is a strong nucleophile, the trend is just the opposite.

HALOALKANES AND HALOCYCLOALKANES

Alkyl fluorides are for practical purposes resistant to reduction. In a few instances, where fluorine in fluoroalkanes was replaced by hydrogen, very vigorous conditions were required and usually gave poor yields [*651, 652*]. For this reason it is not difficult to replace other halogens present in the molecule of chlorofluorocarbons without affecting fluorine, although not in preparative yields [*425, 427*].

Alkyl chlorides are with a few exceptions not reduced by mild catalytic hydrogenation over platinum [*653*], rhodium [*43*], and nickel [*63*], even in the presence of alkali. Metal hydrides and complex hydrides are used more successfully: Various *lithium aluminum hydrides* [*654, 655*], *lithium copper hydrides* [*652*], *sodium borohydride* [*656, 657*], *triethylsilane* in the presence of *tert*-butylhyponitrite and *tert*-dodecanethiol [*161*], and especially different *tin hydrides (stannanes)* [*658, 659, 660, 661*] are the reagents of choice for selective replacement of halogen in the presence of other functional groups. In some cases the reduction is stereoselective. Both *cis*- and *trans*-9-chlorodecalin, on reductions with triphenylstannane or dibutylstannane, give predominantly *trans*-decalin [*660*]. Only tertiary chlorides are reported to be reduced in high yields by *sodium cyanoborohydride* and stannous chloride [*662*] or by *zinc borohydride* [*663*] in ether at room temperature (equation **84**).

84

$C_8H_{17}Cl$	$\xrightarrow{\hspace{4cm}}$	C_8H_{18}
[*654*]	LiAlH$_4$, THF, 25 °C, 24 h	73%
[*654*]	NaAlH$_4$, THF, 25 °C, 24 h	38%
[*654*]	LiAlH(OMe)$_3$, THF, 25 °C, 24 h	19%
[*654*]	LiBHEt$_3$, THF, 25 °C, 24 h	73%
[*655*]	2 LiAlH(OMe)$_3$·CuI, THF, 25 °C, 15 h	96%
[*657*]	NaBH$_4$, DMSO, 45–50 °C, 4 h	42%
[*654*]	NaBH$_4$, diglyme, 25 °C, 24 h	2.5%
[*654*]	LiAlH(OCMe$_3$)$_3$, LiBH$_4$, or BH$_3$·SMe$_2$	0
[*161*]	Et$_3$SiH, *t*-BuONO, *t*-C$_{12}$H$_{25}$SH, C$_6$H$_{14}$, reflux	96%
[*656*]	[a]NaBH$_4$, DMSO, 85 °C, 48 h	67.2%
[*658*]	[a]Bu$_3$SnH, *hv*, 7 h	70%

[a]2-Chlorooctane

Alkyl bromides and especially **alkyl iodides** are reduced faster than chlorides. *Catalytic hydrogenation* is accomplished in good yields using Raney nickel in the presence of potassium hydroxide [63] (Procedure 6, p 298). More frequently, bromides and iodides are reduced by hydrides [659] and complex hydrides in good to excellent yields [652, 656]. Most powerful are *lithium triethylborohydride* and *lithium aluminum hydride* [654]. *Sodium borohydride* reacts much more slowly. Sodium cyanoborohydride in hexamethylphosphoric amide at 70 °C converts decyl iodide to decane in 88–90% yield [1489]. The complex hydrides are believed to react by an S_N2 mechanism [657, 664], so it is not surprising that secondary bromides and iodides react more slowly than the primary ones [654]. The reagent prepared from trimethoxylithium aluminum deuteride and cuprous iodide was found to be stereospecific, giving, surprisingly, *endo-* or *exo-*2-deuteronorbornane from *endo-* or *exo-*2-bromonorbornane, respectively, with complete retention of configuration [655]. Bromo- and iodoalkanes are reduced to alkanes by a radical chain reaction with *trialkylsilanes* in the presence of a free-radical source and an alkanethiol as a catalyst [161]. Reductions with *stannanes* follow a free-radical mechanism [174] and are stereoselective [660]. The order of reducing power of different stannanes was found to be Ph_2SnH_2, $BuSnH_3$ > Ph_3SnH, Bu_2SnH_2 > Bu_3SnH [659] (Procedure 28, p 308). Replacement of bromine or iodine can also be accomplished by reduction with metals such as *magnesium* [207] and *zinc* [677] and with metal salts such as *chromous sulfate* [271] or *samarium diiodide* in good to high yields [304, 307] (equation **85**).

8 5

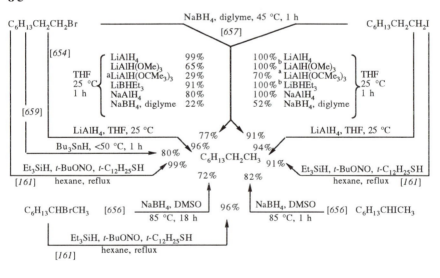

^{a}24 h ^{b}0.25 h

Geminal dihalides undergo partial or total reduction. Total reduction can be achieved *by catalytic hydrogenation* over platinum oxide [665], palladium [665], or Raney nickel [63, 665]. Both partial and total reduction can be accomplished with *lithium aluminum hydride [666]*, *sodium bis(2-methoxy-ethoxy)aluminum hydride [667]*, *tributylstannane [658, 667]*, *electrolytically* [668], with *sodium* in alcohol [669], and with *chromous sulfate [271, 276]*. For partial reduction only, *sodium arsenite [300]*, *sodium sulfite [348]*, and *diethyl phosphite [409]* are used (equation **86**) (Procedure 50, p 317).

8 6

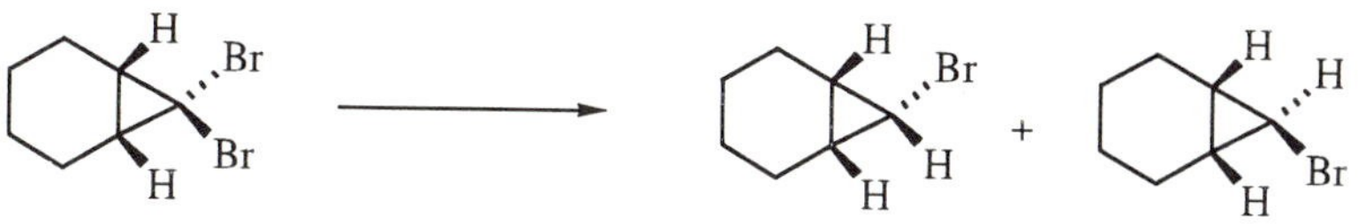

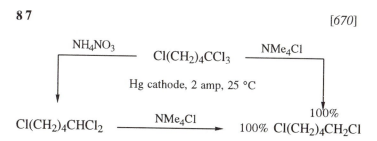

[667]	NaAlH$_2$(OCH$_2$CH$_2$OMe)$_2$	*syn*	17%	*anti*	48%
[667]	Bu$_3$SnH	*syn*	59%	*anti*	23%
[668]	Hg cathode, LiCl,				
	MeOH 90%, H$_2$O 10%		67–76%		13–14%
	AcOH		64–72%		16–18%
	DMF		65–73%		15–17%
[409]	(EtO)$_2$PHO, Et$_3$N, 90 °C		68%		15%

Trigeminal trihalides are completely reduced by *catalytic hydrogenation* over palladium [62] and Raney nickel [63] and partially reduced to geminal dihalides or monohalides by *trichlorosilane [167]*, by *electrolysis* using a mercury cathode [670], by *aluminum amalgam [219]*, and by *sodium arsenite* [300, 671] in good to excellent yields (equation **87**).

8 7   [670]

$$\text{Cl(CH}_2)_4\text{CCl}_3$$

NH$_4$NO$_3$ (left)   NMe$_4$Cl (right)

Hg cathode, 2 amp, 25 °C

$$\text{Cl(CH}_2)_4\text{CHCl}_2 \xrightarrow{\text{NMe}_4\text{Cl}} 100\% \ \text{Cl(CH}_2)_4\text{CH}_2\text{Cl}$$

100%

The trichloromethyl group activated by an adjacent double bond or carbonyl group is partly or completely reduced by *sodium amalgam* and *zinc [672, 673]* (pp 171 and 198).

Selective replacement of halogens is possible if their reactivities are different enough. Examples are the replacement of two chlorine atoms in 1,2-dichlorohexafluorocyclobutane by hydrogens with *lithium aluminum hydride* at 0 °C [674], the reduction of 1-chloro-1-fluorocyclopropane to fluoro-cyclopropane with *sodium* in liquid ammonia at –78 °C [675] and the reduction of 1,1-diphenyl-2-iodo-2-fluorocyclopropane to 1,1-diphenyl-2-fluorocyclo-propane with *samarium diiodide* in tetrahydrofuran and hexamethylphosphoric amide at ambient temperature in 55% yield [307]. Both chlorine and iodine are replaced by hydrogen in 1-chloro-2-iodohexafluorocyclobutane using *lithium aluminum hydride* [676] (62% yield); bromine is replaced by hydrogen in 1-bromo-1-chloro-2,2,2-trifluoroethane with *zinc* in methanol [677] (59% yield); and one bromine atom is replaced by hydrogen in 1,1-dibromo-1-chlorotrifluoroethane with *iron* [348], *sodium sulfite* [348], and *monosodium hypophosphite* [396] (equation 88).

88

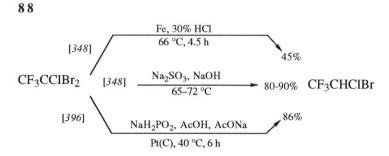

Chlorine is replaced by hydrogen in 1,1,1-trichlorotrifluoroethane [425, 427] and in 1,1,2-trichloropentafluoropropane [678] with *isopropyl alcohol* under irradiation, although not always in good yields (equation 89).

89   [678]

$$CF_3CClFCCl_2F \xrightarrow[\text{reflux 4 h}]{Me_2CHOH, h\nu} CF_3CClFCHClF + CF_3CHFCCl_2F$$
$$\qquad\qquad\qquad\qquad\qquad\qquad\qquad 49.5\% \qquad\qquad\quad 10\%$$

In **vicinal dihalides**, the halogens may be replaced by hydrogen using *hydrogenation* over Raney nickel [63] or *lithium* in *tert*-butyl alcohol [679]. Alkenes are frequently formed using *alkylstannanes* [659, 680]. Such dehalogenations are carried out usually with magnesium and zinc but are not discussed in this book. 1,3-Dihalides are either reduced or converted to cyclopropanes depending on the halogens and the reducing reagents used [681] (equation 90).

HALOALKENES, HALOCYCLOALKENES, AND HALOALKYNES

The ease of hydrogenolytic replacement of halogens in halogenated unsaturated hydrocarbons depends on the relative positions of the halogen and the multiple bond. In *catalytic hydrogenation,* **allylic or propargylic halogen** is replaced very easily, usually with the concomitant saturation of the multiple bond. **Vinylic** and **acetylenic halogens** are hydrogenolyzed less readily than the allylic ones but more easily than the halogens in saturated halides. The multiple bond is usually hydrogenated indiscriminately [63]. Replacement of allylic bromine [682] and vinylic bromine [683] without saturation of the double bond is rare.

The conspicuously high speed of hydrogenolysis of allylic and vinylic halogens as compared with those removed further from the multiple bonds or halogens in alkanes implies that the multiple bond participates in a multicenter

9 0 [681]

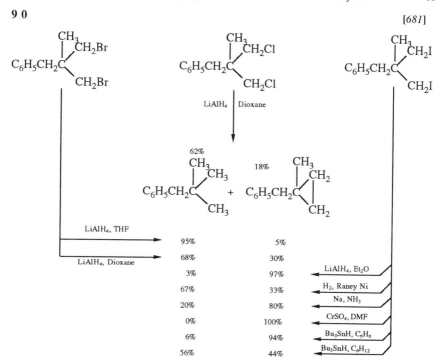

transition state [63]. Such a mechanism would account even for the surprisingly easy hydrogenolysis of allylic and vinylic fluorine [66, 684, 685, 686].

Hydrogenations of all types of unsaturated halides are carried out over *Raney nickel* in the presence of potassium hydroxide at room temperature and atmospheric pressure with yields of 32–92%. Even perhalogenated alkenes such as perchlorocyclopentadiene are reduced to the saturated parent compounds, both using Raney nickel [63], or palladium on silica gel at 330 °C (82% yield) [687].

In 1,2,3,4-tetrachlorobicyclo[2.2.1]-2-heptene, only the vinylic chlorines are hydrogenolyzed over Raney nickel, whereas the double bond is saturated [688] (equation **91**).

In the absence of potassium hydroxide, hydrogenation of vinylic and allylic chlorides to the saturated chlorides (with only partial or no hydrogenolysis of chlorine) is accomplished over 5% platinum, 5% palladium, and best of all, 5% rhodium on alumina [43].

In reductions of allyl and vinyl halides with other reducing agents, the double bond is usually conserved. Allylic halides are readily reduced to alkenes in high yields by treatment with *chromous sulfate* in *N,N*-dimethylformamide [271]. Allylic halogen may be replaced in preference to the vinylic halogen [689]. In 2-fluoro-1,1,3-trichloro-1-propene, *lithium aluminum hydride* reduces only the allylic chlorine and neither of the vinylic halogens [690]. If, however, the vinylic halogen is considerably more reactive, it can be replaced preferentially. In 4,4,4-trichloro-1,1,1-trifluoro-2-iodo-2-butene, reduction with

9 1

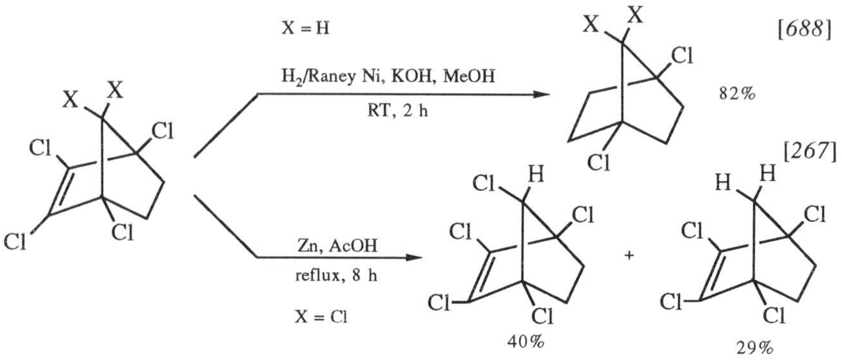

zinc replaces only the vinylic iodine and none of the allylic chlorines, giving 1,1,1- trichloro-4,4,4-trifluoro-2-butene in a 63% yield [*691*]. In reductions with *sodium* or *lithium* in *tert*-butyl alcohol, both the allylic and vinylic chlorines are replaced indiscriminately [*692*].

Surprisingly, *chromous acetate* hydrogenolyzes homoallylic chlorine in preference to both allylic and vinylic chlorines [*277*]. By reduction with *zinc* dust in deuterium oxide in the presence of sodium iodide and cupric chloride, all six chlorines in perchloro-1,3-butadiene are replaced by deuterium in 80–85% yield [*693*]. In 1,2,3,4,7,7-hexachlorobicyclo[2.2.1]-2-heptene, zinc in acetic acid hydrogenolyzes only one or both apical chlorines, while the tertiary and vinylic halogens and the double bond are preserved [*688*] (equation **91**). Similar selectivity is noticed in the reduction using chromous acetate [*267*].

Strongly nucleophilic *lithium aluminum hydride* and *sodium borohydride* in diglyme replace vinylic halogens including fluorine [*674, 694*], sometimes under surprisingly gentle conditions [*674, 695*] (equation **92**).

9 2

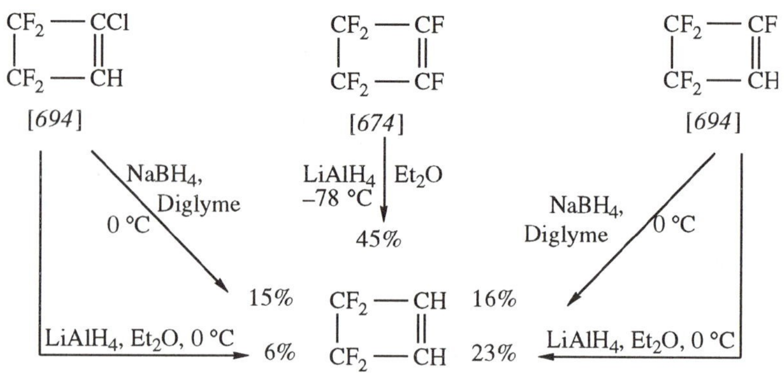

Because the carbon atom linked to fluorine is more electrophilic than that bonded to chlorine, fluorine in 1-chloroperfluorocyclopentene is replaced by hydrogen preferentially to chlorine using sodium borohydride (88% yield) [*694*].

Vinylic iodides and bromides are reduced by *tributylstannane* in the presence of tetrakis(triphenylphosphine)palladium as a catalyst [*696*] (equation **93**).

9 3   [*696*]

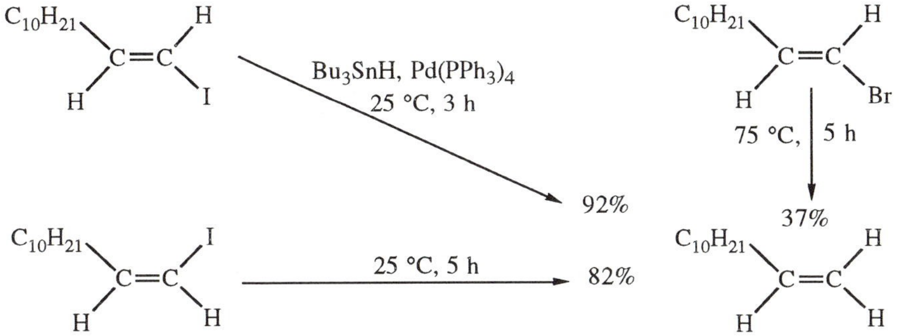

HALOAROMATIC COMPOUNDS

In the aromatic series both **benzylic halogens** and **aromatic halogens** can be replaced by hydrogen by means of *catalytic hydrogenation*; benzylic halogens much faster than aromatic [*653*]. Palladium on calcium carbonate [*62*] and, better still, Raney nickel [*63, 697*] in the presence of potassium hydroxide are the catalysts of choice for hydrogenations at room temperature and atmospheric pressure with good to excellent yields. Also *hydrogen transfer* from ammonium [*438*] or triethylammonium formate [*439*] over palladium on charcoal is successful in replacing aromatic halogen by hydrogen in good yields. In polyhaloaromatic compounds, partial replacement of halogens can be achieved by using amounts of potassium hydroxide equivalent to the number of halogens to be replaced [*697*].

Complete hydrodechlorination of decachlorobiphenyl is achieved by hydrogenation over palladium on silica gel at 330 °C [*687*]. In polychlorobiphenyls, chlorine is replaced by hydrogen on treatment with ammonium formate in the presence of 10% palladium on charcoal in methanol and tetrahydrofuran at ambient temperatures [*438*].

Benzylic halides are reduced very easily using complex hydrides. In α-chloroethylbenzene *lithium aluminum deuteride* replaced the benzylic chlorine by deuterium with inversion of configuration (optical purity 79%) [*698*]. *Borane* replaces chlorine and bromine in chloro- and bromodiphenylmethanes, chlorine in chlorotriphenylmethane, and bromine in benzyl bromide by hydrogen in 90–96% yields. Benzyl chloride, however, is not reduced [*699*]. *Sodium dimethylaminoborohydride* converts α-chloro-, bromo-, and iodonaphthalenes to naphthalene in 53–66% yields [*130*]. Benzylic chlorine and bromine in a *sym-*

triazine derivative are hydrogenolyzed by *sodium iodide* in acetic acid in 55% and 89% yields, respectively [700].

The difference in the reactivity of benzylic versus aromatic halogens makes it possible to reduce benzylic halogens preferentially. *Lithium aluminum hydride* replaces only the benzylic bromine by hydrogen in 2-bromomethyl-3-chloro-naphthalene (75% yield) [701]. *Sodium borohydride* in diglyme reduces, as a rule, benzylic halides but not aromatic halides (except for some iodo derivatives) [657, 702]. *Lithium aluminum hydride* hydrogenolyzes benzyl halides and aryl bromides and iodides. Aryl chlorides and especially fluorides are quite resistant [701, 703]. However, in polyfluorinated aromatic compounds, because of the very low electron density of the ring, even fluorine is replaced by hydrogen using lithium aluminum hydride [704] (equation **94**).

Reductions of fluorobenzene, chlorobenzene, bromobenzene, and iodobenzene are shown in equation **94** [160, 697, 703, 705, 706].

If different halogens are bound to the aromatic rings, selective replacement by hydrogen can take place. Thus *tetracarbonyl hydridoferrate*, $H_2Fe(CO)_4$, prepared in situ from iron pentacarbonyl, potassium carbonate, and carbon monoxide in methanol at 60 °C, converts all three chloroiodobenzenes to 4-chlorobenzene and bromoiodobenzene to bromobenzene in 95–100% and 98% yields, respectively [707].

p-Bromoiodobenzene is also reduced in 91% conversion to bromobenzene on heating at 70 °C in acetonitrile with 1-benzyl-1,4-dihydronicotinamide in the presence of bis(triphenylphosphine)rhodium chloride [89].

9 4

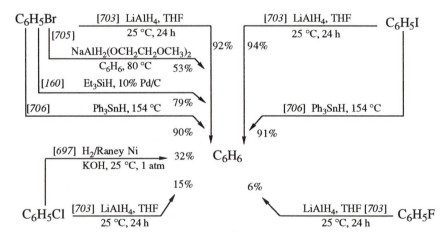

Halogenated naphthalenes are reduced to naphthalene in 80–98% yields [74, 696, 708, 709, 710] (equation **95**). 9-Bromoanthracene and 9-bromophenanthrene are converted to the parent hydrocarbons in 79–97% yields by ethanethiol and aluminum chloride [711].

A specially modified stannane, (*o-dimethylaminomethylphenyl*)*dimethylstannane*, hydrogenolyzes aromatic bromine and iodine in 78% and 85% yields, respectively [174].

9 5

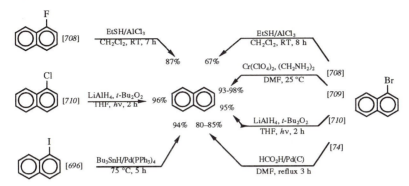

Occasionally metals have been used for the hydrogenolysis of aromatic halides. *Sodium* in liquid ammonia replaced (with hydrogen) even fluorine bonded to the benzene ring [712], and *zinc* activated with potassium hydrogenolyzed bromobenzene [233]. *Magnesium* can be used indirectly for conversion of an aromatic bromide or iodide to a Grignard reagent, which on treatment with water gives the parent compound.

An interesting case is the hydrogenolysis of aryl chlorides, bromides, and iodides in respective yields of 72%, 72%, and 82% on irradiation in *isopropyl alcohol* [431].

Halogens in aromatic heterocycles are replaced by hydrogen in *catalytic hydrogenations* without the reduction of the aromatic rings [713]. *Zinc* in acetic acid replaces both α-bromine atoms but not the β-bromine in 2,3,5-tribromo-thiophene [714].

In pyridine and quinoline and their derivatives, halogens in α and γ positions are replaced especially easily. (This is quite understandable because such halogen derivatives are actually disguised imidoyl chlorides or their vinylogs, and those compounds undergo easy hydrogenolysis.) Catalytic hydrogenation over palladium in acetic acid and alkaline acetate [715, 716, 717] or over Raney nickel in alkaline medium [718, 719] hydrogenolyzes α- and γ-chlorine atoms in yields up to 94%. All three chloropyridines [63] and bromopyridines [697, 720] are converted almost quantitatively to pyridine over Raney nickel or palladium at room temperature and atmospheric pressure.

Halogens in α and γ positions to the nitrogen atoms are distinctly more reactive than those in the β positions toward nucleophilic reagents. Catalytic hydrogenation over palladium converts β-chlorotetrafluoropyridine to tetrafluoropyridine [721] (equation **96**), whereas *lithium aluminum hydride* replaces (with hydrogen) fluorine in the γ position in preference to chlorine in the β position [722]. In pyrimidine derivatives, chlorine α to nitrogen is hydrogenolyzed in 89% yield by refluxing with *zinc dust* [723] and in 78% yield by heating at 100–110 °C with azeotropic *hydriodic acid* [724]. Such chlorine is replaced by hydrogen in preference to bromine and even iodine in the β position [725] (equation **97**).

The reactive chlorine in 3-chloro-4,5-benzopyrazole is replaced with hydrogen by refluxing for 24 h with hydriodic acid and phosphorus (82–86% yield) [726].

96

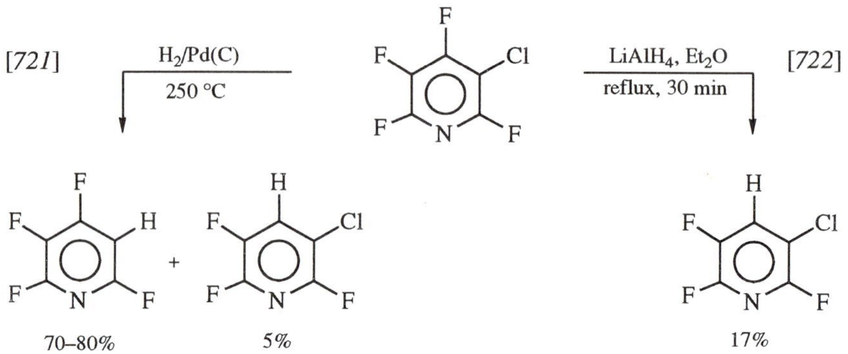

97

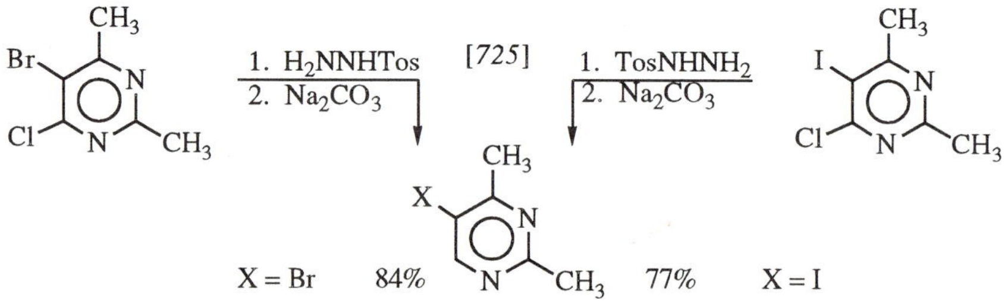

X = Br   84%   77%   X = I

Replacement of halogen in halonitro compounds and in halo compounds with functional groups is discussed in the appropriate chapters.

SUMMARY

Alkyl fluorides and to a certain extent even alkyl chlorides resist catalytic hydrogenation. As the reducibility of halogens increases along the series F << Cl < Br < I, bromine and iodine may be hydrogenolyzed by catalytic hydrogenation. More reliable is reduction of all halides by complex hydrides and especially by stannanes. Vinylic halides are hydrogenolyzed more readily: with lithium aluminum hydride and sodium borohydride, even vinylic fluorine is replaced by hydrogen, and in preference to chlorine. Benzylic halides are reduced more readily than aromatic halides, and these in turn more readily than saturated halides. Halides can also be reduced by alkali metals, aluminum, magnesium, and especially zinc and some metal compounds such as chromous sulfate and samarium diiodide. Nonmetal compounds such as sodium arsenate and sulfite are used for partial reduction of geminal dihalides.

CHAPTER 8

Reduction of Nitro, Nitroso, Diazo, and Azido Derivatives of Hydrocarbons and Basic Heterocycles

Nitrogen-containing substituents such as nitro, nitroso, diazo, and azido groups undergo exceedingly easy reduction by many reagents. Nitro and nitroso groups behave in a manner that depends on whether they are bonded to tertiary or aromatic carbon or to carbons carrying hydrogens. When they are bound to carbons carrying hydrogens, tautomerism generates forms that are reduced in ways different from those of the simple nitro or nitroso groups (equation **98**).

9 8

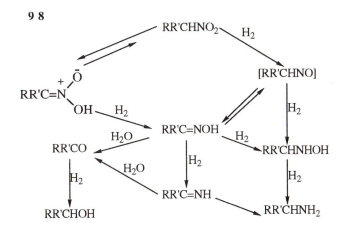

Aliphatic nitro compounds with the nitro group on a tertiary carbon are reduced to amines with *aluminum amalgam* [*220*] or *iron* [*727*]. 2-Nitro-2-methylpropane affords *tert*-butylamine in 65–75% yield [*220*]. Even some secondary nitroalkanes are *hydrogenated* to amines. *trans*-1,4-Dinitro-cyclohexane is converted to *trans*-1,4-diaminocyclohexane with retention of configuration (equation **99**).

9 9

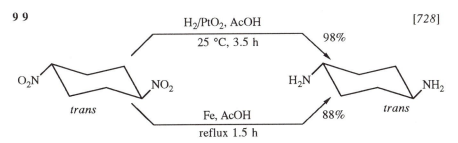

Retention of configuration may be considered as evidence that the intermediate nitroso compound is reduced directly and not after tautomerization to the isonitroso compound [*728*] (equation **98**). All aliphatic nitro compounds can be reduced to hydroxylamino compounds or amines in 72–93% yields by treatment with four or six equivalents of *samarium diiodide*, respectively, in tetrahydrofuran–methanol solutions at room temperature [*310*].

Generally primary nitro compounds are reduced to aldoximes and secondary nitro compounds to ketoximes by metal salts. On reduction with *stannous chloride*, 1,5-dinitropentane gives a 55–60% yield of the dioxime of glutaric dialdehyde [*254*]; 3β,5α-dichloro-6β-nitrocholestane with *chromous chloride* gives a 60% yield of the oxime of 3β-chloro-5α-hydroxycholestan-6-one [*272*]; and nitrocyclohexane is reduced to cyclohexanone oxime *catalytically* [*729*] or with *sodium thiosulfate* [*730*] (equation **100**).

100

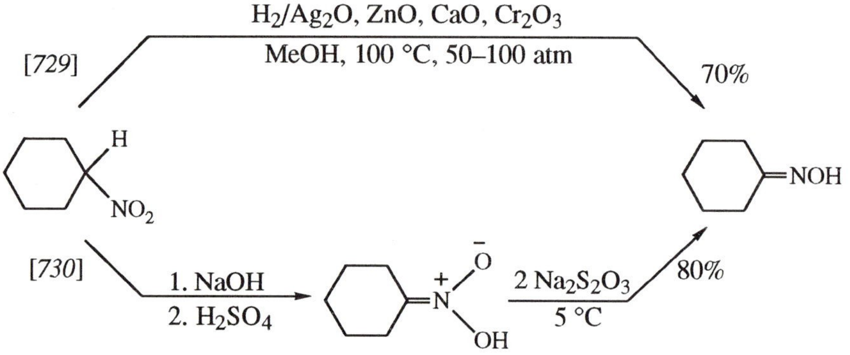

Potassium salts of primary and secondary nitro compounds (potassium nitronates) are reduced with *diborane* to N-substitued hydroxylamines [*731*] (equation **101**).

101   [*731*]

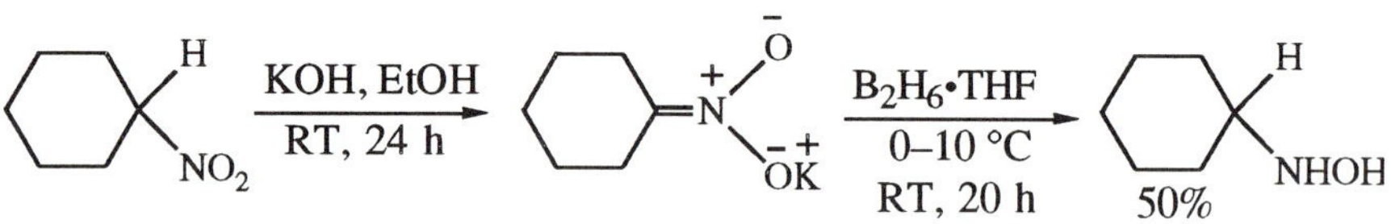

Although primary and secondary nitro compounds may be converted, respectively, to aldehydes and ketones by consecutive treatment with alkalies and sulfuric acid (*the Nef reaction*), the same products can be obtained by reduction with *titanium trichloride* (yields of 45–90%) [*732*] or *chromous chloride* (yields of 32–77%) [*268*]. The reaction seems to proceed through a nitroso rather than an aci-nitro intermediate [*732*] (equation **98**).

Reduction of the nitro group is one of the easiest to accomplish. However, several instances have been recorded of preferential reduction of carbon–carbon double bonds in unsaturated nitro compounds. In 4-nitrocyclohexene with a bulky substituent in position 5, the double bond is saturated by *hydrogenation* over platinum oxide without change of the nitro group [*733*], and even a pyridine ring is hydrogenated in preference to a remote nitro group under special conditions (platinum oxide, acetic acid, room temperature, 3.5 atm, 90% yield) [*734*]. Diazo and azido groups, however, are reduced in preference to amino groups (pp 100).

Hydrodenitration, replacement of a nitro group by hydrogen, can be achieved by catalytic hydrogenation over platinum on silica gel at 150 °C. Nitrohexane is thus converted to hexane in a 91% yield [*735*].

The outcome of the reduction of primary or secondary **nitro compounds with conjugated double bonds** depends on the mode of addition of hydrogen, which in turn depends on the reagents used. 1,4-Addition results in the reduction of the carbon–carbon double bond and leads to a saturated nitro compound. Dimeric compounds formed by coupling of the half-reduced intermediates sometimes accompany the main products [*736*]. 1-Nitrooctane is prepared in an 85% yield by reducing 1-nitrooctene [*736*], and β-nitrostyrene gives 93% β-nitroethylbenzene on reduction with *sodium borohydride* [*737*]. Saturated nitro compounds are also obtained by specially controlled reduction with *lithium aluminum hydride* [*738*], *lithium tris(sec-butyl)borohydride* [*739*], *zinc borohydride* [*740*], *tributylstannane* [*741*], by treatment of the unsaturated nitro derivatives with *formic acid* and triethylamine [*441*], and by heating in benzene at 60 °C with silica gel and *diethyl 1,4-dihydro-2,6-dimethylpyridine-3,5-dicarboxylate* (yields of 55–100%) [*1470*] (equation **102**).

102

$$C_6H_5CH=CHNO_2 \longrightarrow C_6H_5CH_2CH_2NO_2$$

[*739*]	LiBH(*s*-Bu)$_3$, THF, RT, 30 min	78%
[*737*]	NaBH$_4$, SiO$_2$, *i*-PrOH, CHCl$_3$, 25 °C, 25 min	93%
[*740*]	Zn(BH$_4$)$_2$, (CH$_2$OCH$_3$)$_2$, 0 °C, 6 h	88%
[*741*]	Bu$_3$SnH, CH$_2$Cl$_2$, RT, 20–24 h; MeOH, HF	72%

Ring-substituted β-nitrostyrenes are reduced to β-nitroethylbenzenes by homogeneous *catalytic hydrogenation* using tris(triphenylphosphine)rhodium chloride in 60–90% yields [*91*], and to β-aminoethylbenzenes by hydrogenation over 10% palladium on carbon in aqueous hydrochloric acid at 85 °C and 35 atm (82% yield) [*742*]. Reduction of 1-(α-furyl)-2-nitroethylene with an excess of lithium aluminum hydride in refluxing ether gives an 85% yield of 1-(α-furyl)-2-aminoethane [*1498*].

If the addition of hydrogen takes place in a 1,2-mode, the products are oximes [*738, 740, 743, 744, 745*], hydroxylamines [*738, 746*], amines [*738, 745, 747*], and carbonyl compounds resulting from the hydrolysis of the oximes [*738*]. Oximes and carbonyl compounds also result from reductions of α,β-unsaturated nitro compounds with *iron* [*743, 748*], *sodium stannite* [*744*], and *chromium dichloride* [*749*] (equation **103**).

An oxime can be prepared by catalytic hydrogenation of a β-nitrostyrene derivative over palladium in pyridine (89% yield) [*750*]. Depending on the amount of *lithium aluminum hydride* used, 1-(3-pyridyl)-2-nitropropane is

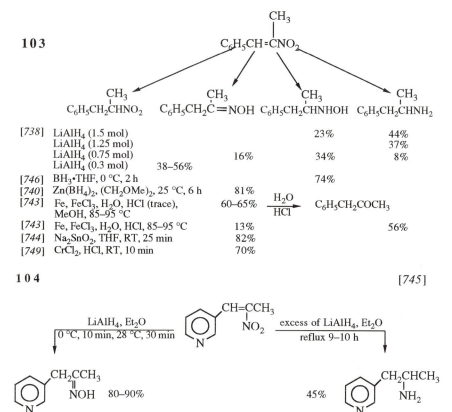

103

$$\underset{\text{C}_6\text{H}_5\text{CH}=\overset{\displaystyle CH_3}{\underset{\displaystyle}{\text{C}}}\text{NO}_2}{}$$

		$\underset{\text{C}_6\text{H}_5\text{CH}_2\overset{CH_3}{\underset{}{\text{CHNO}_2}}}{}$	$\underset{\text{C}_6\text{H}_5\text{CH}_2\overset{CH_3}{\underset{}{\text{C}}}=\text{NOH}}{}$	$\underset{\text{C}_6\text{H}_5\text{CH}_2\overset{CH_3}{\underset{}{\text{CHNHOH}}}}{}$	$\underset{\text{C}_6\text{H}_5\text{CH}_2\overset{CH_3}{\underset{}{\text{CHNH}_2}}}{}$
[738]	LiAlH$_4$ (1.5 mol)			23%	44%
	LiAlH$_4$ (1.25 mol)				37%
	LiAlH$_4$ (0.75 mol)		16%	34%	8%
	LiAlH$_4$ (0.3 mol)	38–56%			
[746]	BH$_3$·THF, 0 °C, 2 h			74%	
[740]	Zn(BH$_4$)$_2$, (CH$_2$OMe)$_2$, 25 °C, 6 h	81%			
[743]	Fe, FeCl$_3$, H$_2$O, HCl (trace), MeOH, 85–95 °C	60–65% $\xrightarrow[\text{HCl}]{\text{H}_2\text{O}}$ C$_6$H$_5$CH$_2$COCH$_3$			
[743]	Fe, FeCl$_3$, H$_2$O, HCl, 85–95 °C	13%			56%
[744]	Na$_2$SnO$_2$, THF, RT, 25 min	82%			
[749]	CrCl$_2$, HCl, RT, 10 min	70%			

104   [745]

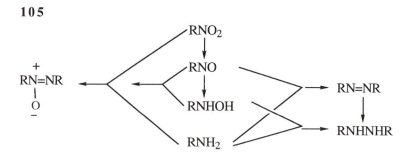

converted to either 3-pyridylacetoxime or to 1-(3-pyridyl)-2-aminopropane [745] (equation **104**).

Aromatic nitro compounds were among the first organic compounds ever reduced. The nitro group is readily converted to a series of functions of various degrees of reduction: very rarely to a nitroso group, more often to a hydroxylamino group, and most frequently to the amino group. In addition azoxy, azo, and hydrazo compounds are formed by combination of two molecules of the reduction intermediates (equation **105**).

105

With the exception of the nitroso stage, all the intermediate stages of the reduction of nitro compounds can be obtained by controlled *catalytic hydrogenation* [751], and all reduction intermediates were prepared by reduction with appropriate *hydrides* or *complex hydrides*. However, the outcome of many hydride reductions is difficult to predict. Therefore, more specific reagents are preferred for partial reductions of nitro compounds.

By controlling the amount of hydrogen and the pH of the reaction, hydroxylamino, azoxy, azo, hydrazo, and amino compounds are obtained in good yields by catalytic hydrogenation over 2% palladium on carbon at room temperature and atmospheric pressure [751].

Nitroso compounds are usually not obtained directly but rather by reoxidation of hydroxylamino compounds or amines. They are reduced to amino compounds with *potassium borohydride and cuprous chloride* in methanol at room temperature in 94% yields [752].

Hydroxylamino compounds are prepared by reduction of nitro compounds with *hydrazine and Raney nickel* at temperatures of 0–10 °C in 65–100% yields [71], by *electrolytic reduction* using a lead anode and a copper cathode [753], by *zinc* in an aqueous solution of ammonium chloride [754], or by *aluminum amalgam* [221], generally in good yields.

Azoxybenzene is synthesized in an 85% yield by reduction of nitrobenzene with *sodium arsenite* [301]. Nitrotoluenes and 2,5-dichloronitrobenzene were converted to the corresponding **azoxy compounds** by heating to 60–90 °C with *hexoses* (up to 74% yields) [437]. Some ring-substituted nitrobenzenes were converted to azoxy compounds and some others to azo compounds by *sodium bis(2-methoxyethoxy)aluminum hydride* [755]. Reduction of azoxy compounds to amino compounds is achieved with *potassium borohydride and cuprous chloride* in methanolic solution at room temperature in quantitative yields [752].

Reduction of nitro compounds with *zinc* in alkali hydroxides gives *azo and hydrazo compounds* [756]. With an excess of zinc, hydrazo compounds are obtained, and they are easily reoxidized to azo compounds (Procedure 37, p 312).

Direct preparation of **azo compounds** in good yields is accomplished by treatment of nitro compounds with *lithium aluminum hydride* [757], *magnesium aluminum hydride* [758], *sodium bis(2-methoxyethoxy)aluminum hydride* [755], *silicon* in alcoholic alkali [759], or *zinc* in strongly alkaline medium [756]. *Hydrazobenzene* is obtained by controlled hydrogenation of nitrobenzene in alkaline medium (80% yield) [751] and by reduction with *sodium bis(2-methoxyethoxy)aluminum hydride* (37% yield) [705].

Reaction conditions of the reduction of nitrobenzene to different products are shown in equation **106**.

Complete reduction of nitro compounds to amines is accomplished by *catalytic hydrogenation*. The difference in the rate of hydrogenation of the nitro group and of the aromatic ring is so large that the hydrogenation may be easily regulated not to reduce the aromatic ring. Catalysts used for the conversion of aromatic nitro compounds to amines are platinum oxide [41, 760], rhodium–platinum oxide [41], palladium [41, 761], Raney nickel [46], copper chromite [51], rhenium sulfide [56], and others. Instead of hydrogen, *hydrazine* [381, 382], *formic acid* [72], *ammonium formate* [78], or *triethylammonium formate* [80] have been used for catalytic hydrogen transfer. Hydrides and complex hydrides are not used for complete reduction to amines because they usually

106

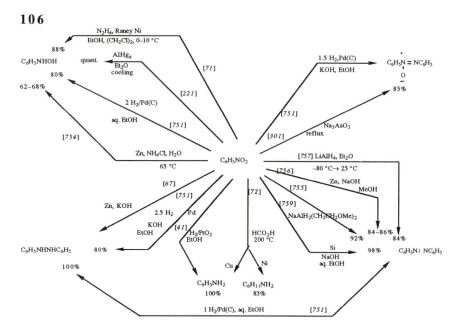

reduce nitro compounds to azo compounds and other derivatives, or else they do not reduce the nitro group at all (like sodium borohydride [762], diborane, lithium hydride, lithium triethylborohydride, and sodium trimethoxyborohydride). In the presence of *palladium on carbon, sodium borohydride* in methanol reduces 2,4-dinitroaniline to 1,2,4-triaminobenzene [763], and in the presence of *cuprous chloride, potassium borohydride* is reported to reduce aromatic nitro compounds to amino compounds in yields of 82–100% [752].

The most popular reducing agent for conversion of aromatic nitro compounds to amines is *iron* [243]. It is cheap and gives good to excellent yields [242, 764]. The reductions are usually carried out in aqueous or aqueous–alcoholic media and require only catalytic amounts of acids (acetic or hydrochloric) or salts such as sodium chloride, ferrous sulfate, or better still, ferric chloride [242]. Thus the reductions are run essentially in neutral media. The rates of the reductions and sometimes even the yields can be increased by using iron in the form of small particles [242]. Iron is also suitable for reduction of complex nitro derivatives because it does not attack many functional groups [765] (equation **107**).

Zinc reduces nitro groups in various ways, depending mainly on the pH. Reduction to amino groups can be achieved in aqueous alcohol in the presence of calcium chloride [766] or nickel dichloride [241]. In strong hydrochloric acid the aromatic ring may be chlorinated in positions *para* or *ortho* to the amino group [767]. The same is true of reduction with *tin* [768]. Other common reducing agents are *titanium trichloride* [280] (Procedure 43, p 314), *stannous chloride* [769], *ferrous sulfate* [298], *hydrogen sulfide* [327] or its salts [329, 333], and *sodium hydrosulfite* (hyposulfite or dithionite) [353, 354].

107

$$C_6H_5NO_2 \xrightarrow{\hspace{6cm}} C_6H_5NH_2$$

[751]	H$_2$/2% Pd(C), aq. EtOH, 25 °C, 1 atm	>90%
[41]	H$_2$/PtO$_2$, EtOH, 25 °C, 1 atm	quant.
[41]	H$_2$/PtO$_2$-RhO$_2$, EtOH, 25 °C, 1 atm	quant.
[56]	H$_2$/Re$_2$S$_7$, EtOH, 25 °C, 63 atm	100%
[51]	H$_2$/CuCr$_2$O$_4$, 175 °C, 250–300 atm	100%
[381]	N$_2$H$_4$·H$_2$O/Raney nickel, PhMe, EtOH, reflux	95%
[72]	HCO$_2$H/Cu, 200 °C	quant.
[78]	HCO$_2$NH$_4$/Pd(C), MeOH, RT	high yield
[767]	Zn, HCl	55%
[767]	SnCl$_2$, HCl	+
[242]	Fe, H$_2$O, FeCl$_3$, 100 °C	46–100%
[241]	NiCl$_2$, Zn, MeOH, reflux	80%

Most of the reagents used for the preparation of aromatic amines from nitro compounds can be used for complete reduction of polynitro compounds to polyamines [243, 764]. However, not all of them are equally suited for partial or selective reductions of just one or one particular nitro group.

Partial reduction of dinitro compounds is accomplished by *catalytic hydrogen transfer* using palladium on carbon and cyclohexene [770] or triethylammonium formate [80]. *Iron* is superior to stannous chloride because it does not require strongly acidic medium, which dissolves the initially formed nitro amine and makes it more susceptible to complete reduction [242, 771] (Procedure 39, p 312). However, *stannous chloride* is frequently used for selective reductions in which one particular nitro group is to be reduced preferentially or exclusively [769]. Other reagents for the same purpose are *titanium trichloride* [772], *hydrogen sulfide in ammonia* [773] or *in pyridine* [327], and *sodium* or *ammonium sulfide* [327, 331]. Surprisingly enough, the sterically less accessible *ortho* nitro group usually is reduced.

Triethylammonium formate in the presence of palladium reduces the less hindered *para* nitro group in 2,4-dinitrotoluene but the more hindered *ortho* nitro group in 2,4-dinitrophenol, 2,4-dinitroanisole, 2,4-dinitroaniline, and 2,4-dinitro-acetanilide in yields of 24–92% [80]. Catalytic hydrogenation over palladium reduces the *para* nitro group in 2,4-dinitroaniline [774] (equation **108**).

Reduction of the aromatic nitro group takes preference over the reduction of the aromatic ring. Under certain conditions, however, even the benzene ring can be reduced. Hydrogenation of nitrobenzene over platinum oxide or rhodium–platinum oxide in ethanol yields aniline, whereas in acetic acid cyclohexylamine is produced [41]. Heating of nitrobenzene with formic acid in the presence of copper at 200 °C gives a 100% yield of aniline; similar treatment in the presence of nickel affords 67% cyclohexylamine [72] (equation **106**). Catalytic

108

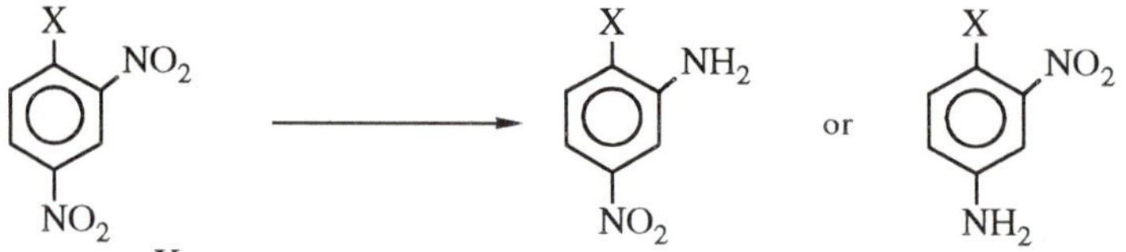

	X			
[80]	Me			HCO$_2$H, Et$_3$N/Pd 92%
[331]	OH	Na$_2$S, NH$_4$OH	64–67%	
[80]	OH	HCO$_2$H•Et$_3$N/Pd	57%	
[80]	OMe	HCO$_2$H•Et$_3$N/Pd	24%	
[80]	NH$_2$	HCO$_2$H•Et$_3$N/Pd	49%	
[763]	NH$_2$	H$_2$/5% Rh(Al$_2$O$_3$) DMF, NH$_4$OH	87%	
[774]	NH$_2$			H$_2$/5% Pt(C) HCl, EtOH, 60–65 °C, 2–4 atm 70%
[80]	NHAc	HCO$_2$H•Et$_3$N/Pd	56%	
[327]	NHMe	H$_2$S, C$_5$H$_5$N	60%	
[769]	NMe$_2$	SnCl$_2$	72%	
[772]	CHO	TiCl$_3$	50%	

hydrogenation of nitrobenzene over platinum on silica gel at 150 °C cleaves the carbon–nitrogen bond and gives cyclohexane in 89% yield [735].

Selective reduction of an aromatic nitro group in aliphatic–aromatic dinitro compounds is achieved with *sodium sulfide*. Depending on the structure of the dinitro compound and on the use of aqueous or anhydrous solution of sodium sulfide, the aliphatic nitro group is left intact or is eliminated. 1-*p*-Nitrophenyl–2-methyl-2-nitropropane and aqueous–ethanolic solution of sodium sulfide hydrate afford 70% 1-*p*-aminophenyl-2-methyl-2-nitropropane and 10% 1-*p*-aminophenyl-2-methyl-1-propene [332]. Reduction of 2-nitro-1-*o*-, *m*- or *p*-nitrophenylpropene by baker's yeast at 33 °C gives 16–61% yields of 2-nitro-1-*o*-, *m*-, or *p*-aminophenylpropane besides 10–24% 2-nitro-1-*o*-, *m*- or *p*-nitrophenylpropane [1485].

Halogenated nitro compounds can be reduced to nitro compounds, amino compounds, or halogenated amino compounds.

Only fairly reactive halogens may be replaced by hydrogen without the reduction of the nitro groups. *p*-Iodonitrobenzene gives a quantitative yield of nitrobenzene on treatment with *1-benzyl-1,4-dihydronicotinamide* in the presence of tris(triphenylphosphine)rhodium chloride [89], and 1-chloro-2,4,6-trinitrobenzene affords 1,3,5-trinitrobenzene on refluxing with *sodium iodide and formic or acetic acid* in 70% or 60% yields, respectively [314]. Chloronitrobenzenes and bromonitrobenzenes are selectively reduced to nitrobenzene by hydrogen with tetrakis(triphenylphosphine)palladium as a catalyst in *N,N*-dimethylformamide at 100 °C in 67–80% yields [775]. Bromonitrobenzenes are reduced to nitrobenzene and to a smaller extent to aniline with *triethylammonium formate* in the presence of palladium on carbon or triarylphosphinepalladium acetate at 50 °C.

2-Iodo-6-nitronaphthalene and 3-iodo-6-nitronaphthalene afford the corresponding iodoaminonaphthalenes in almost quantitative yields on refluxing with aqueous–alcoholic solutions of *sodium hydrosulfite* (hyposulfite or dithionite) [*354*], and 2-iodo-7-nitrofluorene gives 2-iodo-7-aminofluorene in 85% yield by reduction with *hydrazine* in the presence of Raney nickel [*381*].

Bromonitrophenanthrenes reduced with *zinc* and acid, *stannous chloride, ammonium sulfide,* or *hydrazine* in the presence of palladium give aminophenanthrenes. Reduction to bromoaminophenanthrene is accomplished with *iron* in dilute acetic acid [*382*] (equation **109**).

109   [*382*]

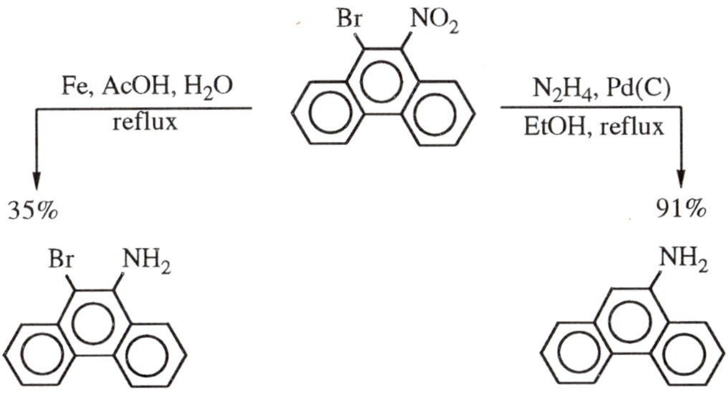

A very reliable reagent for reduction of aromatic nitro compounds in the presence of sensitive functions is *iron,* which reduces nitro groups to amino groups [*765*] and hardly ever affects other functions such as carbonyl in aldehydes, reactive halogens, and amine oxides (p 131, 132 and 144).

Primary and secondary **nitroso compounds** tautomerize to isonitroso compounds—oximes of aldehydes and ketones, respectively. Their reductions are discussed in the sections on derivatives of carbonyl compounds (pp 184).

Tertiary and aromatic nitroso compounds are not readily accessible; consequently, not many reductions have been tried. Nitrosobenzene is converted to azobenzene by *lithium aluminum hydride* (yield 69%) [*776*], and *o*-nitrosobiphenyl is converted to carbazole, probably via a hydroxylamino intermediate, by treatment with *triphenylphosphine* or *triethyl phosphite* (yields of 69% and 76%, respectively) [*407*]. 4-Nitrosothymol is transformed to aminothymol with *ammonium sulfide* (73–80% yield) [*339*], and α-nitroso-β-naphthol is converted to α-amino-β-naphthol with *sodium hydrosulfite* (66–74% yield) [*352*].

Nonfunctionalized **aliphatic diazo compounds** are fairly rare, and so are their reductions. Good examples of the reduction of diazo compounds to either amines or hydrazones are found with α-diazo ketones and α-diazo esters (pp 173 and 222).

Aromatic diazonium compounds, which are prepared readily by diazotization of primary amines, can be *converted either to their parent compounds* by replacement of the diazonium group with hydrogen or to *hydrazines* by reduction of the diazonium group. Both reactions are carried out at room temperature or below using reagents soluble in aqueous solutions and usually give high yields.

Replacement of the diazonium group with hydrogen is accomplished by *sodium borohydride* (48–77% yields) [777], *tin* and hydrochloric acid (51–100% yields) [778], *sodium stannite* (73% yield) [299], *hypophosphorous acid* (52–87% yields) [394, 779] (Procedure 54, p 318), *ethanol* (79–93% yields) [421, 422], *hexamethylphosphoric triamide* (71–94% yields) [780], *formamide* [440], *1-benzyl-1,4-dihydronicotinamide* [781], and others [395]. Reducing agents suitable for the replacement of the diazonium groups by hydrogen usually do not attack other elements or groups, so they can be used for compounds containing halogens [299, 394, 782], nitro groups [421], and other functions. Replacement of the aromatic amino group via diazotization is accomplished even in a compound containing an aliphatic amino group because, at pH lower than 3, only the aromatic amino group is diazotized [779] (equation **110**).

Conversion of diazonium salts to hydrazines is achieved by reduction with *zinc* in acidic medium (85–90% yields) [783], *stannous chloride* (70% yield) [261], and *sodium sulfite* (70–80% yields) [347, 784]. Nitro groups present in the diazonium salts survive the reduction unharmed [261, 347].

110

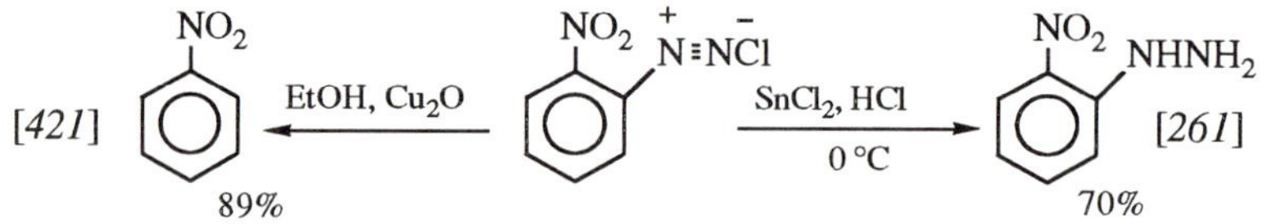

Azido groups are reduced to amino groups most readily. *Catalytic hydrogenation* is carried out over platinum (81% yield) [785] or by hydrogen transfer from ammonium formate over palladium (74–93% yields) [78]. The volume of nitrogen eliminated during the reduction of the azido group equals that of the hydrogen used, so it is impossible to follow the progress of the reduction by measurement of volume or pressure. The easy hydrogenation of the azido group made it possible to reduce an azido compound to an amino compound selectively over 10% palladium on charcoal without hydrogenolyzing a benzyloxycarbonyl group and even in the presence of divalent sulfur in the same molecule [786].

Lithium aluminum hydride reduces β-azidoethylbenzene to β-amino-ethylbenzene in an 89% yield [787]. The azido group is also reduced with *sodium borohydride* [788], *magnesium* or *calcium* in methanol [214], *aluminum amalgam* (71–86% yields) [223], *titanium trichloride* (54–83% yields) [789], *vanadous chloride* (70–95% yields) [297] (Procedure 45, p 314), *hydrogen sulfide* (90% yield) [341], *sodium hydrosulfite* (90% yield) [357], *hydrogen bromide* in acetic acid (84–97% yields) [323], and *1,3-propanedithiol* (84–100% yields) [790]. Unsaturated azides are reduced to unsaturated amines with *aluminum amalgam* [223] and *1,3-propanedithiol* [790] (equation **111**).

111

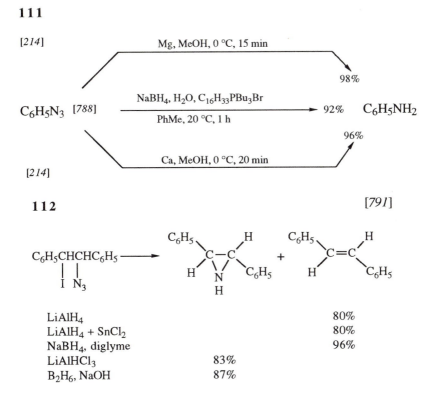

[*214*]

Mg, MeOH, 0 °C, 15 min

98%

$C_6H_5N_3$ [*788*]

$\dfrac{NaBH_4, H_2O, C_{16}H_{33}PBu_3Br}{PhMe, 20 °C, 1 h}$ 92% $C_6H_5NH_2$

96%

Ca, MeOH, 0 °C, 20 min

[*214*]

112 [*791*]

$C_6H_5CHCHC_6H_5$
$\quad |\quad |$
$\quad I\quad N_3$

Reagent	Aziridine	Alkene
LiAlH₄		80%
LiAlH₄ + SnCl₂		80%
NaBH₄, diglyme		96%
LiAlHCl₃	83%	
B₂H₆, NaOH	87%	

Neither aromatic halogens [*323, 790*] nor nitro groups are affected during the reductions of the azido group [*323, 341, 790*]. α-Iodo azides give, on reduction, aziridines or alkenes depending on the substituents and on the reagents used [*791*] (equation **112**).

SUMMARY

Although nitro compounds and azides can be reduced catalytically to amines and nitro compounds by complex hydrides to azo compounds, most reductions, especially of aromatic nitro compounds, nitroso compounds, and diazo compounds, are accomplished by metals and metal and nonmetal reducing agents. For partial and complete reduction of nitro compounds, the following reagents are used: zinc, iron, tin, titanium chloride, stannous chloride, ferrous sulfate, sodium stannite, hydrogen sulfide, sodium and ammonium sulfides, and sodium dithionite. For selective reductions of nitro compounds containing other reducible functions, metal and nonmetal reducing agents are most suitable.

CHAPTER 9

Reduction of Hydroxy Compounds and Their Derivatives

In this chapter, reduction of hydroxy compounds and their substitution and functional derivatives is discussed.

ALCOHOLS, PHENOLS, AND THEIR SUBSTITUTION DERIVATIVES

Reduction of saturated alcohols to the ***parent hydrocarbons*** is not easy and requires high temperatures and high pressures. Catalytic hydrogenation over *molybdenum or tungsten sulfides* at 310–350 °C and 76–122 atm converts tertiary, secondary, and primary alcohols to the corresponding alkanes in yields up to 96% [*792*].

Many hydroxy compounds would not survive such harsh treatment; therefore, other methods must be used. Some alcohols are hydrogenolyzed with *chloroalanes* generated in situ from lithium aluminum hydride and aluminum chloride, but the reaction gives alkenes as byproducts [*793*]. Tertiary alcohols are converted to saturated hydrocarbons on treatment at room temperature with triethyl- or triphenylsilane and trifluoroacetic acid in methylene chloride (41–92% yields). Rearrangements due to carbonium-ion formation occur [*492*]. Sequential treatment of tertiary alcohols with Raney nickel in boiling toluene followed by room-temperature hydrogenation over 5% palladium or carbon gives alkanes and cycloalkanes in 71–99% yields [*794*].

An elegant method for the replacement of hydroxy groups in alcohols by hydrogen is the addition of alcohol to dialkylcarbodiimide followed by catalytic hydrogenation of the intermediate *N,N,O*-trialkylisoureas over palladium at 40–80 °C and 1–60 atm (52–99% yields) [*795*] (equation **113**).

113   [*795*]

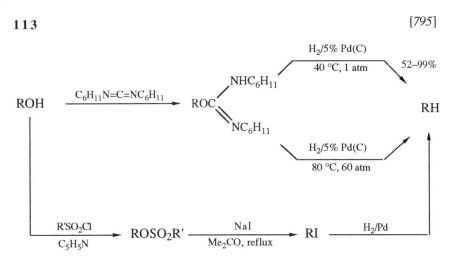

Deoxygenation of alcohols, especially secondary alcohols, can also be accomplished by reduction of their O-phenylthionocarbonates [796], S-methyl dithiocarbonates (methyl xanthates, prepared easily from carbon disulfide [161, 796, 797], or N,N-diethyl thiocarbamates [798] with one of three reducing agents: *borane* and *tributylstannane* [796, 797], *triethyl-* or *tripropylsilane* in the presence of an initiator and *alkanethiol* [161], or *potassium* and a crown ether [798] in yields of 46–93% (equation **114**).

1 1 4

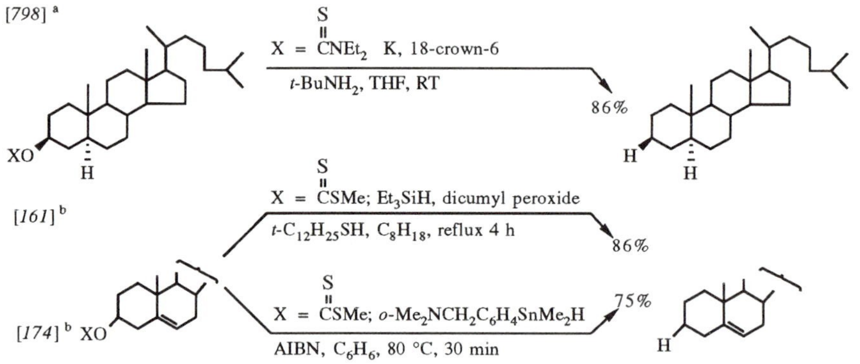

[798] [a]

X = $\overset{\overset{\text{S}}{\|}}{\text{CNEt}_2}$ K, 18-crown-6

t-BuNH$_2$, THF, RT

86%

[161] [b]

X = $\overset{\overset{\text{S}}{\|}}{\text{CSMe}}$; Et$_3$SiH, dicumyl peroxide

t-C$_{12}$H$_{25}$SH, C$_8$H$_{18}$, reflux 4 h

86%

[174] [b]   XO

X = $\overset{\overset{\text{S}}{\|}}{\text{CSMe}}$; o-Me$_2NCH_2C_6H_4$SnMe$_2$H

AIBN, C$_6$H$_6$, 80 °C, 30 min

75%

[a]Cholestanol, cholestane

[b]Cholesterol, 5-cholestene

Another procedure for the hydrogenolysis of a hydroxyl is the conversion of the hydroxy compound to its arenesulfonyl ester, exchange of the sulfonyloxy group for iodine using sodium iodide, and replacement of the iodine by hydrogen via catalytic hydrogenation or other reductions (equation **113**). Decyl tosylate is converted to decane in a 73% yield by refluxing in dimethoxyethane with *sodium iodide and tributylstannane* in the presence of azobisisobutyronitrile [799].

Hydrogenolysis of hydroxyl is somewhat easier in **polyhydric alcohols**. Diols give alcohols on *hydrogenation* over copper chromite at 200–250 °C and 175 atm, after hydrogenolysis of one hydroxyl group [800]. Vicinal diols are converted to alkenes stereospecifically by a special treatment with *trialkyl phosphites* [801]. Similar reaction can also be achieved by heating of vicinal diols with *low-valent titanium* obtained by reduction of titanium trichloride with potassium [286].

Results of the **reduction of unsaturated alcohols** depend on the relative positions of the hydroxyl and the double bond. The hydroxyl group is fairly resistant to hydrogenolysis by *catalytic hydrogenation*, so almost any catalyst working under mild conditions can be used for saturation of the double bond with conservation of the hydroxyl [802]. In addition, *sodium* in liquid ammonia and *lithium* in ethylamine reduce double bonds without affecting the hydroxyl in nonallylic alcohols [802] (equation **115**).

115 [*802*]

H2/Raney Ni
150 atm
85%
CH2CH2OH
Na/NH3, EtOH
95%
CH2CH2OH
83%
Li/EtNH2

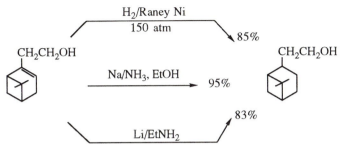

Double bonds in allylic alcohols are saturated on heating of the alcohols with *diimide* generated from hydroxylamine, ethyl acetate, and potassium hydroxide [*366*]. Enantioselective reduction of allylic and homoallylic alcohols is accomplished in 97–100% yields and with 96–98% enantiomeric excess by homogeneous catalytic hydrogenation at 18–20 °C using a chiral catalyst, ruthenium Binap (2,2'–bis(diphenylphosphino)–1,1'–binaphthyl) [*803, 804*].

With few exceptions [*805*], unsaturated alcohols with hydroxyls in allylic positions undergo hydrogenolysis and give alkenes. Such reductions are achieved by *chloroalanes* made from lithium aluminum hydride and aluminum chloride [*806*] and by *sodium* in liquid ammonia [*802, 807*]. Rearrangement of the double bond usually takes place [*807*], and dehydration to a diene may be another side reaction (equations **116** and **117**).

116 [*802*]

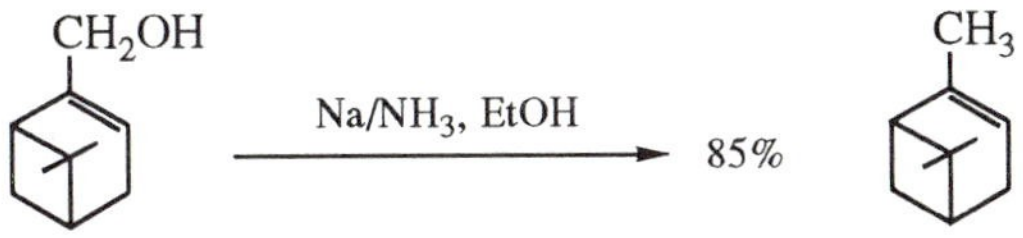

CH2OH
Na/NH3, EtOH
85%
CH3

Replacement of an allylic hydroxyl without saturation or a shift of the double bond is achieved by treatment of some allylic-type alcohols with *triphenyliodophosphorane* (Ph_3PHI), *triphenyldiiodophosphorane* (Ph_3PI_2), or their mixture with triphenylphosphine (24–60% yields) [*808*]. Still another way is the treatment of an allylic alcohol with a *pyridine–sulfur trioxide* complex followed by reduction of the intermediate with *lithium aluminum hydride* in tetrahydrofuran (6–98% yields) [*809*]. With this method saturation of the double bond has taken place in some instances [*809*].

Acetylenic alcohols, usually of propargylic type, are frequently intermediates in the synthesis of compounds such as vitamin A in which a selective reduction of the triple bond to a double bond is desirable. Selective reduction can be accomplished by carefully controlled *catalytic hydrogenation* over deactivated palladium [*39, 513, 514, 515, 517, 519*], by reduction with *lithium aluminum hydride* [*534, 535*], *sodium bis(2-methoxyethoxy)aluminum hydride* [*810*], *zinc* [*535*], and *chromous sulfate* [*273*]. Some partial reductions

1 1 7

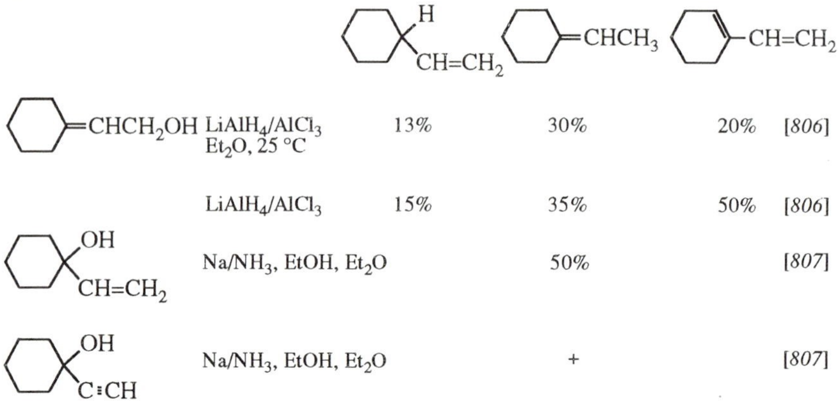

are achieved in alcohols in which the triple bonds are conjugated with one or more double bonds [39, 517, 535] and even aromatic rings [273].

Hydroxy compounds in the aromatic series behave in a manner depending on whether the hydroxy group is phenolic, benzylic, or more remote from the ring.

Phenolic hydroxyl is difficult to hydrogenolyze. It can be replaced by hydrogen if the phenol is first added to dialkylcarbodiimide to form the O-aryl-N,N'-dialkylisourea, which is then hydrogenated over palladium [811]. Alternatively, phenols are converted by treatment with cyanogen bromide to aryl cyanates; these are then treated with diethylamine to give O-aryl-N,N-diethyl-isoureas, which on *hydrogenation over palladium* afford hydrocarbons or hydroxyl-free derivatives in 85–94% yields [812]. Good to high yields are also obtained when phenols are transformed to aryl ethers by reaction with 2-chlorobenzoxazole or 1-phenyl-5-chlorotetrazole, and the aryl ethers are hydrogenated over 5% palladium on carbon in benzene, ethanol, or tetrahydrofuran at 35 °C [813] (equation 118).

Still another way to replace phenolic hydroxyl by hydrogen is treatment of a phenol with octadecanesulfonyl fluoride and triethylamine in N,N-dimethyl-formamide and *hydrogenation* of the phenyl sulfonate with *formic acid* over bis(triphenylphosphine)palladium dichloride. *m*-Methoxyphenol is thus converted to anisole in a 70% yield after 5 h at 80 °C [81], and phenyl trifluoromethanesulfonates are reduced to aromatic parent compounds on heating at 60–65 °C. Conversion of phenols to phenyl trifluoromethane-sulfonates followed by hydrogenation with *triethylammomium formate* over palladium diacetate [814] or with *sodium borohydride* in the presence of tetrakis(triphenylphosphine)palladium [815] seems to be an efficient way for deoxygenation of phenols (equation 119).

Distillation of phenols with zinc dust [816] or with dry lithium aluminum hydride [817] also results in hydrogenolysis of phenolic hydroxyls, but because the reaction requires very high temperatures, hardly any other function in the molecule can survive, and the method has only limited use.

118

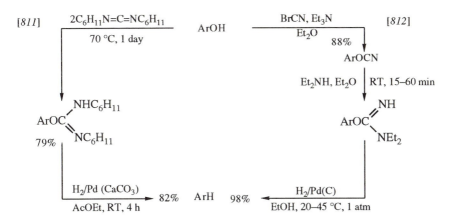

119 [*814, 815*]

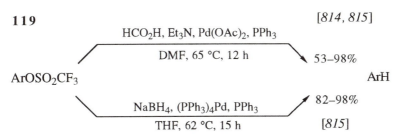

In some phenols, such as α-naphthol and 9-hydroxyphenanthrene, the hydroxyl group is replaced by hydrogen by refluxing with 57% *hydriodic acid* and acetic acid (52% and 96% yields, respectively) [*317*].

In contrast to phenolic hydroxyl, **benzylic hydroxyl** is replaced by hydrogen very easily. In *catalytic hydrogenation* of aromatic aldehydes, ketones, acids, and esters, it is sometimes difficult to prevent the easy hydrogenolysis of the benzylic alcohols that result from the reduction of these functions. A catalyst suitable for preventing hydrogenolysis of benzylic hydroxyl is platinized charcoal [*31*]. Other catalysts, especially palladium [*818*], nickel [*44*], Raney nickel [*818*], and copper chromite [*819*] promote hydrogenolysis. In chiral alcohols such as 2-phenyl-2-butanol, hydrogenolysis takes place with inversion over platinum and palladium, and with retention over Raney nickel (optical purities of 59–66%) [*818*].

Benzylic alcohols are also converted to hydrocarbons by *sodium borohydride* [*820*], *chloroalane* [*821*], *borane* [*822*], *zinc* [*823*], *hydriodic acid* [*316, 824*], and *diphosphorus tetraiodide* [*393*], generally in good to excellent yields. Hydrogenolysis of benzylic alcohols may be accompanied by dehydration (where feasible) [*821*]. A mixture of *titanium trichloride* and 0.33 equivalent of lithium aluminum hydride in 1,2-dimethoxyethane causes coupling of the benzyl residues: Benzyl alcohol thus affords bibenzyl in 78% yield [*284*] (equation **120**).

120

[821]
$LiAlH_4/AlCl_3$
Et_2O, 25 °C
$C_6H_5CHCH_3$
OH
[540]
$LiAlH_4/TiCl_3$
$(CH_2OMe)_2$, reflux
[284]
68%
$C_6H_5CH_2CH_3$ 74%
+
$C_6H_5CH{=}CH_2$ 10%
PtO_2
25 °C
EtOH
AcOH
1 atm
$C_6H_{11}CHCH_3$
OH
86%

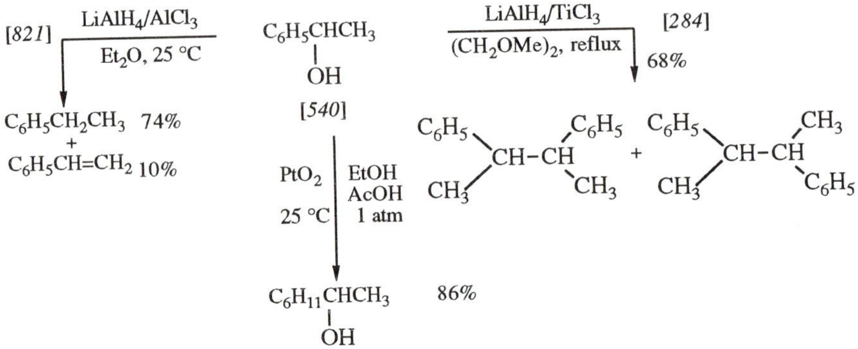

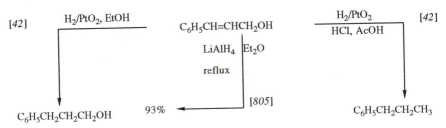

Hydrogenolyses of benzyl-type ethers, amines, and esters are discussed on pp 111, 130, and 211, respectively.

Hydroxylic groups in α positions to heterocyclic aromatic compounds undergo hydrogenolysis in *catalytic hydrogenation*. In furfuryl alcohol, hydrogenation also reduces the aromatic nucleus and easily cleaves the furan ring, giving α-methylfuran, tetrahydrofurfuryl alcohol, and a mixture of pentanediols and pentanols [*41, 800*].

Vinylogs of benzylic alcohols, for example, cinnamyl alcohol, undergo easy saturation of the double bond by catalytic hydrogenation over platinum, rhodium–platinum, or palladium oxides [*42*] or by reduction with *lithium aluminum hydride* [*805*]. In the presence of acids, catalytic hydrogenolysis of the allylic hydroxyl takes place, especially over platinum oxide in acetic and hydrochloric acids [*42*] (equation **121**).

121

[42]
H_2/PtO_2, EtOH
$C_6H_5CH{=}CHCH_2OH$
H_2/PtO_2
HCl, AcOH
[42]
$LiAlH_4$
Et_2O
reflux
[805]
$C_6H_5CH_2CH_2CH_2OH$
93%
$C_6H_5CH_2CH_2CH_3$

Aromatic hydroxy compounds, both phenols and alcohols, can be **reduced in the aromatic rings.** *Catalytic hydrogenation* over platinum oxide [*8*], rhodium–platinum oxide [*41*], nickel [*44*], or Raney nickel [*46*] gives alicyclic alcohols: Phenol yields cyclohexanol by hydrogenation over platinum oxide at room temperature and 210 atm (47% yield) [*8*], by heating with hydrogen and nickel at 150 °C and 200–250 atm (88–100% yield) [*44*], or by hydrogenation over Urushibara nickel at 70–110 °C at 66 atm (79% yield) [*49*]. 2-Phenylethanol gives 2-cyclohexylethanol at 175 °C and 210–260 atm (66–75% yield) [*44*], and hydroquinone yields *cis*-1,4-cyclohexanediol by hydrogenation over nickel at 150 °C and 170 atm [*44*], or over Raney nickel at room temperature at

2–3 atm [8]. Hydrogenation of the aromatic ring in benzylic alcohols is usually preceded by hydrogenolysis of the alcohol group prior to the saturation of the ring. However, under special conditions, using platinum oxide as the catalyst, ethanol with a trace of acetic acid as the solvent, and room temperature and atmospheric pressure, 1-phenylethanol is converted in 86% yield to 1-cyclohexylethanol [540]. β-Naphthol is, under various conditions, hydrogenated in the substituted ring, in the other ring, or in both [8, 46]. The results of partial hydrogenation of naphthols depend on the position of the hydroxyl (α or β) and on the catalyst used (Raney nickel [825, 826, 827] or copper chromite [825]).

Partial reduction of naphthols is accomplished by *sodium* in liquid ammonia. In the absence of alcohols, small amounts of 5,8-dihydro-α-naphthol or 5,8-dihydro-β-naphthol are isolated. In the presence of *tert*-amyl alcohol, α-naphthol gives a 65–85% yield of 5,8-dihydro-α-naphthol, and β-naphthol affords 55–65% β-tetralone [547] (Procedure 30, p 309). *Lithium* in liquid ammonia and ethanol reduces α-naphthol to 5,8-dihydro-α-naphthol in 97–99% yield [828] (equation 122).

1 2 2

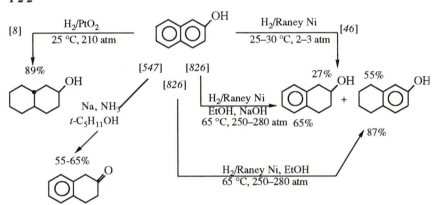

Halogenated saturated alcohols are reduced to *alcohols* by *catalytic hydrogenation* over Raney nickel [63]. Treatment of halogenated tertiary alcohols with Raney nickel in boiling toluene followed by hydrogenation over palladium leads to replacement by hydrogen of the hydroxyl, whereas chlorine survives and bromine is only partially hydrogenolyzed [794] (equation 123).

In polyfluorochloro alcohols, some chlorines are replaced by hydrogen using *isopropyl alcohol* under ultraviolet irradiation [425] (equation 124).

In **halogenated phenols** and naphthols, the halogens are hydrogenolyzed over a nickel catalyst in high yields [697]. 2,4,6-Trichlorophenol is converted to phenol on treatment with *ammonium formate* in the presence of 10% palladium on charcoal in methanol at ambient temperature [438]. 2,4,6-Tribromophenol is reduced to phenol in a 73% yield, and 2,4-bromochlorophenol is reduced to 4-chlorophenol in a 91% yield on stirring with *ethanethiol* and aluminum chloride in dichloromethane at room temperature [336].

123

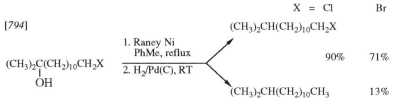

[794]

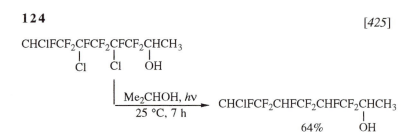

124   [425]

A rather unusual replacement of all three atoms of fluorine by hydrogen in *o*-hydroxybenzotrifluoride is due to the electron-releasing effect of the hydroxy group, which enhances nucleophilic displacement by the hydride anion [829].

In **bromohydrins,** *hydrogenation* over Raney nickel may lead to epoxides rather than to alcohols because regular Raney nickel contains enough alkali to cause dehydrobromination. Pure hydrogenolysis of bromine can be achieved over Raney nickel that has been washed free of alkali by acetic acid [830]. Bromohydrins treated with *titanium dichloride* prepared from lithium aluminum hydride and titanium trichloride give alkenes nonstereospecifically: Both *erythro-* and *threo-*5-bromo-6-decanols give 80:20 and 70:30 mixtures of *trans-* and *cis-*5-decene in 91% and 82% yields, respectively [284].

Selective hydrogenolysis of a secondary bromine in preference to a primary bromine is achieved by catalytic hydrogenation over palladium [831] or on treatment with *sodium iodide* and trifluoroacetic acid [315].

Nitro alcohols are reduced to ***amino alcohols*** by *catalytic hydrogenation* over platinum [832] and with *iron* [727], and nitrosophenols [352] and nitrophenols [353] are reduced to aminophenols with *sodium hydrosulfite, sodium sulfide* [331], or *tin* [253]. Bromine atoms in 2,6-dibromo-4-nitrophenol are not affected [253].

ETHERS AND THEIR DERIVATIVES

Open-chain aliphatic ethers are completely resistant to hydrogenolysis. Cyclic ethers may undergo reductive cleavage under strenuous conditions. Epoxides are discussed on p 113. The tetrahydrofuran ring is cleaved in vigorous hydrogenations over Raney nickel [800] and copper chromite [800] to give, ultimately, alcohols.

Tetrahydrofuran itself is not entirely inert to some hydrides, although it is a favorite solvent for reduction with these reagents. A mixture of lithium aluminum hydride and aluminum chloride produces butyl alcohol on prolonged refluxing in yields corresponding to the amount of *alane* generated [*833*].

Under reasonable conditions the tetrahydropyran ring is not affected. *Catalytic hydrogenation* of 2,3-dihydropyran over Raney nickel at room temperature and 2.7 atm saturates the double bond to give a quantitative yield of tetrahydropyran [*834*].

Vinyl ethers (**enol ethers**) are reductively cleaved by *lithium, sodium, or potassium* in liquid ammonia, especially in the absence of alcohols (except *tert*-butyl alcohol) [*835*]. A mixture of l-methoxy-1,3- and l-methoxy-1,4-cyclo-hexadienes gives in this way first methoxycyclohexene and, on further reduction, cyclohexene [*835*]. 2-Alkyl-4,5-dihydrofurans and 2-alkyl-5,6-di-hydropyrans are reductively opened to unsaturated alcohols by refluxing with *isopropylmagnesium bromide* in tetrahydrofuran in the presence of bis(tributylphosphine)nickel dichloride [*451*] (equation **125**).

125 [*451*]

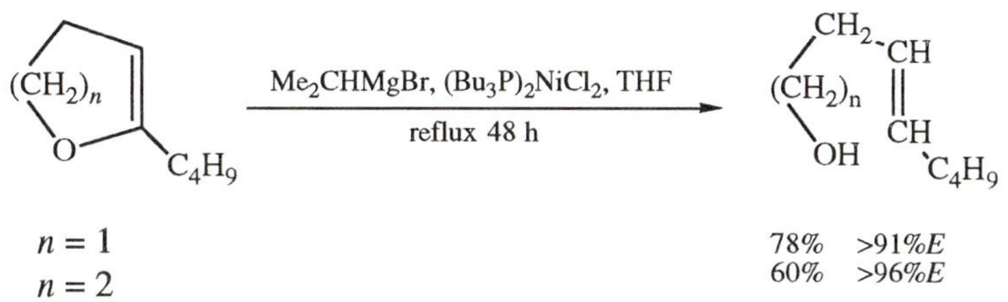

$$n = 1$$
$$n = 2$$

78%   >91%E
60%   >96%E

Reductive cleavage of α-alkoxytetrahydrofurans and pyrans will be discussed in the chapter on acetals (p 145 and 146).

Allyl ethers are cleaved to alcohols and alkenes by *lithium aluminum hydride, lithium diisobutyl aluminum hydride,* and *lithium triethylborohydride* [*836*] (equation **126**).

126 [*836*]

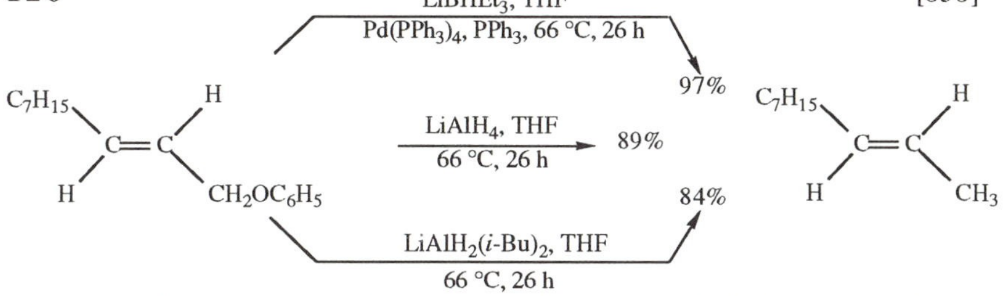

A similar reductive cleavage can be accomplished with *lithium* in ethylamine [*837*] (equation **127**).

Benzyl ethers are hydrogenolyzed easily, even more readily than benzyl alcohols [*838*]. 3,5-Bis(benzyloxy)benzyl alcohol gives toluene and 3,5-di-hydroxybenzyl alcohol on *hydrogenation over palladium on carbon* at room temperature and atmospheric pressure in a quantitative yield [*839*].

127 $\xrightarrow[\text{EtNH}_2, \text{ 5 min}]{\text{Li}}$ [837]

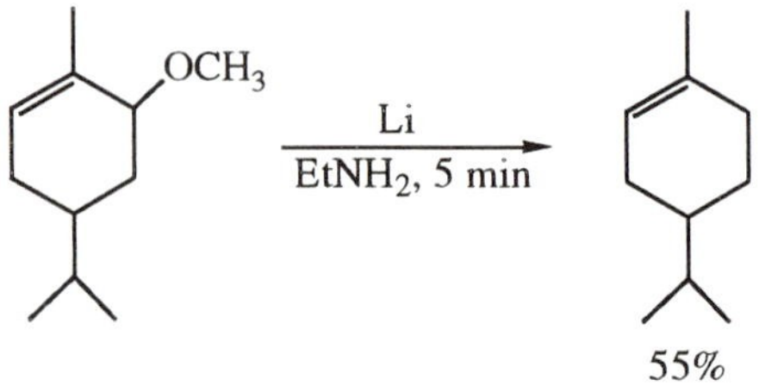

55%

Hydrogenolysis of benzylic ethers can also be achieved by refluxing the ether with cyclohexene (as a source of hydrogen) in the presence of 10% palladium on carbon in the presence of aluminum chloride [840].

Hydrogenolysis of benzyl alkynyl ethers with *calcium* in liquid ammonia occurs in preference to the reduction of the triple bond [211] (equation **128**).

128   [211]

$$CH_3C{\equiv}CCH_2CH_2OCH_2C_6H_5 \xrightarrow[\text{NH}_3]{\text{Ca}} CH_3C{\equiv}CCH_2CH_2OH + C_6H_5CH_3$$

90%

Cleavage of a benzyloxy bond is also accomplished by treatment with *sodium* in liquid ammonia [841] or by refluxing with sodium and butyl alcohol in toluene [842]. On the other hand, aluminum amalgam reduces a nitro group but does not cleave the benzyl ether bond in the dibenzyl ether of 2-nitro-hydroquinone [843].

Ether linkage in **alkyl aryl ethers** is usually stable to reduction, and hydrogenolysis of the central methoxy group in the trimethyl ether of pyrogallol by refluxing with *sodium* in ethanol is an exception [844]. Catalytic hydrogenation does not affect the ether bonds. In 4-[2,3-dimethoxyphenyl]-5-nitrocyclohexene only the double bond (not even the nitro group) is reduced by hydrogenation over platinum oxide [733]. In methyl β-naphthyl ether *hydrogenation* over Raney nickel at 130 °C and 24 atm reduces the substituted ring to give a tetrahydro derivative but does not break the ether bond [827].

Partial reduction of benzene and naphthalene rings in aryl alkyl ethers is accomplished by *electrolysis* in aqueous tetrabutylammonium hydroxide [845] and by *dissolving metal reduction*. Anisole treated with lithium [846] or with sodium [835] and ethanol in ammonia gives first l-methoxy-1,4-cyclohexadiene, which rearranges to the conjugated 1-methoxy-1,3-cyclohexadiene. Further reduction affords l-methoxycyclohexene and cyclohexene [835] (equation **129**).

129   K, NH$_3$, MeOH   [835]

OCH$_3$ $\xrightarrow[\text{NH}_3]{\text{Na}}$ OCH$_3$ $\underset{\text{KNH}_2}{\rightleftarrows}$ OCH$_3$ →   OCH$_3$ 78%   5%

8%

Na, NH$_3$, *t*-BuOH   90%

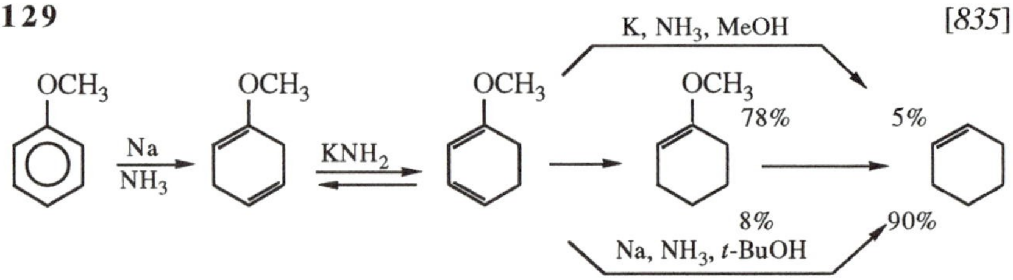

Similar reductions are accomplished in methoxyalkylbenzenes [547]. 2-Ethoxynaphthalene is reduced by sodium in ethanol to 3,4-dihydro-2-ethoxy-naphthalene, which on hydrolysis affords β-tetralone in 40–50% yield [847].

In **alkyl haloaryl ethers**, halogen can be selectively replaced by hydrogen in 68–97% yields on heating at 70 °C with zinc and catalytic amounts of nickel dichloride, sodium iodide, and triphenylphosphine in aqueous *N,N*-dimethylformamide [848].

EPOXIDES, PEROXIDES, AND OZONIDES

Epoxides (oxiranes) can be reduced in various ways. So-called *deoxygenation* converts epoxides to alkenes. Such a reaction is very useful because it is the reversal of epoxidation of alkenes and because both reactions (epoxidation and deoxygenation) combined represent temporary protection of a double bond.

Reduction to alkenes (deoxygenation) is accomplished by refluxing the epoxides with *zinc* dust in acetic acid [849, 850] (69–70% yields) or with a *zinc–copper couple* in ethanol [851] (8–95% yields), by heating at 65 °C with *titanium dichloride* prepared in situ from titanium trichloride and 0.25 mol of lithium aluminum hydride [284] (11–79% yields), by treatment with *titanocene*, prepared in situ from dicyclopentadienyltitanium dichloride and magnesium in tetrahydrofuran [852], with *chromous chloride–ethylenediamine* complex in *N,N*-dimethylformamide at 90 °C [853], with *samarium diiodide* in tetrahydrofuran at room temperature (65–98% yields) [304], with a mixture of *butyllithium* and *ferric chloride* in tetrahydrofuran at room temperature (57–92% yields) [854], or with *alkali O,O'-diethyl phosphorotellurate* prepared in situ from diethyl sodium phosphite and tellurium (39–91% yields) [411]. The latter procedure is faster with terminal than with internal epoxides and with Z than with E compounds and is stereospecific. The other methods give either mixtures of isomers or predominantly *trans*-alkenes. The method is applied to epoxides of sesquiterpenes [851] and steroids [849] (equation **130**).

130

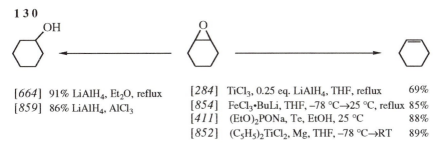

[664] 91% LiAlH₄, Et₂O, reflux	[284] TiCl₃, 0.25 eq. LiAlH₄, THF, reflux	69%
[859] 86% LiAlH₄, AlCl₃	[854] FeCl₃·BuLi, THF, –78 °C→25 °C, reflux	85%
	[411] (EtO)₂PONa, Te, EtOH, 25 °C	88%
	[852] (C₅H₅)₂TiCl₂, Mg, THF, –78 °C→RT	89%

More frequent than the deoxygenation of epoxides to alkenes is **reduction of epoxides to alcohols**. Its regiospecificity and stereospecificity depend on the reducing agents.

Hydrogenolysis of epoxides to alcohols by *catalytic hydrogenation* over platinum requires acid catalysis. 1-Methylcyclohexene oxide is reduced to a mixture of *cis*- and *trans*-2-methylcyclohexanols [855]. Steroidal epoxides usually give axial alcohols stereospecifically, and 4,5-epoxycoprostan-3α-ol affords cholestan-3α,4β-diol [855].

Reagents of choice for reduction of epoxides to alcohols are hydrides and complex hydrides. A general rule of regioselectivity is that the nucleophilic complex hydrides such as *lithium aluminum hydride* [*664, 856*] and *sodium bis(methoxyethoxy)aluminum hydride* [*857*] approach the oxide from the less hindered side and thus give alcohols with hydroxyls on the more substituted carbon. In contrast, hydrides of electrophilic nature such as *alanes* (prepared in situ from lithium aluminum hydride and aluminum halides) [*857, 858, 859, 860, 861*] or *boranes*, especially in the presence of boron trifluoride, open the ring in the opposite direction and give predominantly alcohols with hydroxyls on the less substituted carbon [*862, 863, 864*]. As far as stereoselectivity is concerned, lithium aluminum hydride yields *trans* products [*664*], whereas electrophilic hydrides yield predominantly *cis* products [*862, 864*]. Both these rules have exceptions, as shown by a study of reduction of deuterated styrene-2,2-d_2 oxide with deuteroalanes [*860*].

Sterically hindered epoxides are sometimes easier to reduce by dissolving metals: 2-Methyl-2,3-epoxybutane gives, after treatment with *lithium* in ethylenediamine at 50 °C for 1 h, 74% 3-methyl-2-butanol and 8% 2-methyl-2-butanol [*865*]. A regiospecific reagent is prepared by treating sodium hydride, sodium amylate, and zinc or nickel dichloride in 1,2-dimethoxyethane at 65 °C. The direction of the cleavage of the epoxide depends on the metal salt used [*1488*] (equation **131**).

131

[855]	H$_2$/PtO$_2$, AcOEt, HClO$_4$		42%	58%
[862]	NaBH$_3$CN/BF$_3$·Et$_2$O, THF, 25 °C	3%	84%	trace
[864]	BH$_3$/NaBH$_4$, THF, 0 °C	26%	74%	
[864]	BH$_3$/LiBH$_4$, THF, 0 °C	26%	74%	
[864]	AlH$_3$/2 AlCl$_3$, THF, 0 °C	10%	19%	12%
[865]	Li,(CH$_2$NH$_2$)$_2$, 50 °C	93%		
[1488]	NaH, C$_5$H$_{11}$ONa, ZnCl$_2$, DME, 65 °C	92%		
[1488]	NaH, C$_5$H$_{11}$ONa, NiCl$_2$, DME, 65 °C	19%	76%	

Unsaturated epoxides are reduced preferentially at the double bonds by *catalytic hydrogenation* over 5% palladium on carbon [*866*]. The rate of hydrogenolysis of the epoxides is much lower than that of the addition of hydrogen across the carbon–carbon double bond. In α,β-unsaturated (allylic) epoxides *borane* attacks the conjugated double bond at β-carbon in a *syn* mode with respect to the epoxide ring and gives Z-allylic alcohols [*867*]. *Lithium aluminum hydride* modified with *cuprous cyanide* reduces allylic epoxides by 1,2- and 1,4-addition of hydrogen, giving predominantly homoallylic alcohols. The ratio of the two products depends on the substituents at the double bond [*868*] (equations 132 and 133). Similar complex reduction of epoxides occurs in α-keto epoxides (p 175). Reduction with *chromous chloride* does not affect the double bond or the carbonyl and results in deoxygenation to alkenes [*270, 869*] (equations **132** and **133**).

132 [867]

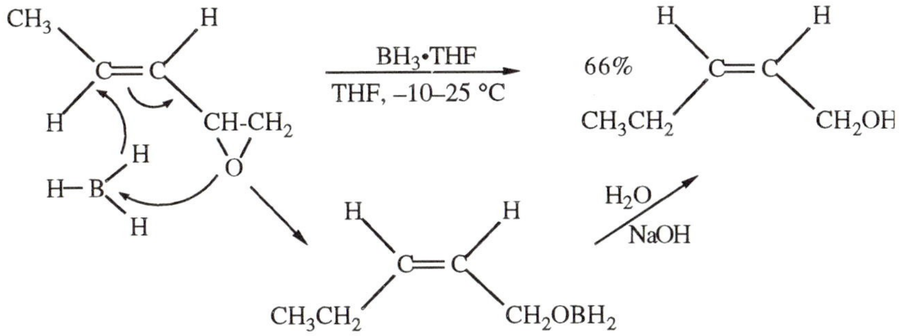

133 [868]

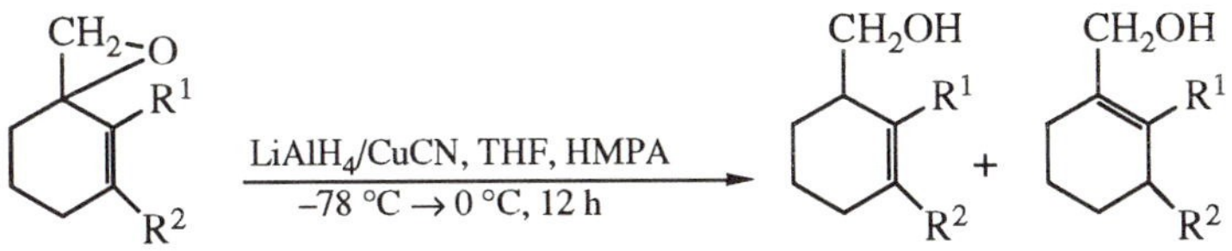

$R^1 = R^2 = H$	31%	62%
$R^1 = R^2 = CH_3$	4%	83%
$R^1 = CH_3, R^2 = C_6H_5$	40%	20%

In **halogenated epoxides**, the halogens can be replaced by hydrogen by *electrolytic reduction* in acetonitrile using a mercury electrode; epoxides are obtained in 80–92% yields [*197*]. Reduction with *lithium aluminum hydride, diisobutylalane, or lithium triethylborohydride* affords halogen-free alcohols in 92–98% yields [*861*].

The oxygen–oxygen bond in peroxides and hydroperoxides is cleaved very easily *by catalytic hydrogenation* over platinum oxide [*870, 871*], palladium [*83, 871, 872, 873*], or Raney nickel [*874*]. Hydroperoxides yield alcohols [*870*]; unsaturated hydroperoxides yield unsaturated and saturated alcohols [*83*]. Unsaturated cyclic peroxides (endoperoxides) give saturated peroxides [*871*], unsaturated diols [*871, 872*], or saturated diols [*873*]. Similar results are obtained by reduction with *zinc* [*875*] (equations **134** and **135**).

A very convenient way of reducing hydroperoxides to alcohols in high yields is treatment with a 20% aqueous solution of *sodium sulfite* at room or elevated temperatures [*345*] or with *tertiary phosphines* in organic solvents [*399, 876*].

Triethyl- and *triphenylphosphines* have been used for "*deoxygenation*" not only of hydroperoxides to alcohols, but also of dialkyl peroxides to ethers; diacyl peroxides to acid anhydrides; peroxy acids and their esters to acids or esters, respectively; and endoperoxides to oxides [*399*] in good to excellent

134

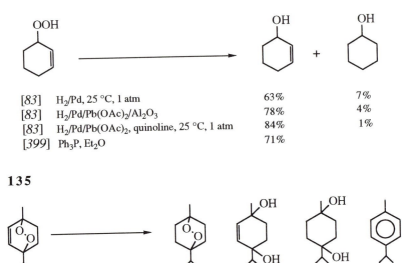

			63%	7%
[83]	H₂/Pd, 25 °C, 1 atm		63%	7%
[83]	H₂/Pd/Pb(OAc)₂/Al₂O₃		78%	4%
[83]	H₂/Pd/Pb(OAc)₂, quinoline, 25 °C, 1 atm		84%	1%
[399]	Ph₃P, Et₂O		71%	

135

[872]	H₂/Pd black		53%		
[871]	H₂/Pd (C)	23%	47%		
[873]	H₂/Pd, MeOH			~100%	
[877]	(EtO)₃P, 160–170 °C				30%

yields. The deoxygenation of ascaridole to 1-methyl-4-isopropyl-1,4-oxido-2-cyclohexene [399] was later challenged (the product is claimed to be *p*-cymene instead [877]).

Apart from the conversion of peroxides to useful products, it is sometimes necessary to reduce peroxides, and especially hydroperoxides formed by autoxidation. Such compounds are formed especially in hydrocarbons containing branched chains, double bonds, or aromatic rings, and in ethers such as diethyl ether, diisopropyl ether, tetrahydrofuran, and dioxane. **Most peroxidic compounds decompose violently at higher temperatures and could cause explosion and fire; therefore, it is necessary to remove them from the liquids they contaminate.** Water-immiscible liquids can be stripped of peroxides by shaking with an aqueous solution of sodium sulfite or ferrous sulfate. A simple and efficient way of removing peroxides is treatment of the contaminated compounds with 0.4-nm molecular sieves [878].

Reduction of ozonides is very useful, especially when aldehydes are the desired products. Ozonides, prepared in solutions in ethyl acetate, dichloromethane, or methanol and usually not isolated, are easily *hydrogenolyzed over palladium* [879, 880], or reduced by *zinc* in acetic acid [881] in good yields. Ozonolysis of methyl oleate followed by hydrogenation over 10% palladium on charcoal gives a 58% yield of nonanal and a 60% yield of methyl 9-oxononanoate [879], whereas reduction of the ozonide with zinc dust in acetic acid affords the respective yields of 81% and 85%. Ozonolysis of

6-chloro-2-methyl-2-heptene in acetic acid followed by treatment with zinc dust in ether gives a 55% yield of 4-chloropentanal [*881*].

An elegant "one-pot" reduction of ozonides consisting of the treatment of a crude product of ozonization of an alkene in methanol with *dimethyl sulfide* (36% molar excess) gives 62–97% yields of very pure aldehydes [*338*]. On the other hand, treatment of ozonides with the *borane–dimethyl sulfide* complex affords alcohols in 67–98% yields [*882*] (equation **136**).

136

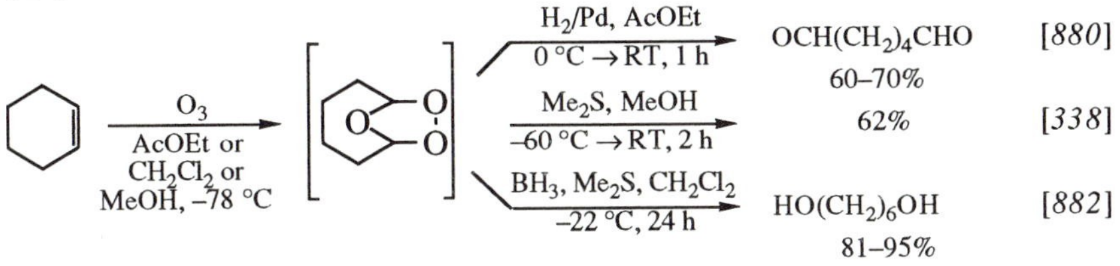

$$\text{H}_2/\text{Pd, AcOEt}$$
$$\overline{0\,^{\circ}\text{C} \rightarrow \text{RT, 1 h}} \quad \text{OCH(CH}_2)_4\text{CHO} \quad [880]$$
$$60\text{–}70\%$$

$$\text{Me}_2\text{S, MeOH}$$
$$\overline{-60\,^{\circ}\text{C} \rightarrow \text{RT, 2 h}} \quad 62\% \quad [338]$$

$$\text{BH}_3, \text{Me}_2\text{S, CH}_2\text{Cl}_2$$
$$\overline{-22\,^{\circ}\text{C, 24 h}} \quad \text{HO(CH}_2)_6\text{OH} \quad [882]$$
$$81\text{–}95\%$$

$$\text{O}_3 \atop \overline{\text{AcOEt or} \atop \text{CH}_2\text{Cl}_2 \text{ or} \atop \text{MeOH, } -78\,^{\circ}\text{C}}$$

SUMMARY

The carbon–oxygen bond in most alcohols and phenols is very resistant to any reduction. Only benzylic alcohols are an exception, and their hydroxyl can be relatively easily replaced by hydrogen in catalytic hydrogenation and complex hydride reductions. Dialkyl and alkyl aryl ethers resist reductions, with the exception of benzylic ethers, which are reduced even more readily than benzylic alcohols by catalytic hydrogenation and alkali metal reduction.

Epoxides are deoxygenated by metals and metal compounds and reductively cleaved by complex hydrides. Peroxides are deoxygenated by catalytic hydrogenation and with sodium sulfite, and ozonides are cleaved by catalytic hydrogenation, zinc reduction, and treatment with dimethyl sulfide.

CHAPTER 10

Reduction of Sulfur Compounds (Except Sulfur Derivatives of Aldehydes, Ketones, and Acids)

Reduction of sulfur compounds is discussed in this chapter in the sequence divalent, tetravalent, and hexavalent sulfur derivatives.

DIVALENT SULFUR DERIVATIVES

Sulfur analogs of hydroxy compounds, **thiols (mercaptans and thiophenols),** behave differently from their oxygen counterparts. Whereas hydrogenolysis of alcohols and especially phenols to parent hydrocarbons is far from easy, conversion of thiols to the parent hydrocarbons is achieved readily by several reagents. Benzyl mercaptan is desulfurized to toluene by treatment with *triethyl phosphite [883]* (Procedure 56, p 319). Refluxing with *iron* in ethanol and acetic acid desulfurizes the mercapto group in 2-mercapto-6-methylbenzothiazole and gives an 82% yield of 6-methylbenzothiazole. The sulfidic sulfur in the heterocyclic ring remains intact [*248*]. A general method for the desulfurization of thiols is treatment with *Raney nickel [884]*. Such reaction applies not only to thiols but also to sulfides and almost any sulfur-containing organic compound (equation **137**).

Raney nickel and *nickel boride* somehow extract sulfur from sulfur-containing compounds to form nickel sulfide, and because they contain adsorbed hydrogen at their surfaces, they even cause hydrogenation. The reaction is achieved by refluxing sulfur-containing compounds with a large excess of Raney nickel or nickel boride in ethanol and generally gives excellent yields [*884*].

Aliphatic and aromatic sulfides undergo desulfurization with *Raney nickel* [*884*], *nickel boride* [*884*], *lithium aluminum hydride* in the presence of cupric chloride [*885*], *titanium dichloride* [*886, 887*], and *triethyl phosphite* [*888*]. In saccharides, benzylthioethers are not desulfurized but are reduced to toluene and mercaptodeoxysugars using *sodium* in liquid ammonia [*889*]. This reduction has general application where catalytic hydrogenation would not work [*838*] (equation **137**).

Divalent sulfur is a poison for most noble metal catalysts, so that catalytic hydrogenation of sulfur-containing compounds poses serious problems (p 11). However, allyl phenyl sulfide is *hydrogenated over tris(triphenylphosphine)-rhodium chloride* in benzene to give a 93% yield of phenyl propyl sulfide [*890*].

Cyclic sulfides treated with *triethyl phosphite* eject sulfur and form a new ring with one fewer member [*888*]. From compounds containing sulfur in three-membered rings **(thiiranes)** alkenes are formed in high yields [*400, 403*]. The same reaction can be achieved with *triphenylphosphine* [*400*] (equation **138**).

Reductive cleavage of 1,3-dithiolanes is accomplished in low yields with *sodium* in liquid ammonia and gives either dithiols or mercapto sulfides depending on the substituents in position 2 [*891*] (equation **139**).

137

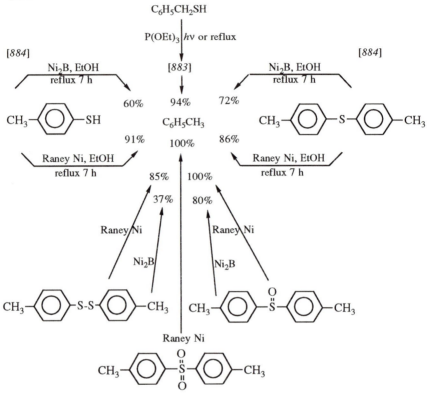

138   [400]

139   [891]

Alkyl thiocyanates are reduced with *lithium aluminum hydride* to mercaptans (81% yield), and aryl thiocyanates are reduced to thiophenols (94% yield) [*892*].

Disulfides can be either reduced to *two thiols or desulfurized*. Reduction to two thiols is achieved in high yields using *lithium aluminium hydride* [*893*]; *lithium triethylborohydride* [*133*]; *sodium borohydride* [*894*]; *aluminum* in liquid ammonia [*229*]; and *hydrazine, methylhydrazine,* and *1,1-dimethyl-hydrazine* [*390*].

Partial desulfurization of **disulfides to sulfides** is accomplished by treatment with *hexamethylphosphorous triamide* [*415*] or *hexaethylphosphorous triamide* [*414*] in good yields. 1,2-Dithiacyclohexane is thus quantitatively converted to thiophane (tetrahydrothiophene) at room temperature [*414*]. Complete desulfurization to hydrocarbons results when disulfides are refluxed in ethanol with *Raney nickel* or *nickel boride* (86 and 72% yields, respectively) [*884*] (equation **140**).

1 4 0

$$[415] \quad \xrightarrow[\text{C}_6\text{H}_6,\ \text{reflux}]{\text{(Me}_2\text{N)}_3\text{P}} \quad \text{C}_6\text{H}_5\text{CH}_2\text{SSCH}_2\text{C}_6\text{H}_5 \quad \xrightarrow[\text{25 °C, 4 h}]{\text{Al/I}_2,\ \text{NH}_3} \quad [229]$$

$$\text{C}_6\text{H}_5\text{CH}_2\text{SCH}_2\text{C}_6\text{H}_5 \quad 97\text{--}100\% \qquad\qquad\qquad \text{C}_6\text{H}_5\text{CH}_2\text{SH} \quad 65\%$$

Reductive cleavage of disulfides is very easy and can be accomplished without affecting some other readily reducible groups such as nitro groups.

If, on the other hand, a nitro group must be reduced to an amino group without reducing the sulfidic bond, *hydrazine* is the reagent of choice [*895*].

TETRAVALENT SULFUR DERIVATIVES

Dialkyl and diaryl sulfoxides are reduced (*deoxygenated*) to sulfides by several methods. As a little surprise, sulfides are obtained from sulfoxides by *catalytic hydrogenation* over 5% palladium on charcoal in ethanol at 80–90 °C and 65–90 atm in yields of 59–99% [*896*]. With *p*-tolyl β-styryl sulfoxide, deoxygenation is achieved even without reduction of the double bond, giving 66% *p*-tolyl β-styryl sulfide and only 16% *p*-tolyl β-phenylethyl sulfide [*896*].

Chemical deoxygenation of sulfoxides to sulfides is carried out by reduction with *boron tribromide, dimethylboron bromide,* or *9-BBN-Br* (9-bromo-9-borabicyclo[3.3.1]nonane) [*416*]; with *methyltrichlorosilane and sodium iodide* [*419*]; by refluxing in aqueous–alcoholic solutions with *stannous chloride* (62–93% yields) [*263*] (Procedure 41, p 313); with *titanium trichloride* (68–91% yields) [*283*]; by treatment at room temperature with *molybdenum trichloride* (prepared by reduction of molybdenum oxychloride, MoOCl_3, with zinc dust in tetrahydrofuran) (78–91% yields) [*296*]; by heating with *vanadium dichloride* in aqueous tetrahydrofuran at 100 °C (74–88% yields) [*296*]; by refluxing with *samarium diiodide* in tetrahydrofuran (77–90% yields) [*304*]; and by refluxing in aqueous methanol with *chromium dichloride* (24% yield) [*268*]. A very impressive method is the conversion of dialkyl and diaryl sulfoxides to sulfides by treatment in acetone solutions for a few minutes with 2.4 equivalents of

141

$$\underset{C_6H_5CH_2\overset{\overset{\displaystyle O}{\|}}{S}CH_2C_6H_5}{}\qquad\longrightarrow\qquad C_6H_5CH_2SCH_2C_6H_5$$

[896]	H$_2$/5% Pd(C), EtOH, 86–90 °C, 65–78 atm, 4 days	90%
[263]	SnCl$_2$•2 H$_2$O, HCl, MeOH, reflux 2 h	82%
[283]	TiCl$_3$, H$_2$O, MeOH, CHCl$_3$, reflux	78%
[296]	MoCl$_3$, THF, H$_2$O, Zn, 25 °C, 1 h	89%
[296]	VCl$_2$, H$_2$O, THF, 40 mm, 100 °C, 8 h	82%
[268]	CrCl$_2$, H$_2$O, MeOH, reflux 2 h	24%
[897]	NaI, (CF$_3$CO)$_2$O, Me$_2$CO, 25 °C, few min	93%
[416]	BBr$_3$, CH$_2$=CHMe, CH$_2$Cl$_2$, –23 °C, 30 min	91%
[416]	Me$_2$BBr, CH$_2$=CHMe, CH$_2$Cl$_2$, –23 °C, 30 min	96%
[416]	9-BBN-Br, CH$_2$=CHMe, CH$_2$Cl$_2$, –23 °C, 30 min; 0 °C, 10 min	92%
[419]	MeSiCl$_3$, NaI, MeCN, 25 °C, 10 min	85%

sodium iodide and 1.2–2.6 equivalents of trifluoroacetic anhydride (90–98% isolated yields) [897] (equation **141**).

In addition to deoxygenation, sulfoxides undergo *reductive cleavage* at the carbon–sulfur bond when treated with tert-*butyllithium* and methanol at –78 °C [898] (equation **142**) or heated in tetrahydrofuran with *aluminum amalgam*.

142   [898]

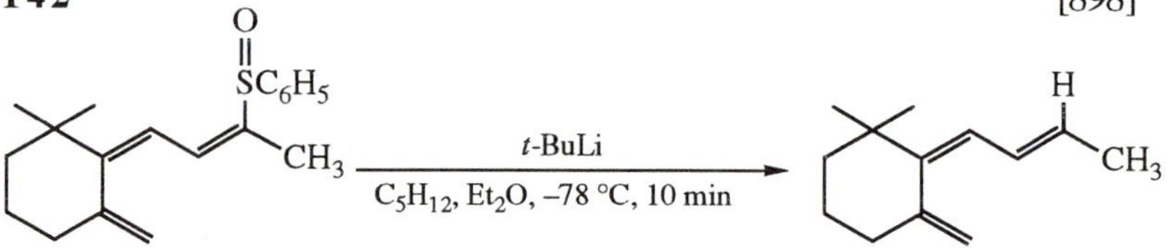

72%

Keto sulfoxides are thus converted to ketones in high yields (the keto group remains intact) [215] (p 173). Both sulfur–oxygen and sulfur–carbon bonds are cleaved by *electrolytic reduction* of 1-methylthio-1-methylsulfinyl-2-phenyl-ethylene [198] (equation **143**).

143   [198]

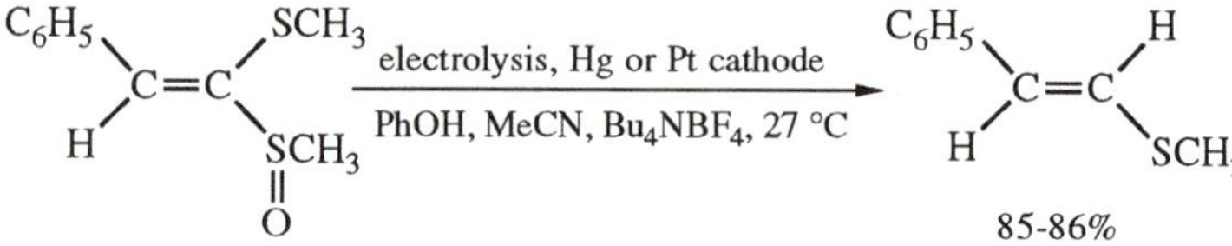

85-86%

HEXAVALENT SULFUR DERIVATIVES

The sulfonyl group in *sulfones* resists catalytic hydrogenation. Double bonds in α,β-unsaturated sulfones are reduced by hydrogenation over palladium on charcoal (94% yield) [*899, 900*] or over Raney nickel (62% yield) without the sulfonyl group being affected [*899*]. In *p*-thiopyrone-1,1-dioxide, both double bonds are reduced with *zinc* in acetic acid, but the keto group and the sulfonyl group survive [*901*]. *Raney nickel* may desulfurize sulfones to hydrocarbons [*884*].

Reduction of sulfones to sulfides is accomplished by *lithium aluminum hydride* [*900*], *diisobutyl aluminum hydride* [*902*], and sometimes by *zinc* dust in acetic acid [*900*]. Diisobutyl aluminum hydride (Dibal-H) is superior to both lithium aluminum hydride and zinc. At least 2.2 mol of the hydride is necessary for the reduction carried out in mineral oil at room temperature or in refluxing toluene over 18–72 h. Five-membered cyclic sulfones are reduced most readily, and aliphatic sulfones least readily. Yields are 12–77% [*902*].

Reductions of five-membered cyclic sulfones with lithium aluminum hydride are run in refluxing ether, and reductions of other sulfones are run in refluxing ethyl butyl ether (92 °C) (12–92% yields). Benzothiophene-1,1-dioxide is reduced at the double bond as well, giving 2,3-dihydrobenzothiophene in 79% yield after 18 h of refluxing in ether [*900*].

In vinyl sulfones containing vinylic halogen, the halogen in replaced by hydrogen, but the double bond and the sulfonyl group are not affected. (Z)-β-chloro-β-phenylsulfonylstyrene is converted to phenyl (Z)-β-styryl sulfone in 80% yield by refluxing with *zinc* dust in acetic acid [*903*].

An entirely different type of reduction occurs if sulfones are treated with *lithium aluminum hydride* in the presence of cupric chloride (1:2) [*226*], *sodium* in liquid ammonia [*904*], *aluminum amalgam* [*215, 226*], *sodium amalgam* [*905, 906, 907*], *samarium diiodide* in hexamethylphosphoric triamide [*311*], or *sodium dithionite* [*358*]. The carbon–sulfur bond is cleaved to give a sulfinic acid and a compound resulting from replacement of the sulfone group by hydrogen. Such hydrogenolysis can be achieved even without reducing a nitro group in nitro sulfones [*358*] (equation **144**), and a keto group in keto sulfones (2-methylsulfonylcyclohexanone on heating at 65 °C with aluminum amalgam in aqueous tetrahydrofuran affords an 89% yield of cyclohexanone [*215*]), and without reducing a double bond conjugated with aromatic rings in α,β-unsaturated sulfones [*226*] (equation **145**). Similar carbon–sulfur bond cleavage is observed with *alkyl aryl sulfoximines* [*227*] (equation **146**).

144 [*358*]

$$C_8H_{17}\underset{\underset{NO_2}{|}}{C}HSO_2C_6H_5 \xrightarrow[\text{octylviologen, } CH_2Cl_2, \text{ 35 °C, 3 h}]{Na_2S_2O_4, \text{ } K_2CO_3, \text{ } H_2O} C_8H_{17}CH_2NO_2$$

76%

Sulfenyl chlorides, sulfinic acids, and sulfinyl chlorides are reduced in good yields by *lithium aluminum hydride* to disulfides [*892*]. The same products are obtained from sodium or lithium *salts of sulfinic acids* on treatment with *sodium hypophosphite* or *ethyl hypophosphite* [*412*]. Sulfoxy sulfones are intermediates in this reaction [*412*] (equation **147**).

145   [226]

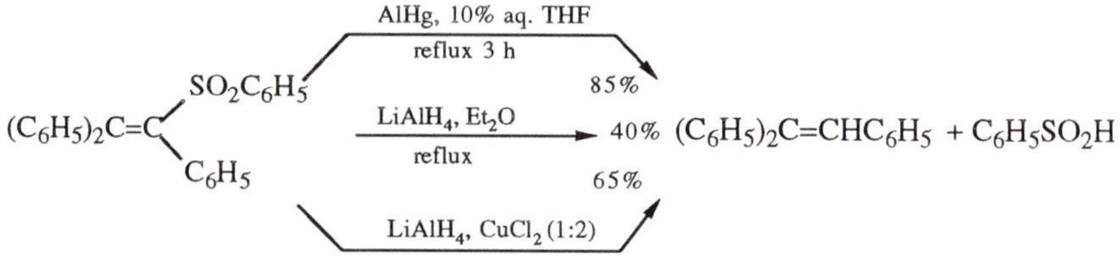

146   [227]

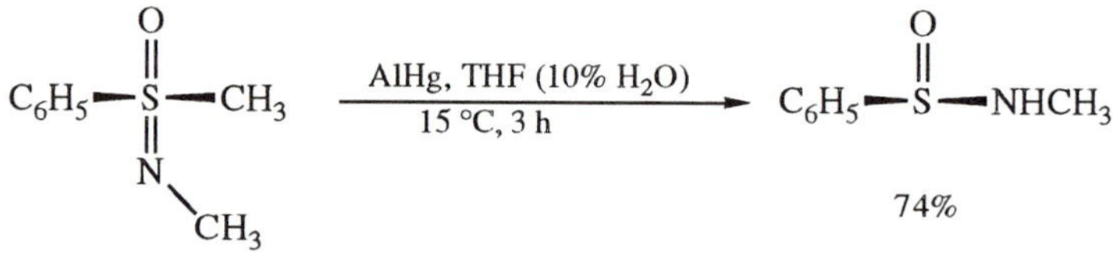

147

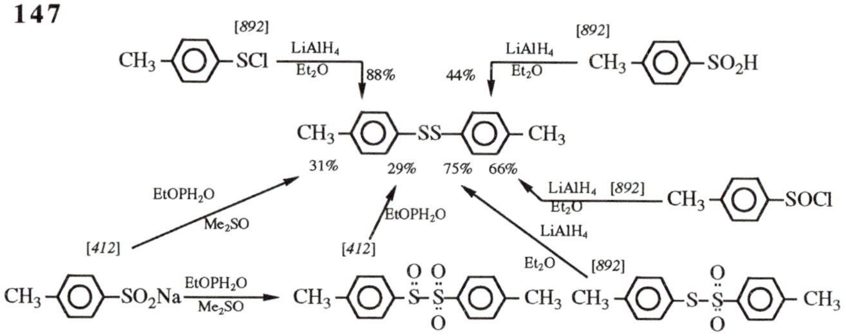

Sulfonic acids are completely resistant to any reductions, and other functions contained in their molecules can be easily reduced, for example, nitro groups with *iron* [908].

Chlorides of sulfonic acids can be reduced either partially to sulfinic acids, or completely to thiols. Both reductions are accomplished in high yields with *lithium aluminum hydride.* An inverse addition technique at a temperature of −20 °C is used for the preparation of sulfinic acids, whereas the preparation of thiols is carried out at the boiling point of ether [909].

More reliable reagents for the *preparation of sulfinic acids* are *zinc* [910, 911], *sodium sulfide* [343], and *sodium sulfite* [346]. These reagents not only stop the reduction at the stage of the sulfinic acids (in the form of their salts), but they also do not reduce other functions present in the molecules. In the reduction of anthraquinone-1,5-disulfonyl chloride with sodium sulfide below 40 °C, anthraquinone-1,5-disulfinic acid is obtained in 83.5% yield [343], and p-cyano-benzenesulfonyl chloride is reduced to *p*-cyanobenzenesulfinic acid in 87.4% yield [*346*].

Complete reduction of sulfonyl chlorides to thiols can be achieved by *lithium aluminum hydride* [*892, 909*], *zinc* [*911, 912*], and *hydriodic acid* generated in situ from iodine and red phosphorus [*321*] (equation **148**).

148

[*892*]	LiAlH$_4$, Et$_2$O	90%
[*909*]	LiAlH$_4$, Et$_2$O, –20 °C	93%
[*909*]	LiAlH$_4$, Et$_2$O, reflux	89%
[*911*]	Zn dust, 70–90 °C; NaOH, Na$_2$CO$_3$; H$^+$	64%
[*321*]	P, I$_2$, AcOH, reflux 2–3 h	90%

m-Nitrobenzenesulfonyl chloride, however, is reduced not to the thiol but to bis(*m*-nitrophenyl) disulfide by hydriodic acid in 86–91% yield [*913*].

Esters of sulfonic acids give products that depend on the reducing agent; the structure of the parent sulfonic acid; and especially the structure of the hydroxy compound, alcohol, or phenol.

Catalytic hydrogenation over Raney nickel converts benzenesulfonates of both alcohols and phenols to ***parent hydroxy compounds***, benzene, and nickel sulfide. *p*-Toluenesulfonates of alcohols are reduced similarly, whereas *p*-toluenesulfonates of phenols give nickel *p*-toluenesulfinates and ***aromatic hydrocarbons***. The yields of the hydroxy compounds range from 25 to 96% [*914*].

Similarly, *lithium aluminum hydride* gives various products. β-Naphthyl *p*-toluenesulfonate affords *p*-thiocresol and β-naphthol, and phenyl methanesulfonate gives methyl mercaptan and phenol. On the other hand, propyl *p*-toluenesulfonate yields *p*-toluenesulfonic acid and propane, and cetyl methanesulfonate and cetyl *p*-toluenesulfonate give hexadecane in 92% and 96% yields, respectively [*892*].

Lithium aluminum hydride frequently leads to mixtures of hydrocarbons and alcohols from alkyl *p*-toluenesulfonates, but *lithium triethylborohydride* (Super-Hydride) in tetrahydrofuran reduces *p*-toluenesulfonates of primary and secondary alcohols to the corresponding ***hydrocarbons*** at room temperature in high yields [*915*]. Also, *sodium borohydride* in dimethyl sulfoxide gives hydrocarbons: dodecane and cyclododecane from dodecyl and cyclododecyl *p*-toluenesulfonates in 86.5% and 53.9% yields, respectively [*656*]. Dodecane is also obtained by heating dodecyl tosylate with sodium cyanoborohydride in hexamethylphosphoramide at 80 °C for 12 h (73–78% yield) [*1489*].

Replacement of sulfonyloxy groups by hydrogen is also achieved by refluxing the methanesulfonyl and *p*-toluenesulfonyl esters of alcohols with *sodium iodide and zinc dust* in wet dimethoxyethane. Evidently iodine displaces sulfonyloxy group and is replaced by hydrogen by means of zinc. Yields range from 26% to 65% [*916*] (Procedure 38, p 312).

Alkyl [917] and aryl [814, 815] trifluoromethanesulfonates are reduced to the corresponding hydrocarbons by heating with *sodium borohydride* or *trialkylammonium formate* in the presence of palladium and triphenylphosphine [917] (equation **119**, p 107).

Cleavage of the sulfonyl esters to the **parent alcohols** is accomplished in yields of 60–100% by treatment of the *p*-toluenesulfonates with 2–6 equivalents of *sodium naphthalene* in tetrahydrofuran at room temperature (60–100% yields) (equation **149**).

149

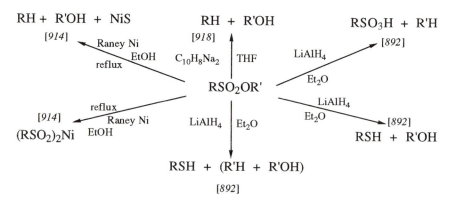

Sodium naphthalene is prepared by stirring sodium with an equivalent amount or a slight excess of naphthalene in tetrahydrofuran for 1 h at room temperature under an inert gas [918]. Benzenesulfonates and bromobenzenesulfonates are also cleaved to the parent alcohols, whereas alkyl methanesulfonates are reduced also to hydrocarbons [918].

In saccharides, *sodium amalgam* is used for regeneration of sugars from their sulfonates, but only in low yields [919].

Sulfonamides can be **reduced to amines** and either to **sulfinic acids** or **thiols**. Heating *p*-toluenesulfonamide with fuming *hydriodic acid* and phosphonium iodide at 100 °C affords 85% *p*-thiocresol [920].

Substituted sulfonamides are reductively cleaved to amines and sulfinic acids, thiols, or even parent hydrocarbons of the sulfonic acid [921]. For the regeneration of amines, *sodium naphthalene* proves very useful because it gives excellent yields (68–96%, isolated) under very gentle conditions (room temperature, 1 h) [922] (Procedure 32, p 310). Other reagents usually require refluxing. Thus *sodium* in refluxing amyl alcohol gives 78–93.5% yields, *zinc* dust in refluxing acetic and hydrochloric acids gives 24.5–87% yields, and *stannous chloride* in the same system gives 61.8–96% yields of the amines [921]. The other reduction products are toluene in the reductions with sodium, and *p*-thiocresol (74%) in the reductions with zinc [921] (equation **150**).

Reduction of both sulfoesters and sulfonamides has as its main objective preparation of hydrocarbons, alcohols, phenols, and amines. Reduction products of the sulfur-containing parts of the molecules are only corollaries.

150

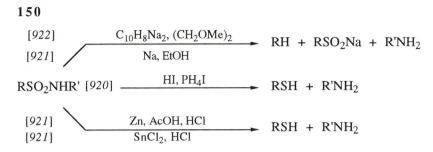

$$[922] \xrightarrow{C_{10}H_8Na_2, (CH_2OMe)_2} RH + RSO_2Na + R'NH_2$$

$$[921] \xrightarrow{Na, EtOH}$$

$$RSO_2NHR' [920] \xrightarrow{HI, PH_4I} RSH + R'NH_2$$

$$[921] \xrightarrow{Zn, AcOH, HCl} RSH + R'NH_2$$

$$[921] \xrightarrow{SnCl_2, HCl}$$

SUMMARY

Most reductions of compounds of divalent sulfur are aimed at the removal of the sulfur atom. Such desulfurizations are most reliably performed by Raney nickel or nickel boride. Thiiranes are desulfurized to alkenes by trialkylphosphine or trialkyl phosphites, and disulfides are converted to sulfides by tris-(diethylamino)phosphine. Sulfones resist reduction, but sulfoxides can be reduced to sulfides by metal chlorides. Most useful reduction is that of alkyl and aryl sulfonates and sulfonamides. Alkyl sulfonates are reduced to hydrocarbons and alcohols by sodium naphthalene, lithium aluminum hydride, or Raney nickel. Sulfonamides are redued also by zinc and stannous chloride.

CHAPTER 11

Reduction of Amines and Their Derivatives and of Phosphorus Compounds

AMINES AND THEIR DERIVATIVES

Like saturated alcohols, **saturated amines** are fairly resistant to reduction. The carbon–nitrogen bond is hydrogenolyzed in the *catalytic hydrogenation* of 2,2-dimethylaziridine, which, on treatment with hydrogen over Raney nickel at 60 °C and 4.2 atm, affords *tert*-butylamine in 75–82% yields [*923*]. Evidently the strain of the aziridine ring contributes to the relative ease of the hydrogenolysis, because other cyclic amines, such as piperidine, require heating to 200–220 °C for their hydrogenolysis to alkanes [*924*]. Excellent yields of hydrocarbons are obtained by hydrogenation of both aliphatic and aromatic amines over a specially activated platinum on silica gel at 150–225 °C [*735*].

Primary amines can be reduced to their parent compounds by treatment with *hydroxylamine-O-sulfonic acid* and aqueous sodium hydroxide at 0 °C. Substituted hydrazines and diimides are believed to be the reaction intermediates in the ultimate replacement of the primary amino group by hydrogen in yields of 49–72% [*925*].

Unsaturated amines are hydrogenated at the multiple bonds by *catalytic hydrogenation* over any catalyst. The double bond in indole is saturated in catalytic hydrogenation over platinum dioxide in ethanol containing fluoroboric acid, and indoline is obtained in greater than 85% yield [*606*]. **Allylic amines** such as allylpiperidine are also reduced by *sodium* in liquid ammonia in the presence of methanol (75% yield) [*926*].

Vinylamines (enamines) are reduced by *alane, chloroalane,* and *dichloroalane* to saturated amines and hydrogenolyzed to amines and alkenes [*927*]. Reduction is favored with dichloroalane, whereas hydrogenolysis is favored with alane. Alane, chloroalane, and dichloroalane give the following results with 1-(1-cyclohexenyl)pyrrolidine (1-*N*-pyrrolidinylcyclohexene): *N*-pyrrolidinylcyclohexane in 13, 15, and 22% yields, and cyclohexene in 80, 75, and 75% yields, respectively [*927*]. Saturated amines are also obtained by treatment of enamines with *sodium borohydride* [*928*], *sodium cyanoborohydride* [*138, 929*] (Procedure 26, p 307), *zinc cyanoborohydride* (prepared in situ from sodium cyanoborohydride and anhydrous zinc chloride at room temperature in yields of 73–90%) [*930*] and by heating for 1–2 h at 50–70 °C with 87% or 98% *formic acid* (37–89% yields) [*444*].

Aromatic amines are hydrogenated in the rings by *catalytic hydrogenation.* Aniline yields 17% cyclohexylamine and 23% dicyclohexylamine over platinum dioxide in acetic acid at 25 °C and 125 atm [*8*]. A better yield (90%) of cyclohexylamine is obtained by hydrogenation over nickel at 175 °C and 180 atm [*44*]. Reduction to saturated amines is also achieved *by lithium in ethylamine:* Dimethylaniline gives dimethylcyclohexylamine in 44% yield [*931*]. Reduction with *sodium* in liquid ammonia in the presence of ethanol converts dimethylaniline and dimethyltoluidines to dihydro and even tetrahydro derivatives or the products of their hydrolysis, cyclohexenones and cyclohexanones, in 58–72% yields [*932*]. Heating with *lithium aluminum hydride* replaces the amino group by hydrogen, albeit in low yields (17–28%) [*817*].

In naphthylamines reduction with *sodium* in refluxing pentanol affects exclusively the substituted ring, giving 51–57% 2-amino-1,2,3,4-tetrahydronaphthalene from β-naphthylamine [*933*].

N-Alkyl or *N*-aryl derivatives of *benzylamine* are readily hydrogenolyzed to toluene and primary or secondary amines by *catalytic hydrogenation* [*838*], the best catalysts being palladium oxide [*934*] or palladium on charcoal [*935, 936*]. The same type of cleavage takes place in 3-dimethylaminomethylindole, which gives 3-methylindole in 83.9% yield on treatment with hydrogen over palladium on charcoal in ethanol at room temperature and atmospheric pressure [*937*]. Hydrogenolysis over palladium is easier than over platinum and takes place in preference to saturation of double bonds [*937*].

Quaternary ammonium salts, especially iodides, in the aromatic series are converted to tertiary amines. *Electrolysis* of trimethylanilinium iodide using a lead cathode gives 75% benzene and 77% trimethylamine, and electrolysis of methylethylpropylanilinium iodide gives a 47.5% yield of methylethylpropylamine [*938*]. 2,3-Dimethylbenzyltrimethylammonium iodide is reduced by *sodium amalgam* to 1,2,3-trimethylbenzene in 85–90% yield [*939*].

If the quaternary nitrogen is a member of a ring, the ring is cleaved. 3-Benzyl-2-phenyl-*N*,*N*-dimethylpyrrolidinium chloride is cleaved by *hydrogenation* over Raney nickel at 20–25 °C almost quantitatively to 2-benzyl-4-dimethylamino-1-phenylbutane [*940*]. Reduction of methylpyridinium iodide (and its methyl homologs) with *sodium aluminum hydride* gives 24–89% yields of 5-methylamino-1,3-pentadiene (and its methyl homologs) in addition to *N*-methyldihydro- and tetrahydropyridines [*598*].

Amines are fairly resistant to reduction, and consequently **halogens in halogenated amines** behave toward reducing agents as if the amino group were not present. Occasionally, however, the amino group may help hydrogenolysis of halogen. In *o*-aminobenzotrifluoride, all three fluorine atoms are replaced by hydrogen in 80–100% yields on refluxing with *lithium aluminum hydride* [*829*]. This rather unusual hydrogenolysis is due to the electron-releasing effect of the amino group, which favors nucleophilic displacement of benzylic halogens by hydride ions.

Similarly, **nitroso groups** and **nitro groups in amines** are reduced independently. *p*-Nitrosodimethylaniline gives 64% *N*,*N*-dimethyl-*p*-phenylenediamine on reduction with *chromous chloride* [*268*], *o*-nitroaniline gives 74–85% *o*-phenylenediamine on refluxing with *zinc* in aqueous ethanolic sodium hydroxide [*941*], and 5-amino-2,4-dinitroaniline yields 49–52% 1,2,5-triamino-4-nitrobenzene on reduction with *sodium sulfide* [*942*].

As in the hydrogenation of glycols, in vigorous hydrogenations of **hydroxy amines**, a carbon–carbon bond cleavage can occur along with hydrogenolysis of carbon–oxygen and carbon–nitrogen bonds.

N-Ethyl-*N*-(2-hydroxyethyl)aniline affords 52.5% *N*-ethyl-*N*-[2-hydroxyethyl]cyclohexylamine and 14% *N*-methyl-*N*-ethylcyclohexylamine upon *hydrogenation* over Raney nickel at 150 °C and 65 atm in ethanol [*943*].

Cleavage of a carbon–nitrogen bond in a position β to a keto group takes place during catalytic hydrogenation of Mannich bases and gives ketones in yields of 51–96% [*944*]. The carbon–nitrogen bond next to a carbonyl group in α-amino ketones is cleaved by *zinc* [*945*] (pp 164).

Cleavage of the carbon–nitrogen bond occurs in *benzyloxycarbonylamino compounds* as a result of decarboxylation of the corresponding free carbamic acids resulting from hydrogenolysis of benzyl residues [*946, 947, 948*] (p 211).

In **N-nitrosoamines** the more polar nitrogen–oxygen bond is usually reduced preferentially to the nitrogen–nitrogen bond with *lithium aluminum hydride* so that the products are disubstituted *asym*-hydrazines. *N*-Nitroso-*N*-methylaniline gives 77% *N*-methyl-*N*-phenylhydrazine, *N*-nitrosodicyclohexylamine gives 48% *N,N*-dicyclohexylhydrazine, and *N*-nitrosopiperidine gives 75% *N,N*-pentamethylenehydrazine [*949*] (equation **151**). *N*-Nitrosodimethylamine is reduced to *N,N*-dimethylhydrazine with lithium aluminum hydride in 78% yield [*950*], and with *zinc* in acetic acid in 77–83% yields [*951*]. However, reduction of *N*-nitrosodiphenylamine with lithium aluminum hydride using the "inverse technique" yields 74% diphenylamine. When 1 mol of *N*-nitrosodiphenylamine is added to 2.6 mol of lithium aluminum hydride, cleavage of the nitrogen–nitrogen bond takes place, giving diphenylamine [*950*]. A 95% yield of *N,N*-diphenylhydrazine is obtained on reduction of *N*-nitrosodiphenylamine by *titanium dichloride*, prepared in situ from titanium tetrachloride and magnesium at room temperature [*281*].

N-nitramines, too, are reduced to disubstituted hydrazines. *Electrolysis* in 10% sulfuric acid over copper or lead cathodes reduces *N*-nitrodimethylamine to *N,N*-dimethylhydrazine (69% yield), *N*-nitro-*N*-methylaniline to *N*-methyl-*N*-phenylhydrazine (54% yield), *N*-nitropiperidine to *N,N*-pentamethylenehydrazine (52% yield) [*952*] (equation **151**), and nitrourea to semicarbazide (61–69% yield) [*953*].

151

| [*949*] | 75% | | 52% | [*952*] |

N-**Amine oxides** can be reduced (*deoxygenated*) to tertiary amines. Such a reaction is very desirable, especially in aromatic nitrogen-containing heterocycles in which conversion to amine oxides makes possible electrophilic substitution of the aromatic rings in different positions than in the parent heterocyclic compounds. The reduction is very easy and is accomplished by *catalytic hydrogenation* over palladium [*954, 955*], by hydrogen transfer from *sodium hypophosphite* in the presence of 5% palladium on carbon (83–90% yields) [*956*], and by hydrogen transfer from *ammonium formate* in the presence of 10% palladium on carbon (85–98% yields) [*957*]. With the exception of *borane* [*958*], the reducing agents *alane, complex hydrides, electroreduction,* and *sodium* not only deoxygenate the pyridine oxide but also reduce the pyridine ring [*959*]. More selective are reductions with *iron* in acetic acid (80–100%

yields) [247], with *titanium trichloride* (71–97% yields) [278], *titanium dichloride* (75–98% yields) [960], *chromous chloride* (39–96% yields) [269], *sulfur dioxide* in water or dioxane (21–78% yields) [961], *sulfur dioxide–trialkylamine complex* (60–81% yields) [962], *carbon disulfide* (69–92% yields) [963], *formic–acetic anhydride* (69–100% yields) [964], or *trimethyl phosphite* (95–100% yields) [408] (equations **152** and **153**). *Trimethyl* and *triethyl phosphites* also deoxygenate *nitrile oxides* to nitriles in 80–98% yields [965].

152

153

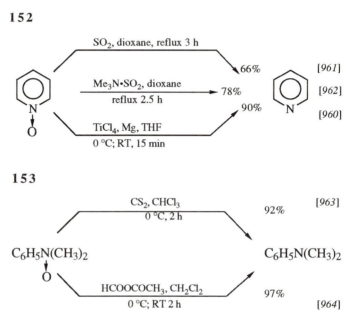

In pyridine oxides containing unsaturated side chains or aromatic chlorine, hydrogenation over palladium saturates the carbon–carbon double bonds and hydrogenolyzes chlorine prior to the reduction of the amine oxide to the amine [966].

In **nitroamine oxides** the nitro group may be reduced preferentially, but usually both functions are affected. 5-Ethyl-2-methyl-4-nitropyridine *N*-oxide is converted quantitatively to 4-amino-5-ethyl-2-methylpyridine by *hydrogen over 30% palladium* on charcoal in ethanolic solution [955]. The outcome of the hydrogenation of a nitroamine oxide may be influenced by reaction conditions [954] (equation **154**).

Iron in acetic acid at 100 °C reduces 4-nitropyridine oxide quantitatively to 4-aminopyridine [247] and 3-bromo-4-nitropyridine oxide to 3-bromo-4-amino-pyridine in an 80% yield [247]. Deoxygenation of nitropyridine oxides, nitroquinoline oxides, and pyrazine oxides with *chromous chloride* takes place in 41–96% yields [269].

A bond between nitrogen and oxygen in **alkyl hydroxylamines** is hydrogenolyzed by *microorganisms*. Chiral 3-hydroxylamino-2,4,4-trimethyl-1-pentene is reduced to the corresponding amine in 81–86% yields with 50–86% enantiomeric excesses with *Clostridium thermoaceticum* [480]. In a cyclic derivative of hydroxylamine, 2-methyl-6-[β-pyridyl]-1,2-oxazine, treatment with *zinc* dust in acetic acid gives, upon ring cleavage, 84.5% 1-hydroxy-4-methyl-amino-1-[β-pyridyl]butane [*967*].

154   [*954*]

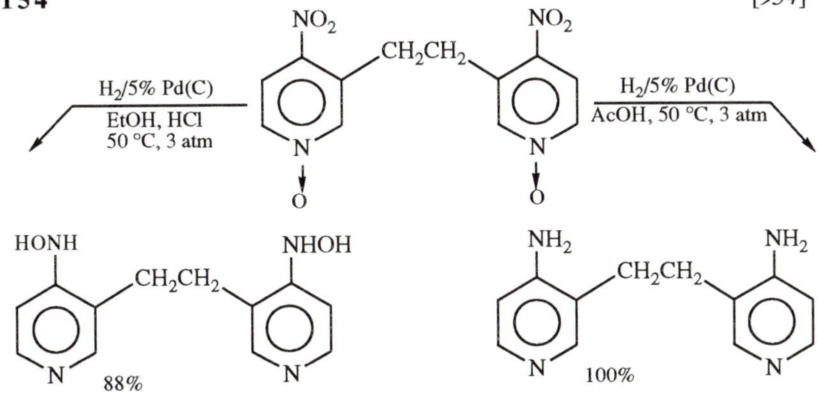

Hydrazo compounds are hydrogenolyzed to primary amines by *catalytic hydrogenation* over palladium [*968*]. Hydrazo compounds are intermediates in the reductive cleavage of azo compounds to amines, so it is very likely that all the reducing agents converting azo compounds to amines also cleave the hydrazo compounds.

Hydrazines are hydrogenolyzed *catalytically over Raney nickel* [*969*] and other catalysts. Nitrogen–nitrogen bonds in tertiary symmetrical hydrazines are cleaved by *sodium* in liquid ammonia or, better still, by Raney nickel, *zinc*, or *aluminum amalgam* [*970*] (equation **155**). Hydrogenolysis of hydrazones and azines is discussed elsewhere (pp 186).

155   [*970*]

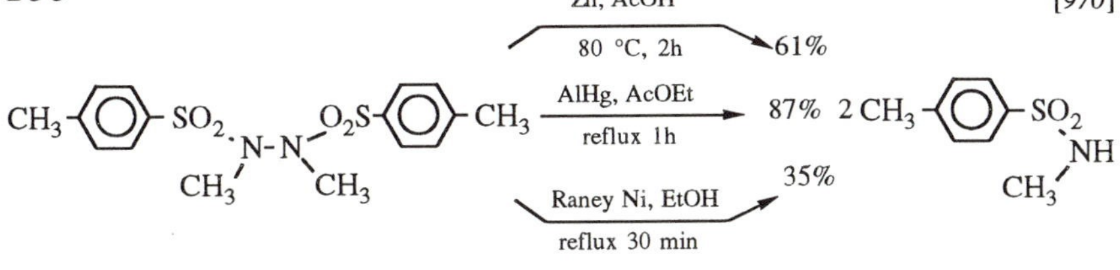

Azo compounds can be reduced either to hydrazo compounds or else cleaved to two molecules of amines.

Conversion of azo compounds to hydrazo compounds is achieved by controlled *catalytic hydrogenation* [*751, 968, 971*], by treatment with tributylstannane (74–92% yields) [*972*], with *diimide* (generated in situ from dipotassium azodicarboxylate and acetic acid) [*364, 973*], by *sodium amalgam* (90% yield) [*974*], by *aluminum amalgam* [*222*], and by *zinc* in alcoholic ammonia (70–85% yield) [*975*]. *Sodium hydrogen sulfide* has been used for selective reduction of a nitro group to an amino group with conservation of both azo groupings in mono-*p*-nitro-*m*-phenylenebisazobenzene (79% yields) [*333*]. This reaction is unusual because hydrogen sulfide, under slightly different conditions, reduces the azo group to the hydrazo group [*976*] as well as azo compounds to primary amines [*340*] (equation **156**).

The **reductive cleavage of azo compounds to amines** is accomplished in good yields by *catalytic hydrogenation* [*968, 977*], with *potassium borohydride and cuprous chloride* in 100% yields [*752*], *sodium hydrosulfite* [*356*], and O,O-*dialkyl dithiophosphoric acid* (23–93% yields) [*978*].

156

$$C_6H_5N=NC_6H_5 \longrightarrow C_6H_5NHNHC_6H_5 + 2\ C_6H_5NH_2$$

[*971*]	1 H$_2$/(py)$_3$RhCl$_3$-NaBH$_4$, 25 °C, 1 atm		+
[*968*]	1 H$_2$/Pd, 25 °C, 1 atm, 5 min		+
[*751*]	1 H$_2$/2% Pd(C), 83%, EtOH, 25 °C	100%	
[*972*]	Bu$_3$SnH, C$_6$H$_6$, reflux 4 h	92%	
[*364*]	KO$_2$CN=NCO$_2$K, DMSO, 25 °C, 20 h	75%	
[*222*]	AlHg, EtOH, 100 °C, few min	100%	
[*968*]	H$_2$/Pd, 25 °C, 1 atm, 4.5 h		+
[*971*]	H$_2$/(py)$_3$RhCl$_3$-NaBH$_4$, 25 °C, 1 atm		+
[*752*]	KBH$_4$/Cu$_2$Cl$_2$, MeOH, RT, 15 h		100%
[*978*]	(EtO)$_2$P(S)SH (10 eq.), 70 °C, 24 h		40%

Azoxy compounds are cleaved to amines by *potassium borohydride* and *cuprous chloride* in quantitative yields [*752*].

Alkyl isothiocyanates are reduced to *N*-alkylthioformamides with *tributylstannane* in ether at 25 °C in 10–89% yields [*979*].

Reduction of **diazo and azido compounds** is discussed on pp 99 and 100.

PHOSPHORUS COMPOUNDS

Phosphine oxides are reduced by *phenylsilane* and by *samarium diiodide*. Triphenylphosphine oxide, the most common byproduct of the Wittig reaction, is reduced to triphenylphosphine in a 75% yield by heating for 16 h at 65 °C with *samarium diiodide* in tetrahydrofuran in the presence of hexamethylphosphoric triamide [*309*].

Phenylsilane reduces phosphine oxides to phosphines in 85–96% yields on heating for 1 h at 80–100 °C. The chiral methylphenylpropylphosphine oxide is thus converted in 96% yield to methylphenylpropylphosphine with complete retention of configuration [*164*].

Chlorides of diesters of phosphoric acid and of **monoesters of alkane- and arenephosphonic acids** are reduced with *sodium borohydride* in refluxing dioxane to derivatives of phosphorous acid in 53–99% yields [*980*] (equations **157** and **158**).

157

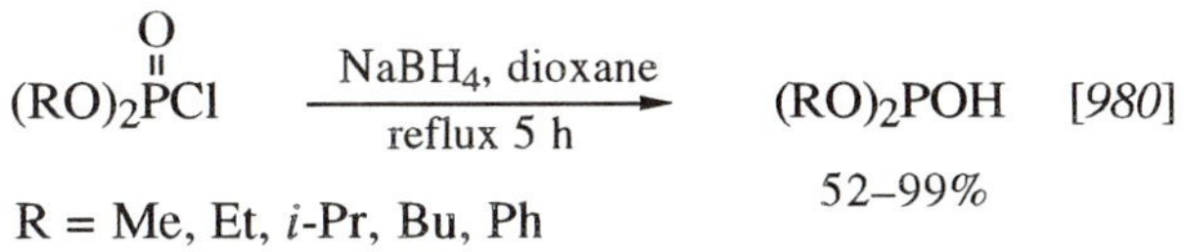

$$(RO)_2\overset{\overset{\displaystyle O}{\|}}{P}Cl \xrightarrow[\text{reflux 5 h}]{\text{NaBH}_4,\ \text{dioxane}} (RO)_2POH \quad [980]$$

R = Me, Et, *i*-Pr, Bu, Ph   52–99%

158

$$R-\overset{\overset{\displaystyle O}{\|}}{\underset{\underset{\displaystyle OR'}{|}}{P}}Cl \xrightarrow[\text{reflux 5 h}]{\text{NaBH}_4,\ \text{dioxane}} R-P\overset{OR'}{\underset{OH}{\diagdown}} \quad [980]$$

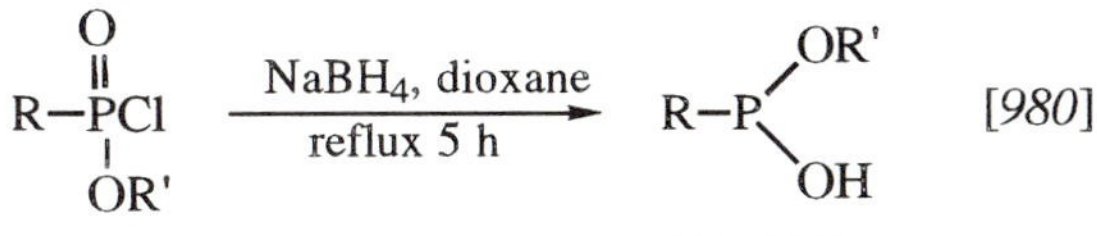

R = Me, Ph; R' = Et, *i*-Pr   76–81%

SUMMARY

Carbon–nitrogen bonds in saturated and aromatic amines resist reduction. Cleavage occurs only in allylic and benzylic amines that are cleaved to amines and the nitrogen-free residues by catalytic hydrogenation and with sodium, and in quaternary ammonium salts that are cleaved by electroreduction or sodium amalgam. Amine oxides are deoxygenated by catalytic hydrogenation and by reduction with metals and metal salts. Azo compounds are converted to hydrazo compounds by catalytic hydrogenation or metal amalgam reduction, and to amines by catalytic hydrogenation and sodium hydrosulfite. Hydrazo compounds and hydrazines are cleaved to amines by catalytic hydrogenation.

Phosphine oxides are deoxygenated by phenylsilane or by samarium diiodide, and ester–chlorides of phosphonic acids are deoxygenated to phosphinic esters by sodium borohydride.

CHAPTER 12

Reduction of Aldehydes and Their Derivatives

Reduction of aldehydes to primary alcohols is very easy and can be accomplished by a legion of reagents. Most of them reduce ketones as well, although at a slower rate. Some reagents, for example, *tetrabutylammonium triacetoxyborohydride* [*132*], *potassium triphenylborohydride* [*136*], *9-borabicyclo[3.3.1]nonane–pyridine* complex [*981*], and *samarium diiodide* [*304*] reduce predominantly only aldehydes. The real challenge is to reduce aldehydes containing other functional groups selectively or to reduce the other function in preference to the aldehyde group (p 176).

ALIPHATIC ALDEHYDES

Saturated aliphatic aldehydes are readily reduced to alcohols by *catalytic hydrogenation* over almost any catalyst, for example, platinum oxide, especially in the presence of iron ions, which accelerate the reduction of the carbonyl group [*982*], and Raney nickel (after washing away strong alkalinity, which could cause side reactions) [*46*]. High yields of primary alcohols are obtained by reductions with *lithium aluminum hydride* [*110*] (Procedure 14, p 302), *lithium borohydride* [*983*], and *sodium borohydride* [*984*], as well as with more sophisticated complex hydrides such as *lithium trialkoxyaluminum hydrides* [*985*], *sodium bis(2-methoxyethoxy)aluminum hydride* [*705*], *tetrabutyl-ammonium cyanoborohydride* [*139*], and *tributylstannane* [*172, 986, 987*]. *B*-3-Pinanyl-9-borabicyclo[3.3.1]nonane (prepared from α-pinene and 9-borabicyclo[3.3.1]nonane or 9-BBN) is suitable for asymmetric reduction of 1-deuteroaldehydes to chiral 1-deuteroalcohols (enantiomeric excess of 64–83%) [*150*]. Aliphatic aldehydes are also reduced by treatment with a suspension of *sodium* or *lithium hydride* in a solution of ferrous or ferric chloride in tetrahydrofuran [*988*], by *zinc and nickel dichloride* in methanol [*241*], by *iron* and acetic acid [*989*], by refluxing with *sodium dithionite* (hydrosulfite) in aqueous dioxane or *N,N*-dimethylformamide [*361*], and by the *Meerwein–Ponndorf* reaction using isopropyl alcohol and aluminum isopropoxide [*430*] or alumina [*990*] (Procedure 58, p 319) (equation **159**).

Several reagents reduce aldehydes in preference to ketones in mixtures of both (p 176).

Reduction of saturated aliphatic aldehydes to alkanes is carried out by refluxing with amalgamated *zinc* and hydrochloric acid (the *Clemmensen reduction*) [*236, 991*] (p 142) or by heating with *hydrazine* and potassium hydroxide (the *Wolff–Kizhner reduction*) [*384, 992*] (p 142). Heptaldehyde gives heptane in a 72% yield by the Clemmensen reduction and in a 54% yield by the Wolff–Kizhner reduction.

Other possibilities of **converting the aldehyde group to a methyl group** are *desulfurization of the mercaptal* (p 145) and reduction of azines, hydrazones, and tosylhydrazones (p 147 and 148).

An interesting reduction of aldehydes takes place on treatment with a reagent prepared from *titanium trichloride and potassium* [*286*] *or magnesium*

159

RCHO			RCH$_2$OH
	R=		
[705]	C$_3$H$_7$	NaAlH$_2$(OCH$_2$CH$_2$OMe)$_2$, C$_6$H$_6$, 30–80 °C	97%
[984]	C$_3$H$_7$	NaBH$_4$, H$_2$O	85%
[46]	C$_5$H$_{11}$	H$_2$/Raney Ni(W6), 25 °C, 2 h	~100%
[985]	C$_5$H$_{11}$	LiAlH(OCMe$_3$)$_3$, THF, –78 °C or 0 °C	≥90%
[361]	C$_5$H$_{11}$	Na$_2$S$_2$O$_4$, H$_2$O, reflux 4 h	63%
[50]	C$_6$H$_{13}$	H$_2$/Ni, EtOH, Et$_3$N, 25 °C, 1 atm, 14 h	88%
[110]	C$_6$H$_{13}$	LiAlH$_4$, Et$_2$O, reflux	86%
[983]	C$_6$H$_{13}$	LiBH$_4$, Et$_2$O	83%
[172]	C$_6$H$_{13}$	Bu$_3$SnH, TiCl$_3$, Zn/Cu, C$_6$H$_6$, reflux	64%
[989]	C$_6$H$_{13}$	Fe, AcOH, H$_2$O, 100 °C, 6–7 h	75-81%
[987]	C$_7$H$_{15}$	SiO$_2$, Bu$_3$SnH, hexane, 25 °C, 1 h	90%
[241]	C$_7$H$_{15}$	Zn, NiCl$_2$, MeOH, reflux 5 h	73%
[988]	C$_7$H$_{15}$	FeCl$_3$, LiH, THF, 25 °C, 24–48 h	79%
[139]	C$_8$H$_{17}$	Bu$_4$NBH$_3$CN, HMPA, 25 °C, 1 h	84%
[990]	C$_9$H$_{19}$	Al$_2$O$_3$, Me$_2$CHOH, CCl$_4$, 25 °C, 2 h	84%

[287]. In tetrahydrofuran, propionaldehyde gives a 60% yield of a mixture of 30% cis- and 30% trans-3-hexene [287].

A method for the **conversion of unsaturated aliphatic aldehydes to** *saturated aldehydes* is a gentle *catalytic hydrogenation.* Palladium is more selective than nickel. Hydrogenation over palladium reduced with sodium borohydride in methanol at room temperature and 2 atm reduces crotonaldehyde to butyraldehyde but does not hydrogenate butyraldehyde [34]. Nickel prepared by reduction with sodium borohydride is less selective: It effects reduction of crotonaldehyde to butyraldehyde but also reduction of butyraldehyde to butyl alcohol, though at a slower rate [34]. Hydrogenation of 2.2-dimethyl-4-pentenal over 5% palladium on alumina at 25 °C and 2 atm gives 88.5% 2,2-dimethylpentanal, whereas over Raney nickel at 125 °C and 100 atm, 2,2-dimethylpentanol is obtained in a 94% yield [993]. 3-Cyclohexenecarbox–aldehyde affords an 81% yield of cyclohexanecarboxaldehyde on hydrogenation over 5% palladium on charcoal at 75–80 °C and 14 atm [994]. Reduction using cobalt hydrocarbonyl HCo(CO)$_4$ at 25 °C and 1 atm converts crotonaldehyde to butyraldehyde (80% yield) with only a trace of butyl alcohol [995]. A very selective reduction of a double bond in an unsaturated aldehyde is achieved by *homogeneous catalytic hydrogenation* (Procedure 10, p 300) and by *catalytic transfer of hydrogen* using 10% palladium on charcoal and triethylammonium formate (equation **160**). In citral, only the α,β-double bond is reduced, yielding 91% citronellal (3,7-dimethyl-6-octenal) [439].

Unsaturated aliphatic aldehydes are selectively *reduced to unsaturated alcohols* by specially controlled *catalytic hydrogenation.* Citral treated with hydrogen over platinum dioxide in the presence of ferrous chloride or sulfate

160

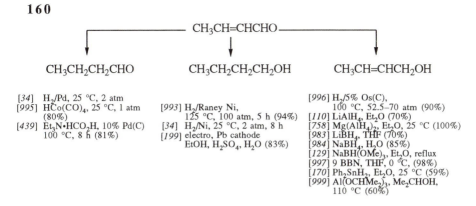

CH₃CH=CHCHO

CH₃CH₂CH₂CHO   CH₃CH₂CH₂CH₂OH   CH₃CH=CHCH₂OH

[*34*] H₂/Pd, 25 °C, 2 atm
[*995*] HĊo(CO)₄, 25 °C, 1 atm
       (80%)
[*439*] Et₃N•HCO₂H, 10% Pd(C)
       100 °C, 8 h (81%)

[*993*] H₂/Raney Ni,
       125 °C, 100 atm, 5 h (94%)
[*34*] H₂/Ni, 25 °C, 2 atm, 8 h
[*199*] electro, Pb cathode
       EtOH, H₂SO₄, H₂O (83%)

[*996*] H₂/5% Os(C),
       100 °C, 52.5–70 atm (90%)
[*110*] LiAlH₄, Et₂O (70%)
[*758*] Mg(AlH₄)₂, Et₂O, 25 °C (100%)
[*983*] LiBH₄, THF (70%)
[*984*] NaBH₄, H₂O (85%)
[*129*] NaBH(OMe)₃, Et₂O, reflux
[*997*] 9 BBN, THF, 0 °C, (98%)
[*170*] Ph₂SnH₂, Et₂O, 25 °C (59%)
[*999*] Al(OCHMe₂)₃, Me₂CHOH,
       110 °C (60%)

and zinc acetate at room temperature and 3.5 atm is reduced only at the carbonyl group and gives geraniol (3,7-dimethyl-2,6-octadienol) [*59*], and crotonaldehyde on hydrogenation over 5% osmium on charcoal yields crotyl alcohol [*996*].

Complex hydrides can be used for the selective reduction of the carbonyl group, although some of them, especially lithium aluminum hydride, may reduce the α,β-conjugated double bond as well. Crotonaldehyde is converted to crotyl alcohol by reduction with *lithium aluminum hydride* [*110*], *magnesium aluminum hydride* [*758*], *lithium borohydride* [*983*], *sodium borohydride* [*984*], *sodium trimethoxyborohydride* [*129*], *diphenylstannane* [*170*], and *9-borabicyclo[3.3.1]nonane* [*997*]. *Sodium borohydride* in the presence of *cerium trichloride hexahydrate* or *erbium trichloride* in 40% aqueous solution of ethanol at –15 °C reduces conjugated unsaturated aldehydes to unsaturated alcohols in preference to nonconjugated aldehydes in ratios of 8:1 [*998*].

A dependable way to convert α,β-unsaturated aldehydes to unsaturated alcohols is the *Meerwein–Ponndorf reduction* [*999*] (equation **160**).

Reduction of unsaturated aliphatic aldehydes *to saturated alcohols* is effected by catalytic hydrogenation over nickel catalysts [*34, 993*] and by electroreduction [*199*] (equation **160**).

AROMATIC ALDEHYDES

Reduction of aromatic aldehydes can give three kinds of product: primary alcohols of benzylic type, methylated aromatic compounds, or methylated alicyclic (or heterocyclic) compounds with fully hydrogenated rings. Benzyl-type alcohols are easily hydrogenolyzed, and thus reduction to the alcohol sometimes requires careful control or a specially chosen reducing agent to prevent reduction of the aldehyde group to a methyl group. Total reduction requires rather intensive hydrogenation and occurs only after the hydrogenolysis of the carbon–oxygen bond.

Conversion of aromatic aldehydes to alcohols can be accomplished in a variety of ways. *Catalytic hydrogenation* over almost any catalyst gives good yields of alcohols provided it is monitored to prevent deeper hydrogenation. Platinum and palladium catalysts tend to hydrogenolyze the alcohols and even to hydrogenate the rings if used in acidic media. Some reviewers of hydrogenations consider platinum better than palladium, but others have found the opposite. At any rate, good yields of alcohols are obtained with either catalyst, especially when the hydrogenation is stopped after absorption of the amount of hydrogen required for the hydrogenation to the alcohol [33, 982, 1000, 1001]. Addition of ferrous chloride in hydrogenations using platinum oxide is advantageous [982].

Nickel, Raney nickel, and copper chromite are other catalysts suitable for hydrogenation of aldehydes to alcohols with little if any further hydrogenolysis. Benzaldehyde is hydrogenated to benzyl alcohol over nickel [44], Raney nickel [46], and copper chromite [51] in excellent yields. With copper chromite, the benzyl alcohol is accompanied by 8% toluene [51]. Reduction of benzaldehyde by *hydrogen transfer* using formic acid and copper gives 56% benzyl alcohol and 18% toluene [72]. Salicylaldehyde is reduced by nascent hydrogen generated by dissolving *Raney nickel alloy* (50% nickel and 50% aluminum) in 10% aqueous sodium hydroxide. At a temperature of 10–20 °C the product is saligenin (salicyl alcohol) (77% yield), whereas at 90 °C a 76% yield of *o*-cresol is obtained [1002] (equation 161).

161

$$C_6H_5CHO \xrightarrow{\hspace{4cm}} C_6H_5CH_2OH \quad \text{or} \quad C_6H_5CH_3$$

[1000]	H$_2$/Pt, EtOH, 20 °C, 2 atm, 5 h	(~100%)	
[1000]	H$_2$/Pt, AcOH, 20 °C, 2 atm, 10 h		80%
[50]	H$_2$/Ni, EtOH, 25 °C, 1 atm, 1.6 h	91%	
[72]	HCO$_2$H/Cu, 200 °C	56%	18%
[988]	NaH/FeCl$_3$, THF, 25 °C, 24 h	85%	
[110]	LiAlH$_4$, Et$_2$O, exothermic	85%	
[1003]	LiAlH$_4$/AlCl$_3$, Et$_2$O, reflux 30 min	60%	
[983]	LiBH$_4$, Et$_2$O, exothermic	91%	
[129]	NaBH(OMe)$_3$, Et$_2$O, reflux 4 h	78%	
[130]	NaBH$_3$NMe$_2$, THF, RT, 30 min	90%	
[1004]	Bu$_4$NBH$_4$, CH$_2$Cl$_2$, 0→25 °C, 24 h	91%	
[150]	*B*-3-Pinanyl-9-BBN, enantiomeric excess	70%	
[1006]	Et$_3$SiH/BF$_3$, CH$_2$Cl$_2$, 0 °C, 11 min		52%
[987]	Bu$_3$SnH, SiO$_2$, cyclohexane	81%	
[986]	Bu$_3$SnH, HMPA, 60 °C, 2 h	88%	
[172]	Bu$_3$SnH, TiCl$_3$, Zn/Cu, C$_6$H$_6$, reflux	45%	
[170]	Ph$_2$SnH$_2$, Et$_2$O, 25 °C	62%	
[1010]	Li/NH$_3$, THF, *t*-BuOH or NH$_4$Cl		90%
[241]	Zn, NiCl$_2$, MeOH, reflux 1 h	76%	
[361]	Na$_2$S$_2$O$_4$, H$_2$O, dioxane	84%	
[1012]	N$_2$H$_4$, KOH, 80–100 °C		79%
[990]	Al$_2$O$_3$, Me$_2$CHOH, CCl$_4$, 25 °C, 2.5 h	77%	
[999]	Al(OCHMe$_2$)$_3$, Me$_2$CHOH, reflux	55%	

Chemical reduction of aromatic aldehydes to alcohols is accomplished with *lithium aluminum hydride* [110], *alane* [1003], *lithium borohydride* [983], *sodium borohydride* [984], *sodium trimethoxyborohydride* [129], *sodium dimethylaminoborohydride* [*1 3 0*], *tetrabutylammonium borohydride* [*1004*], *tetrabutylammonium cyanoborohydride* [*139*], B-*3-pinanyl-9-borabicyclo-*[*3.3.1*]*nonane* [*150*], *zinc borohydride* [*1005*], *tributylstannane* [*172, 986, 987*], *diphenylstannane* [*170*], *zinc and nickel dichloride* [*241*], *sodium dithionite* [*361*], *isopropyl alcohol* [*990*], *formaldehyde* (crossed Cannizzaro reaction) [*434*], and others (equation **161**).

Selective reduction of aromatic aldehydes in the presence of saturated or nonconjugated aldehydes is achieved by *sodium borohydride* in the presence of *chromium trichloride, cerium trichloride,* or *erbium trichloride* in aqueous ethanol with the selectivity of 8–100 :1 [*998*].

The chiral reagent prepared in situ from (+)-α-pinene and 9-borabicyclo-[3.3.1]nonane reduces benzaldehyde-1-*d* to benzyl-1-*d* alcohol in 81.6% yield and 95% enantiomeric excess [*150*] (equation **162**).

162   [*150*]

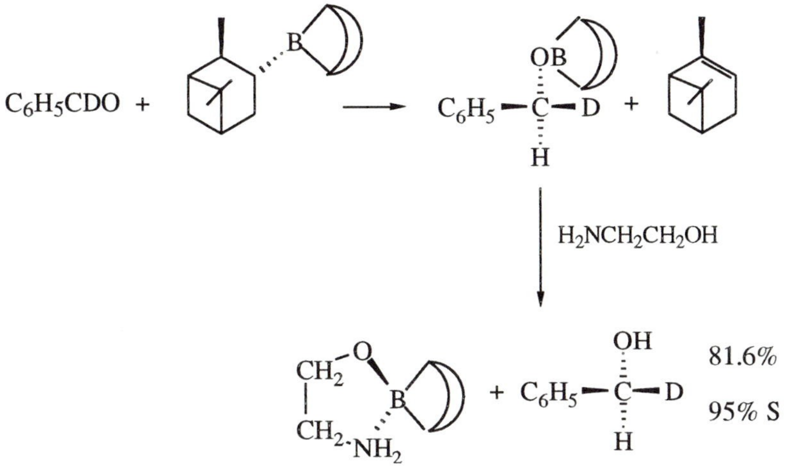

Most of these reagents do not reduce aromatic aldehydes beyond the stage of alcohols. Reagents that reduce via carbocation intermediates, *alane* [*1003*] and *triethylsilane* with boron trifluoride [*1006*], for example, tend to reduce the carbonyl group to the methyl group, especially in aldehydes containing electron-releasing groups that stabilize the carbonium intermediates. On the contrary, electron-withdrawing substituents favor reduction to alcohols only (equation **163**).

Reduction of the carbonyl group in aromatic aldehydes *to the methyl group* is achieved by *catalytic hydrogenation* over palladium in acetic acid (86% yield) [*1007*] or by intensive hydrogenation over nickel [*44*] or copper chromite [*1008*]. Fluorene-1-carboxaldehyde gives 86% 1-methylfluorene on hydrogenation over palladium on charcoal in acetic acid at room temperature and atmospheric pressure [*1007*], and fural yields 90–95% α-methylfuran over copper chromite at 200–230 °C at atmospheric pressure [*1008*].

Triethylsilane in the presence of boron trifluoride [*1006*] or trifluoroacetic acid [*1009*] also reduces the aldehyde group to a methyl group.

163

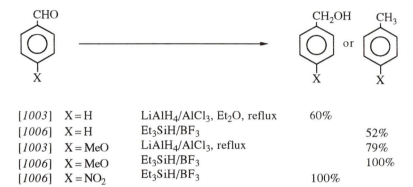

[1003]	X = H	LiAlH$_4$/AlCl$_3$, Et$_2$O, reflux	60%	
[1006]	X = H	Et$_3$SiH/BF$_3$		52%
[1003]	X = MeO	LiAlH$_4$/AlCl$_3$, reflux	79%	
[1006]	X = MeO	Et$_3$SiH/BF$_3$		100%
[1006]	X = NO$_2$	Et$_3$SiH/BF$_3$	100%	

Electrolytic reduction using a lead cathode in 20% sulfuric acid converts pyridine α-carboxaldehyde to a mixture of 41% α-picoline, 25% α-pipecoline, and 11% 2-methyl-1,2,3,6-tetrahydropyridine [591].

Reduction of benzaldehyde and *p*-alkylbenzaldehydes to the corresponding hydrocarbons is carried out by *lithium* in liquid ammonia and tetrahydrofuran in the presence of *tert*-butyl alcohol or ammonium chloride (90–94% yields) [1010].

Like any aldehydes, aromatic aldehydes undergo *Clemmensen reduction* [991, 1011] and *Wolff–Kizhner* reduction [992, 1012] and give the corresponding methyl compounds, generally in good yields. The same effect is accomplished by conversion of the aldehydes to *p*-toluenesulfonylhydrazones followed by reduction with *lithium aluminum hydride* (p 148).

Reduction of aromatic aldehydes to pinacols using *sodium amalgam* is quite rare. Equally rare is conversion of aromatic aldehydes *to alkenes* formed by deoxygenation and coupling and accomplished by treatment of the aldehyde with a reagent obtained by reduction of *titanium trichloride with lithium* in 1,2-dimethoxyethane. Benzaldehyde thus affords *trans*-stilbene in a 97% yield [286, 289].

Hydrogenation of the nuclei in aromatic aldehydes is possible by *catalytic hydrogenation* over noble metals in acetic acid and is rather slow [1000].

A typical example of an **unsaturated aromatic aldehyde** is cinnamaldehyde. Hardly any other aldehyde has become so extensively employed for testing new reagents for their selectivity. The α,β-double bond is conjugated not only with the aldehyde carbonyl group but also with the aromatic ring, which makes it readily reducible. Consequently, the development of a selective method for reducing cinnamaldehyde to hydrocinnamaldehyde or to cinnamyl alcohol is a great challenge.

Reduction of only the double bond is achieved by *catalytic hydrogenation* over palladium prepared by reduction with sodium borohydride. This catalyst does not promote hydrogenation of the aldehyde group [34]. Thus nickel reduced with sodium borohydride is used for conversion of cinnamaldehyde to hydrocinnamaldehyde [34]. *Homogeneous hydrogenation* over tris(triphenylphosphine)rhodium chloride gives 60% hydrocinnamaldehyde and 40% ethylbenzene [*91*]. Raney nickel, by contrast, catalyzes total reduction to hydrocinnamyl alcohol [*46*]. Total reduction of both the double bond and the carbonyl group is also accomplished by *electrolysis* [*199*] and by refluxing with *lithium aluminum hydride* in ether [*757, 805*] (equation **164**).

164

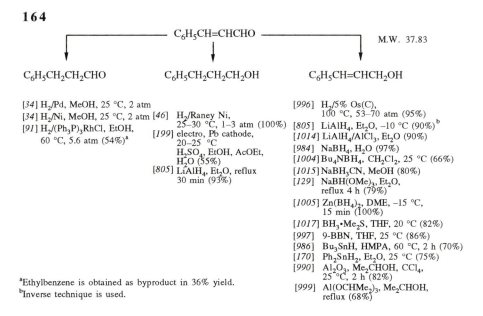

C₆H₅CH=CHCHO   M.W. 37.83

C₆H₅CH₂CH₂CHO   C₆H₅CH₂CH₂CH₂OH   C₆H₅CH=CHCH₂OH

[*34*] H₂/Pd, MeOH, 25 °C, 2 atm
[*34*] H₂/Ni, MeOH, 25 °C, 2 atm [*46*] H₂/Raney Ni,
[*91*] H₂/(Ph₃P)₃RhCl, EtOH,   25–30 °C, 1–3 atm (100%)
    60 °C, 5.6 atm (54%)ᵃ   [*199*] electro, Pb cathode,
                 20–25 °C
                 H₂SO₄, EtOH, AcOEt,
                 H₂O (55%)
       [*805*] LiAlH₄, Et₂O, reflux
                 30 min (93%)

[*996*] H₂/5% Os(C),
       100 °C, 53–70 atm (95%)
[*805*] LiAlH₄, Et₂O, –10 °C (90%)ᵇ
[*1014*] LiAlH₄/AlCl₃, Et₂O (90%)
[*984*] NaBH₄, H₂O (97%)
[*1004*] Bu₄NBH₄, CH₂Cl₂, 25 °C (66%)
[*1015*] NaBH₃CN, MeOH (80%)
[*129*] NaBH(OMe)₃, Et₂O,
       reflux 4 h (79%)
[*1005*] Zn(BH₄)₂, DME, –15 °C,
       15 min (100%)
[*1017*] BH₃·Me₂S, THF, 20 °C (82%)
[*997*] 9-BBN, THF, 25 °C (86%)
[*986*] Bu₃SnH, HMPA, 60 °C, 2 h (70%)
[*170*] Ph₂SnH₂, Et₂O, 25 °C (75%)
[*990*] Al₂O₃, Me₂CHOH, CCl₄,
       25 °C, 2 h (82%)
[*999*] Al(OCHMe₂)₃, Me₂CHOH,
       reflux (68%)

ᵃEthylbenzene is obtained as byproduct in 36% yield.
ᵇInverse technique is used.

The **acetylenic aromatic aldehyde,** tolane-4-carboxaldehyde, is reduced to the unsaturated aldehyde, *cis*-stilbene-4-carboxaldehyde, by treatment with formic acid, triethylamine, and palladium on charcoal [*78*].

Many more examples exist for **reduction of only the carbonyl.** Over an osmium catalyst [*996*] or platinum catalyst activated by zinc acetate and ferrous chloride [*1013*], cinnamaldehyde is *hydrogenated* to cinnamyl alcohol. The same product is obtained by gentle reduction with *lithium aluminum hydride* at –10 °C using the inverse technique [*805*] and by reduction with *alane* (prepared in situ from lithium aluminum hydride and aluminum chloride) [*1014*], *sodium borohydride* [*984*], *tetrabutylammonium borohydride* [*1004*], *sodium trimethoxyborohydride* [*129*], *sodium cyanoborohydride* [*1015*], *zinc borohydride* [*1016, 1005*], *9-borabicyclo[3.3.1]nonane* [*997*], *borane–dimethyl sulfide complex* [*1017*], *tributylstannane* [*986*], *diphenylstannane* [*170*], and dehydrated alumina soaked with *isopropyl alcohol* [*990*]. Reduction of furfurylidene acetaldehyde (3-(α-furyl)acrolein) with ethanol and aluminum ethoxide in xylene at 100 °C for 2 h gives 60–70% 3-(α-furyl)allyl alcohol [*1018*].

On rare occasions the aldehyde group in an α,β-unsaturated aromatic aldehyde is *reduced to a methyl group.*

Reduction of unsaturated aromatic aldehydes to unsaturated hydrocarbons poses a serious problem, especially if the double bond is conjugated with the benzene ring, the carbonyl, or both. 6-Benzyloxyindole-3-carboxaldehyde is transformed to 6-benzyloxy-3-methylindole by *sodium* *borohydride* in isopropyl alcohol in the presence of 10% *palladium* on charcoal (89% yield) [*1019*]. In the *Clemmensen reduction* the α,β-unsaturated double bond is usually reduced [*236*], and in the *Wolff–Kizhner reduction* a cyclopropane derivative may be generated as a result of decomposition of pyrazolines formed by intramolecular addition of the intermediate hydrazones across the double bonds [*384*]. The only way to convert unsaturated aromatic aldehydes to unsaturated hydrocarbons is the reaction of the aldehyde with *p*-toluenesulfonylhydrazide to form *p*-toluenesulfonylhydrazone and its subsequent reduction with sodium triacetoxyborohydride [*1020*] or bis(benzoyloxy)borane [*1021*]. The reduction is accompanied by a double-bond shift.

Conversion of unsaturated aromatic aldehydes to saturated hydrocarbons can be realized by the *Clemmensen reduction* [*236*].

DERIVATIVES OF ALDEHYDES

Halogenated aldehydes are usually reduced without the loss of the halogen [*1012*]. In many reductions the rate of reduction of the carbonyl group is higher than that of the hydrogenolysis of most carbon–halogen bonds. Use of *alkali metals* and *zinc* for the reduction may involve some risk, especially if a reactive halogen is bonded in the position α to a carbonyl [*236*]. In such cases even the *Meerwein–Ponndorf* reduction may lead to complications because of the reaction of the halogen with the alcoholic group in alkaline medium [*430*].

Halogenated unsaturated aldehydes such as 3-chloro-2-methyl-3-*p*-tolyl-propenal are reduced to saturated aldehydes, such as 2-methyl-3-*p*-tolylpropanal, by hydrogenation over 5% palladium on carbon in the presence of potassium carbonate at 50–60 °C and 1.5–3.5 atm (75% yield) [*1022*].

In **nitro aldehydes** both the nitro group and the aldehyde group are readily reduced by *catalytic hydrogenation.* It may be difficult, if not impossible, to hydrogenate either function separately. More dependable methods are reduction by *alane* [*1023*], *tributylstannane* in the presence of hexamethylphosphoramide [*986*], *sodium sulfide* on alumina [*1024*], or *isopropyl alcohol* and aluminum isopropoxide (*Meerwein–Ponndorf reduction*) [*1025*] *to nitro alcohols,* and by *stannous chloride* [*1026, 1027*], *titanium trichloride* [*772*], *sodium sulfide* [*1024*], *or ferrous sulfate* [*298*] *to amino aldehydes* (Procedure 43, p 314) (equation **165**).

Hydroxy aldehydes are reduced to ***hydroxy alcohols*** without complications. In benzyl ethers of phenol aldehydes, the benzyl group is hydrogenolyzed in preference to the reduction of the carbonyl by *catalytic hydrogenation* over 10% palladium on charcoal at room temperature and 2 atm to give phenol aldehyde [*1028*].

A peculiar reduction occurs on treatment of a dialdehyde, biphenyl-2,2'-di-carboxaldehyde, with tris(dimethylamino)phosphine at room temperature: Phenanthrene-9,10-oxide is formed in an 81–89% yield [*413*].

Acetals of aldehydes are usually stable to lithium aluminum hydride but are reduced to ethers with *alane* prepared in situ from lithium aluminum hydride and aluminum chloride in ether. Butyraldehyde diethyl acetal gives a 47% yield of butyl ethyl ether, and benzaldehyde dimethyl acetal and diethyl acetal afford

165

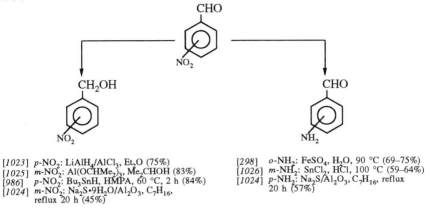

[1023] *p*-NO$_2$: LiAlH$_4$/AlCl$_3$, Et$_2$O (75%)
[1025] *m*-NO$_2$: Al(OCHMe$_2$)$_3$, Me$_2$CHOH (83%)
[986] *p*-NO$_2$: Bu$_3$SnH, HMPA, 60 °C, 2 h (84%)
[1024] *m*-NO$_2$: Na$_2$S·9H$_2$O/Al$_2$O$_3$, C$_7$H$_{16}$, reflux 20 h (45%)

[298] *o*-NH$_2$: FeSO$_4$, H$_2$O, 90 °C (69–75%)
[1026] *m*-NH$_2$: SnCl$_2$, HCl, 100 °C (59–64%)
[1024] *p*-NH$_2$: Na$_2$S/Al$_2$O$_3$, C$_7$H$_{16}$, reflux 20 h (57%)

benzyl methyl ether and benzyl ethyl ether in 88% and 73% yields, respectively [*1029*].

2-Tetrahydrofuranyl and 2-tetrahydropyranyl alkyl ethers, being acetals, resist lithium aluminum hydride but are reduced by *alane* prepared from lithium aluminum hydride and aluminum chloride or boron trifluoride etherate. Exocyclic reductive cleavage gives tetrahydrofuran or tetrahydropyran and alcohols (5–90% yields), whereas endocyclic reductive cleavage affords 4-hydroxybutyl or 5-hydroxypentyl alkyl ethers, respectively (10–79% yields). Which of the cleavages takes place depends on the structure of the alkyl groups and is affected both by electronic and steric effects. Use of aluminum chloride gives better yields with tertiary alkyl groups, whereas boron trifluoride etherate favors better yields with secondary and especially primary alkyl groups [*1030*] (equation **166**).

Sulfur analogs, 2-tetrahydrofuranyl and 2-tetrahydropyranyl **thioethers,** are reduced by *alane* to alkyl 4- or alkyl 5-hydroxyalkyl thioethers resulting from the preferential reductive cleavage of the carbon–oxygen (rather than the carbon–sulfur) bond. Thus refluxing for 2 h with alane in ether converts 2-alkylthiotetrahydrofurans to alkyl 4-hydroxybutyl thioethers in 63–72% yields, and 2-alkylthiotetrahydropyrans to alkyl 5-hydroxypentyl thioethers in 58–82% yields [*1031*] (equation **167**).

Cyclic five-membered **mono- and dithioacetals** are reduced with *calcium* in liquid ammonia to give β-alkoxy- and β-thioalkoxymercaptans, respectively: RCH$_2$OCH$_2$CH$_2$SH and RCH$_2$SCH$_2$CH$_2$SH. Benzaldehyde ethylene-dithioacetal, on the other hand, gives toluene (75%), and phenylacetaldehyde-ethylenedithioacetal affords ethylbenzene (71% yield) [*1032*] (equation **168**).

Mercaptals (dithioacetals) and cyclic mercaptals (prepared from aldehydes and 1,2-ethanedithiol or 1,2-propanedithiol) are easily **desulfurized** by refluxing with an excess of *Raney nickel* in ethanol and give *hydrocarbons* in good yields [*1033, 1034*]. If the molecule of the mercaptal contains a function reducible by hydrogen, Raney nickel must be refluxed for several hours in acetone prior to the desulfurization to remove the hydrogen adsorbed on its surface [*47*]. Desulfurization of mercaptals can also be achieved by heating of the mercaptal

166

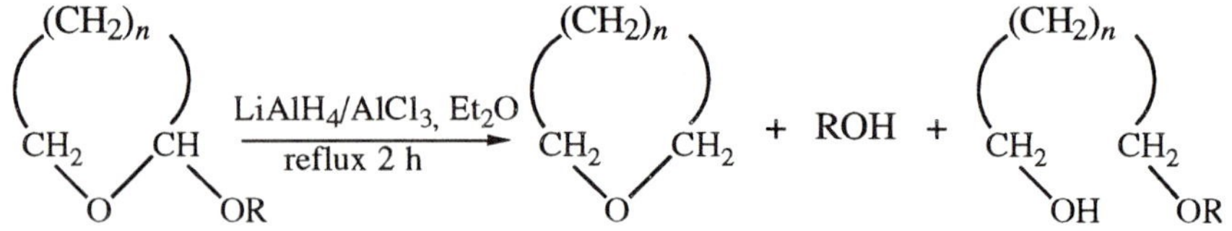

[1030] n = 2, R = Me$_3$C; AlCl$_3$ NOTE   58%
                            BF$_3$·Et$_2$O   16%
       R = C$_6$H$_{13}$;   AlCl$_3$ 27%
                            BF$_3$·Et$_2$O   66%
       n = 3, R = Me$_3$C; AlCl$_3$ 72%
                            BF$_3$·Et$_2$O   46%
       R = C$_6$H$_{13}$;   AlCl$_3$ 11%
                            BF$_3$·Et$_2$O   41%

Note. The column under ether and alcohols is blank because the yields of
ethers + alcohols are the balance in 65–100% recovery.

167   [1031]

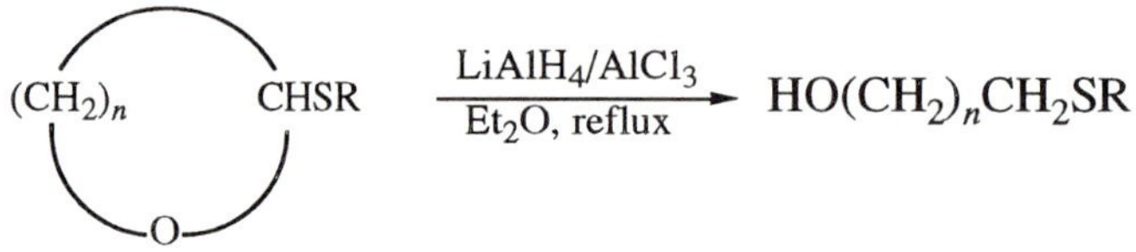

n = 3   63–72%
n = 4   58–82%

with 3–4 equivalents of *tributylstannane* and a catalytic amount of azobis(isobutyronitrile) for 1.5 h at 80 °C and distilling the mixture in vacuo [1035], or on treatment of the mercaptal with *sodium* in liquid ammonia [1036].

These reactions represent a gentle method for conversion of aldehydes to hydrocarbons in yields of 40–95% and are frequently used, especially in saccharide chemistry [1033, 1034, 1035] (equations **167** and **168**).

Cinnamaldehyde diacetate is reduced by heating at 100 °C with *iron* in 50% acetic acid to cinnamyl acetate in a 60.7% yield [1037].

2-Alkylaminotetrahydropyrans may be considered aminoacetals of aldehydes. They are reductively cleaved by *lithium aluminum hydride* to *amino alcohols*. Thus 2-(*N*-piperidyl)tetrahydropyran affords 5-piperidino-1-pentanol in an 82% yield [1038] (equation **169**).

Alkyl or aryl aldimines (Schiff's bases) give on reduction secondary amines. Reduction is carried out by *catalytic hydrogenation* over platinum [1039] and nickel in good yields [1040, 1041]; *lithium aluminum hydride* in refluxing ether or tetrahydrofuran [1042]; *sodium aluminum hydride, lithium borohydride,* and *sodium borohydride* [1042]; *sodium amalgam* [1043]; and *potassium–graphite* [1044], generally in fair to excellent yields. In the reduction

168

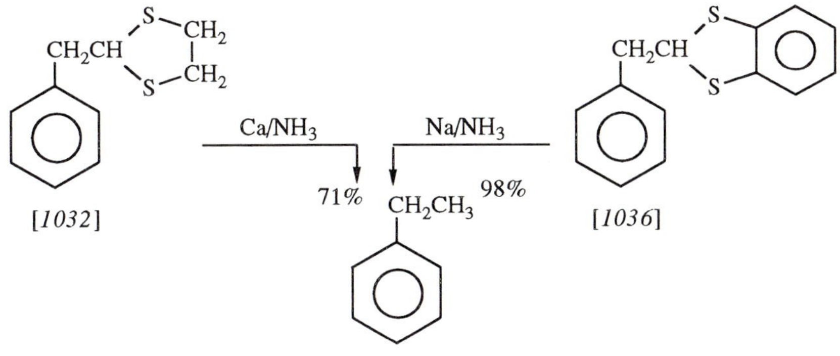

[*1032*]   Ca/NH$_3$ Na/NH$_3$ [*1036*]

71%   CH$_2$CH$_3$ 98%

169   [*1038*]

2 LiAlH$_4$, Et$_2$O
reflux 2 h   HO(CH$_2$)$_5$N⟨⟩

82%

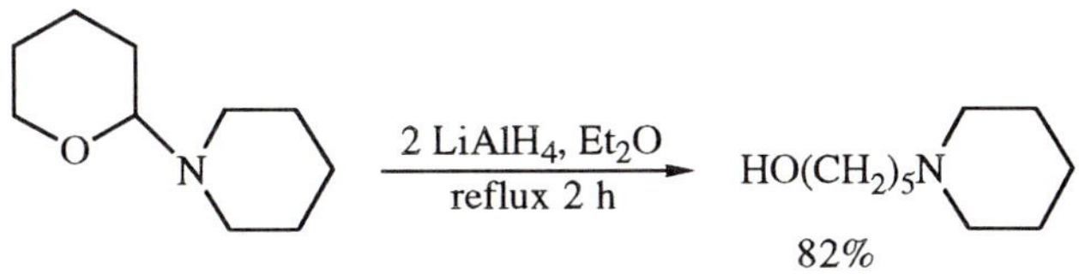

of *N*-cyclopropylbenzaldimine the energetic action of lithium aluminum hydride not only reduces the carbon–nitrogen double bond but also cleaves the three-membered ring and yields benzylpropylamine [*1042*] (equation **170**).

170   [*1042*]

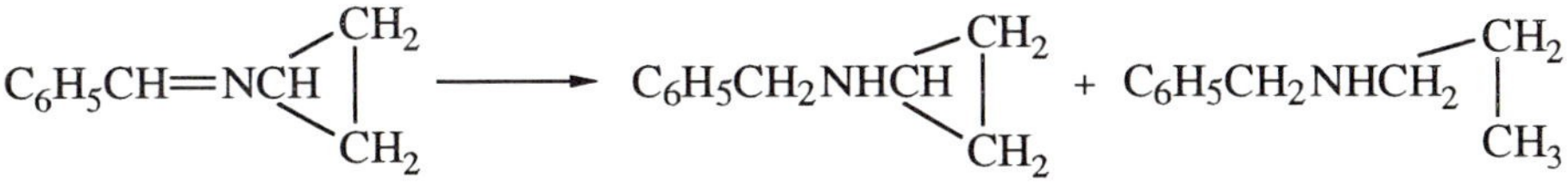

H$_2$/Pt, EtOH, 2 atm, 12 h	72%	
LiAlH$_4$, Et$_2$O, reflux 12 h		80%
LiAlH$_4$, Et$_2$O, reflux 4 h	59%	
NaAlH$_4$, THF, reflux 24 h	23.5%	23.5%
LiBH$_4$, THF, reflux 24 h	52%	
NaBH$_4$, THF, reflux 24 h	67%	

Aldimines derived from aromatic aldehydes suffer hydrogenolysis in *hydrogenation* over palladium at 117–120 °C at 20 atm and give products in which the original aldehyde group is transformed to methyl group (62–72% yields) [*1045*].

Aldoximes yield *primary amines by catalytic hydrogenation:* Benzaldehyde gives benzylamine in a 77% yield over nickel at 100 °C and 100 atm [*1040*], with *lithium aluminum hydride* (47–79% yields) [*1046*], with *sodium* in refluxing ethanol (60–73% yields) [*1047*], and with other reagents. **Hydrazones**

of aldehydes are intermediates in the Wolff–Kizhner reduction of the aldehyde group to a methyl group but are hardly ever reduced to amines.

p-**Toluenesulfonylhydrazones of aldehydes** prepared from aldehydes and *p*-toluenesulfonylhydrazide are reduced by *lithium aluminum hydride* [*1048, 1049*], *sodium borohydride* [*1020, 1048*], *sodium cyanoborohydride* [*1050*], or *bis(benzyloxy)borane* prepared in situ from benzoic acid and borane–tetrahydrofuran complex [*1021*]. The aldehyde group is converted to the methyl group. If sodium cyanoborohydride is used as the reducing agent, the tosylhydrazones may be prepared in situ because the cyanoborohydride does not reduce the aldehydes prior to their conversion to the tosylhydrazones. Thus this sequence represents a method for the reduction of aldehydes to hydrocarbons under very gentle reaction conditions. It is therefore suitable even for reduction of α,β-unsaturated aldehydes, which cannot be reduced by Clemmensen or Wolff–Kizhner reductions. Rearrangement of the double bond accompanies the reduction of tosylhydrazones of α,β-unsaturated aldehydes [*1020, 1021*] (equations **171** and **172**).

Selective reduction of keto groups in the presence of aldehydes is dealt with on p 177.

171

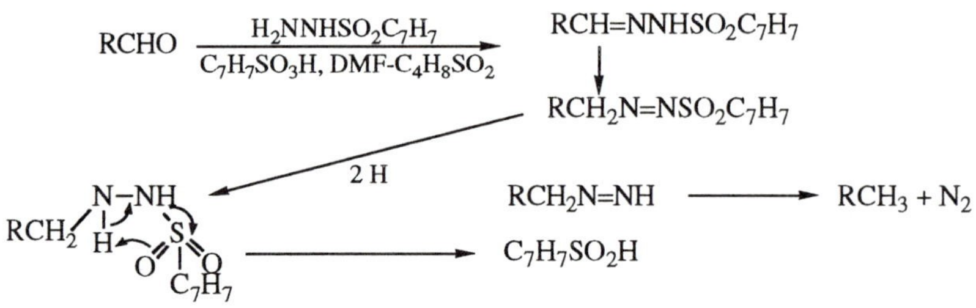

R=
[*1021*] C_9H_{19}; $BH_3 \cdot THF$, BzOH, $CHCl_3$, 0 °C, 30 min   78%
[*1050*] C_9H_{19}; $NaBH_3CN$, DMF–$C_4H_8SO_2$, 100–105 °C, 4 h   66%
[*1048*] $C_{11}H_{23}$; $NaBH_4$, MeOH   60–70%
[*1049*] β-naphthyl; $LiAlH_4$, THF, reflux 12 h   70%
[*1049*] β-indolyl; $LiAlH_4$, THF, reflux 12 h   35–40%

172

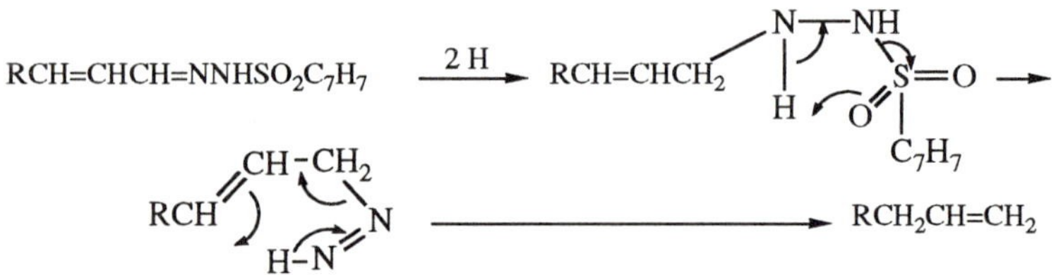

R=
[*1020*] CH_3; $NaBH_4$, AcOH, 25 °C, 1 h, 70 °C, 1.5 h   42–56%
[*1021*] C_6H_5; $BH_3 \cdot THF$, BzOH, $CHCl_3$, 0 °C, 1 h   85%

CHAPTER 13

Reduction of Ketones and Their Derivatives

The immense number of reductions performed on ketones dictates subdivision of this topic into reductions of aliphatic, alicyclic, aromatic, and unsaturated ketones. However, differences in behavior toward reduction are very small between aliphatic and alicyclic ketones, so the section on alicyclic ketones includes specific examples due only to the stereochemistry of the ring systems.

ALIPHATIC KETONES

Reduction of saturated ketones to alcohols is very easy by many means but is distinctly slower than that of comparable aldehydes. It is frequently possible to carry out selective reductions of the two types (p 176).

Alcohols are obtained by *catalytic hydrogenation*. Palladous oxide proves ineffective [*41, 1051*], but hydrogenation over platinum oxide and especially rhodium–platinum oxide [*41*] and over platinum, rhodium, and ruthenium is fast and quantitative at room temperature and atmospheric pressure [*41, 1051*]. Under comparable conditions Raney nickel and nickel prepared by reduction of nickel acetate with sodium hydride in the presence of sodium alkoxides [*50*] have also been used and give excellent yields. Normal nickel [*44*] as well as copper chromite [*51*] catalysts require high temperatures (100–150 °C) and high pressures (100–165 atm) and give almost quantitative yields. Moderate to good yields are obtained by reduction of ketones with 50% nickel–aluminum alloy in 10% aqueous sodium hydroxide [*1002*]. This result is somewhat surprising considering that the reduction is carried out under conditions favoring aldol condensation.

Transformation of ketones to alcohols has been accomplished by many hydrides and complex hydrides: *lithium aluminum hydride* [*110*], *magnesium aluminum hydride* [*117*], *lithium tris*(tert-*butoxy*)*aluminum hydride* [*1052*], *sodium dimethylaminoborohydride* [*130*], *dichloroalane* prepared from lithium aluminum hydride and aluminum chloride [*1053*], *lithium borohydride* [*983*], *lithium triethylborohydride* [*133*], *sodium borohydride* [*984, 1054*], *sodium trimethoxyborohydride* [*129*], *tetrabutylammonium borohydride* [*1004*], *tetrabutylammonium cyanoborohydride* [*139*], *zinc borohydride* [*121, 1005*], *zinc cyanoborohydride* [*930*], *dibutyl-* and *diphenylstannane* [*170*], *tributylstannane* [*987*], and others (Procedures 24 and 25, pp 306 and 307).

As a consequence of the wide choice of hydride reagents, the classical methods such as reduction with *sodium* in ethanol almost fell into oblivion [*1055*]. Nevertheless some old reductions were resuscitated [*1056*]. *Sodium dithionite* is an effective reducing agent [*361*], and the reduction *by alcohols* [*430*] was modified to cut down on the temperature [*990*] or the time required [*1057*], or to furnish chiral alcohols in good yields and excellent optical purity by using optically active pentyl alcohol and its aluminum salt [*1058*].

Good to high optical yields of **chiral alcohols** are obtained by using chiral *diisopinocampheylborane* [*1059*], chiral *K-glucoride [9-O-(1,2:5,6)-di-O-*

isopropylidene-α-D-glucofuranosyl)-9-borabicyclo[3.3.1]nonane] [1060], or *borane* and chiral complexing compounds such as *diphenyloxazaborolidine* [154] or *oxazaphospholidine* [153] (equation **173**).

173

$$RCOCH_3 \longrightarrow RCH(OH)CH_3$$

	R=		
[46]	CH$_3$	H$_2$/Raney Ni, 25–30 °C, 1–3 atm, 38 min	100%
[44]	CH$_3$	H$_2$/Ni, 125 °C, 165 atm, 1.5 h	85–100%
[51]	CH$_3$	H$_2$/CuCr$_2$O$_4$, 150 °C, 100–150 atm, 42 min	100%
[110]	C$_2$H$_5$	LiAlH$_4$, Et$_2$O, reflux	80%
[983]	C$_2$H$_5$	LiBH$_4$, Et$_2$O, reflux	77%
[984]	C$_2$H$_5$	NaBH$_4$, MeOH–H$_2$O	87%
[1059]	C$_2$H$_5$	Diisopinocampheylborane, THF, –30 °C	73%[a]
[1058]	C$_2$H$_5$	EtMeCHCH$_2$OH,Al(OCH$_2$CHMeEt)$_3$, reflux 3 h	94.6%[b]
[1002]	C$_3$H$_7$	Ni–Al, NaOH, 80–90 °C, 1 h	54%
[153]	(CH$_3$)$_2$CH	BH$_3$, THF, oxazaphospholidine, PhMe, 110 °C, 5 min	75–80% (99% ee)
[1004]	(CH$_3$)$_3$C	Bu$_4$NBH$_4$, CHCl$_3$, reflux 5.5 h	91%
[1057]	(CH$_3$)$_3$C	i-PrOH, Al(O-i-Pr)$_3$, reflux 1 h	36%
[154]	(CH$_3$)$_3$C	BH$_3$, diphenyloxazaborolidine, THF, 25 °C, 1 min	100% (92% ee)
[1062]	(CH$_3$)$_2$CH-CH$_2$CH$_2$	electro, Cd cathode, dild. H$_2$SO$_4$, EtOH	83.5%
[588]	C$_6$H$_{13}$	LiAlH$_4$, 4 C$_5$H$_5$N, 25 °C, 12 h	69%
[1054]	C$_6$H$_{13}$	NaBH$_4$, H$_2$O, Al$_2$O$_3$	90%
[361]	C$_6$H$_{13}$	Na$_2$S$_2$O$_4$, DMF–H$_2$O, reflux 4 h	75%
[139]	C$_9$H$_{19}$	Bu$_4$NBH$_3$CN, HMPA, 25 °C	84%
[1055]	C$_9$H$_{19}$	Na, EtOH, reflux	75%

[a] Optical purity 16.5%
[b] Optically active

A good method for the preparation of optically active hydroxy compounds is biochemical reduction using *microorganisms*. The yields are not always very good, but the enantiomeric excesses are high. Cyclohexyl methyl ketone treated with *baker's yeast* gives a 22–35% yield of (S)-cyclohexylmethylcarbinol with 97–98% enantiomeric excess [454], and ω,ω,ω-trifluoroacetophenone yields 71% (S)-1-phenyl-2,2,2-trifluoroethanol of 99% enantiomeric excess [470] (Procedure 63, p 321). Such reductions frequently apply to polyfunctional derivatives such as unsaturated ketones, hydroxyketones, and keto acids, and are dealt with in the appropriate sections.

Some reducing agents are chemoselective. Thus *potassium triphenylboro-hydride* reduces 2-heptanone in preference to 4-heptanone [136] (equation **174**).

174 [136]

$$CH_3COC_5H_{11} + (C_3H_7)_2CO \xrightarrow[-78\ °C,\ 6\ h;\ 0\ °C,\ 2\ h]{KBHPh_3,\ THF} \underset{\underset{94\%}{OH}}{CH_3CHC_5H_{11}} + \underset{4\%}{(C_3H_7)_2CHOH}$$

For the **reduction of aliphatic ketones to hydrocarbons**, several methods are available: reduction with *triethylsilane and boron trifluoride* [1006]; the *Clemmensen reduction* [236, 991]; the *Wolff–Kizhner reduction* [384, 385, 992]; reduction of *p*-toluenesulfonylhydrazones with *sodium borohydride* [1020], *sodium cyanoborohydride* [1050], or *bis(benzoyloxy)borane* [1021]; *desulfurization of dithioketals* (mercaptoles) [1036, 1061] (pp 182 and 183); and *electroreduction* [1062] (equation **175**).

175

$$RCOCH_3 \xrightarrow{\hspace{6cm}} RCH_2CH_3$$

	R=		
[992]	C_6H_{13}	$N_2H_4 \cdot H_2O$, $HO(CH_2O)_3H$, AcOH, 130–180 °C; MeONa, $HO(CH_2CH_2O)_3H$, 190–200 °C, 0.5–1.5 h	66%
[1021]	C_6H_{13}	$TosNHNH_2 \cdot EtOH$; $BH_3 \cdot THF$, BzOH, $CHCl_3$, 1 h	78%
[1036]	C_8H_{17}	$o\text{-}C_6H_4(SH)_2$, TosOH; Na/NH_3	98%
[1006]	C_9H_{19}	Et_3SiH/BF_3, CH_2Cl_2, 1 h	80%
[991]	C_9H_{19}	ZnHg, HCl, reflux 24 h	87%
[1050]	C_9H_{19}	$TosNHNH_2$, $DMF\text{-}C_4H_8SO_2$, $NaBH_3CN$, 100–105 °C, 4 h	86%

Somewhat less frequent than the reductions of aliphatic ketones to secondary alcohols and to hydrocarbons are one-electron *reductions to pinacols*. These are accomplished by metals such as *sodium,* but better still by *magnesium* or *aluminum.* Acetone gives a 43–50% yield of pinacol on refluxing with magnesium amalgam in benzene [210], and 45% and 51% yields on refluxing with aluminum amalgam in methylene chloride or tetrahydrofuran, respectively [1063].

An interesting **deoxygenation of ketones** takes place on treatment with *diisobutylalane and aluminum chloride or bromide* [1064] and especially with *low-valence-state titanium.* Reagents prepared by treatment of titanium trichloride in tetrahydrofuran with lithium aluminum hydride [285], potassium [286], magnesium [287], or in 1,2-dimethoxyethane with lithium [286] or zinc–copper couple [289] convert **ketones to alkenes** formed by coupling of the ketone carbon skeleton at the carbonyl carbon. Diisopropyl ketone thus gives tetraisopropylethylene (37% yield) [286], and cyclic and aromatic ketones afford much better yields of symmetrical or mixed coupled products [286, 287, 289]. The formation of the alkene may be preceded by pinacol coupling. In some cases a pinacol was actually isolated and reduced by low-valence-state titanium to the alkene [286] (equation **179**, p 156).

AROMATIC KETONES

Like aliphatic ketones, **aromatic and aromatic–aliphatic ketones are reduced to alcohols** very easily. The secondary alcoholic group adjacent to the benzene ring is easily hydrogenolyzed; and therefore special precautions must be taken to prevent the reduction of the keto group to the methylene group, especially in catalytic hydrogenations.

Catalytic hydrogenation of aromatic ketones is successful over a wide range of catalysts under mild conditions (25 °C, 1 atm). Of the noble metal family, some metals more than others tend to hydrogenolyze the alcohol formed and/or saturate the ring. Evaluation of some metal catalysts is based on hydrogenation of acetophenone as a representative [41, 1051].

Palladium is ineffective in reducing aliphatic and alicyclic ketones, but it is the catalyst of choice for reduction of acetophenone: It does not hydrogenate the ring, and at proper conditions, reduction may stop at the stage of the alcohol without hydrogenolysis. Over platinum oxide in ethanol, reduction to 1-phenylethanol is fast and proceeds further only slowly, whereas in acetic acid up to 84% of completely hydrogenolyzed and saturated product, ethylcyclohexane, is obtained. Rhodium–platinum oxide in ethanol or acetic acid facilitates quantitative reduction to 1-cyclohexylethanol with a slight hydrogenolysis (6%) [41]. Use of palladium and platinum oxides in acetic acid with added hydrochloric acid enhances hydrogenation of acetophenone to ethylbenzene [41]. Compared with palladium, platinum favors more extensive hydrogenolysis, and rhodium favors both hydrogenolysis and hydrogenation of the ring [1051].

Very good yields (86–92%) of alcohols are obtained by catalytic hydrogen transfers using ammonium formate and Raney nickel [77] or palladium on carbon [76] in methanol at room temperature or 70 °C, respectively (at 100 °C, palladium catalysts hydrogenolyze the hydroxyl [76]).

In contrast to hydrogenation over noble metals, hydrogenation of acetophenone over different nickel catalysts and over copper chromite results in the formation of 1-phenylethanol without hydrogenolysis [44, 46, 50, 51] (equation **176**).

Hydrogenation of the acetophenone analogs, isomeric acetylpyridines, is more complicated. 2-Acetylpyridine hydrogenated over 5% palladium on charcoal in 95% ethanol gives, at room temperature and 3 atm, a 79% yield of 2-(1-hydroxyethyl)pyridine, also obtained over Raney nickel. Hydrogenations over platinum oxide or rhodium afford, in addition to 60–70% 2-(1-hydroxyethyl)pyridine, 23–25% of the starting material and varying amounts of 2-(1-hydroxyethyl)piperidine resulting from the ring hydrogenation. 3-Acetyl-pyridine is hydrogenated over 5% palladium on charcoal in ethanol to a mixture of 70% 3-acetyl-1,4,5,6-tetrahydropyridine and 7.2% 3-acetylpiperidine. Treatment of 4-acetylpyridine with hydrogen over platinum oxide in 95% ethanol yielded 62% 4-(1-hydroxyethyl)pyridine and 15% of a pinacol, 2,3-bis(4-pyridyl)-2,3-butanediol. The pinacol is also obtained in a 60% yield in hydrogenation over palladium on charcoal or over rhodium on alumina, in a 75–80% yield over platinum oxide, and in a 90% yield over Raney nickel in alcohol [1065]. This formation of a pinacol in catalytic hydrogenation is unique.

Surprisingly good results in the reduction of aromatic ketones are obtained by treating the ketones with a 50% nickel–aluminum alloy in 10–16% aqueous sodium hydroxide at temperatures of 20–90 °C (65–90% yields) [*1002, 1066*].

Very dependable reduction of aromatic ketones to secondary alcohols is accomplished by hydrides and complex hydrides: *lithium aluminum hydride* [*110, 1067*], *lithium tetrakis(N-dihydropyridyl)aluminate* (diaryl ketones only) [*588*], *lithium borohydride* [*983*], *lithium triethylborohydride* [*133*], *sodium bis(2-methoxyethoxy)aluminum hydride* [*705*], *sodium borohydride* [*984, 1054*], *sodium trimethoxyborohydride* [*129*], *sodium dimethylaminoborohydride* [*130*], *tetrabutylammonium borohydride* [*1004*], *tetrabutylammonium cyanoborohydride* [*139*], *diphenylstannane* [*170*], *dimethylphenylsilane* [*166*], and others, generally in good to excellent yields with practically no hydrogenolysis of the alcohol. As a consequence, reduction with metals almost ceased to be used, and other reducing reagents are used only sporadically. Some of these are *sodium dithionite* [*361*], *alcohols* (especially isopropyl alcohol) in the *Meerwein–Ponndorf reduction* [*430*] or its modifications such as shorter periods of heating without the time-consuming removal of acetone [*1057, 1068*], and *aldehydes* (formaldehyde or salicylaldehyde, suitable only for purely aromatic ketones) [*435*]. Reduction of aromatic ketones *by Grignard reagents* [*448*] lacks practical importance (equation **176**).

176

C₆H₅COCH₃ ⟶ C₆H₅CH(OH)CH₃

[*50*]	H₂/Ni–NaH, EtOH, 25 °C, 1 atm, 2.5 h	92%
[*46*]	H₂/Raney Ni, 25–30 °C, 1–3 atm, 22 min	100%
[*44*]	H₂/Ni, EtOH, 175 °C, 160 atm, 8 h	60%
[*51*]	H₂/CuCr₂O₄, 150 °C, 100–150 atm, 30 min	100%
[*76*]	HCO₂NH₄, Pd(C), MeOH, 70 °C	91%
[*77*]	HCO₂NH₄, Raney Ni, MeOH, RT, 3 h	91%
[*1003*]	LiAlH₄/AlCl₃, Et₂O, reflux 30 min	94%
[*705*]	NaAlH₂(OCH₂CH₂OMe)₂, C₆H₆, 20–40 °C, 6 min	95%
[*129*]	NaBH(OMe)₃, Et₂O, reflux 4 h	82%
[*1004*]	Bu₄NBH₄, CHCl₃, reflux 2 h	91%
[*1002*]	Ni–Al alloy, 10% aq. NaOH, 10–20 °C	75%
[*241*]	Zn/NiCl₂, MeOH, reflux 1 h	76%
[*1058*]	MeEtCHCH₂OH, Al(OCH₂CHMeEt)₃, reflux 10 h	80%
[*361*]	Na₂S₂O₄, DMF–H₂O, reflux 4 h	94%

As in catalytic hydrogenation, in reduction with metals, **alkyl pyridyl ketones** make a complex picture.

2-Acetyl- and 4-acetylpyridines reduced by refluxing for 8 h with amalgamated zinc in hydrochloric acid (the *Clemmensen reduction*) give 83% and 86% yields of pure 1-(2-pyridyl)ethanol and 1-(4-pyridyl)ethanol, respectively. On the other hand, reduction by refluxing for 8 h with *zinc* dust in 80% formic acid yields, respectively, 86% and 64% pure 2-ethyl- and 4-ethyl-pyridines. Refluxing with zinc dust and acetic acid for 9 h gives mixtures of the alcohols and alkylpyridines. 3-Acetylpyridine affords 14% 1-(3-pyridyl)ethanol by the first method, 65% 3-ethylpyridine by the second method, and mainly the alcohol and also a pinacol, 2,3-bis(3-pyridyl)-2,3-butanediol [1069], by refluxing with zinc dust in acetic acid.

Special attention should be directed to reductions of pro-chiral ketones by *chiral reducing agents* that lead to chiral secondary alcohols. Such reagents are prepared, for example, by treatment of *(2S,3S)- and (2R,3R)-1,4-bis(diethylamino)-2,3-butanediol* with *lithium aluminum hydride* [1070] or by adding *borane* [148, 1059] or 9-bora[3.3.1]bicyclononane [150, 1071] to optically active α-pinene. Reductions with such *boranes* give alcohols of varying optical purities.

Diisopinocampheylchloroborane affords alkylarylcarbinols in 45–90% yields with 78–100% enantiomeric excesses [149]. High optical purity is also obtained from reductions with *K-glucoride* [1060], *chiral oxazaborolidines* [154], and *chiral oxazaphospholidines* [153]. One of the best chiral reducing agents is 2,2'-dihydroxy-1,1'-binaphthylethoxyaluminum hydride (Binal-H) [156, 1072] (p 23) (equation **177**).

177

C6H5COCH3		→	C6H5CH(OH)CH3		
			Yield	Config.	ee
[1073]	Catalyst[a], Me2CHOH, RT, 1–2 h		74%		96%
[149]	Diisopinocampheylchloroborane, THF, –25 °C, 7 h		72%	S	98%
[154]	BH3, THF, (S)-5,5-diphenyloxazaborolidine, 23 °C, 1 min		100%	R	97%
[153]	BH3, THF, oxazaphospholidine, PhMe, 110 °C, 5 min		75–80%	R	99%
[156]	R-Binal-H, THF, –100 °C, 1 h, –78 °C, 2 h		60%	R	95%
[1074]	Cryptococcus macerans		90%	?	90%
[469]	Pseudomonas sp. PED, NAD, RT		34%	R	97%

$$
\begin{array}{c}
\text{Ph} \\
| \\
\text{CH}_2 \\
\text{Ph··,} \diagdown \overset{|}{\underset{|}{N}} \diagup \text{Ph} \\
\end{array}
$$

[a]The catalyst   O–Sm–O   is prepared from chiral PhCHCH2NCH2CHPh,
                                                                                          | |
                                                                                        OH   CH2Ph
                           I
BuLi, SmI2, and ICH2CH2I in THF.   OH   CH2Ph

Another way of preparing optically active alcohols in good yields and excellent optical purity is based on reduction of pro-chiral ketones with the aluminum salt of optically active amyl alcohol (pentyl alcohol) [1058] or by the *Meerwein–Ponndorf–Verley* reduction at room temperature using a chiral samarium complex as a catalyst [1073] (equation **177**).

Alcohols of high enantiomeric excesses, usually in not very high conversions, are obtained by *biochemical reductions* using microorganisms. Acetophenone gives a 90% yield of (–)-(S)-1-phenylethanol after incubation with *Cryptococcus macerans* [1074]. Other ketones are reduced by *Sporobolomyces pararoseus* [1074], *baker's yeast* [454], *Pseudomonas* [469], *Lactobacillus kefir* [468], and others. α-, β- and γ-Acetylpyridines are similarly reduced by *Crypto-*

coccus macerans to the corresponding (–)-(*S*)-1-pyridylethanols of 79–85% enantiomeric excesses [*1075*]. The *S* configuration predicted by Prelog's rule (if the ketone is placed with the larger group on the observer's left, the hydroxyl group formed is closer to the observer [*1076*]) was confirmed by chemical means [*1075*]. Treatment of 1,2-di(β-pyridyl)-2-methyl-1-propanone with *Botryodiplodia theobromae* Pat. affords (–)-1,2-di(β-pyridyl)-2-methyl-1-propanol of 99% optical purity in a 90% yield [*1077*].

Different microorganisms can reduce ketones to alcohols of opposite configurations (equation **178**).

178

$$C_6H_5COCF_3 \longrightarrow C_6H_5CH(OH)CF_3$$

		Conversion	Config.	ee
[*469*]	*Pseudomonas sp. PED*	37%	S	92%
[*468*]	*Lactobacillus kefir*	71%	S	99%
[*468*]	*Thermoanaerobicum brockii*		R	95%

In ketones having a chiral cluster next to the carbonyl carbon, reduction with *lithium aluminum hydride* gives one of the two possible diastereomers, *erythro* or *threo*, in larger proportions. The outcome of the reduction is determined by the approach of the reducing agent from the least hindered side (steric control of asymmetric induction) [*1067*]. With lithium aluminum hydride, as much as 80% of one diasteromer can be obtained. This ratio is higher with more bulky reducing hydrides such as *diisopinocampheylborane* [*1078*] or *dimethylphenylsilane* [*166*].

One-electron reduction of **aromatic ketones** gives ***pinacols***. Occasionally, such a reduction can be achieved by catalytic hydrogenation of 3-acetylpyridine [*1065*] (p 153). Common reducing agents for the synthesis of pinacols are metals: *aluminum amalgam* [*218*] (Procedure 34, p 311), *zinc* in acetic acid [*1079*], and *magnesium* with magnesium iodide [*1080*]. An excellent yield of tetraphenylpinacol is obtained by *irradiation* of benzophenone dissolved in *isopropyl alcohol* [*429*].

Titanium in a low valence state, as prepared by treatment of solutions of titanium trichloride with potassium [*286*] or magnesium [*287*] in tetrahydrofuran or with lithium in dimethoxyethane [*286*], *deoxygenates ketones and effects coupling* of two molecules at the carbonyl carbon to form alkenes, usually a mixture of both stereoisomers. If a mixture of acetone with other ketones is treated with titanium trichloride and lithium, the alkene formed by combination of acetone with the other ketone predominates over the symmetrical alkene produced from the other ketone [*286*] (equation **179**) (Procedure 44, p 314).

Reduction of aromatic ketones to hydrocarbons occurs very easily as the carbonyl group is directly attached to an aromatic ring. In these cases reduction produces benzylic-type alcohols that are readily hydrogenolyzed to hydrocarbons [*1081, 1082*]. This happens during catalytic hydrogenations as well as in chemical reductions (equation **180**).

Carbonyl groups adjoining an aromatic ring have been converted to methylene in good to excellent yields by *hydrogenation* over palladium and

179

$$C_6H_5COC_6H_5 \longrightarrow (C_6H_5)_2C{=}C(C_6H_5)_2 + \underset{\underset{HO}{|}}{(C_6H_5)_2C}{-}\underset{\underset{OH}{|}}{C(C_6H_5)_2} + C_6H_5CH_2C_6H_5$$

[286]	TiCl$_3$, Li, (CH$_2$OMe)$_2$ reflux 16.5 h	96%	
[286]	TiCl$_3$, K, THF, RT, reflux 16.5 h	80%	
[1079]	Zn, AcOH, heat		92%	
[1080]	Mg, MgI$_2$, Et$_2$O, C$_6$H$_6$		99.6%	
[429]	Me$_2$CHOH, $h\upsilon$, 3–5 h		93–94%	
[385]	N$_2$H$_4$·H$_2$O, NaOH, HO(CH$_2$CH$_2$O)$_3$H			83.3%

180

$$C_6H_5COCH_3 \longrightarrow C_6H_5CH_2CH_3$$

[1081]	LiAlH$_4$/P$_2$I$_4$, C$_6$H$_6$, reflux 2 h	100%[a]
[1003]	LiAlH$_4$/AlCl$_3$, Et$_2$O, reflux 30 min	94%
[1064]	iBu$_2$AlH, C$_7$H$_{16}$, 70 °C, 4 h; AlBr$_3$, (C$_5$H$_5$)$_2$Ti, 90 °C, 22 h	75%
[1082]	NaBH$_3$CN/ZnI$_2$, (CH$_2$Cl)$_2$, 83 °C, 48 h	64%
[181]	Et$_3$SiH, CF$_3$CO$_2$H, 55 °C, 15 h	70%
[1021]	TosNHNH$_2$; Bz$_2$BH	68%

[a] The yield is based on gas-liquid chromatography.

platinum catalysts in acetic and hydrochloric acids at room temperature and atmospheric pressure [1083], over Raney nickel [575] or copper chromite [575] at 130–175 °C and 70–350 atm, and over molybdenum sulfide at 270 °C and 100 atm [58]. Aromatic ketones containing pyrrole [575] and pyridine rings [1065] must not be hydrogenated under too vigorous conditions if the aromatic rings are to be preserved. With 2,4-diacetyl-3,5-dimethylpyrrole, temperature during the hydrogenation over Raney nickel or copper chromite should not exceed 180 °C [575].

Reduction of carbonyl to methylene in aromatic ketones is also achieved *by alane* prepared from lithium aluminum hydride and aluminum chloride [1003], *dihaloisobutylalanes* [1064], *lithium aluminum hydride* and *diphosphorus tetraiodide* [1081], *sodium borohydride* in trifluoroacetic acid [1084], *sodium cyanoborohydride* and *zinc iodide* [1082], *triethylsilane* in trifluoroacetic acid [181, 485, 1009], *sodium* in refluxing ethanol [1085], *zinc* in hydrochloric acid [1086], and *hydrogen iodide* and *phosphorus* [318], generally in good to high yields.

However, the most frequently used methods for reduction of carbonyl groups adjacent to aromatic rings to methylene groups are the *Clemmensen reduction* [236, 237, 991, 1086, 1087] (Procedure 36, p 312), *Wolff–Kizhner* reduction [*384, 385, 386, 992, 1012, 1088*] (Procedure 53, p 317), or reduction of *p*-toluenesulfonylhydrazones of the ketones with *lithium aluminum hydride* [*1048, 1049*] or with *borane and benzoic acid* [*1021*] (equation **180**).

Deoxygenation of ketones to unsaturated hydrocarbons takes place with low-valence *titanium compounds* [*285, 286, 287*] (equation **179**).

Reduction of carbocyclic rings in aromatic ketones can be accomplished by *catalytic hydrogenation* over platinum oxide or rhodium–platinum oxide and takes place only after the reduction of the carbonyl group, either to the alcoholic group or to a methylene group [*41*].

In **heterocyclic aromatic ketones** containing pyrrole and pyridine rings, partial or total reduction of the ring may precede hydrogenation of the keto group. *N*-Methyl-2-pyrrylacetone is hydrogenated over platinum oxide in acetic acid at 20 °C and 1 atm to hygrine (*N*-methyl-2-pyrrolidinylacetone) in a 60% yield [*579*]. On the other hand, in 2,4-diacetyl-3,5-dimethylpyrrole, hydrogenation of the pyrrole ring requires temperatures of 180 °C for Raney nickel and 250 °C for copper chromite to give 2,4-diethyl-3,5-dimethyl-pyrrolidine in 70% and 50% yields, respectively [*575*]. In the hydrogenation of 3-acetylpyridine over 5% palladium on charcoal in ethanol, the keto group survives the reduction to 3-acetyl-1,4,5,6-tetrahydropyridine (70%) and 3-acetylpiperidine (7.2%) [*1065*]. In 4-acetylquinoline the pyridine ring is reduced to the 1,2-dihydro- and 1,2,3,4-tetrahydro derivatives, but only after the conversion of the acetyl group to an ethyl group [*443*].

CYCLIC KETONES

All reducing agents used for reductions of **aliphatic and aromatic ketones** can be used for reduction of cyclic ketones to *secondary alcohols* (pp 157-163). In fact, reduction of cyclic ketones is sometimes easier than that of aliphatic and aromatic ketones [*361*]. In the reduction of an equimolar mixture of cyclohexanone and 4-heptanone with *borane*-tert-*butylamine complex*, the ratio of cyclohexanol to 4-heptanol is 96:4 [*1089*]; with *potassium triphenyl-borohydride*, the ratio is 99.4:0.6 [*136*]. The latter reagent shows *chemo-selectivity* even in reductions of cyclanones of different sizes. A mixture of cyclohexanone and cyclopentanone gives the two alcohols in the ratio of 97:3 at −78 °C [*136*]. An important aspect in reductions of cyclic ketones is stereoselectivity of the reduction and stereochemistry of the products.

According to the older literature *catalytic hydrogenation* tends to favor *cis* isomers (where applicable), whereas reduction with *metals* gives *trans* isomers [*1090*]. But even catalytic hydrogenation may give either *cis* isomers, when carried out over platinum oxide [*1091*], or *trans* isomers, if Raney nickel is used [*1091*]. An acidic medium favors alcohols with axial hydroxyls. Thus 4-*tert*-butylcyclohexanone is hydrogenated over platinum dioxide in acetic acid to *cis*-4-*tert*-butylcyclohexanol [*1092*].

Similarly reductions with metal hydrides, metals, and other compounds may give predominantly one isomer. The stereochemical outcome depends strongly on the structure of the ketone and on the reagent and may be affected by the solvents.

The structure of the cyclic ketone is of utmost importance. Reduction of cyclic ketone by complex hydrides is started by a nucleophilic attack at the carbonyl function by a complex hydride anion (equation **181**).

181

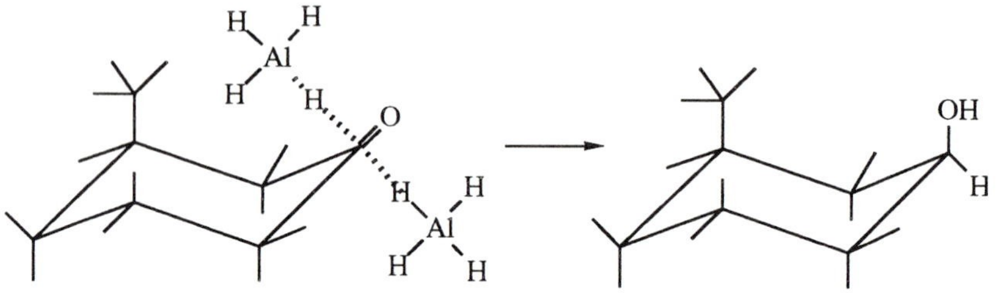

The approach of the nucleophile takes place from the less crowded side of the molecule (**steric approach or steric strain control**) (equation **181**), leading usually to the less stable alcohol. In ketones with no steric hindrance (no substituents flanking the carbonyl group or bound in position 3 of the ring), usually the more stable (equatorial) hydroxyl is generated (**product development or product stability control**) [*1093, 1094, 1095, 1096*]. The contribution of product development to the stereochemical outcome of the reduction has been disputed [*1093*]. The reductions with hydrides and complex hydrides are very fast; hence it is assumed that the transition states develop early in the reaction, and consequently the steric approach is more important than the stability of the product. What is sometimes difficult to estimate is what causes stronger steric hindrance: interactions between the incoming nucleophile and axial substituents in position 3 of the ring or torsional strain between the nucleophile and a hydrogen or a substituent flanking the carbonyl carbon. In other words, the determination of which side of the molecule is more accessible is at times difficult (see equation **183** and **184**, p 160).

Reduction of 2-alkylcycloalkanones having four- to six-membered rings with *lithium aluminum hydride* in tetrahydrofuran gives less of the less stable *cis*-2-alkylcyclanol with axial hydroxyl, whereas 2-methylcycloheptanone and 2-methylcyclooctanone yield predominantly the *cis*-isomer [*1078*] (equation **182**).

The effect of reducing agents on the stereochemistry of the reduction of 2-methylcyclohexanone is demonstrated for the reduction with *lithium aluminum hydride*, *zinc borohydride* [*1005*], *borane* in tetrahydrofuran, *di-sec-amylborane* in tetrahydrofuran, *dicyclohexylborane* in diglyme, and *diisopinocampheylborane* in diglyme [*1078*] (equation **182**). Distribution of products of reduction of 4-*tert*-butylcyclohexanone with complex hydrides is shown in equation **183** (*1053, 1093, 1095, 1097, 1098, 1099*).

Different stereoselectivities caused by solvent effects are demonstrated in the reduction of dihydroisophorone (3,3,5-trimethylcyclohexanone) with *lithium aluminum hydride* and its derivatives [*1100*] and with *sodium borohydride* and its modifications, which give less stable *trans*-3,3,5-trimethylcyclohexanol (with axial hydroxyl) by reduction in anhydrous isopropyl alcohol (55–56%), anhy-

182

R = H; *n* =

2	[*1005*]	Zn(BH$_4$)$_2$, (CH$_2$OMe)$_2$, –15 °C, 30 min	100%
3	[*1005*]	Zn(BH$_4$)$_2$, (CH$_2$OMe)$_2$, –78 °C, 5 min	100%
3	[*170*]	Ph$_2$SnH$_2$, Et$_2$O, RT	82%
4	[*1005*]	Zn(BH$_4$)$_2$, (CH$_2$OMe)$_2$, –15 °C, 30 min	100%[a]

R = CH$_3$; LiAlH$_4$, THF; *n* =	*cis*	*trans*	[*1078*]
1	25	75	
2	21	79	
3	25	75	
4	73	27	
5	73	27	
3; BH$_3$•THF	26	74	
3; BH(CHMeCHMe$_2$)$_2$	79	21	
3; BH(C$_6$H$_{11}$)$_2$, MeO(CH$_2$CH$_2$O)$_2$Me	94	6	
3; diisopinocampheylborane, diglyme	94	6	

[a] The yields were determined by NMR spectroscopy.

drous *tert*-butyl alcohol (55%), 65% aqueous isopropyl alcohol (59.5%), anhydrous ethanol (67%), and 71% aqueous methanol (73%) (the balance to 100% being the more stable *cis* isomer with equatorial hydroxyl) [*1053, 1093, 1097, 1100, 1101*] (equation **184**).

The effect of steric hindrance can be nicely demonstrated in the reduction of two bicyclic ketones, norcamphor and camphor. On reduction with complex hydrides, the relatively accessible norcamphor yields predominantly (the less stable) *endo* norborneol, whereas sterically crowded camphor is reduced by the same reagents predominantly to the less stable *exo* compound, isoborneol [*1078*]. The numerous examples show that the stereoselectivity increases with increasing bulkiness (with some exceptions), and that it is affected by the nucleophilicity of the reagent and by the solvent.

Similar stereoselectivity can be noticed in reductions of ketones by *alkali metals* in liquid ammonia and alcohols. 4-*tert*-Butylcyclohexanone gives almost exclusively the more stable equatorial alcohol (equation **183**), norcamphor affords 68–91% of the less stable *endo*-norborneol, and camphor yields a mixture of borneol and isoborneol [*1102*] (equations **185** and **186**).

183

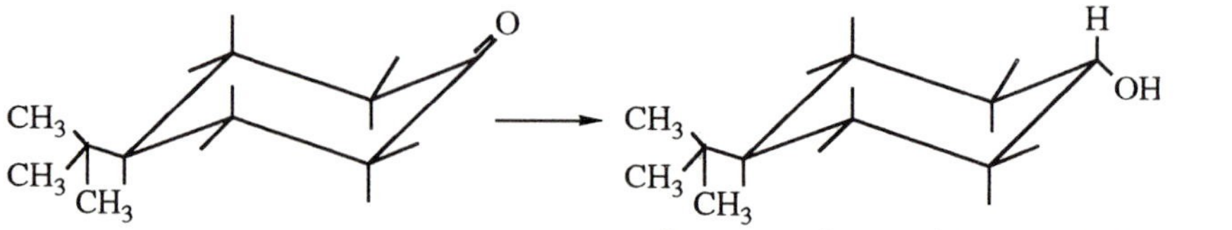

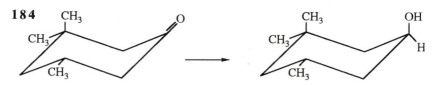

Percentage of equatorial (*trans*) alcohol

Reference	[*1053*]	[*1093*]	[*1095*]	[1097]	[*1098*]	[*1099*]
LiAlH$_4$, Et$_2$O	89	90–91				
LiAlH$_4$, THF			42	92		
LiAlH(i-Bu)$_2$Bu, Et$_2$O,C$_6$H$_{14}$					61	
LiAlH(i-Bu)$_2s$-Bu, Et$_2$O,C$_6$H$_{14}$					58	
LiAlH(i-Bu$_2$)t-Bu, Et$_2$O,C$_6$H$_{14}$					51	
LiAlH(OMe)$_3$, THF			59	36		
LiAlH(OCMe$_3$)$_3$, THF			90	46	89.7	
AlHCl$_2$(LiAlH$_4$/AlCl$_3$), Et$_2$O	80–99.5					
NaBH$_4$, MeOH			85–86			
NaBH(OMe)$_3$, MeOH			75–76			
NaBH(OCHMe$_2$)$_3$, i-PrOH			75–80			
i-PrOH-Al(O-i-Pr)$_3$	79					
(MeO)$_3$P, H$_2$O, i-PrOH, IrCl$_4$, HCl						4

184

Percentage of axial (*trans*) alcohol

Reference	[*1053*]	[*1100*]	[*1093*]	[*1097*]	[*1101*]
LiAlH$_4$,Et$_2$O	55	52–55	58–63		
LiAlH$_4$,THF		72–74		87.8	
LiAlH(OMe)$_3$, Et$_2$O		75			
LiAlH(OMe)$_3$, THF		92	98		
LiAlH(OEt)$_3$, Et$_2$O		83			
LiAlH(OCHMe$_2$)$_3$, Et$_2$O		54			
LiAlH(OCHMe$_2$)$_3$, THF		63–73			
AlHCl$_2$(LiAlH$_4$/AlCl$_3$),Et$_2$O	85[a],0[b]	85–86			

Continued on next page

184 *Continued*

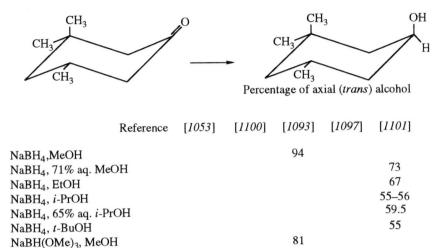

Percentage of axial (*trans*) alcohol

	Reference	[1053]	[1100]	[1093]	[1097]	[1101]
NaBH₄,MeOH				94		
NaBH₄, 71% aq. MeOH						73
NaBH₄, EtOH						67
NaBH₄, *i*-PrOH						55–56
NaBH₄, 65% aq. *i*-PrOH						59.5
NaBH₄, *t*-BuOH						55
NaBH(OMe)₃, MeOH				81		
NaBH(OMe)₃, *i*-PrOH						65
NaBH(OCHMe₂)₃, *i*-PrOH					80–83	
i-PrOH, Al(O-*i*-Pr)₃		6				

ᵃKinetic product. ᵇThermodynamic product.

185

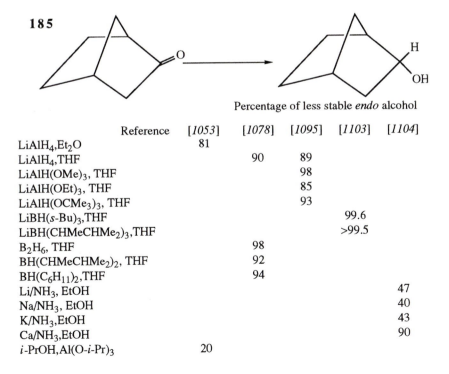

Percentage of less stable *endo* alcohol

	Reference	[1053]	[1078]	[1095]	[1103]	[1104]
LiAlH₄,Et₂O		81				
LiAlH₄,THF			90	89		
LiAlH(OMe)₃, THF				98		
LiAlH(OEt)₃, THF				85		
LiAlH(OCMe₃)₃, THF				93		
LiBH(*s*-Bu)₃,THF					99.6	
LiBH(CHMeCHMe₂)₃,THF					>99.5	
B₂H₆, THF			98			
BH(CHMeCHMe₂)₂, THF			92			
BH(C₆H₁₁)₂,THF			94			
Li/NH₃, EtOH						47
Na/NH₃, EtOH						40
K/NH₃,EtOH						43
Ca/NH₃,EtOH						90
i-PrOH,Al(O-*i*-Pr)₃		20				

186

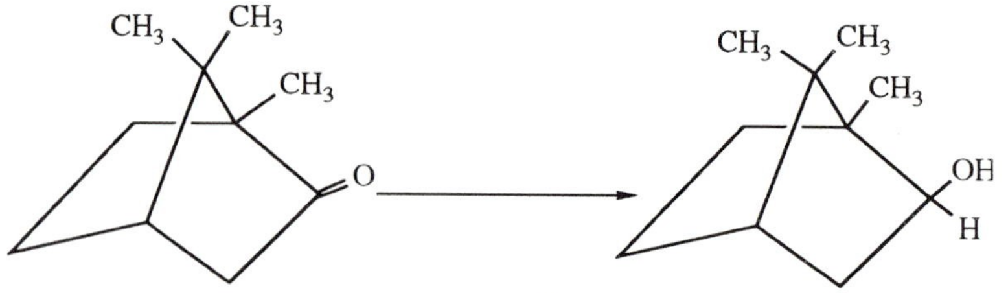

Percentage of less stable *exo* alcohol

Reference	[1053]	[1078]	[1095]	[1103]	[1104]	[1105]
LiAlH$_4$,Et$_2$O	90					
LiAlH$_4$,THF		91	92			
LiAlH(OMe)$_3$, THF			99			
LiAlH(OCMe$_3$)$_3$, THF			93			
AlHCl$_2$ (LiAlH$_4$/AlCl$_3$), Et$_2$O	73					
LiBH(s-Bu)$_3$,THF				99.6		
LiBH(C$_6$H$_{11}$)$_3$, THF				99.3		
B$_2$H$_6$, THF			52			
BH(CHMeCHMe$_2$)$_2$, THF			65			
BH(C$_6$H$_{11}$)$_2$,THF			93			
Li/NH$_3$, EtOH					21–23	
Na/NH$_3$, EtOH					42	
K/NH$_3$,EtOH					60–70	
Ca/NH$_3$,EtOH					28	
iso-PrOH,Al(O-iso-Pr)$_3$	29					70

Stereoselective reductions of cyclic ketones have immense importance in the chemistry of steroids from which either α or β epimers can be obtained [50]. A few examples demonstrate that *catalytic hydrogenation* [50] and catalytic *hydrogen transfer* (the *Henbest reduction*) [1106] favor formation of α epimers, whereas reduction with *hydrides* and *complex hydrides* is more stereoselective and gives predominantly β epimers [139, 1053] (equation **187**).

Cyclic ketones are reduced *diastereoselectively* or *enantioselectively* to *cyclic alcohols* by chemical and biochemical methods. Very high yields and diastereomeric excesses are obtained by reduction with chiral *K-glucoride* [1060]. A chiral reagent, prepared by successive treatment of lithium aluminum hydride with (–)-(*1R, 2S*)-*N*-methylephedrine and 2-ethylaminopyridine, reduces α-tetralone to (*R*)-α-tetralol in a 93% yield with 96% enantiomeric excess, and β-tetralone to (*R*)-β-tetralol in an 84% yield with 93% enantiomeric excess [1107]. The same cyclanols are reduced by *Rhodotorula rubra* in the respective conversions of 53 and 100% and enantiomeric excesses of 96% and 90% [476].

187

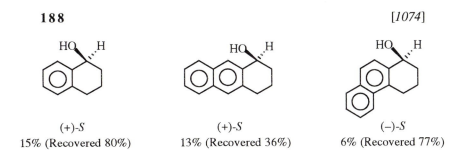

X = CHC$_8$H$_{17}$ (3-Cholestanone)

		α	β
[50]	H$_2$/Ni(NaH), 25 °C, 1 atm, 3.8 h (96%)	67%	33%
[1053]	LiAlH$_4$, Et$_2$O	9%	91%
[1053]	AlHCl$_2$(LiAlH$_4$/AlCl$_3$), Et$_2$O	0–17%	83–100%
[139]	Bu$_4$NBH$_3$CN, HMPA, 25 °C (94.5%)	17%	83%
[1053]	i-PrOH, Al(O-i-Pr)$_3$, reflux	16%	84%

X = CO (Androstane-3,17-dione)

[50]	H$_2$/Ni(NaH), 25 °C, 1 atm, 70 min (97%)	47%	53%
[1106]	H$_2$IrCl$_6$, (MeO)$_3$P, 90% i-PrOH	94%	2%
	reflux 94 h (Henbest reduction)		

Incubation of 2,3-benzosuberone with *Cryptococcus macerans* gives 27% conversion to the optically active (–)-(*S*)-2,3-benzo-1-cycloheptanol. Other hydroaromatic ketones are reduced, although in low conversions (2–41%), to *S*-alcohols (according to Prelog's rule, p 50) [*1074*] (structure group **188**). Biochemical reductions are especially important in steroids.

188   [*1074*]

(+)-*S*	(+)-*S*	(–)-*S*
15% (Recovered 80%)	13% (Recovered 36%)	6% (Recovered 77%)

Reductions of cyclic ketones to cycloalkanes, pinacols, and alkenes are accomplished by the same reagents that are used for similar reductions of aliphatic and aromatic ketones (pp 151 and 155) (equation **189**).

Reduction of the **carbonyl group to methylene** is carried out by the *Clemmensen reduction* [*236, 991*], the *Wolff–Kizhner reduction* [*384, 385, 386*], or its modifications: decomposition of hydrazones with potassium *tert*-butoxide in dimethyl sulfoxide at room temperature in yields of 60–90% [*1088*], or reduction of *p*-toluenesulfonylhydrazones with *sodium borohydride* (yields 65–80%) [*1048*] (p 187).

In keto steroids the reductions can also be achieved by *electrolysis* in 10% sulfuric acid and dioxane using a divided cell with lead electrodes (85–97%

yields) [*1108*], by specially activated *zinc* dust in anhydrous solvent (ether or acetic anhydride saturated with hydrogen chloride) (50–87% yields) [*231, 1109*], and by the aforementioned reduction of tosylhydrazones with *sodium borohydride* (60–75% yields) [*1048*].

189

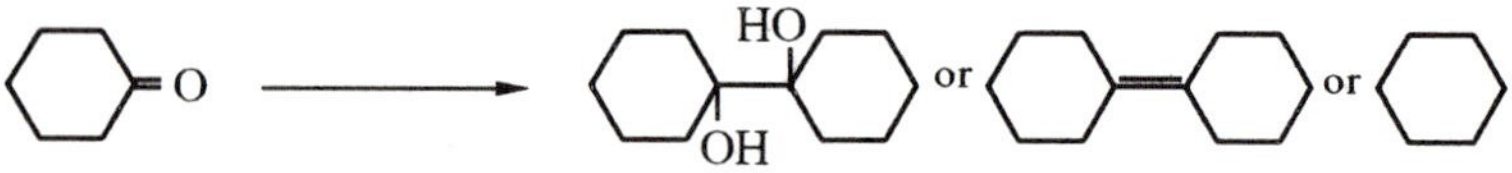

[*1063*] AlHg, CH_2Cl_2, reflux 1–4 h   55%
[*287*] $TiCl_3$, Mg, THF   45%
[*286*] $TiCl_3$, Li, $(CH_2OMe)_2$, reflux 16 h   79%
[*286*] $TiCl_3$, K, THF, reflux 16 h   85%
[*385*] N_2H_4•H_2O, KOH, $HO(CH_2CH_2O)_3H$,   80.4%
            reflux 1 h
[*1048*] $C_7H_7SO_2NHNH_2$; $NaBH_4$, dioxane   70–80%

Heterocyclic ketones having sulfur or nitrogen atoms in the ring in a position β to the carbonyl group are cleaved during Clemmensen reduction between the α-carbon and the hetero atom. Recyclization during the reaction leads to ring contractions [*235, 1110*]. Reduction of *N*-alkyl-α-pyridones, which are amides "in disguise", is discussed elsewhere (p 235) (equation **190**).

190

X=
S   [*1110*]   ZnHg, HCl, reflux 1 h   21%
NMe   [*235*]   ZnHg, HCl, reflux 12 h   60%
NMe   [*235*] N_2H_4•H_2O, KOH,   45%
                 $HO(CH_2CH_2O)_3H$, reflux 1 h

For the **reduction to pinacols,** *aluminum amalgam* [*1063*] or *low-valence-state titanium chloride* [*287*] can be used. Under various conditions the titanium reagent **deoxygenates the ketones and couples them** at the carbonyl carbon to alkenes [*286*] (equation **189**).

UNSATURATED KETONES

Unsaturated ketones of all kinds can be converted to saturated ketones, unsaturated alcohols, saturated alcohols, alkenes, and alkanes. If the double bond is conjugated with the carbonyl group, reduction usually takes place more readily and may give, in addition to the products just listed, 1,6-diketones formed by coupling at β-carbons.

Unsaturated ketones having double bonds not conjugated with the carbonyl group can be ***reduced to saturated ketones*** by controlled *catalytic hydrogenation*. However, if another, conjugated, double bond is present in the molecule of the ketone, it may be reduced preferentially. In 3-ethylenedioxy-5,16-pregnadiene-20-one, only the conjugated double bond in position 16 is reduced by hydrogen over 10% palladium on calcium carbonate in ethyl acetate (91% yield) [*1111*], and in 4,6,22-ergostatriene-3-one, only the double bond in the position γ,δ to the carbonyl is reduced over 5% palladium on charcoal in methanol containing potassium hydroxide (70% yield) [*1112*]. The isolated double bonds in these two compounds are not hydrogenated provided the reduction is stopped after absorption of 1 mol of hydrogen. The same results are obtained in heterogeneous hydrogenation of eremophilone (5,10-dimethyl-3-isopropenyl-1-oxo-1,2,3,4,5,6,7,10-octahydronaphthalene) over palladium catalysts. However, homogeneous hydrogenation using tris(triphenyl-phosphine)rhodium chloride in benzene gives a 94% yield of 12,13-dihydro-eremophilone in which the conjugated double bond survives [*866*]. Homogeneous hydrogenation of carvone reduces either the isolated or the conjugated double bond depending on the catalyst used [*94, 176*] (equation **191**).

191

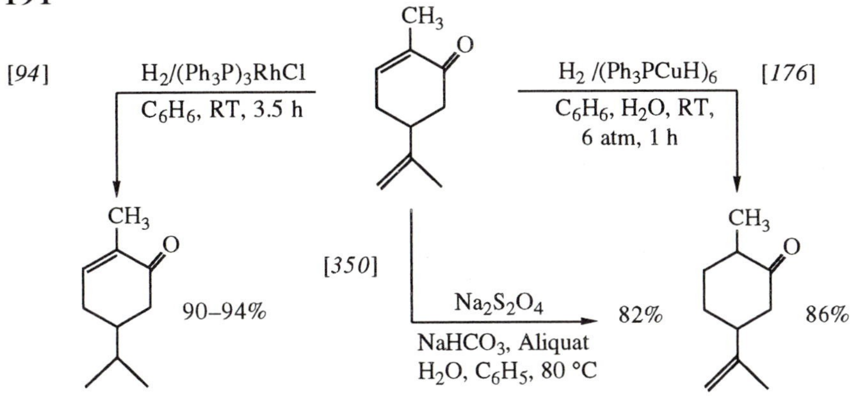

α,β-**Unsaturated ketones (enones) are reduced to saturated ketones** by *catalytic hydrogenation* provided it is stopped after absorption of 1 mol of hydrogen [*1051*]. Platinum [*1000*], platinum oxide [*37, 1113*], platinum on carbon [*1051*], palladium on carbon [*1051, 1114*], rhodium on carbon [*1051*], and tris(triphenylphosphine)rhodium chloride [*91, 1115*] are used as the catalysts. *Nickel–aluminum alloy* in 10% aqueous sodium hydroxide [*1002*] and *zinc-reduced nickel* in aqueous medium [*1116*] also reduce only the conjugated double bonds.

In cyclic α,β-unsaturated ketones such as 2-oxo-2,3,4,5,6,7,8,10-octahydro-naphthalene, where the reduction may lead to two stereoisomers, hydrogenation over 10% palladium on charcoal in aqueous ethanol gives 93% *cis*- and 7% *trans*-2-decalone in the presence of hydrochloric acid, and 62% *cis*- and 38% *trans*-2-decalone in the presence of sodium hydroxide [*1114*]. A whole spectrum of ratios of the two stereoisomers is obtained over different catalysts in solvents of different polarities and different pHs [24, 1114].

Lithium aluminum hydride preferentially reduces the carbonyl function, but *alanes* prepared by reactions of aluminum hydride with two equivalents of isopropyl or *tert*-butyl alcohol or of diisopropylamine reduce the conjugated double bonds with high regioselectivity in quantitative yields [1117] (p 167).

Triphenylstannane reduces the α,β double bond in dehydro-β-ionone in 84% yield [1118]. Complex *copper hydrides* prepared in situ from lithium aluminum hydride and cuprous iodide in tetrahydrofuran at 0 °C [1119] or from lithium trimethoxyaluminum hydride or sodium bis(methoxyethoxy)aluminum hydride and cuprous bromide [1120] in tetrahydrofuran at 0 °C reduce the α,β double bonds selectively in yields from 40 to 100%. Similar selectivity is found with a complex *sodium bis(irontetracarbonyl)hydride*, $NaHFe_2(CO)_8$ [1121].

Reduction of α,β-unsaturated to saturated ketones is further achieved by *electrolysis* in a neutral medium using copper or lead cathodes (55–75% yields) [199], *lithium* in propylamine (40–65% yields) [1122], *potassium–graphite* clathrate, C_8K (57–85% yields) [1044], and *zinc* in acetic acid (87% yield) [901]. Reduction with amalgamated zinc in hydrochloric acid *(Clemmensen reduction)* usually reduces both functions [1123]. A very selective reagent for reduction of the conjugated double bond is *sodium hydrosulfite* (dithionite) [350, 351] (equation **191**) (p 165).

Biochemical reduction of α,β-unsaturated ketones using microorganisms (the most efficient is *Beauveria sulfurescens*) takes place only if there is at least one hydrogen in the β position and the substituents on the α carbons are not too bulky. The main product is the saturated ketone, and only a small amount of the saturated alcohol is formed, especially in slightly acidic medium (pH 5–5.5). The carbonyl is attacked from the equatorial side. Results of biochemical reduction of 5-methylcyclohex-2-en-1-one are illustrative of the biochemical reduction by incubation with *Beauveria sulfurescens:* after 24 h 74% of the enone is reduced to 3-methylcyclohexanone and 26% is converted to 3-methylcyclohexanol containing 55% *cis* and 45% *trans* isomer. After 48 h the numbers are 70% and 30%, and 78% and 22%, respectively [1124]. *Corynebacterium equii* reduces the conjugated double bond without affecting the carbonyl group and a nitro group [473].

Reduction of unsaturated ketones to unsaturated alcohols is best carried out with *complex hydrides*. α,β-Unsaturated ketones may suffer reduction even at the conjugated double bond [997, 1125]. Usually only the carbonyl group is reduced, especially if the inverse technique is applied. Such reductions are accomplished in high yields with *lithium aluminum hydride* [1125, 1126, 1127, 1128]; *lithium trimethoxyaluminum hydride* [997]; *alane* [1125]; *diisobutylalane* [1129]; *lithium butylborohydride* [1130]; *sodium borohydride* [984], especially in the presence of *trichlorides of cerium, samarium, and europium* [1131, 1132]; *sodium cyanoborohydride* [1015, 1133]; *zinc borohydride* [1016]; *diisopropoxytitanium borohydride* [1134]; *9-borabicyclo[3.3.1]nonane (9-BBN)* [997]; and *isopropyl alcohol* and *aluminum isopropoxide (Meerwein–Ponndorf reduction)* [430, 1135]. Many α,β-unsaturated ketones, for example, cyclopentenones and cyclohexenones and their homologs, are reduced by these reagents [997, 1015, 1017, 1125, 1130] (equation **192**).

192

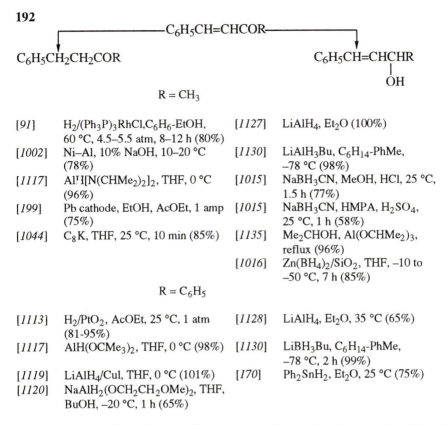

	$R = CH_3$		
[91]	$H_2/(Ph_3P)_3RhCl,C_6H_6$-EtOH, 60 °C, 4.5–5.5 atm, 8–12 h (80%)	[1127]	$LiAlH_4$, Et_2O (100%)
[1002]	Ni–Al, 10% NaOH, 10–20 °C (78%)	[1130]	$LiAlH_3Bu$, C_6H_{14}-PhMe, –78 °C (98%)
[1117]	$Al^rI[N(CHMe_2)_2]_2$, THF, 0 °C (96%)	[1015]	$NaBH_3CN$, MeOH, HCl, 25 °C, 1.5 h (77%)
[199]	Pb cathode, EtOH, AcOEt, 1 amp (75%)	[1015]	$NaBH_3CN$, HMPA, H_2SO_4, 25 °C, 1 h (58%)
[1044]	C_8K, THF, 25 °C, 10 min (85%)	[1135]	Me_2CHOH, $Al(OCHMe_2)_3$, reflux (96%)
		[1016]	$Zn(BH_4)_2/SiO_2$, THF, –10 to –50 °C, 7 h (85%)

	$R = C_6H_5$		
[1113]	H_2/PtO_2, AcOEt, 25 °C, 1 atm (81-95%)	[1128]	$LiAlH_4$, Et_2O, 35 °C (65%)
[1117]	$AlH(OCMe_3)_2$, THF, 0 °C (98%)	[1130]	$LiBH_3Bu$, C_6H_{14}-PhMe, –78 °C, 2 h (99%)
[1119]	$LiAlH_4/CuI$, THF, 0 °C (101%)	[170]	Ph_2SnH_2, Et_2O, 25 °C (75%)
[1120]	$NaAlH_2(OCH_2CH_2OMe)_2$, THF, BuOH, –20 °C, 1 h (65%)		

A regioselective reduction of an unsaturated steroidal triketone is achieved by yeast [1497].

Reduction of unsaturated ketones to saturated alcohols is achieved by *catalytic hydrogenation* using a nickel catalyst [50], or a copper chromite catalyst [51, 1136] or by treatment with a *nickel–aluminum alloy* in sodium hydroxide [1137]. If the double bond is conjugated, complete reduction can also be obtained with some hydrides. 2-Cyclopentenone is reduced to cyclopentanol in an 83.5% yield with *lithium aluminum hydride* in tetrahydrofuran [997], *lithium tris*(tert-*butoxy)aluminum hydride* (88.8% yield) [997], and *sodium borohydride* in ethanol at 78 °C (100% yield) [997]. Most frequently, however, only the carbonyl is reduced, especially with application of the inverse technique (p 29) (equation **193**).

Reduction of α,β-unsaturated ketones to unsaturated hydrocarbons is rather rare and is frequently accompanied by a shift of the double bond. Occasionally, such reductions are achieved by catalytic hydrogenation over palladium [1138]. Most reductions are accomplished in good to high yields by treatment of the *p*-toluenesulfonylhydrazones of the unsaturated ketones with *sodium borohydride* [1020], *borane* [1021], or *catecholborane* [1139], or by the

Wolff–Kizhner reduction or its modifications [*1140*]. However, *complete reduction to saturated hydrocarbons* may also occur during the Wolff–Kizhner reduction [*1141*] as well as during the Clemmensen reduction [*236*].

193

$$[51] \quad \overset{H_2/CuCr_2O_4}{\underset{175\ °C,\ 100–150\ atm}{\big|}} \quad C_6H_5CH = CHCOCH_3 \quad \overset{1.\ H_2NNHO_2SC_7H_7}{\underset{\substack{2.\ NaBH_4,\ AcOH \\ 25\ °C,\ 1\ h;\ 70\ °C, \\ 3\ h}}{\big|}} \quad [1020]$$

$$C_6H_5CH_2CH_2\underset{|}{\overset{}{CH}}CH_3 \quad 100\% \qquad\qquad 54\%\ \ C_6H_5CH_2CH=CHCH_3$$
$$\qquad\qquad\quad OH$$

One-electron reduction of α,β-unsaturated ketones yields, instead of pinacols, **1,6-diketones** formed by coupling of the semireduced species at the β carbons [*1142*]. The practical usefulness of this reaction is minimized by low yields (16% with benzalacetone).

Ketones containing acetylenic bonds are reduced selectively at the triple bond *by catalytic hydrogenation.* Over 5% palladium on calcium carbonate in pyridine (which decreases the activity of the catalyst), 17-ethynyltestosterone is reduced to 17-vinyltestorone in a 95% yield, but over palladium on charcoal in dioxane, 80% 17-ethyltestosterone is obtained; the carbonyl in position 3 and the conjugated 4,5-double bond remain intact [*537*]. *Chromous chloride* and *chromous sulfate* convert α,β-acetylenic ketones to *trans*-olefinic ketones in yields of 51–85% [*274*].

Reduction of α,β-acetylenic ketones to *acetylenic alcohols* is accomplished by o-*(dimethylaminomethyl)phenyldimethylstannane* in an 81% yield [*174*]. If the triple bond is conjugated with the carbonyl, reduction to the corresponding *allylic alcohols* occurs with *sodium cyanoborohydride* or *tetrabutylammonium cyanoborohydride*. Thus 4-phenyl-3-butyn-2-one gives 4-phenyl-3-buten-2-ol in 70–89% yields [*1015*]. If the same ketone is converted to its *p*-toluenesulfonylhydrazone and this is reduced with *bis(benzoyloxyborane)*, 1-phenyl-1,2-butadiene is obtained in a 21% yield [*1021*].

Reduction with *chiral borane NB*-Enanthrane prepared by addition of 9-borabicyclo[3.3.1]nonane to the benzyl ether of nopol (p 21) yields optically active acetylenic alcohols in 74–84% yields and 91–96% enantiomeric excess (ee) [*151*]. Similar results are obtained by using chiral B-*3-pinanyl-9-borabicyclo[3.3.1]nonane (B-3-pinanyl-9-BBN)* [*1143, 1144*] (59–98% yields, 71–92% ee), Binal-H (R = OCH_3 or OC_2H_5) [*1145*] (64–90% yields, 57–94% ee) and B-*chlorodiisopinocampheylborane* [*1146*] (70–92% yields, mostly 96–99% ee) (equation **194**).

Another route to optically active acetylenic alcohols is reduction with a reagent prepared from lithium aluminum hydride and (+)-(2S,3R)-4-dimethylamino-3-methyl-1,2-diphenyl-2-butanol. At –78 °C mainly *R* alcohols are obtained in 62–99% yields and 34–90% enantiomeric excesses [*1147*].

A peculiar reduction takes place when 4-methyl-1,1,1-triphenyl-5-undecyn-2-one is treated with *lithium triethylborohydride* in tetrahydrofuran. Without

194

$$HC \equiv CCOC_5H_{11} \longrightarrow HC \equiv CC\underset{\underset{OH}{\vdots}}{\overset{\overset{H}{\vdots}}{C}}C_5H_{11}$$

[151]	NB Enanthrane, THF, 25 °C, 48 h	74%	95% ee	
[1143]	B-3-pinanyl-9-BBN, THF, reflux; 2.5 h, RT, 48 h	65%	92%ee	R
[1144]	B-3-pinanyl-9-BBN, THF, reflux; H₂O₂, NaOH, 40 °C, 3 h	86%	86%ee	R
[1145]	Binal-H (R = OMe), THF, –100 °C, 1 h; –78 °C, 2 h	80%	96%ee	S

affecting the triple bond, the reagent hydrogenolyzes the triphenylmethyl group and reduces the carbonyl [1148] (equation **195**).

195 [1148]

$$(C_6H_5)_3CCOCH_2\underset{\underset{CH_3}{|}}{C}HC \equiv CC_5H_{11} \xrightarrow[\text{–45 °C→RT, 40 h}]{\text{LiBHEt}_3, \text{THF}} HOCH_2CH_2\underset{\underset{CH_3}{|}}{\underset{77\%}{C}}HC \equiv CC_5H_{11}$$

$$+ (C_6H_5)_3CH$$

SUBSTITUTION DERIVATIVES OF KETONES

Halogenated ketones can be *reduced to halogenated alcohols* with complex hydrides. Although these reagents are capable of hydrogenolyzing the carbon–halogen bond, reduction of the carbonyl group takes preference. 1,3-Dichloro-acetone is reduced with *lithium aluminum hydride* to 1,3-dichloro-2-propanol in ether at –2 °C in 77% yield, even if an excess of the hydride is used [1149], and 4-bromo-2-butanone yields 4-bromo-2-butanol on reduction with *sodium borohydride* [1150]. In the reduction of halogenated ketones containing reactive halogens, the inverse technique and mild conditions are advisable. Cyclohexyl trifluoromethyl ketone gives the optically active fluorinated alcohol by reduction with *baker's yeast* in 86–88% yield and 99% enantiomeric excess [454]. o-*(Dimethylaminomethyl)phenyldimethylstannane* reduces halogenated ketones to halogenated alcohols in 61–87% yields [174].

Catalytic hydrogenation in basic medium may **hydrogenolyze halogen even without reducing the keto group**: 8-chloro-5-methoxy-1-tetralone gives a 53–56% yield of 5-methoxy-1-tetralone on treatment with hydrogen over 10% palladium on charcoal in ethanol containing methylamine at room temperature and atmospheric pressure [1151]. Other reagents capable of hydrogenolyzing reactive halogens without affecting the carbonyl group are *hexamethyldisilane* with tetrakis(triphenylphosphine)palladium [417], *zinc* in acetic acid and tetramethylenediamine at 0–25 °C (78–86% yields) [1152], *sodium dithionite*

196

$$C_6H_5COCH_2X \xrightarrow[\text{X = Cl}]{} C_6H_5COCH_3$$

[302]	NaI, 50% aq. THF, RT, 3 h	53%
[349]	$Na_2S_2O_4$, $NaHCO_3$, H_2O, DMF, 70 °C, 1 h	90%
[363]	$HOCH_2OSONa$, EtOH, reflux 0.5 h	83%
[1153]	Ph_3PHI, MeCN, RT, 4 h	83%

X = Br

[417]	$Me_3SiSiMe_3$, $(Ph_3P)_4Pd$, C_6H_6, 150 °C, 17 h	50%
[302]	NaI, 50% aq. THF, RT, 24 min	74%
[349]	$Na_2S_2O_4$, $NaHCO_3$, H_2O, DMF, RT, 5 h	95%
[363]	$HOCH_2OSONa$, EtOH, reflux <0.5 h	79%
[1153]	Ph_3PHI, MeCN, RT, 3 h	85%

[349], *sodium formaldehyde sulfoxylate* [363], and *triphenylphosphonium iodide* [1153] (equation **196**).

Isopropyl alcohol under ultraviolet iradiation reduces α-fluoro-cyclohexanone to cyclohexanone in 60–65% yield [428]. A particularly reactive halogen like the bromine in 1-bromo-1,1-dibenzoylethane is replaced quantitatively by hydrogen on refluxing for 5 min with a solution of *potassium iodide* and dilute *hydrochloric acid* in acetone [320]. *Sodium iodide* in 50% aqueous tetrahydrofuran or dioxane dehalogenates α-chloro-, α-bromo-, and α-iodoketones to ketones [302]. Steroidal α-iodo ketones are reduced to ketones with *chromous chloride* [265] (Procedure 42, p 313).

The *Meerwein–Ponndorf reduction* of 2-bromocyclohexanone gives 2-bromocyclohexanol (30%) and cyclohexanol (33%) [432]. Better results are obtained by using triphenylphosphonium iodide, which reduces α-chloro- and α-bromocyclohexanones to cyclohexanone at room temperature in respective yields of 72 and 80% [1153].

Treatment of **α-halocyclohexanones** with *hydrazine* in alkaline medium affords **unsaturated hydrocarbons**. The reaction is carried out by refluxing the α-halo ketone with hydrazine and potassium acetate in dimethoxyethane or cyclohexene. Cyclohexene is used as an acceptor of hydrogen produced by the decomposition of hydrazine to diimide and further to nitrogen and hydrogen in order to prevent the formation of saturated hydrocarbons. This reaction gives good yields only with ketones containing the carbonyl group in a six-membered ring and has found use in transformations of steroids [1154] (equation **197**).

In α-bromo-α,β-unsaturated cyclic ketones the *replacement of vinylic bromine* by hydrogen is accompanied by a shift of the double bond into the β,γ position [1155] (equation **198**).

Vinylic bromine in 2-bromocoprost-1-en-3-one is hydrogenolyzed in high yield by refluxing with *zinc* in ethanol without the double bond or carbonyl group being affected [1156]. A trichloromethyl ketone, treated with zinc in acetic acid, gives a methyl ketone. With *amalgamated zinc* in hydrochloric acid even the carbonyl is reduced [*673*] (equation **199**). Bromine in α-bromo ketones is selectively replaced by hydrogen by means of *chromous chloride* [*1157, 1158*] *or vanadous chloride* (80–98% yields) [*294*].

197 [*1154*]

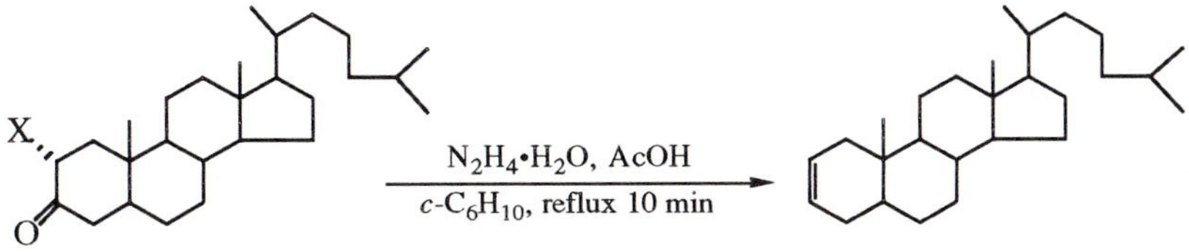

X	F	Cl	Br	I
Yield	71%	68%	62%	54%

198 [*1155*]

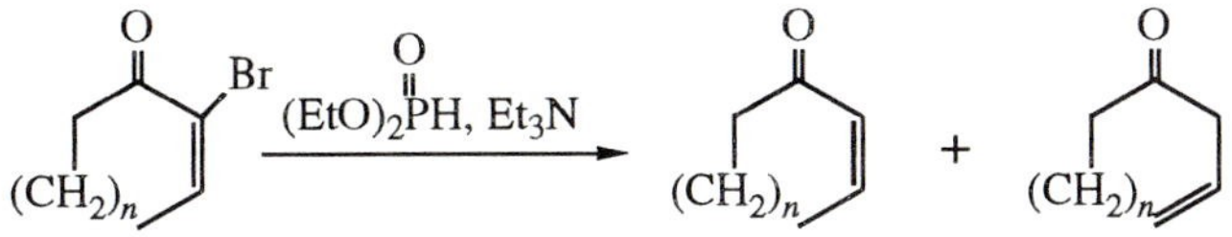

n			
2	RT, 15 h	3%	63%
3	80 °C, 3.5 h	9%	38%
6	80 °C, 1 h	8%	76%

199 [*673*]

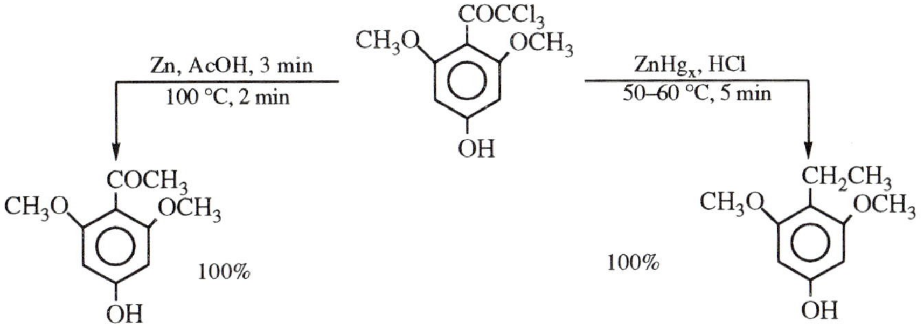

Halogen, especially chlorine, in aromatic rings of ketones usually survives reduction of the carbonyl group. Concomitant hydrogenolysis of any halogen and reduction to alcohols occurs in the treatment of **haloaryl ketones** with *Raney nickel* or *Raney copper alloys* [*1159*] (equation **200**).

Denitration of α-nitro ketones to ketones is accomplished in 61–86% yields by treatment with *triethylsilane and sodium hydrosulfite* in aqueous hexamethylphosphoric triamide at room temperature [355].

The **reduction of ketones containing nitro groups to nitro alcohols** is best carried out by borohydrides. 5-Nitro-2-pentanone is converted to 5-nitro-2-pentanol in an 86.6% yield by reduction with *sodium borohydride* at 20–25 °C. Other nitro ketones give 48.5–98.7% yields, usually higher than were obtained by the *Meerwein–Ponndorf reduction* [1160]. 2-Acetamido-3-(*p*-nitrophenyl)-1-hydroxypropan-3-one is reduced with *calcium borohydride* at –30 °C to 70% *threo-* and 10% *erythro-*2-acetamido-3-(*p*-nitrophenyl)propane-1,3-diol, whereas *sodium borohydride* affords a mixture of these isomers in 25% and 47% yields, respectively [1161]. Highly enantioselective reduction of nitro ketones is accomplished by *baker's yeast.* 4-Nitro-2-butanone and 5-nitro-2-butanone give optically active nitro alcohols in 40–60% yields and 94–98% enantiomeric excesses [1162].

200   [1159]

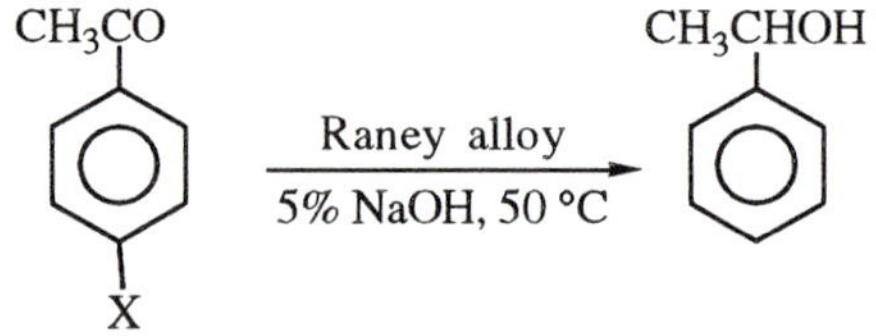

X	NiAl	CuAl
F	86%	93% [a]
Cl	88%	85%
Br	71%	91%

[a] Irradiation or ultrasound was applied.

The **reduction of a dinitro ketone to an azo ketone** is best achieved with glucose. 2,2'-Dinitrobenzophenone treated with *glucose* in methanolic sodium hydroxide at 60 °C affords 82% dibenzo[*c*,*f*][1,2]diazepin-11-one, whereas *lithium aluminum hydride* yields 24% bis(*o*-nitrophenyl)methanol [436].

Conversion of aromatic nitro ketones with a nitro group in the ring into amino ketones is achieved by *zinc and nickel dichloride* in refluxing methanol in an 87% yield [241] and also by means of *stannous chloride,* which reduces 4-chloro-3-nitroacetophenone to 3-amino-4-chloroacetophenone in a 91% yield [255]. A more dependable reagent for this purpose is *iron,* which, in acidic medium, reduces *m*-nitroacetophenone to *m*-aminoacetophenone in an 80% yield and *o*-nitrobenzophenone to *o*-aminobenzophenone in an 89% yield (stannous chloride was unsuccessful in the latter case) [1163]. Iron has also been used for the reduction of *o*-nitrochalcone, 3-(*o*-nitrophenyl)-1-phenyl-2-propen-1-one, to 3-(*o*-aminophenyl)-1-phenyl-2-propen-1-one in an 80% yield [765].

Catalytic hydrogenation over palladium in acetic acid and sulfuric acid at room temperature and 2.5 atm reduces nitroacetophenones and their homologs and derivatives all the way through to the alkylanilines in yields of 78.5–95% [*1164*].

By reduction combined with hydrolysis, 5-nitro-2-heptanone is converted to 2,5-heptanedione on treatment with *titanium trichloride* in aqueous glycol monomethyl ether (85% yield) [*1165*].

α-Diazoketones yield various products depending on their structures and on the reducing agents. Alicyclic α-diazo ketones such as α-diazocamphor, on *hydrogenation* over palladium oxide, after absorption of 1 mol of hydrogen gives the corresponding hydrazone (camphorquinone hydrazone, 83.5%) [*1167*]. Aromatic and aliphatic–aromatic α-diazo ketones over the same catalyst after absorption of 2 mol of hydrogen yield α-amino ketones (which condense to 2,5-disubstituted pyrazines) [*1166, 1167*]. In the presence of copper oxide or in hydrochloric acid, nitrogen is replaced by hydrogen in hydrogenation with palladium oxide or palladium on charcoal, and ketones are formed [*1166, 1167*]. The same result is obtained (without the copper oxide) by reduction with *hydrogen iodide* (68–100% yields) [*322, 1168*]. Also *ethanol* on irradiation causes replacement of nitrogen by hydrogen, giving up to an 85% yield of ω-(*p*-toluenesulfonyl)acetophenone from ω-(*p*-toluenesulfonyl)diazoacetophenone [*423*]. *Lithium aluminum hydride* in ether reduces ω-diazoacetophenone to 2-amino-1-phenylethanol in 93% yield [*1166*] (equation **201**).

201

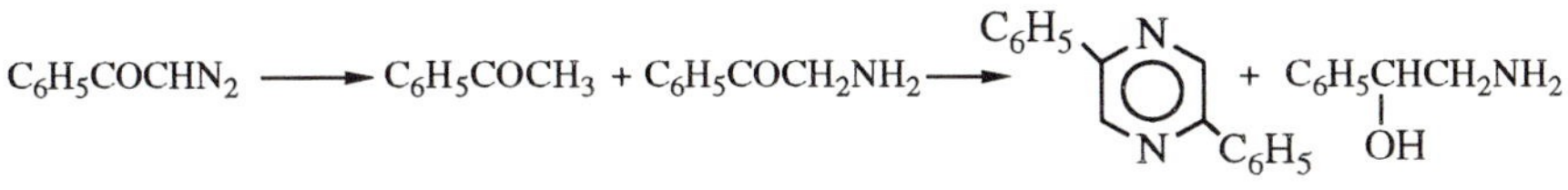

$$C_6H_5COCHN_2 \longrightarrow C_6H_5COCH_3 + C_6H_5COCH_2NH_2 \longrightarrow$$

[structure of 2,5-diphenylpyrazine] $+ \ C_6H_5\underset{\underset{OH}{|}}{CH}CH_2NH_2$

[*1166*] H$_2$/PdO, AcOEt 25 °C, 1 atm			70%
[*1166*] H$_2$/PdO, AcOEt, AcOH		55%	
[*1166*] H$_2$/PdO, AcOH, HCl	30%		
[*1167*] H$_2$/Pd(C), EtOH	50%		
[*1167*] H$_2$/Pt	15%		45%
[*1167*] AlHg, HCl, or Zn, HCl	21%		
[*1167*] LiAlH$_4$, Et$_2$O			93%

In **aromatic ketones containing diazonium groups,** *hypophosphorous acid* replaces nitrogen by hydrogen. Diazotized 4-amino-3,5-dichloro-acetophenone thus affords 3,5-dichloroacetophenone in an 80% yield [*394*].

In **azido ketones** both functions are reduced with *lithium aluminum hydride* in refluxing ether; ω-azidoacetophenone gives a 49.5% yield of 2-amino-1-phenylethanol [*787*].

Sulfoxide-containing ketones with the sulfoxide group in α positions are reductively cleaved to ketones with *aluminum amalgam* [*215*] (equation **202**).

202   [*215*]

$$RCOCH_2SCH_3 \xrightarrow[\text{65 °C, 1-1.5 h}]{\text{AlHg}_x, \text{ aq. THF}} RCOCH_3 \quad 89-100\%$$
$$\underset{O}{\|}$$

HYDROXY AND AMINO KETONES

Hydroxy ketones and hydroxy-α,β-unsaturated ketones in steroids such as estrone and testosterone, respectively, can be *reduced to diols biochemically.* Estrone acetate gives a 68% yield of α-estradiol on incubation with *baker's yeast* at room temperature after 5 days [*1169*]. Testosterone is reduced by bacteria to the saturated hydroxy ketones, etiocholan-17-ol-3-one and androstan-17-ol-3-one, and further to the diols, *epi*-etiocholane-3,17-diol and the epimeric isoandrostane-3,17-diol, both in low yields [*1170*].

Bacillus polymyxa in a hydrogen atmosphere reduces (*R,S*)-acetoin to *erythro*- and *threo*-butane-2,3-diol in 100% yield [*1171*], and *Saccharomyces cerevisiae* converts 3,3-dimethyl-1-hydroxybutan-2-one to (–)-(*R*)-3,3-dimethyl-butane-1,2-diol in a 66% yield [*1172*].

Acyloins are converted to mixtures of stereoisomeric *vicinal diols* by *catalytic hydrogenation* over copper chromite [*1173*]. More frequently they are reduced to *ketones*: by *zinc* (77% yield) [*1174, 1175*], by *zinc amalgam* (50–60% yields) [*1176*], by *tin* (86–92% yields) [*250*], or by refluxing with 47% *hydriodic acid* in glacial acetic acid (70–90% yields) [*1177*], by treatment with *lithium diphenylphosphide* (Ph_2PLi) in tetrahydrofuran at room temperature (52–86% yields) [*398*], with *red phosphorus and iodine* in carbon disulfide at room temperature (80–90% yields) [*1178*] (Procedure 47, p 316), or with *iodotrimethylsilane* in acetonitrile at room temperature (27–90% yields) [*418*]. Acyloins are readily accessible by reductive condensation of esters (p 212), and hence these reductions provide a very good route to ketones and the best route to macrocyclic ketones [*1174*] (equation **203**).

203

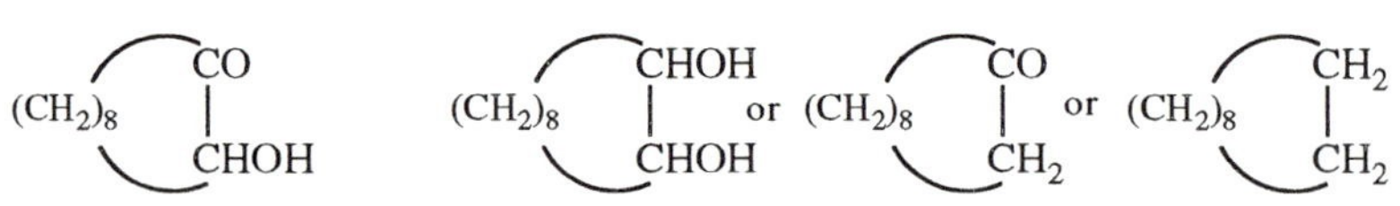

[*1173*]	H_2/$CuCr_2O_4$, 150 °C, 135 atm	48–52% *cis* 27–32% *trans*		
[*1174*]	Zn wool, HCl		77%	
[*1174*]	Zn wool, HgCl$_2$, HCl, AcOH, reflux 88 h			79%
[*1177*]	HI, AcOH, reflux 2 h		90%	

In **steroidal α-hydroxy ketones** with the keto group in position 20 and the hydroxylic group in position 17, the hydroxy, and better still, acetoxy group is replaced by hydrogen on refluxing for 24 h with *zinc* dust and acetic acid.

3β,17α-Diacetoxyallopregnan-20-one gives an 89% yield of 3β-acetoxy-allopregnane, whereas its 17β-epimer's yield is only 46% [*1179*].

Complete reduction of acyloins to hydrocarbons is accomplished by the *Clemmensen reduction*. Sebacoin affords cyclodecane in a 79% yield [*1174*], and benzoin gives 1,2-diphenylethane in an 84% yield [*991*]. A stereoselective reduction of *polyhydroxycyclanones* is achieved by *catalytic hydrogenation* over Raney nickel. Thus *allo*-inositol is prepared in better than 90% stereoselectivity [*1180*] (equation **204**).

In **ketones containing oxirane rings** in positions α,β to their carbonyls, only the oxiranes are *deoxygenated to α,β-unsaturated ketones* by refluxing with *zinc* dust in acetic acid for 3 h [*849*] or on treatment with *chromous chloride* [*270, 869*]. The keto groups remain intact, even if there are other keto groups with conjugated double bonds in the same molecule. Accordingly 16,17-epoxycorticosterone-21-acetate is converted to 16,17-dehydrocorticosterone-21-acetate in a 94% yield [*270*]. α,β-Epoxy ketones may also be converted to allylic alcohols. On heating with *hydrazine hydrate* to 90 °C and refluxing for 15 min, 4β,5-epoxy-3-coprostanone gives 3-coprostene-5-ol in a 68% yield. This reaction can be carried out even at room temperature using alcoholic solutions of α,β-epoxy ketones, 2–3 equivalents of hydrazine hydrate, and 0.2 equivalent of acetic acid and gives good yields of the rearranged allylic alcohols [*1181*] (equation **205**).

204 [*1180*]

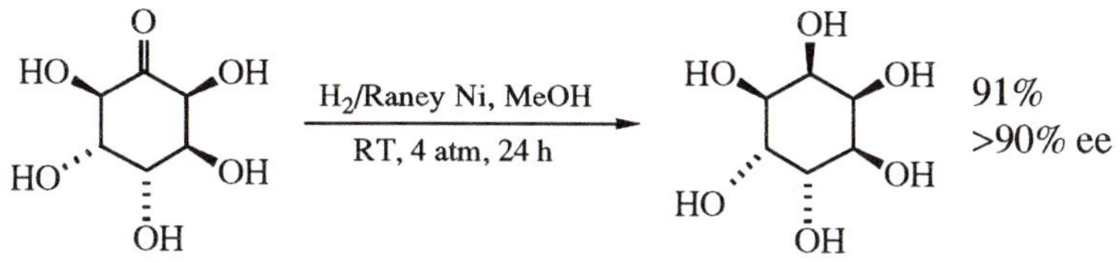

205 [*1181*]

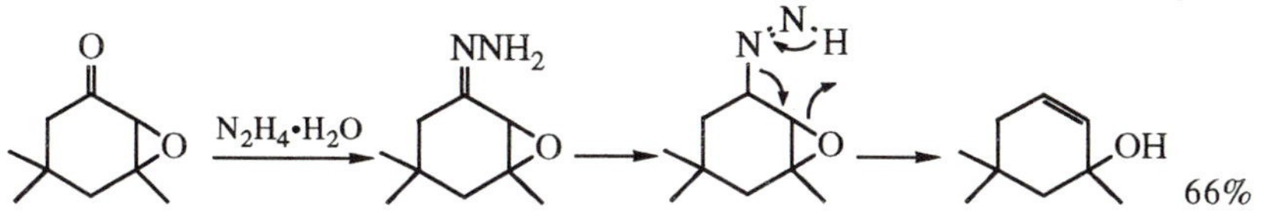

Ketones containing sulfur or nitrogen atoms bound to carbons suffer carbon–sulfur or carbon–nitrogen bond cleavage under the conditions of the *Clemmensen reduction* [*235, 1110*]. A **ketosulfone** is reduced to a ketone with *aluminum amalgam* in aqueous tetrahydrofuran in an 89% yield [*215*] and to a sulfone-alcohol with *zinc* in refluxing 80% acetic acid in a 70% yield [*1182*].

Amino ketones are reduced to *amino alcohols* with modified lithium aluminum hydrides. The best stereoselective results are obtained with *lithium tris(tert-butoxy)aluminum hydride* or *diisobutylaluminum hydride* in the presence of anhydrous zinc chloride [*1183*] (equation **206**). Possible ambiguities in the designation of stereoisomers with more chiral centers can be avoided by using a special nomenclature, an extension of Cahn–Ingold–Prelog specification [*1184*].

Selective reduction of **aldehydes in the presence of ketones** can be accomplished with *tetrabutylammonium triacetoxyborohydride*, which gives 80–90% yields of primary alcohols [*132*], with *aminoborane* and *tert-butylaminoborane* (94–98% yields) [*1185*], or *tributylstannane* [*987*], especially in the presence of *low-valent titanium* (43–80% yields) [*172*]. High selectivity is also achieved with *lithium trialkoxyaluminum hydrides* [*985*] and *potassium triphenylborohydride*, both at –78 °C [*136*] (equation **207**).

206   [*1183*]

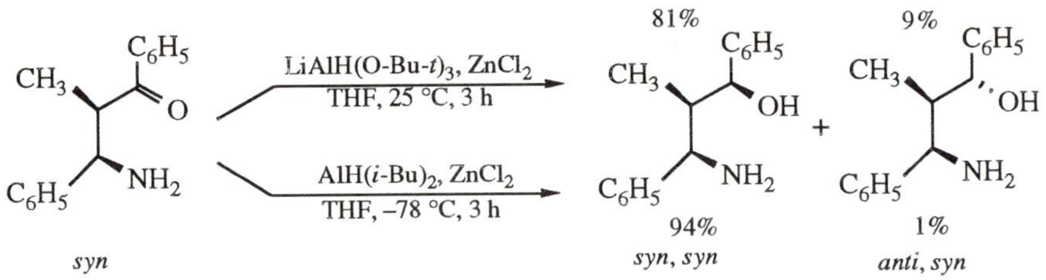

207   [*172*]

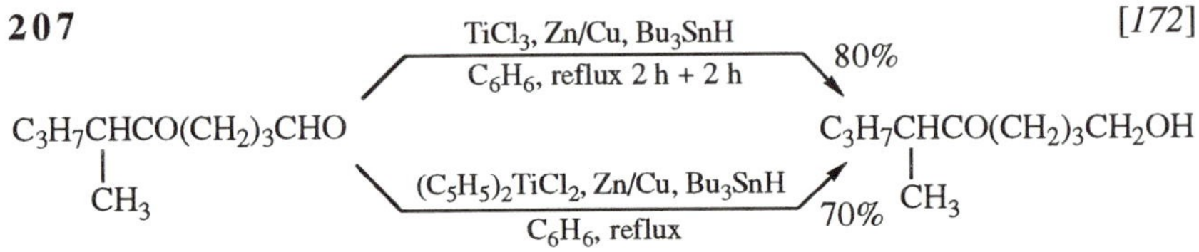

In the system hexanal–cyclohexanone the ratio of the primary to the secondary alcohol is 87:13 at 0 °C and 91.5:8.5 at –78 °C with *lithium tris(*tert-butoxy*)aluminum hydride* [*985*], and 93.6:6.4 at 0 °C and 99.6:0.4 at –78 °C with *lithium tris(triethylmethoxy)aluminum hydride* [*985*].

Also dehydrated aluminum oxide soaked with *isopropyl alcohol* and especially *diisopropylcarbinol* reduces aldehydes in preference to ketones [*990*].

Because of large differences in the rate of reduction of various ketones, *amine boranes* allow for selective reductions of ketones [*1185*]. Thus cyclohexanone is reduced in preference to acetophenone and aliphatic ketones.

Selective reduction of **ketones in preference to aldehydes** is achieved by *sodium borohydride* in the presence of *lanthanide chlorides* such as *cerium trichloride*, and *erbium trichloride* in aqueous ethanol. Most likely, these metals coordinate with hydrated aldehydes, thus allowing the reducing reagent to reduce ketones that do not form hydrates so readily [*998, 1186, 1187*]. This "one-pot" procedure is preferable to the classical three-step process converting aldehydes to acetals, reducing the ketones, and deprotecting the aldehydes. The method is especially suitable for the selective reduction of ketoaldehydes [*998, 1186*] (equation **208**).

208 [*998, 1186*]

DIKETONES AND QUINONES

The reduction of diketones is very complex. They can be partially reduced to ketols (hydroxy ketones), ketones or diols, or completely reduced to hydrocarbons. Depending on the mutual distance of the two carbonyl groups and reagents used, carbon–carbon bond cleavage may occur and may be followed by recyclizations or rearrangements. Some reactions may result in the formation of alkenes. Quinones react in their own specific way.

α-**Diketones** (1,2-diketones) *are reduced to* α-*hydroxy ketones or ketones.* Diacetyl (butane-2,3-dione) is quantitatively converted to acetoin (3-hydroxy-2-butanone) by heating at 100 °C with granulated *zinc in dilute sulfuric acid* [*1188*]; benzils are quantitatively converted to benzoins by refluxing with *zinc dust* in aqueous *N,N*-dimethylformamide [*240*], by treatment with *vanadous chloride* [*295*], by heating with *benzpinacol* [*1189*], and by treatment at 0 °C with *hydrogen sulfide* in piperidine–*N,N*-dimethylformamide [*328*] (Procedure 48, p 316). Cyclotetradecane-1,2-dione is reduced to 2-hydroxycyclotetradecanone by refluxing for 1 h with *triethyl phosphite* in benzene (78% yield) [*405*] (equation **209**).

Ketones are obtained from α-**diketones** by reduction with *hydrogen sulfide* in a pyridine–methanol solution [*328*], by refluxing with 47% *hydriodic acid* in acetic acid (80% yield) [*1177*], and by *decomposition of monohydrazones* with alkali [*1190*]. **Reduction of** α-**diketones to hydrocarbons** is achieved by *decomposition of bis(hydrazones)* by alkali [*1190*] (equation **209**).

The Clemmensen **reduction of** β-**diketones** (1,3-diketones) is rather complicated. The first step in the reaction of 2,4-pentanedione with *zinc amalgam* is an intramolecular pinacol reduction leading to a cyclopropanediol. Next the cyclopropane ring is opened in the acidic medium, and a rearrangement followed by a reduction gives the final product, a ketone, with a changed carbon skeleton [*1191, 1192*]. The ketone is usually accompanied by small amounts of the corresponding hydrocarbon [*1191*] or an α-hydroxy ketone [*1192*] (equation **210**).

γ-**Diketones** (1,4-diketones) are reduced to *hydroxy ketones* by *catalytic hydrogenation* using ruthenium on silica gel at 20 °C and 6.2 atm. 1,4-Cyclohexanedione thus gives 70% yield of 4-hydroxycyclohexanone [*1193*]. The *Clemmensen reduction* is very complex. In addition to products of partial reduction of hexane-2,5-dione, 2-hexanone, 2-hexanol, and a mixture of *cis-* and *trans*-2-hexen-4-ol, a small amount of 3-methyl-2-pentanone is isolated [*1194*]. This compound results from a rearrangement of the initially formed 1,2-dimethyl-1,2-cyclobutanediol to 1-acetyl-1-methylcyclopropane that ultimately

209

$$C_6H_5COCOC_6H_5 \longrightarrow C_6H_5\underset{\underset{OH}{|}}{C}HCOC_6H_5 \text{ or } C_6H_5CH_2COC_6H_5 \text{ or } C_6H_5CH_2CH_2C_6H_5$$

[240]	Zn dust, 80% aq. DMF, reflux 5–6 h	93%	
[295]	TiCl$_3$, THF, 25 °C	88%	
[295]	VCl$_2$, THF, 25 °C	100%	
[328]	H$_2$S, C$_5$H$_{11}$N-DMF, 0 °C, 1 h	100%	
[1189]	Ph$_2$C(OH)CPh$_2$(OH), 160–170 °C	85%	
[328]	H$_2$S, C$_5$H$_5$N–MeOH, 0 °C, 4 h		100%
[1190]	1. N$_2$H$_4$, H$_2$O; 2. dil. NaOH, reflux 7.5 h		73%
[1190]	1. 2N$_2$H$_4$; 2. 4 N KOH in MeOH, reflux 40 h			100%
[481]	Xanthomonas oryzae, 30 °C, 3 days R (opt. pure)	86%	

210   [1191]

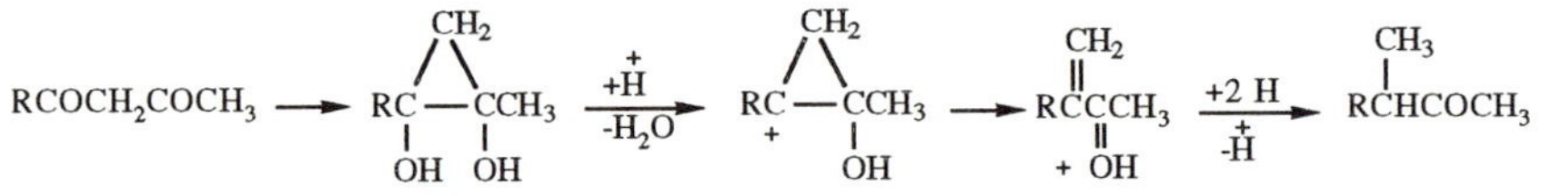

$$RCOCH_2COCH_3 \longrightarrow \underset{\underset{OH\ \ OH}{|\ \ \ |}}{RC-CCH_3}\overset{CH_2}{\diagup\diagdown} \xrightarrow[-H_2O]{+H} \underset{\underset{OH}{|}}{RC-CCH_3}\overset{CH_2}{\diagup\diagdown} \longrightarrow \underset{+\ OH}{RCCCH_3}\overset{CH_2}{\|} \xrightarrow[-H]{+2H} \underset{}{RCHCOCH_3}\overset{CH_3}{|}$$

R = Me   ZnHg, HCl, reflux   35–50%
R = Ph   65%

rearranges to 3-methyl-2-pentanone [1194]. 1-Phenyl-1,4-pentanedione is reduced to 5-phenyl-2-pentanone (76% yield) and 1-phenylpentane (12%) (equation **211**).

In the Clemmensen reduction of 1,4-cyclohexanedione, all the products isolated from the reduction of 2,5-hexanedione are found in addition to 2,5-hexanedione (20%) and 2-methylcyclopentanone (6%). The presence of the two latter compounds reveals the mechanism of the reduction. In the first stage the carbon–carbon bond between carbons 2 and 3 ruptures, and the product of the cleavage, 2,5-hexanedione, partly undergoes aldol condensation, partly its own further reduction [1194]. The cleavage of the carbon–carbon bond in 1,4-diketones occurs during the treatment of 1,2-dibenzoylcyclobutane, which affords, on short refluxing with zinc dust and zinc chloride in ethanol, an 80% yield of 1,6-diphenyl-1,6-hexanedione [232].

211 [*1194*]

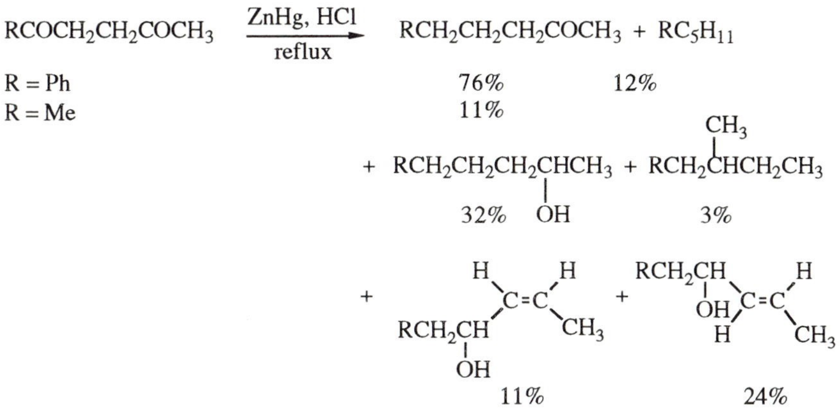

$$RCOCH_2CH_2COCH_3 \xrightarrow[\text{reflux}]{\text{ZnHg, HCl}} RCH_2CH_2CH_2COCH_3 + RC_5H_{11}$$

R = Ph   76%   12%

R = Me   11%

$$+ \; RCH_2CH_2CH_2\underset{\underset{OH}{|}}{CH}CH_3 \; + \; RCH_2\overset{\overset{CH_3}{|}}{C}HCH_2CH_3$$

                32%   3%

$$+ \quad \underset{\underset{OH}{|}}{RCH_2CH}\diagdown_{}^{H}C=C\diagup^{H}_{CH_3} \quad + \quad \underset{H}{\overset{RCH_2CH}{\underset{OH}{|}}}\diagdown C=C\diagup^{H}_{CH_3}$$

                11%   24%

An interesting reaction takes place when diketones with the keto groups in positions 1,4 or more remote are refluxed in dimethoxyethane with *titanium dichloride* prepared by reduction of titanium trichloride with a zinc–copper couple. By deoxygenation and intramolecular coupling, cycloalkenes with up to 22 members in the ring are obtained in yields of 50–95% (equation **212**). Other examples are preparations of 1-methyl-2-phenylcyclopentene in a 70% yield from 1-phenyl-1,5-hexanedione, and of 1,2-dimethylcyclohexadecene in a 90% yield from 2,17-octadecanedione [*286, 290*].

212 [*286*]

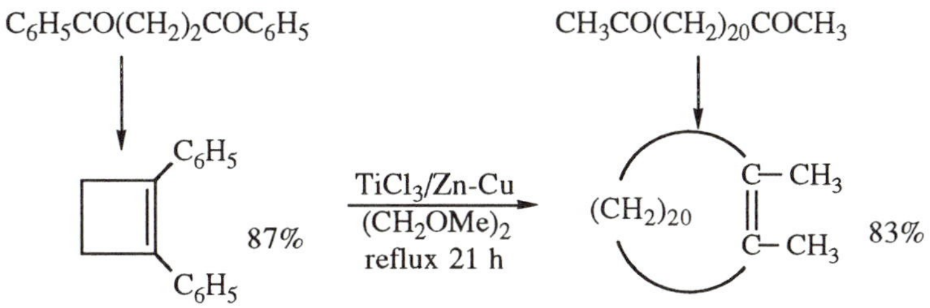

$$C_6H_5CO(CH_2)_2COC_6H_5 \qquad\qquad CH_3CO(CH_2)_{20}COCH_3$$

Reduction of diketones *to optically active hydroxy ketones or diols* can be carried out by catalytic hydrogenation using chiral catalysts or by microbial reduction. Microbial reduction of benzoin with the bacterium *Xanthomonas oryzae* affords an 86% yield of optically pure (*R*)-benzoin [*481*] (equation **209**).

Aliphatic β-diketones are converted predominantly to *anti*-diols in 84–98% yields and 94–99% enantiomeric excesses by homogeneous hydrogenation using chiral Binap catalyst in methanol at 50 °C and 3.5 atm. Under the same conditions 1-phenyl-1,3-butanedione is converted to 1-hydroxy-1-phenylbutan-3-one in an 89% yield and 98% enantiomeric excess [*104*] (equation **213**).

2 1 3   [*104*]

$$R^1\diagdown\underset{CO}{C}\diagup\underset{CO}{C}\diagdown CH_2 \diagup R^2 \xrightarrow[\text{MeOH, 50 °C, 3.5 atm}]{H_2/Ru_2Cl_4(R\text{-Binap})_2, Et_3N}$$

R¹ = CH₃, R² = C₂H₅	84% (94% ee)	5%
R¹ = CH₃, R² = C₆H₅	2%	5%
R¹ = CH₃, R² = C₆H₅ 100 °C, 7 atm	52%	5% 89% (98% ee) 1%

Biochemical reduction of 2,4-diones gives 2-hydroxy-4-alkanones in 75–100% conversions and 74–99% enantiomeric excesses with *baker's yeast (Saccharomyces cerevisiae), Geotrichum candidum,* and *Aspergillus niger.* Baker's yeast leads to the *S* configuration, whereas the other two microorganisms give mainly the *R* configuration [*1195*]. 3,5-Diones give both 3-hydroxy- and 5-hydroxy ketones, mainly with *R* configurations [*1195*].

Bicyclic δ-diketones such as 2,7-decalindiones are reduced to hydroxy ketones of the same configuration by *Rhodotorula rubra* [*477*] (59% yield of high optical purity) and by *horse liver alcohol dehydrogenase* [*470*] in an 89% yield and 98% enantiomeric excess (Procedure 63, p 321).

In **enediones** in which two carbonyl groups of a diketone are linked by an ethylenic bond, *tin* [*251*] and *chromous chloride* [*275*] reduce only the double bond and none of the conjugated carbonyl groups. A double bond conjugated with only one carbonyl group is not reduced. Refluxing cholest-4-ene-3,6-dione with chromous chloride in tetrahydrofuran yields 49% 5β-cholestane-3,6-dione, and a similar reduction of cholesta-1,4-diene-3,6-one gives 5β-cholest-1-ene-3,6-dione [*275*].

Quinones constitute one of the ***most easily reducible systems***. They can be reduced to hydroquinones, ketones, diols, or hydrocarbons.

p-Benzoquinone and its derivatives are *catalytically hydrogenated* to ***hydroquinones*** under very mild conditions. At room temperature and atmospheric pressure, hydrogenation of *p*-benzoquinone stops after absorption of just 1 mol of hydrogen when platinum in acetic acid or platinum or palladium in ethanol are used as catalysts. The reduction over palladium is faster than over platinum. However, platinum proves more efficient because, in the presence of a mineral acid, the hydrogenation proceeds further to the stage of cyclohexanol [*1196*]. Palladium is the best catalyst for the hydrogenation of *p*-benzoquinone to hydroquinone [*1196*].

Lithium aluminum hydride reduces *p*-benzoquinone to hydroquinone (70% yield) [*757*] and anthraquinone to anthrahydroquinone in a 95% yield [*757*]. *Tin* reduces *p*-benzoquinone to hydroquinone in an 88% yield [*251*] (Procedure 40, p 313). *Stannous chloride* converts tetrahydroxy-*p*-benzoquinone to hexa-hydroxybenzene in a 70–77% yield [*1197*], and 1,4-naphthoquinone to 1,4-di-hydroxynaphthalene in a 96% yield [*257*]. Other reagents suitable for reduction of quinones are *titanium trichloride* [*1198*], *chromous chloride* [*264*], *hydrogen sulfide* [*342*], and *sulfur dioxide* [*344*] (equation **214**).

214

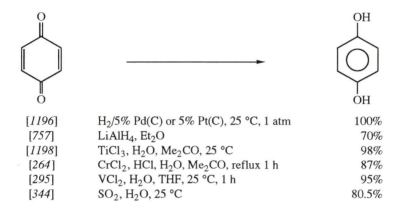

[*1196*]	H₂/5% Pd(C) or 5% Pt(C), 25 °C, 1 atm	100%
[*757*]	LiAlH₄, Et₂O	70%
[*1198*]	TiCl₃, H₂O, Me₂CO, 25 °C	98%
[*264*]	CrCl₂, HCl, H₂O, Me₂CO, reflux 1 h	87%
[*295*]	VCl₂, H₂O, THF, 25 °C, 1 h	95%
[*344*]	SO₂, H₂O, 25 °C	80.5%

Yields are usually good to excellent. Some of the reagents reduce the quinones selectively in the presence of other reducible functions. Thus hydrogen sulfide converts 2,7-dinitrophenanthrenequinone to 9,10-dihydroxy-2,7-dinitrophenanthrene in a 90% yield [*342*].

Anthraquinone is usually reduced to anthrahydroquinone, but refluxing with *tin,* hydrochloric acid, and acetic acid transforms anthraquinone to anthrone in a 62% yield [*252*].

Further reduction of quinones, acquisition of four or more hydrogens per molecule, is achieved with *lithium aluminum hydride*, which reduces in yields lower than 10% 2-methyl-1,4-naphthoquinone to 1,2,3,4-tetrahydro-1,4-dihydroxy-2-methylnaphthalene and 1,2,3,4-tetrahydro-4-hydroxy-1-oxo-2-methylnaphthalene [*1199*]. *Lithium aluminum hydride* [*1199*], *sodium borohydride, lithium triethylborohydride,* and *9-borabicyclo[3.3.1]nonane* [*133*] convert anthraquinone to 9,10-dihydro-9,10-dihydroxyanthracene in respective yields of 67, 65, 77, and 79%.

Complete deoxygenation of quinones to hydrocarbons is accomplished in yields of 80–85% by heating with a mixture of *zinc,* zinc chloride, and sodium chloride at 210–280 °C [*1200*]. Refluxing with *stannous chloride* in acetic and hydrochloric acid followed by refluxing with *zinc* dust and 2 N sodium hydroxide reduces 4'-bromobenzo[5'.6':1.2]anthraquinone to 4'-bromobenzo-[5'.6':1.2]anthracene in a 95% yield [*258*], and heating with iodine, phosphorus, and 47% *hydriodic acid* at 140 °C converts 2-chloroanthraquinone to 2-chloroanthracene in a 75% yield [*313*]. Also *aluminum* in dilute sulfuric acid can be used for reductions of the same kind [*225*].

Reductions of keto acids, keto esters, keto amides, and keto nitriles are discussed in the appropriate sections.

KETALS AND THIOKETALS

Ketals of acetone and cyclohexanone with methyl, butyl, isopropyl, and cyclohexyl alcohols are ***hydrogenolyzed to ethers and alcohols*** by *catalytic hydrogenation.* Platinum and ruthenium are inactive, and palladium is only partly active, but 5% rhodium on alumina is the best catalyst. In the presence of a mineral acid, it converts the ketals to ethers and alcohols in yields of 70–100% [1201].

The reduction is believed to be preceded by an acid-catalyzed reversible cleavage of the ketals to alcohols and unsaturated ethers, which are subsequently hydrogenated. Mineral acid is essential. Best yields and fastest reductions are found with ketals of secondary alcohols. The hydrogenation proceeds at 2.5–4 atm at room temperature with ketals of secondary alcohols, and at 50–80 °C with ketals of primary alcohols. Acetone and cyclohexanone diisopropyl ketals give 75% and 90% yields of diisopropyl and cyclohexyl isopropyl ether at room temperature after 1 and 2.5 h, respectively [1201] (equation 215).

The reagents of choice for the reduction of ketals to ethers are *alane* prepared in situ from lithium aluminum hydride and aluminum chloride in ether [1029, 1202], *diisobutylalane* [1203], and *chloroborane* complexes with ether or dimethyl sulfide [1204]. At room temperature ethers are obtained in 61–92% yields [1029, 1202]. On hydrogenolysis at 0 °C or room temperature, cyclic ketals prepared from ketones and 1,2- or 1,3-diols afford alkyl β- or γ-hydroxyalkyl ethers in 65-92% yields [1029, 1203, 1204].

215

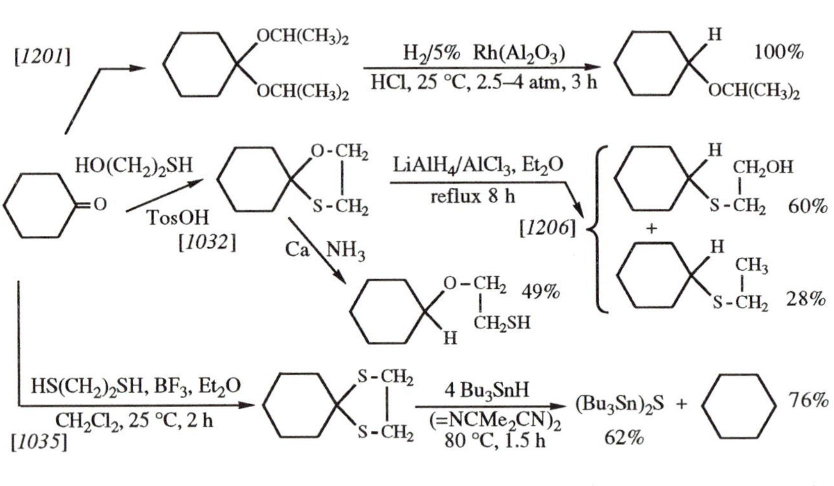

Cyclic five-membered **monothioketals are desulfurized by Raney nickel** mainly to their parent ketones (53–55% yields) and several byproducts [1205]. After stirring for 2 h at 25 °C in benzene with W-2 *Raney nickel*, 4-*tert*-butylcyclohexanone ethylene monothioketal affords 53–55% 4-*tert*-butylcyclohexanone, 12–14% *cis*-4-*tert*-butylcyclohexyl ethyl ether, 6–11% 4-*tert*-butylcyclohexanone diethyl ketal, 16% 4-*tert*-butylcyclohex-1-enyl ethyl ether, and 7–12% 4-*tert*-butylcyclohexene [1205].

Alane formed by the reaction of lithium aluminum hydride and aluminum chloride in ether cleaves exclusively the carbon–oxygen bond in cyclic monothioketals derived from ketones and 2-mercaptoethanol, and on refluxing with a 100% excess for 2 h produces β-hydroxyethyl sulfides (yields 66–91%);

on prolonged heating with the reagent these β-hydroxyethyl sulfides are further reduced to the corresponding ethyl sulfides (thioethers) (28–81% yields) [*1206*] (equation **215**).

Reduction of cyclic five-membered ethylene monothioketals with *calcium* in liquid ammonia cleaves the bond between carbon and sulfur and yields *alkyl β-mercaptoethyl ethers* (7–88%) [*1032*]. Cyclic five-membered *ethylene dithioketals* (ethylenemercaptoles) afford analogously *alkyl β-mercaptoethyl thioethers* (85% yields) [*1032*] (equation **215**).

Mercaptoles (dithioketals) can be partly desulfurized to thioethers (sulfides) or completely desulfurized to hydrocarbons.

Partial *reduction to thioethers* is accomplished by treatment of the dithioacetals with pyridine, *borane*, aluminum chloride, and trifluoroacetic acid in dichloromethane with cooling in 72–88% [*1207*] or by treatment with *thiophenol* and sodium hydride in dimethoxyethane in 54–90% yields [*337*].

Most frequently **mercaptoles (dithioketals) are completely desulfurized to hydrocarbons** by *Raney nickel* [*1033, 1034*]. As with aldehydes, conversion of ketones to mercaptoles followed by desulfurization to hydrocarbons represents the most gentle reduction of the carbonyl group to methylene. The desulfurization is accomplished by refluxing of the mercaptole with a large excess of Raney nickel in ethanol or other solvents. If reducible functions are present in the molecule of the mercaptole, Raney nickel must be stripped of hydrogen by refluxing with acetone for a few hours before use [*47*].

Nickel prepared by reduction of nickel chloride with sodium borohydride is used for desulfurization of the diethylmercaptole of benzil. Partial desulfurization using 2 mol of nickel per mol of the mercaptole gives a 71% yield of ethylthiodesoxybenzoin, whereas treatment with a 10-fold molar excess of nickel over the mercaptole affords a 61% yield of desoxybenzoin (benzyl phenyl ketone) [*1208*] (equation **216**).

216 [*1208*]

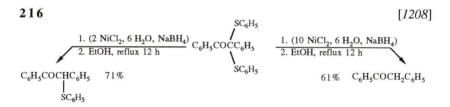

Mercaptoles of ketones are best prepared by treatment of ketones with ethanedithiol or 1,3-propanedithiol in the presence of anhydrous zinc chloride or boron trifluoride etherate. Many desulfurizations have been carried out with these cyclic mercaptoles, especially in steroids. Yields of the hydrocarbons range from 50 to 95% [*1034*].

Alternative desulfurizations can be achieved using *tributylstannane* or hydrazine. Reaction with tributylstannane is carried out by heating the mercaptole for 1.5 h at 80 °C with 3–4 equivalents of tributylstannane and azobis(isobutyronitrile) as a catalyst and distilling the product and the byproduct, bis(tributyltin) sulfide, in vacuo. Yields are 74–95% [*1035*] (equation **215**).

The reaction with *hydrazine* consists of heating the mercaptole with 3–5 parts by volume of hydrazine hydrate, 1.5–2.5 parts by weight of potassium hydroxide, and 8–20 parts by volume of diethylene or triethylene glycol to 90–190 °C until the evolution of nitrogen ceases. Times required are 0.5–3 h, and yields range from 60% to 95% [*1209*].

KETIMINES, KETOXIMES, AND HYDRAZONES

Ketimines are reduced to amines very easily by *catalytic hydrogenation,* by complex hydrides, and by formic acid. They are intermediates in reductive amination of ketones (p 187). An example of the reduction of a ketimine is conversion of 3-aminocarbonyl-2,3-diphenylazirine to the corresponding aziridine by *sodium borohydride, potassium borohydride,* and *sodium bis(2-methoxyethoxy)aluminum hydride* [*1210*] (equation **217**).

217   [*1210*]

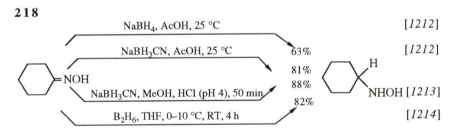

Very good yields of amines are obtained by reduction with *isopropyl alcohol* in the presence of aluminum isopropoxide and *Raney nickel* [*1211*].

Aldoximes and ketoximes are reduced to **N-substituted hydroxylamines** by *sodium borohydride* [*1212*], *sodium cyanoborohydride* [*1212, 1213*], or *diborane* [*1214*] (equation **218**).

218

Treatment of ketoximes with *sodium borohydride* and acetic acids at 25 °C gives *N*-alkylhydroxylamines. Heating of the oximes with sodium borohydride and carboxylic acids at 55 °C affords products of **reductive alkylations,** *N,N*-disubstituted hydroxylamines [*1212*]. Thus cyclohexanone oxime and sodium borohydride heated at 55 °C with acetic acid gives *N*-cyclohexyl-*N*-ethyl-hydroxylamine, and with propionic acid at 55 °C, *N*-cyclohexyl-*N*-propyl-hydroxylamine is obtained [*1212*] (equation **219**).

219 [*1212*]

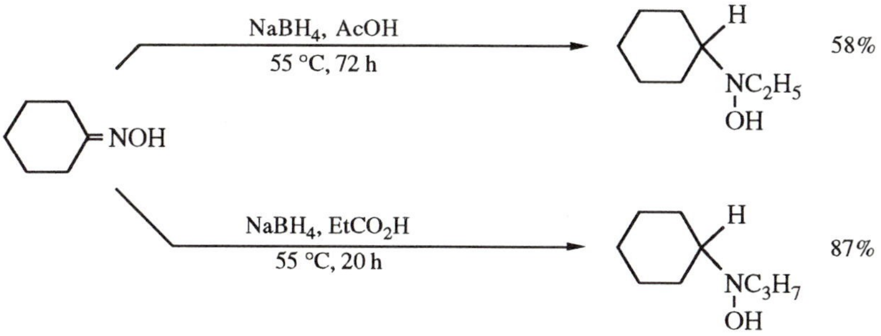

Reduction of ketoximes is one of the most useful routes *to primary amines.* Many methods are available: *Catalytic hydrogenation* over 5% rhodium on alumina converts cycloheptanone oxime to cycloheptylamine in an 80% yield at 20–60 °C and 0.75–1 atm [*1215*], and treatment with hydrogen and Raney nickel at 25–30 °C and 1–3 atm gives cyclohexylamine from cyclohexanone oxime in a 90% yield [*46*]. 2-Alkylcyclohexanone oximes yield *cis-* or *trans*-2-alkylcyclohexylamines depending on the method and reaction conditions used. Catalytic hydrogenation over platinum in acetic acid or over palladium in ethanol and hydrochloric acid affords *cis*-2-alkylcyclohexylamine, whereas hydrogenation over Raney nickel in ammonia or reduction with *sodium* in ethanol yields *trans*-2-alkylcyclohexylamine, and reduction with *sodium amalgam* yields a mixture of both isomers [*614, 1216*]. Similarly, hydrogenation of 11-oximino-5β-pregnane-3α,17α,20β-triol over platinum oxide in acetic acid at 60 °C and 56 atm gives 86% β-11-amino-5β-pregnane-3α,17α,20β-triol, and reduction with sodium in refluxing propanol gives the 75% α-isomer [*1217, 1218*]. Hydrogenation of methyl ethyl ketoxime to *sec*-butylamine in a 76% yield is achieved by treatment with *hydrazine hydrate* and *Raney nickel* W4 at room temperature for 4 h [*1219*]. A very gentle reduction of oximes to amines is carried out by treatment with *samarium diiodide* in tetrahydrofuran and methanol at –40 °C in 64–82% yields [*1220*].

Reduction of cyclohexanone oxime with *lithium aluminum hydride* in tetrahydrofuran gives cyclohexylamine in a 71% yield [*1046*], and reduction of ketoximes with *sodium* in methanol and liquid ammonia [*1221*] or in boiling ethanol [*1222*] affords alkylamines, usually in good to high yields (equations **220** and **221**).

Stannous chloride in hydrochloric acid at 60 °C reduces the dioxime of 9,10-phenanthraquinone to 9,10-diaminophenanthrene in a 90% yield [*262*].

Somewhat less usual is *reductive cleavage* of ketoximes to ketones. It can be accomplished in yields of 56–100% by treatment of the ketoximes with *hydrogen, Raney nickel*, methanol, acetone, and boric acid at room temperature [*1223*] or in 90% yields by reduction with *titanium trichloride* in methanol or dioxane in the presence of sodium acetate [*279*].

220

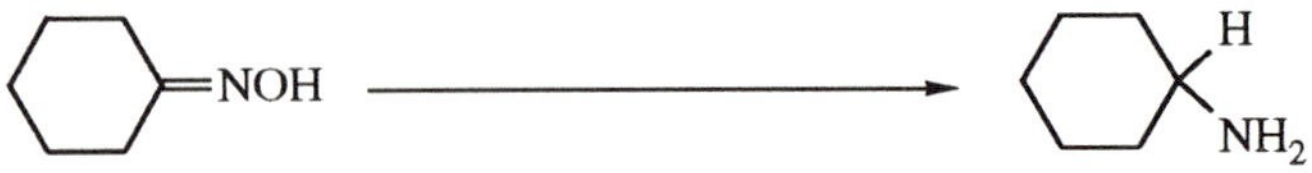

[46]   Raney Ni, EtOH, 25–30 °C, 1–3 atm, 45 min   90%
[1219] N$_2$H$_4$•H$_2$O, Raney Ni, 25 °C, 4 h   65%
[1046] LiAlH$_4$, Et$_2$O, reflux 30 min   71%
[1221] Na, NH$_3$, MeOH   91%

221

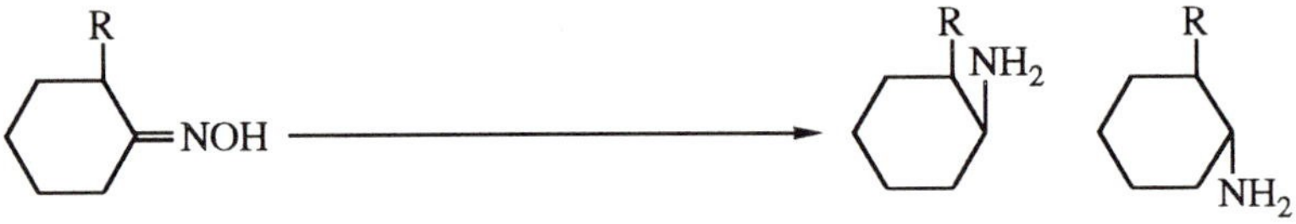

R = CH$_3$ [1216] H$_2$/Pt, AcOH   70–75%
             [1216] Na, ROH   a
R = C$_2$H$_5$ [614] H$_2$/Pd(C), EtOH, 25 °C, 1 atm   13%
             [614] H$_2$/Raney Ni, 130 °C, 83 atm   79%
             [614] Na, EtOH, reflux   74.5%

^{a} Yield is not indicated.

Oximes of α,β-unsaturated ketones yield *unsaturated amines, saturated amines*, and sometimes *aziridines* in fair yields [1224] on reduction with *lithium aluminum hydride,* depending on the structure of the ketoxime and on the reaction conditions.

Whereas *diborane* in tetrahydrofuran reduces oximes only at 105–110 °C, oxime ethers and oxime esters are reduced to amines and alcohols at room temperature in good yields. For example the *p*-nitrobenzoyl ester of cyclohexanone oxime gives a 67% yield of cyclohexylamine and an 81% yield of *p*-nitrobenzyl alcohol [1225].

The nitro group in an oxime is reduced in preference to the oximino group with *ammonium sulfide* [330]. In a monoxime of a diketone, the oximino group is reduced to an amino group to the exclusion of the carbonyl group by *catalytic hydrogenation* over platinum oxide in methanolic hydrochloric acid: 9-Keto-10-oximino-1,2,3,4-tetrahydrophenanthraquinone affords 10-amino-9-keto-1,2,3,4-tetrahydrophenanthrene in a 78% yield [1226].

Hydrazones of ketones may be *reduced to hydrazines, amines, and hydrocarbons* or reconverted *to parent ketones.*

The phenylhydrazone of acetone gives on *hydrogenation* over colloidal platinum a 90% yield of *N*-isopropyl-*N'*-phenylhydrazine [1227], and the semicarbazone of benzil on *electroreduction* affords a 70% yield of *N*-amino-carbonyl-*N'*-(1,2-diphenyl-2-oxoethyl)hydrazine [1228].

The phenylhydrazone of *N*-acetylisopelletierine (*N*-acetyl-2-piperidyl-acetone) is hydrogenolyzed over platinum oxide in acetic acid at 25 °C and 3 atm to 1-(*N*-acetyl-2-piperidyl)-2-aminopropane in a 92% yield [*1229*], and the phenylhydrazone of levulinic acid is reduced with *aluminum amalgam* to 4-aminovaleric acid in a 60% yield [*228*].

Hydrazones treated with alkalis decompose to nitrogen and hydrocarbons [*1088, 1190*] (the *Wolff–Kizhner reduction*) (p 151 and 156), and *p*-toluene-sulfonylhydrazones are reduced to hydrocarbons by *lithium aluminum hydride* [*1048*], *sodium borohydride* [*1020, 1048*], or *sodium cyanoborohydride* [*1050*] (p 151). *Titanium trichloride* hydrogenolyzes the nitrogen–nitrogen bond in phenylhydrazones and forms amines and ketimines that are hydrolyzed to the parent ketones. Thus the 2,4-dinitrophenylhydrazone of cycloheptanone affords cycloheptanone in a 90% yield [*282*].

REDUCTIVE ALKYLATION (REDUCTIVE AMINATION)

Treatment of aldehydes or ketones with ammonia, primary amines, or secondary amines in reducing media is called reductive alkylation (of ammonia or amines) or reductive amination (of aldehydes or ketones). Reducing agents are most frequently hydrogen in the presence of catalysts such as platinum, palladium, nickel or Raney nickel [*1230*], complex borohydrides [*138, 1231, 1232*], formaldehyde, or formic acid [*447*].

Reductive alkylation of ammonia should give primary amines, reductive alkylation of primary amines should give secondary amines, and reductive alkylation of secondary amines should give tertiary amines. In reality, secondary and even tertiary amines are almost always present to varying extents because the primary amines formed in the reaction of the carbonyl compounds with ammonia react with the carbonyl compounds to give secondary amines, and the secondary amines similarly afford tertiary amines, according to equation **222**. In addition, secondary amines may be formed, especially at higher temperatures, by additional reactions shown in equation **223**. Depending on the ratios of the carbonyl compounds to ammonia or amines, different classes of amines predominate.

For example, hydrogenation of benzaldehyde with 1 mol of ammonia gives 89.4% primary and 7.1% secondary amine, whereas the reaction with 0.5 mol of ammonia affords 11.8% primary and 80.8% secondary amine [*1233*].

Consequently, by choosing proper conditions, especially the ratios of the carbonyl compound to the amino compound, very good yields of the desired amines can be obtained [*447, 1230*]. In *catalytic hydrogenations*, alkylation of amines is also achieved by alcohols under the conditions when they may be dehydrogenated to the carbonyl compounds [*1040*]. The reaction of aldehydes and ketones with ammonia and amines in the presence of hydrogen is carried out on catalysts: platinum oxide [*1234*], nickel [*1040, 1235*], or Raney nickel [*1233, 1236, 1237*]. Yields range from low (23–35%) to very high (93%). An alternative route is the use of complex borohydrides: *sodium borohydride* [*1231*], *lithium cyanoborohydride* [*1232*], and *sodium cyanoborohydride* [*138, 1213,1238,1239*] in aqueous–alcoholic solutions of pH 5–8 (equations **224** and **225**).

222

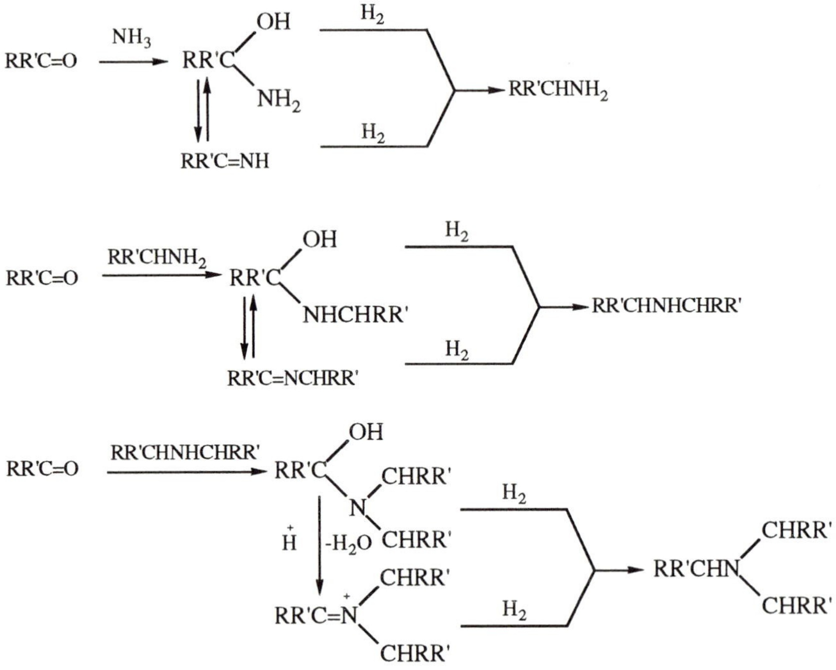

R,R' = H, alkyl, aryl

223   [*1040*]

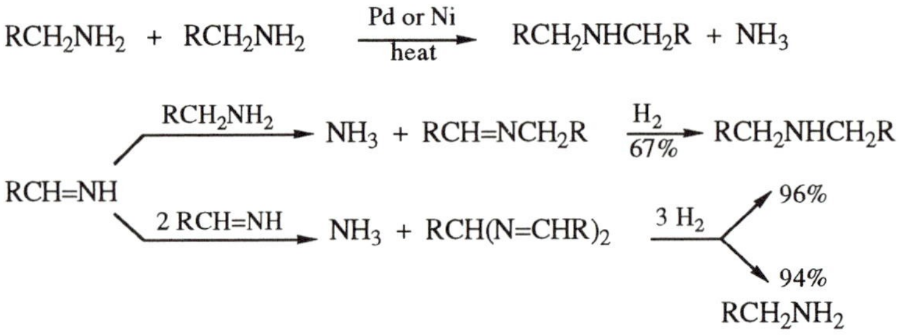

224

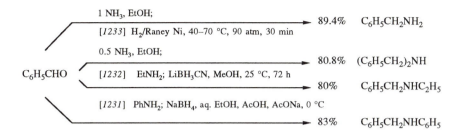

225

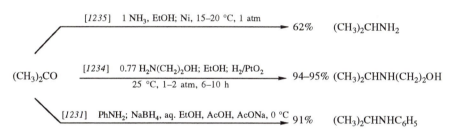

Heating of primary aliphatic amines with *formaldehyde* and an excess of *formic acid* at 100 °C results in the formation of tertiary amines in 80–92% yields [*445*].

Reductive alkylation can also be accomplished by heating carbonyl compounds at 150–250 °C with 4–5 mol of *ammonium formate, formamide, or formates or formamides* prepared by heating primary or secondary amines with formic acid at 180–190 °C (*the Leuckart reaction*) [*447*]. An excess of 85–90% formic acid is frequently used. Formyl derivatives of primary or secondary amines are sometimes obtained as products and have to be hydrolyzed to the corresponding amines by refluxing with 30% sodium hydroxide or with 10%, 20%, or concentrated hydrochloric acid. The reaction requires 4–30 h, and the yields range from some 25% to almost 100% [*446, 447, 1240*] (Procedure 59, p 320) (equations **226** and **227**).

Reductive methylation is achieved by the reaction of *formaldehyde* with ammonium chloride. It is carried out by heating the components at 100–120 °C and gives mono-, di-, and trimethylamines in high yields (*the Eschweiler reaction*) [*433, 1241*]. No catalyst is needed; part of the formaldehyde provides the necessary hydrogen, and the other part is oxidized to formic acid. The same reaction can be applied to methylation of primary and secondary amines [*1241*]. Reductive alkylation can also be accomplished by reducing mixtures of amines with acids that are first reduced to aldehydes (p 236).

226

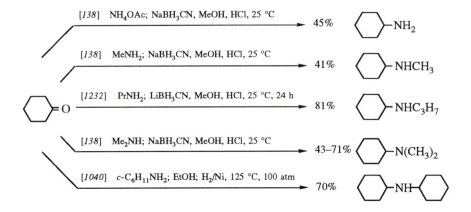

227

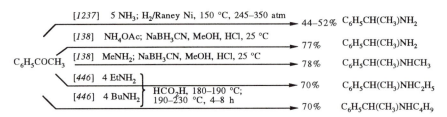

SUMMARY

Both aldehydes and ketones are readily reduced to alcohols by catalytic hydrogenation and by complex hydrides. Aldehydes are reduced more easily than ketones. Thus a selective reduction of aldehydes is feasible. Reduction of ketones in preference to aldehydes requires specially modified hydrides. Whereas α,β-unsaturated ketones can be selectively reduced at the double bond by catalytic hydrogenation and even by specially modified complex hydrides, and at the carbonyl with complex hydrides, selective reduction of α,β-unsaturated aldehydes by the same methods is more difficult.

For the reduction of aldehydes and ketones to hydrocarbons, the Clemmensen and Wolff–Kizhner reductions are the methods of choice. Only aromatic aldehydes are reduced to the corresponding hydrocarbons by catalytic hydrogenation, triethylsilane, and lithium in liquid ammonia.

Of the carbonyl compound derivatives, reduction of oximes and reductive alkylation (amination) are most important. Oximes are reduced to amines by catalytic hydrogenation, with lithium aluminum hydride, with sodium in alcohols, and by metal chlorides; reductive amination is best carried out by catalytic hydrogenation, reduction with sodium cyanoborohydride, and heating with ammonium formate or formamide (the Leuckart reaction).

CHAPTER 14

Reduction of Carboxylic Acids

The carboxyl group in carboxylic acids can be reduced to an aldehyde group, an alcoholic group, and even a methyl group. Unsaturated acids and aromatic acids can be reduced at the multiple bonds or aromatic rings, respectively, without or with concomitant reduction of the carboxyl groups. Other functions in the molecules of carboxylic acids may or may not be affected by various reducing agents. Most frequently, products of the reduction of carboxylic acids are alcohols.

ALIPHATIC CARBOXYLIC ACIDS

Saturated aliphatic acids are converted to aldehydes by *aminoalanes* prepared in situ from alane and two molecules of a secondary amine. One of the best reagents is obtained by adding 2 mol of *N*-methylpiperazine to 1 mol of alane in tetrahydrofuran at 0–25 °C. Refluxing hexanoic acid, octanoic acid, or palmitic acid for 6 h with the solution of the aminoalane affords the corresponding aldehydes in respective yields of 63%, 69%, and 77% [*1242*].

The ideal reagents for the conversion of carboxylic acids to aldehydes are *thexylhaloboranes* (dimethylisopropylmethylhaloboranes). *Thexylchloroborane* is prepared by treatment of thexylborane–dimethyl sulfide with one equivalent of hydrogen chloride or by hydroboration of 2,3-dimethyl-2-butene with monochloroborane–dimethyl sulfide [*144*]. The reagent reduces aliphatic and alicyclic carboxylic acids at room temperature in 15 min in 80–93% yields and aromatic acids after 24 h in 51–86% yields. Selective reduction of aliphatic acids in the presence of aromatic acids can be accomplished. Half esters of dicarboxylic acids are reduced to ester aldehydes (Procedure 23, p 306). Even more selective is *thexylbromoborane*, obtained by hydroboration of 2,3-dimethyl-2-butene with bromoborane–dimethyl sulfide in methylene chloride [*145*]. Unsaturated acids are reduced without the danger of hydroboration of the double bonds [*145*], and nitro carboxylic acids are reduced without reduction of the nitro groups [*145*] (equation **228**).

The best way of isolating the aldehydes is to treat the reaction mixtures with sodium bisulfite [*144, 145*]. Aliphatic acids containing 5–14 carbon atoms are

228 [*144*]

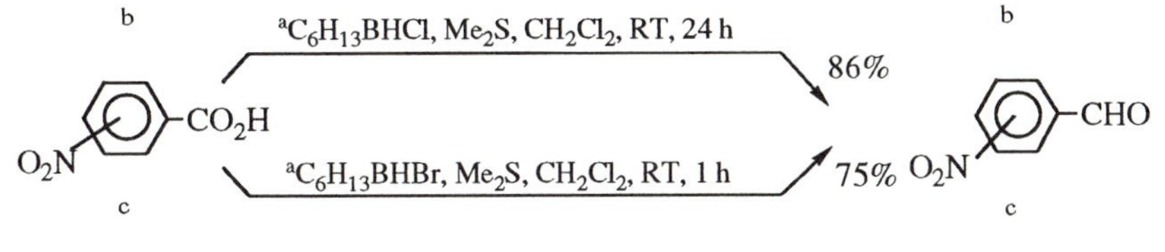

$$O_2N \overbrace{}^{\text{a}C_6H_{13}BHCl, Me_2S, CH_2Cl_2, RT, 24 h}_{\text{a}C_6H_{13}BHBr, Me_2S, CH_2Cl_2, RT, 1 h} \overset{86\%}{\underset{75\%}{\longrightarrow}}$$

b
O$_2$N—⟨⟩—CO$_2$H
c

b
O$_2$N—⟨⟩—CHO
c

^{a}C$_6$H$_{13}$ = (CH$_3$)$_2$CHC(CH$_3$)$_2$–(thexyl) ^{b}*para* ^{c}*meta* [*145*] reduced to the corresponding aldehydes in 61–84% yields by treatment with *lithium* in methylamine followed by hydrolysis of the intermediates, *N*-methylaldimines [*1243*]. The carboxyl group is also reduced to the aldehyde group by *electrolysis* using lead electrodes [*1244*] and on reduction with *sodium amalgam* [*206*] (p 194).

More frequent than the reduction to aldehydes is **reduction of carboxylic acids to alcohols.** *Catalytic hydrogenation* requires special catalysts, high temperatures (140–420 °C), and high pressures (150–990 atm). Under such vigorous conditions the resulting alcohols are accompanied by esters resulting from esterification of the alcohols with the parent acids. The catalysts used for such hydrogenations are copper and barium chromate [*1245*], ruthenium dioxide or ruthenium on carbon [*54*], and especially rhenium heptoxide [*55*] and rhenium heptasulfide [*56, 1246*]. Although yields of alcohols are in some cases very high, this method is of limited use.

The reduction of free acids to alcohols became practical only after the advent of complex hydrides. *Lithium aluminum hydride* reduces carboxylic acids to alcohols in ether solution very rapidly in an exothermic reaction. Because of the presence of acidic hydrogen in the carboxylic acid, an additional equivalent of lithium aluminum hydride is needed beyond the amount required for the reduction. The stoichiometric ratio is 4 mol of the acid to 3 mol of lithium aluminum hydride (equation **44**, p 26). A special technique is recommended for the reduction of esters that contain nitrogen—quenching of the reaction mixture with triethanolamine prior to treatment with water [*1247*] (Procedure 14, p 302). Trimethylacetic acid is reduced to neopentyl alcohol in a 92% yield, and stearic acid is converted to 1-octadecanol in a 91% yield. Dicarboxylic sebacic acid is reduced to 1,10-decanediol even if less than the needed amount of lithium aluminum hydride is used [*1248*].

Another reagent, *sodium bis(2-methoxyethoxy)aluminum hydride* (Vitride, Red-Al), is used to reduce nonanoic acid to 1-nonanol in refluxing benzene in a 92% yield [*1249*]. The same reagent converts sodium or bromomagnesium salts of acids to alcohols: sodium stearate to 1-octadecanol at 80 °C in a 96% yield, and bromomagnesium octanoate to 1-octanol at 80 °C in an 85% yield [*1250*].

Sodium borohydride does not reduce the free carboxyl group, but *borane* prepared from sodium borohydride and boron trifluoride etherate [*1251, 1252*] or iodine [*1253*] in tetrahydrofuran converts aliphatic acids and their salts [*1254*] to alcohols at 0–25 °C in 89–100% yields.

The reagent is suitable for selective reduction of a free carboxyl group in the presence of halogens (equation **229**, p 193), nitro groups (p 198), and ester groups [*1251, 1253, 1255*] (p 227). Another reagent capable of reducing carboxylic acids to alcohols is *zinc borohydride* in trifluoroacetic acid in dimethoxyethane at room temperature (80–92% yields) [*1256*].

Carboxylic acids containing double bonds are easily *converted to saturated acids* by *catalytic hydrogenation* over common catalysts. If a new chiral center is generated in the reduction process, homogeneous hydrogenation over a chiral catalyst gives a 40–45% enantiomeric excess of one enantiomer [*19*].

Better results are obtained with other chiral catalysts, especially in α-acylamino-α,β-unsaturated acids, precursors in the synthesis of optically active amino acids (p 218).

229 [*1253*]

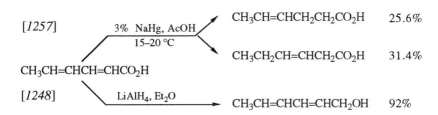

p-Cl; X = H; NaBH₄, I₂, THF, RT, 1 h → 98%

o-Cl; X = Na; BH₃, THF, RT, 10 h → 95%

[*1254*]

Unsaturated carboxylic acids with double bonds conjugated with the carboxyl are reduced at the double bond by *sodium amalgam* [*1257*]. Sorbic acid (2,4-hexadienoic acid) containing a conjugated system of two double bonds is reduced by *catalytic hydrogenation* over palladium or Raney nickel preferentially at the more distant double bond, giving 83–90% 2-hexenoic acid [*1258*]. *Sodium amalgam* adds hydrogen to the 1,2- as well as the 1,4-position and affords a mixture of 4- and 3-hexenoic acids [*1257*] (equation **230**).

230

[*1257*] 3% NaHg, AcOH 15–20 °C	$CH_3CH=CHCH_2CH_2CO_2H$	25.6%
	$CH_3CH_2CH=CHCH_2CO_2H$	31.4%
$CH_3CH=CHCH=CHCO_2H$		
[*1248*] LiAlH₄, Et₂O	$CH_3CH=CHCH=CHCH_2OH$	92%

Maleic and fumaric acids are converted to succinic acid with *chromous sulfate* in aqueous solution at room temperature in 30 and 60 min in 86% and 91% yields, respectively [*1259*]. Chiral reductions of unsaturated carboxylic acid derivatives are described on p 218.

Lithium aluminum hydride **exclusively reduces the carboxyl** group, even in unsaturated acids with α,β-conjugated double bonds. Sorbic acid affords a 92% yield of sorbic alcohol [*1248*] (equation **230**), and fumaric acid gives a 78% yield of *trans*-2-butene-1,4-diol [*1260*]. The monoethyl ester of fumaric acid is reduced to ethyl γ-hydroxycrotonate with *borane* in tetrahydrofuran at –10 °C to room temperature in a 61% yield [*1255*]. If, however, the α,β-conjugated double bond of an acid is at the same time conjugated with an aromatic ring, it is reduced (p 196).

Carboxylic acids containing triple bonds are converted to *cis* olefinic acids by *catalytic hydrogenation* over Raney nickel [*507*] and to ***trans*** acids by reduction with *sodium* in liquid ammonia [*507*] (equation **231**). Acetylene-dicarboxylic acid affords fumaric acid in a 94% yield on treatment with *chromous sulfate* at room temperature [*273*].

Lithium aluminum hydride reduces acetylenic acids containing conjugated triple bonds to olefinic alcohols. Acctylenedicarboxylic acid gives an 84% yield of *trans*-2-butene-l,4-diol at room temperature after 16 h [*1260*].

231   [*507*]

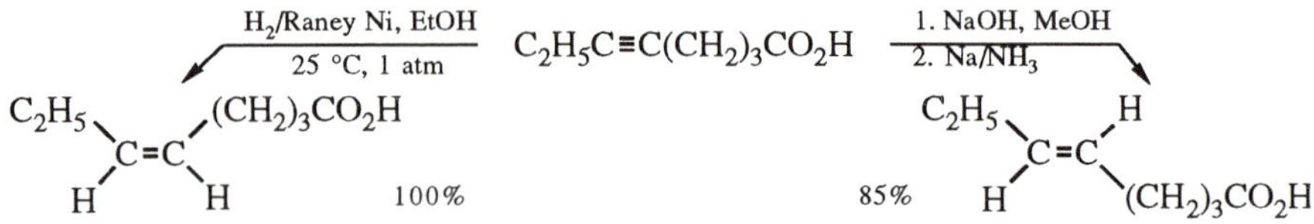

AROMATIC CARBOXYLIC ACIDS

Aromatic carboxylic acids can be reduced to aldehydes or alcohols. In addition, a carboxylic group linked to an aromatic ring can be converted to a methyl, and the aromatic ring can be partially or totally hydrogenated.

Aldehydes are obtained from benzoic acid on refluxing for 6 h and from nicotinic acid on standing at room temperature for 24 h with *bis(N-methylpiperazino)alane* in tetrahydrofuran in 86% and 75% yields, respectively [*1242*]. Thexylchloroborane [*144*] and thexylbromoborane [*145*] reduce carboxylic acids to aldehydes without affecting other reducible groups (equation **232**).

232

$$C_6H_5CO_2H \longrightarrow C_6H_5CHO \ \text{or} \ C_6H_5CH_2OH$$

[*1245*]	AlH(-N⁀NMe)$_2$, THF, reflux 3 h	86%	4%
[*1248*]	LiAlH$_4$, Et$_2$O		81%
[*1249*]	NaAlH$_2$(OCH$_2$CH$_2$OMe)$_2$, C$_6$H$_6$, 80 °C		97%
[*144*]	^{a}C$_6$H$_{13}$BHCl, Me$_2$S, CH$_2$Cl$_2$, RT, 24 h	59%	
[*145*]	^{a}C$_6$H$_{13}$BHBr, Me$_2$S, CH$_2$Cl$_2$, RT, 9 h	49%	
[*1251*]	NaBH$_4$, BF$_3$·Et$_2$O, THF, 0–25 °C, 1 h		89%
[*1253*]	NaBH$_4$, I$_2$, THF, RT, 1 h		93%

^{a}C$_6$H$_{13}$ = (CH$_3$)$_2$CHC(CH$_3$)$_2$–(thexyl).

Reduction of 3-fluorosalicylic acid with 2% *sodium amalgam* in aqueous solution containing sodium chloride, boric acid, and *p*-toluidine gives, at 13–15 °C, a Schiff base, which, on hydrolysis with hydrochloric acid and steam distillation, affords 3-fluorosalicylaldehyde in a 57% yield [*206*]. The purpose of *p*-toluidine is to react with the aldehyde as it is formed and protect it from further reduction.

Reduction of aromatic carboxylic acids to alcohols can be achieved by hydrides and complex hydrides, for example, *lithium aluminum hydride* [*1248*], *sodium aluminum hydride* [*115*], and *sodium bis(2-methoxyethoxy)aluminum hydride* [*705, 1249, 1250*]. This reduction can also be achieved with *borane* (diborane) [*1261*] prepared from sodium borohydride and boron trifluoride etherate [*1251, 1252*] or aluminum chloride [*958, 1262*] in diglyme or borane prepared from sodium borohydride and iodine in tetrahydrofuran [*1253*]. Sodium borohydride alone does not reduce free carboxylic acids. Anthranilic acid is reduced to the corresponding alcohol by *electroreduction* in sulfuric acid at 20–30 °C in a 69–78% yield [*1263*].

Reduction of aromatic carboxylic acids having hydroxy or amino groups in *ortho* or *para* positions is accompanied by *hydrogenolysis* when carried out with *sodium bis(2-methoxyethoxy)aluminum hydride* in xylene at 141–142 °C: *o*- and *p*-cresols [*1264*] and *o*- and *p*-toluidines [*1265*], respectively, are obtained in yields ranging from 71% to 92%. *meta*-Substituted benzoic acids under the same conditions give *m*-hydroxybenzyl alcohol [*1264*] and *m*-aminobenzyl alcohol [*1265*] in 72% yields.

Conversion of **aromatic acids to hydrocarbons** is accomplished with *trichlorosilane*. A mixture of 0.1 mol of an acid and 0.6 mol of trichlorosilane in 80 mL of acetonitrile is refluxed for 1 h, then treated under cooling with 0.264 mol of tripropylamine and refluxed for 16 h. After dilution with 850 mL of ether, removal of tripropylamine hydrochloride, and evaporation of the ether, the residue is refluxed for 20 h with 1 mol of potassium hydroxide in 120 mL of 80% aqueous methanol to give a 78–94% yield of the corresponding hydrocarbon [*168, 1266*]. Hydrogenolysis to hydrocarbons may also occur in catalytic hydrogenations over nickel or copper chromite under very drastic conditions.

Saturation of the aromatic rings in aromatic carboxylic acids takes place during high-temperature–high-pressure *hydrogenation*. Gallic acid (3,4,5-trihydroxybenzoic acid) hydrogenated over 5% rhodium on alumina in 95% ethanol at 90–100 °C and 150 atm gives, after 8–12 h, a 45–51% yield of all-*cis*-3,4,5-trihydroxycyclohexanecarboxylic acid. Palladium, platinum, rhodium on other supports, and Raney nickel are less satisfactory as catalysts [*1267*].

Sodium in liquid ammonia and ethanol reduces benzoic acid to 1,4-dihydrobenzoic acid. Reduction of *p*-toluic acid is more complicated and affords a mixture of *cis*- and *trans*-1,2,3,4-tetrahydro-*p*-toluic acids and *cis*- and *trans*-1,4-dihydrotoluic acids. *m*-Methoxybenzoic acid yields 1,2,3,4-tetrahydro-5-methoxybenzoic acid, and 3,4,5-trimethoxybenzoic acid gives 1,4-dihydro-3,5-dimethoxybenzoic acid in an 87% yield (after hydrogenolysis of the methoxy group *para* to the carboxyl) [*1268*]. With 4'-methoxybiphenyl-4-carboxylic acid, sodium in isoamyl alcohol at 130 °C completely reduces only the ring with the carboxylic group, thus giving a 92% yield of 4-(*p*-methoxyphenyl)cyclohexanecarboxylic acid [*1269*].

1-Naphthoic acid is reduced with sodium in liquid ammonia to 1,4-dihydro-1-naphthoic acid, which, after heating on a steam bath with 20% sodium hydroxide for 30 min, isomerizes to 3,4-dihydro-1-naphthoic acid (63% yield) [*547*]. 2-Naphthoic acid treated with 4 equivalents of *lithium* in liquid ammonia and ethanol gives a 69% yield of 1,2,3,4-tetrahydro-2-naphthoic acid. With 7 equivalents of *lithium*, 1,2,3,4,5,8-hexahydro-2-naphthoic acid is obtained in an 82% yield [*1270*].

Partial reduction of the aromatic ring is especially easy in anthracene-9-carboxylic acid, which is reduced to 9,10-dihydroanthracene-9-carboxylic acid with 2.5% *sodium amalgam* in aqueous sodium carbonate at 10 °C in an 80% yield [*1271*]. Aromatic carboxylic acids with hydroxyl groups in the *ortho* positions suffer ring cleavage during reductions with *sodium* in alcohols and are converted to dicarboxylic acids after fission of the intermediate β-keto acids. Thus salicylic acid on reduction with sodium in boiling isopentyl alcohol affords pimelic acid in a 43–50% yield [*1272*] (equation **233**).

233   [*1272*]

43–50%

Unsaturated aromatic carboxylic acids with double bonds conjugated with both the carboxyl and the aromatic ring undergo easy *saturation of the double bond* by *catalytic hydrogenation* over colloidal palladium [*1273*], Raney nickel [*46*], or copper chromite [*51*]. Homogeneous hydrogenation over a chiral catalyst, tris(triphenylphosphine)rhodium chloride, produces hydrocinnamic acid in an 85% yield [*91*]. A chiral diphosphine–rhodium catalyst converts atropic acid (2-phenylacrylic acid) to (*S*)-hydratropic acid (2-phenylpropionic acid) in a quantitative yield and 63% optical purity [*1274*]. Higher enantiomeric excess is obtained by hydrogenation over (*R*)-1-[*N*-methyl-*N*'-(2-piperidinoethyl)]-1-(*S*)-bis(diphenylphosphino)ferrocenylaminoethane [*106*]. 3-Methyl-2-phenyl-2-butenoic acid gives 3-methyl-2-phenylbutanoic acid in 96–98% enantiomeric excess [*106*]. Such reductions are especially useful for the preparation of optically active amino acids from their precursors, acylamino-α,β-unsaturated acids (p 218). Double bonds are also saturated by treatment with *nickel–aluminum alloy* in 10% sodium hydroxide at 90–100 °C (80–95% yields) [*1275, 1276*]. Cinnamic acid is reduced to hydrocinnamic acid not only by catalytic hydrogenation but also by *electroreduction* [*1277*], by reduction with *sodium amalgam* [*67*] (Procedure 33, p 310), and with *chromous sulfate* [*1259*]. *Lithium aluminum hydride* reduces acylamino-α,β-unsaturated acids completely at the carboxyl as well as at the double bond [*1248*]. On the contrary, *sodium borohydride* and iodine reduce cinnamic acid to cinnamyl alcohol at 0 °C in a 97% yield [*1253*] (equation **234**).

Acetylenic aromatic acids having the triple bond flanked by a carboxyl group and an aromatic ring are partially reduced to olefinic aromatic acids by

234

$$C_6H_5CH=CHCO_2H \longrightarrow C_6H_5CH_2CH_2CO_2H \quad or \quad C_6H_5CH_2CH_2CH_2OH$$

[*46*]	H_2/Raney Ni, 25–30 °C, 1–3 atm, 10 min	100%
[*51*]	H_2/$CuCr_2O_4$, 175 °C, 100–150 atm, 18 min	100%
[*91*]	H_2/$(Ph_3P)_3RhCl$, EtOH, 60 °C, 4–6 atm, 8–12 h	85%
[*1277*]	Electroredn., NaOH, Hg cathode, 30 V, 5–10 amp	80–90%
[*67*]	1. NaOH; 2. NaHg	80%
[*1259*]	$CrSO_4$, DMF–H_2O, 25 °C, 40 h	89%
[*1248*]	LiAlH₄, Et₂O	85%

chromous sulfate in aqueous *N,N*-dimethylformamide at room temperature in high yields. Phenylpropiolic acid affords *trans*-cinnamic acid in a 91% yield [*273*]. Its sodium salt in aqueous solution gives, on *catalytic hydrogenation* over colloidal platinum at room temperature and atmospheric pressure, an 80% yield of *cis*-cinnamic acid if the reaction is stopped after absorption of 1 mol of hydrogen. Otherwise 3-phenylpropanoic acid is obtained in a 75–80% yield [*1278*].

CARBOXYLIC ACIDS CONTAINING SUBSTITUENTS OR OTHER FUNCTIONAL GROUPS

Carboxylic acids containing halogens are easily *reduced to halogenated alcohols* by alanes or boranes. Lithium aluminum hydride could endanger more reactive halogens (p 221). *Borane* in tetrahydrofuran converts chloroacetic acid to chloroethanol at 0–25 °C in 30 min in a quantitative yield, and 2-bromo-dodecanoic acid to 2-bromododecanol in 1 h in a 92% yield without hydrogenolyzing the reactive halogens in positions α to carboxyls [*1251*]. 3-Bromopropanoic acid is reduced to 3-bromopropanol with an ethereal solution of *alane* (prepared from lithium aluminum hydride and aluminum chloride) at 35 °C in a 50% yield [*1279*] and with *lithium aluminum hydride* in ether by the inverse technique at –15 °C in a 26% yield [*1279*]. Alane in tetrahydrofuran at 10 °C converts 3-chloropropanoic acid to 3-chloropropanol in an 89% yield (61% isolated) and 3-bromobutanoic acid to 3-bromobutanol in an 87% yield, both in 15 min [*1280*]. *p*-Chlorobenzoic acid affords *p*-chlorobenzyl alcohol on reduction with borane in diglyme in 64–88% yields [*958, 1252*]. Partial replacement of one atom of iodine takes place during reduction of 3,4,5-triiodo-benzoic acid, which affords 3,5-diiodobenzyl alcohol in a 60% yield on treatment with lithium aluminum hydride. Similar reduction occurs with other polyiodinated benzene derivatives [*1281*].

o-Chlorobenzoic acid and *p*-bromobenzoic acid are transformed into *o*-chlorotoluene and *p*-bromotoluene, respectively, in 94% yields on treatment with *trichlorosilane* and tripropylamine [*1266*]. On the other hand, *o*-, *m*-, and *p*-chlorobenzoic acids are converted to benzoic acid in respective yields of 91%, 64%, and 82% *by catalytic hydrogenation* over Raney nickel in methanolic potassium hydroxide at room temperature [*697*]. *o*-Bromobenzoic acid is reduced quantitatively to benzoic acid over colloidal palladium in aqueous sodium hydroxide [*1282*]. In *N*-acetyl-β-(2-bromobenzo-3-furyl)alanine the bromine in the furan ring is replaced by hydrogen over Raney nickel at room temperature in a 91% yield [*713*].

In **halogenated unsaturated acids** having chlorine or bromine linked to sp^2 carbons, catalytic hydrogenation may cause replacement of the halogen without saturation of the double bond. β-Chlorocrotonic acid is converted to crotonic acid by hydrogenation over 10% palladium on barium sulfate at room temperature and atmospheric pressure [*1282*]. On the other hand, on hydrogenation over 10% palladium on charcoal at room temperature and atmospheric pressure, difluoromaleic acid affords predominantly succinic acid [*66*].

Homogeneous hydrogenation of 2-fluoro-2-alkenoic acids over a chiral catalyst, ruthenium Binap, gives 2-fluoroalkanoic acids in high conversions and 78–91% enantiomeric excesses [*1283*] (equation **235**). Vinylic fluorines in fluorobutenedioic acids are hydrogenolyzed with surprising ease. Products are always saturated acids, with or without fluorine [*66*] (equation **236**).

235   [*1283*]

$$\begin{array}{c} C_3H_7 \\ \diagdown \\ H \diagup \end{array} C=C \begin{array}{c} F \\ \diagup \\ \diagdown CO_2H \end{array} \quad \xrightarrow[\text{MeOH} \ | \ 50\ ^{\circ}\text{C, 5 atm}]{(R)\text{-Ru}_2\text{Cl}_4(\text{Binap})_2(\text{Et}_3\text{N})} \quad \begin{array}{c} C_3H_7 \\ \diagdown \\ H \diagup \end{array} C=C \begin{array}{c} CO_2H \\ \diagup \\ \diagdown F \end{array}$$

$$C_3H_7CH_2\overset{F}{\underset{\vdots}{\overset{|}{C}}}CO_2H$$

90% ee   *R*-   H   83% ee

236   [*66*]

$$\begin{array}{c} HO_2C \\ \diagdown \\ F \diagup \end{array} C=C \begin{array}{c} CO_2H \\ \diagup \\ \diagdown F \end{array} \quad \xrightarrow{H_2,\ 1\ atm} \quad HO_2CHFCHFCO_2H + HO_2CCHFCH_2CO_2H + HO_2CCH_2CH_2CO_2H$$
                                                                meso

10% Pd(C), Et$_2$O, –70 °C	37.5%	41.2%	2.5%
10% Pd(C), Et$_2$O, 25 °C	19.0%	70.0%	11.0%
10% Pd(C), H$_2$O, 25 °C	0	67.0%	33.0%
5% Rh(C), H$_2$O, 25 °C	0	11.0%	89%

Refluxing with *zinc* in ethanol reduces α-bromocinnamic acid to cinnamic acid in an 80% yield [*1284*]. Allylic chlorines in γ,γ,γ-trichlorocrotonic acid are hydrogenolyzed by successive use of *zinc* and *sodium amalgam* to give 31% γ,γ-dichlorocrotonic acid and ultimately 74% crotonic acid [*672*]. Hydrogenolysis of allylic bromine in α,β-unsaturated esters with zinc in acetic acid gives predominantly β,γ-unsaturated esters in 65–97% yields [*1285*].

The biochemical reduction of α,β-unsaturated β-haloaliphatic acids by means of *Clostridium kluyveri* yields halogen-free saturated acids. The same products are obtained from saturated α-halo acids. However, the same microorganism converts α,β-unsaturated α-halo acids to saturated α-halo acids with *R* configuration. Yields of reduction of α-fluoro-, α-chloro-, and α-bromocrotonic acids range from 30% to 100% [*465*].

In **carboxylic acids containing nitro groups**, *boranes* reduce only the carboxyl group [*144, 145, 958, 1251, 1252*], whereas *ammonium sulfide* reduces only the nitro group [*329*] (equation **237**).

The **diazonium group** in *o*-carboxybenzenediazonium salts is ***replaced by hydrogen*** using *sodium borohydride* in methanol [*777*], or *ethanol* and ultraviolet irradiation [*422*]. Yields of benzoic acid are 77% and 92.5%, respectively.

In **2,3-epoxybutyric acid,** *sodium borohydride* opens the epoxide ring without affecting the carboxyl. Varying ratios of 2- and 3-hydroxybutyric acids are obtained, depending on the reaction conditions. Sodium borohydride in alkaline solution gives 18% α- and 82% β-hydroxybutyric acids, but in the presence of lithium bromide, the two isomers are obtained in 60:40 percentage ratio [*1286*].

237

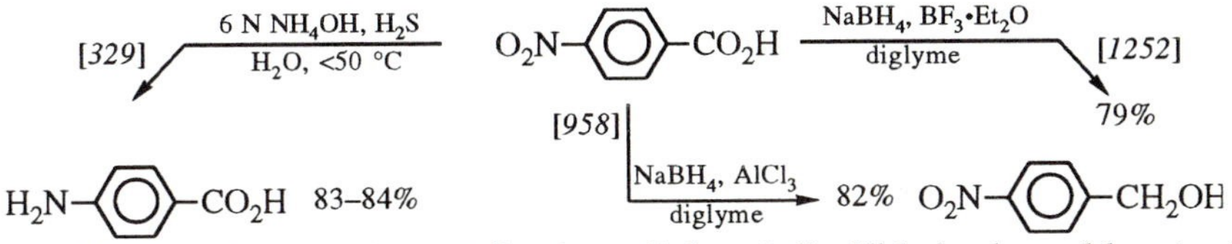

The *sulfidic bond* in *o,o'*-dicarboxydiphenyl disulfide is cleaved by *zinc* in refluxing acetic acid, giving *o*-carboxythiophenol in a 71–84% yield [*1287*].

α-**Amino acids** and their *N*-acyl derivatives are reduced to *amino alcohols* and *N*-alkylamino alcohols, respectively, with *sodium borohydride* and *iodine* or *chlorine* in refluxing tetrahydrofuran in 45–94% yields [*1288*] (equation **238**).

238   [*1288*]

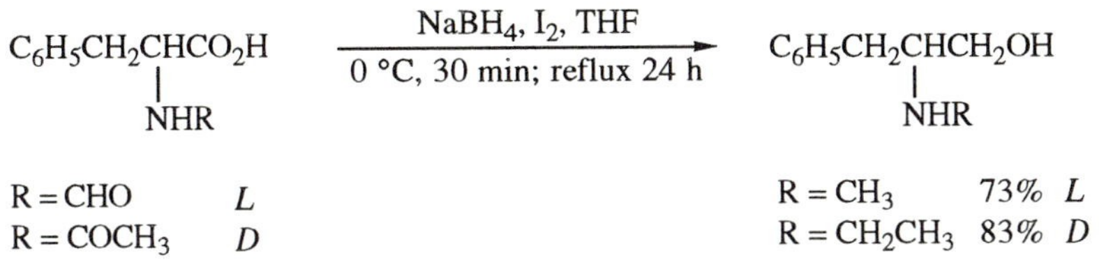

$$C_6H_5CH_2CHCO_2H \xrightarrow[\text{0 °C, 30 min; reflux 24 h}]{\text{NaBH}_4, \text{I}_2, \text{THF}} C_6H_5CH_2CHCH_2OH$$

with NHR below first, NHR below second.

R = CHO   *L*   R = CH$_3$ 73% *L*
R = COCH$_3$ *D*   R = CH$_2$CH$_3$ 83% *D*

In **keto acids,** carboxyl is reduced preferentially to the carbonyl with *borane* in tetrahydrofuran. 4-Phenyl-4-oxobutanoic acid affords 4-phenyl-4-oxobutanol in a 60% yield [*1251*] (equation **239**).

239

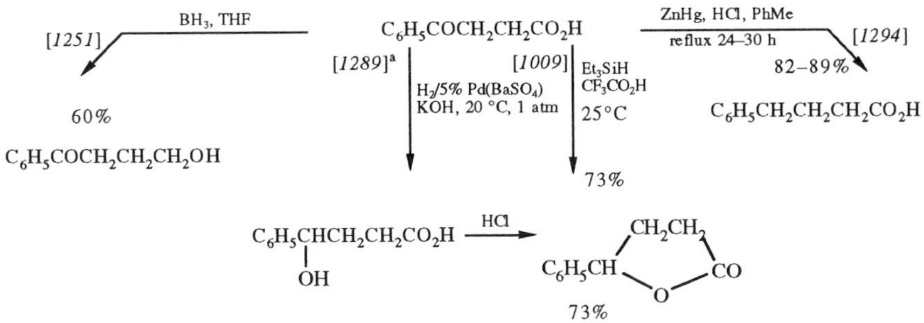

Keto acids are reduced most frequently to ***hydroxy acids***. On *catalytic hydrogenation* over Raney nickel, 4-oxocyclohexanecarboxylic acid gives an 86% yield of *cis*-4-hydroxycyclohexanecarboxylic acid, and on reduction with 4% *sodium amalgam*, it affords a 65–70% yield of the *trans* product [1090]. 4-Oxo-4-phenylbutanoic acid is hydrogenated over palladium on barium sulfate to 4-hydroxy-4-phenylbutanoic acid, which lactonizes to a γ-lactone (73% yield) [1289]. The same compound is obtained in a 73% yield on treatment with *triethylsilane* and trifluoroacetic acid [1009]. *Borane* in tetrahydrofuran reduces the acid to 4-oxo-4-phenyl-1-butanol [1251] (equation **239**).

In 4-oxodecanoic acid, the keto group is reduced *biochemically* using *baker's yeast* to a hydroxy group, which lactonizes to give an 85% yield of the corresponding γ-lactone [1290]. Phenylglyoxylic acid is transformed by heating at 100 °C with *amalgamated zinc* in hydrochloric acid to mandelic acid (α-hydroxyphenylacetic acid) in a 70% yield [1291]. Refluxing of 9-oxofluorene-1-carboxylic acid with *zinc* dust and copper sulfate in aqueous potassium hydroxide for 2.5 h affords 9-hydroxyfluorenecarboxylic acid in a 94% yield [1292].

Reduction with *sodium borohydride* in aqueous methanol at 0–25 °C converts 5-oxopiperidine-2-carboxylic acid to *trans*-5-hydroxypiperidine-2-carboxylic acid (equation **240**). On the other hand, reduction of *N*-benzyloxy-carbonyl-5-oxopiperidine-2-carboxylic acid gives *N*-benzyloxycarbonyl-*cis*-5-hydroxypiperidine-2-carboxylic acid under the same conditions [1293].

240   [1293]

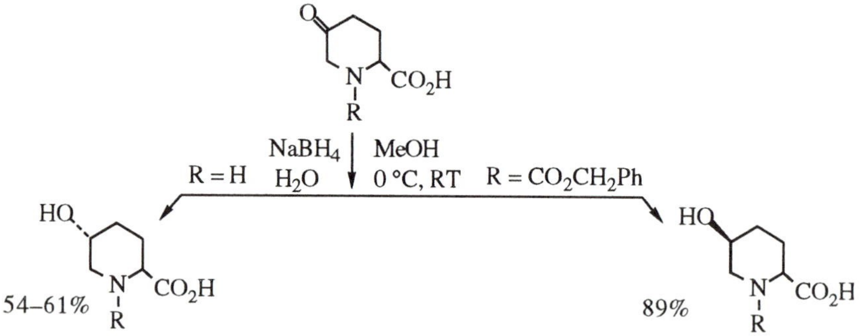

Stereoselective reduction of keto acids to hydroxy acids is accomplished by *homogeneous hydrogenation* using chiral rhodium [108] or ruthenium catalysts [1295] or by *biochemical reduction* [455, 471, 472]. *o*-Acetylbenzoic acid is reduced to (*R*)-*o*-α-hydroxyethylbenzoic acid by hydrogenation over ruthenium Binap catalyst in ethanol at 20–32 °C and 43 atm in a 100% yield and 92% enantiomeric excess [1295].

A roundabout method of enantioselective reductions is based on the introduction of a *chiral auxiliary group*. α-Ketoacyl chloride is converted to a chiral *N*-arylsulfonylaminobornyl ester. This is reduced with *L- or K-Selectride* or better still, with *lithium tris(triethylmethoxy)hydride* to the corresponding hydroxy ester, which is then hydrolyzed to the hydroxy acid by treatment with lithium hydroxide in aqueous tetrahydrofuran at room temperature. The yields of the individual reactions are high, and enantiomeric excesses are in the range of 93–98% [*126*].

Better enantioselectivity is obtained by reduction of α-keto acids with *lactate dehydrogenase* from *Bacillus stearothermophilus* [*471*] or isolated from various animal tissues [*472*]. Yields of α-hydroxy acids and their enantiomeric excesses in the former case are 66–98% and 97–99%, in the latter case 94–99% and >99%, respectively.

Reduction of the keto group in keto acids to a methylene group is accomplished by means of the *Clemmensen reduction*. 4-Phenyl-4-oxobutanoic acid, on refluxing with amalgamated zinc, hydrochloric acid, and toluene, affords 4-phenylbutanoic acid [*1294*] (equation **239**). The *Wolff–Kizhner reduction* or its modifications are used for the conversion of 6-oxoundecanedioic acid to undecanedioic acid (87–93% yield) [*1296*] and for the transformation of *p*-phenoxy-4-phenyl-4-oxobutanoic acid to *p*-phenoxy-4-phenylbutanoic acid (95–96% yield) [*385*]. *o*-Benzoylbenzoic acid is reduced to diphenylmethane-*o*-carboxylic acid in 85% yield by refluxing for 120 h with ethanolic aqueous *hydriodic acid* and phosphorus [*319*], and *p*-benzoylbenzoic acid gives *p*-benzyl-benzoic acid on reduction with *triethylsilane* and trifluoroacetic acid in a 50% yield (50% recovered) [*1009*]. The latter reagent reduces only keto acids with the keto group adjoining the benzene ring, and not without exceptions [*1009*].

CHAPTER 15

Reduction of Acyl Chlorides and Acid Anhydrides

ACYL CHLORIDES

Reduction of acyl chlorides is exceptionally easy. Depending on the reagents and reaction conditions, it can lead to aldehydes or to alcohols. *Catalytic hydrogenation to aldehydes* can be achieved by the *Rosenmund reduction* [38] by simply passing hydrogen through a solution of an acyl chloride in the presence of a deactivated catalyst. Such a catalyst is made by treating palladium on barium sulfate by sulfur compounds like the so-called quinoline-S, prepared by boiling quinoline with sulfur [38], or by deactivating platinum oxide with thiourea [1297]. Reduction yields range from 50% to 97% [1297, 1298]; 2,4,6-trimethylbenzoyl chloride is converted to 2,4,6-trimethylbenzaldehyde in a 70–80% yield [1299].

A variation of the Rosenmund reduction is heating of an acyl chloride at 50 °C with an equivalent of *triethylsilane* in the presence of 10% palladium on charcoal. Yields of aldehydes obtained by this method range from 45% to 75% [88].

Disadvantages of the Rosenmund reduction are high temperature, sometimes necessary to complete the reaction, and long reaction times. In this respect, reduction with complex hydrides offers considerable improvement. Special, not too efficient reagents must be used; otherwise the reduction proceeds further and gives alcohols (p 205). One of the most suitable complex hydrides proved to be *lithium tris(*tert-*butoxy)aluminum hydride*, which, because of its bulkiness, does not react at low temperatures with the aldehydes formed by the reduction. The highest yields of aldehydes (52–85%) are obtained if the reduction is performed in diglyme at −75 to −78 °C for 1 h [125, 1300] (Procedure 16, p 303). *Sodium borohydride* in ether in the presence of N,N-dimethylformamide and pyridine reduces acyl chlorides to aldehydes in 67–70% yields with only 7–8% yields of alcohols [1301].

Hydrides prepared from a mixture of cuprous chloride and triphenylphosphine, trimethyl phosphite, or triisopropyl phosphite in chloroform and an ethanolic solution of sodium borohydride reduce acyl chlorides to aldehydes in acetone solutions at room temperature in 15–90 min in yields ranging from 57% to 83% [175] (equation 241).

Both aforementioned complex hydrides have been successfully used for the **preparation of unsaturated aldehydes from unsaturated acyl chlorides** (48–71% yields) [1300] and for the synthesis of *p*-nitrobenzaldehyde from *p*-nitrobenzoyl chloride [175, 1300], a reduction that could hardly be achieved by applying catalytic hydrogenation.

α,β-Unsaturated acyl chlorides are also converted to α,β-unsaturated aldehydes in 94–100% yields by treatment with *triethyl phosphite* and subsequent reduction of the diethyl acylphosphonates with *sodium borohydride* at room temperature followed by alkaline hydrolysis [1302].

Complete reduction of acyl chlorides to primary alcohols is not nearly as important as the reduction to aldehydes because alcohols are readily obtained by reduction of more accessible compounds such as aldehydes, free carboxylic acids, or their esters [110, 1248]. Because aldehydes are the primary products of the reduction of acyl chlorides, strong reducing agents convert acyl chlorides directly to alcohols.

241

$$C_6H_5COCl \xrightarrow{\hspace{4cm}} C_6H_5CHO$$

[1298]	H$_2$/5% Pd(BaSO$_4$), xylene, reflux	97%
[1298]	H$_2$/Ni, xylene, reflux 4 h	95%
[1297]	H$_2$/PtO$_2$, CS(NH$_2$)$_2$, PhMe, reflux 6–12 h	96%
[88]	SiHEt$_3$/10% Pd(C), 60–80 °C	31–70%
[1300]	LiAlH(OCMe$_3$)$_3$, diglyme, –78 °C, 1 h; 25 °C, 1 h	81%
[175]	(Ph$_3$P)$_2$Cu$\overset{\text{H}\;\;\text{H}}{\underset{\text{H}\;\;\text{H}}{\diagup\!B\!\diagdown}}$, Me$_2$CO, 25 °C, 1 h	83%
[175]	([MeO]$_3$P)$_2$Cu$\overset{\text{H}\;\;\text{H}}{\underset{\text{H}\;\;\text{H}}{\diagup\!B\!\diagdown}}$, Me$_2$CO, 25 °C, 15 min	82%

Catalytic hydrogenation is hardly ever used for this purpose because the reaction byproduct, hydrogen chloride, poses some inconveniences in the experimental procedures. Most transformations of acyl chlorides to alcohols are effected by hydrides or complex hydrides. Addition of acyl chlorides to ethereal solutions of *lithium aluminum hydride* under gentle refluxing produces alcohols from aliphatic, aromatic, and unsaturated acyl chlorides in 72–99% yields [110]. The reaction is suitable even for the preparation of halogenated alcohols. Dichloroacetyl chloride is converted to dichloroethanol in a 64–65% yield [1303]. *Sodium aluminum hydride* converts acyl chlorides to alcohols in 94–99% yields [115], and *sodium bis(2-methoxyethoxy)aluminum hydride* achieves this conversion in 52–99% yields [1249].

High yields (76–81%) of alcohols are also obtained by adding solutions of acyl chlorides in anhydrous dioxane or bis(2-ethoxyethyl)ether (diethylcarbitol) to a suspension of *sodium borohydride* in dioxane and brief heating of the mixtures on the steam bath [984], by stirring solutions of acyl chlorides in ether for 2–4 h at room temperature with aluminum oxide (activity I) impregnated with a 50% aqueous solution of sodium borohydride (Alox) (80–90% yields) [1304], by refluxing acyl chlorides with ether solutions of *sodium trimethoxy-borohydride* [129], and by treatment of acyl chlorides in dichloromethane solutions with *tetrabutylammonium borohydride* at –78 °C [1004] or with *zinc cyanoborohydride* prepared from sodium borohydride and anhydrous zinc chloride in ether [930]. A 94% yield of neopentyl alcohol is obtained by the reaction of trimethylacetyl chloride with tert-*butylmagnesium chloride* [449] (equation 242).

For the **reduction of α,β–unsaturated acyl chlorides** *to unsaturated alcohols*, a reverse technique is used [705, 1249]. Reduction of acyl chlorides to alcohols can also be accomplished using *alanes* generated in situ from lithium aluminum hydride and aluminum chloride in ethereal solutions. These reagents are especially suitable for **reduction of halogenated acyl chlorides**, which, in reductions with lithium aluminum hydride, sometimes yield halogen-free alcohols [*1279*]. Reduction of 3-bromopropanoyl chloride with alane gives higher yields (76–90%) than reduction with lithium aluminum hydride (46–87%), even when a normal (not reverse) technique is used [*1023, 1279*].

242

$$C_6H_5COCl \longrightarrow C_6H_5CH_2OH$$

[*110*]	LiAlH$_4$, Et$_2$O, reflux	72%
[*984*]	NaBH$_4$, dioxane, 100 °C	76%
[*1304*]	NaBH$_4$/Al$_2$O$_3$, 25 °C, 2–4 h	90%
[*129*]	NaBH(OMe)$_3$, Et$_2$O, reflux 4 h	66%
[*1004*]	Bu$_4$NBH$_4$, CH$_2$Cl$_2$, –78 °C, 15 min	97%
[*930*]	Zn(BH$_3$CN)$_2$, Et$_2$O, RT, 12 h	97%

ACID ANHYDRIDES

Reductions of anhydrides of monocarboxylic acids to alcohols are very rare but can be accomplished by complex hydrides [*110, 129*]. *Sodium borohydride* in tetrahydrofuran and methanol reduces anhydrides to 75–97% alcohols and 70–100% acids at room temperature within an hour [*1305*].

More frequent are **reductions of cyclic anhydrides of dicarboxylic acids**, which give *lactones*. Such reductions are carried out by catalytic hydrogenation, complex hydrides, and metals.

Hydrogenation of phthalic anhydride over copper chromite affords an 82.5% yield of the lactone, phthalide, and 9.8% *o*-toluic acid resulting from hydrogenolysis of a carbon–oxygen bond [*1306*]. Homogeneous hydrogenation of α,α-dimethylsuccinic anhydride over tris(triphenylphosphine)ruthenium dichloride gives 65% α,α-dimethyl- and 7% β,β-dimethylbutyrolactone [*1307*]. The results of the reduction of α,α-dimethyl- and α,α-diphenylsuccinic anhydrides with *sodium borohydride* and *potassium tris*(sec-*butyl*)*borohydride* are shown in equation **243** [*1308, 1309*].

Under controlled conditions, especially avoiding an excess of *lithium aluminum hydride* and performing the reaction at –55 °C, cyclic anhydrides are converted to lactones in high yields (equation **244**). Both *cis-* and *trans*-1-methyl-1,2,3,6-tetrahydrophthalic anhydride and *cis-* and *trans*-hexahydrophthalic anhydrides are reduced to lactones at the carbonyl group adjoining the methyl-carrying carbon in yields of 75–88.6% [*1310*]. Similar results, including the stereospecificity (where applicable), are obtained by reduction with *sodium borohydride* in refluxing tetrahydrofuran or *N,N*-dimethylformamide at 0–25 °C in yields of 51–97% [*1305,1311, 1312*].

Other reagents used for the preparation of lactones from acid anhydrides are *lithium borohydride* [*1312*], *lithium triethylborohydride* (Super-hydride) [*133, 1312*], and *lithium tris*(sec-*butyl*)*borohydride* (L-Selectride) [*1312*]. Of the three complex borohydrides, the last one is most stereoselective in the reduction of 3-methylphthalic anhydride, 3-methoxyphthalic anhydride, and 1-methoxynaphthalene-2,3-dicarboxylic anhydride. It reduces the less sterically hindered carbonyl group with 85–90% stereoselectivity and an 83–91% yield [1312]. With nonsymmetrical anhydrides, lithium aluminum hydride [1307] and sodium borohydride [1309] reduce preferentially the sterically hindered carbonyl, whereas *potassium tris(sec-butyl)borohydride* (K-Selectride) reduces predominantly the hindered carbonyl group [1308] (equation 243).

243

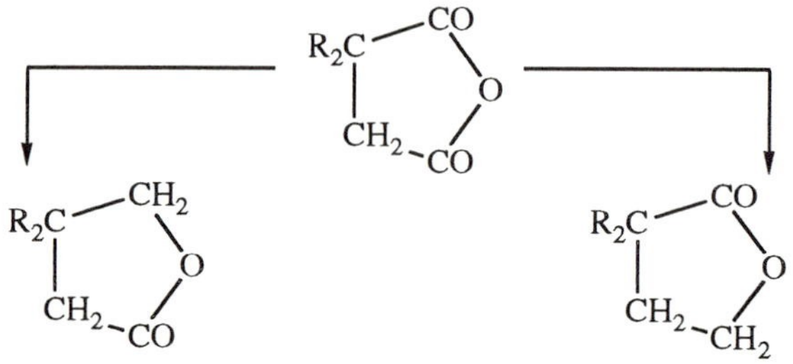

$$R_2C \overset{CO}{\underset{CH_2-CO}{\big|}} O$$

$$R_2C \overset{CH_2}{\underset{CH_2-CO}{\big|}} O \qquad\qquad R_2C \overset{CO}{\underset{CH_2-CH_2}{\big|}} O$$

R = CH₃

[1307]	H₂/(Ph₃P)₃RuCl₂	7%	65%
[1307]	LiAlH₄, THF, –55 °C	66.5%	3.5%
[1308]	KBH(CHMeEt)₃ THF, –70 °C, 1.5–2 h	6%	36%

R = C₆H₅

[1309]	NaBH₄, i-PrOH, 0 °C	67%	
[1308]	KBH(CHMeEt)₃ THF, –70 °C, 1.5–2 h		67%

Results of reductions of cyclic anhydrides to lactones with *sodium amalgam* or *zinc* [1313] are inferior to those achieved by complex hydrides (equation **244**).

244

[1306]	H₂/CuCr₂O₄, C₆H₆, 260 °C, 215 atm	82.5%	
[110]	LiAlH₄, Et₂O, reflux		87%
[1249]	NaAlH₂ (OCH₂CH₂OMe)₂, C₆H₆, 80 °C, 1.5 h		88.5%
[1311]	NaBH₄, DMF, 0–20 °C, 1 h	97%	
[133]	LiBHEt₃, THF, 0 °C, 1 h	76%	
[1313]	Zn, AcOH	30–35%	

Powerful complex hydrides such as *lithium aluminum hydride* in refluxing ether [110] or refluxing tetrahydrofuran [1310] reduce cyclic anhydrides to diols. Phthalic anhydride is thus transformed to phthalyl alcohol (o-hydroxy-methylbenzyl alcohol) [110]. Similar yields of phthalyl alcohol are obtained from phthalic anhydride and *sodium bis(2-methoxyethoxy)aluminum hydride* [705, 1249] (equation **244**).

CHAPTER 16

Reduction of Esters and Lactones of Carboxylic Acids

Reduction of esters of carboxylic acids is complex and takes place in stages and by different routes. In the first stage, hydrogen adds across the carbonyl double bond and generates a hemiacetal (route a in equation **245**), or else it can hydrogenolyze the alkyl–oxygen bond and form an acid and a hydrocarbon (route b). Route b is very rare and is mainly limited to esters of tertiary and benzyl-type alcohols. The hemiacetal is further reduced so that hydrogen replaces either the hydroxylic group (route c) and gives an ether, or else replaces the alkoxy group and yields an alcohol (route d). Alternatively, the hemiacetal may eliminate one molecule of alcohol and afford an aldehyde (route e), which is subsequently reduced to the same alcohol that was formed by route d. The hemiacetals can be isolated in the reductions of esters of carboxylic acids containing electron-withdrawing substituents. One-electron reduction achieved by metals leads to acyloins (route f). The difference in the reaction pathways depends mainly on the reducing agents (equation **245**).

245

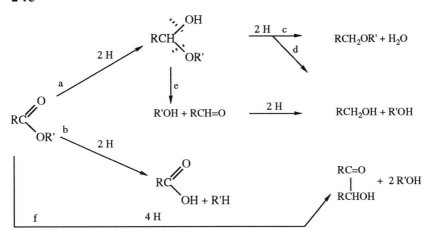

By the use of aluminum hydrides and complex hydrides, **esters can be reduced to aldehydes** in respectable yields. The reductions are usually carried out at subzero temperatures, frequently under cooling with dry ice. *Lithium aluminum hydride* is too powerful and even at –78 °C reduces methyl hexanoate in tetrahydrofuran not only to hexanal (49%) but also to 1-hexanol (22%) [*116*]. *Potassium aluminum hydride* in diglyme at –15 °C produces a 54% yield of butanal from ethyl butyrate [*116*], and *sodium aluminum hydride* in tetrahydrofuran at –45 °C to –60 °C affords an 88% yield of 3-phenylpropanal from methyl hydrocinnamate [*116*]. *Diisobutylaluminum hydride* in ether, hexane, or toluene at –70 °C and *sodium diisobutylaluminum hydride* in ether at –70 °C give, respectively, 78–90% and 60–88% yields of aldehydes from methyl and ethyl esters, respectively, of aliphatic as well as aromatic carboxylic acids [*1314*].

Aromatic esters require lower temperatures and longer times and give 10–15% lower yields than the aliphatic ones [*116, 1314*]. Diisobutylaluminum hydride is frequently used for reduction of esters to aldehydes in syntheses of natural products [*1315*]. Diethyl octafluoroadipate in tetrahydrofuran is reduced quantitatively with *sodium bis(2-methoxyethoxy)aluminum hydride* in toluene at –70 to –50 °C to the bis-hemiacetal, 2,2,3,3,4,4,5,5-octafluoro-1,6-diethoxy-1,6-hexanediol, which on distillation with phosphorus pentoxide gives a 65% yield of octafluorohexanedial [*67, 1316*]. With less powerful hydrides such as *lithium tris(tert-butoxy)aluminum hydride* [*1317*], *lithium bis(diethylamino)aluminum hydride* [*1318*], *bis(N-methylpiperazino)aluminum hydride* [*1319*], and similar amino hydrides, the reductions are carried out at 0 °C, room temperature, and even higher temperatures. *Lithium tris(tert-butoxy)aluminum hydride* converts phenyl esters to aldehydes in tetrahydrofuran at 0 °C in yields of 33–77% [*1317*]; *bis(N-methylpiperazino)aluminum hydride* reduces alkyl esters in tetrahydrofuran at 25 °C or on refluxing in 56–86% yields [*1319*] (equations **246** and **247**).

246

$$C_5H_{11}CO_2CH_3 \quad\longrightarrow\quad C_5H_{11}CHO$$

[*116*]	LiAlH$_4$, THF, –78 °C	49%[a]
[*1318*]	LiAlH$_4$, Et$_2$NH, THF, RT, 1 h[b]	84%
[*116*]	NaAlH$_4$, THF, –60 to –45 °C, 3 h	85%
[*131*]	NaAlHEt$_2$NC$_5$H$_{10}$, THF, 0 °C, 1 h	85%
[*1314*]	NaAlH$_2$ (CH$_2$CHMe$_2$)$_2$, Et$_2$O, –70 °C	72%
[*1314*]	AlH(CH$_2$CHMe$_2$)$_2$, PhMe or C$_6$H$_{14}$, –70 °C, 0.5–1 h	85%
[*1319*]	AlH(N◠NMe)$_2$, THF, reflux 6 h[b]	78%

[a]Hexanol is formed in 22% yield. [b]Ethyl ester is reduced.

247

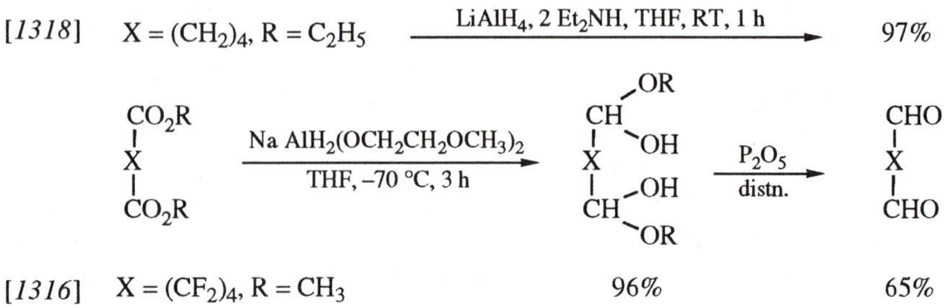

[*1318*]   X = (CH$_2$)$_4$, R = C$_2$H$_5$ $\xrightarrow{\text{LiAlH}_4,\ 2\ \text{Et}_2\text{NH, THF, RT, 1 h}}$ 97%

CO$_2$R   CH⟨OR⟩OH   CHO
| Na AlH$_2$(OCH$_2$CH$_2$OCH$_3$)$_2$ | P$_2$O$_5$ |
X $\xrightarrow{\text{THF, –70 °C, 3 h}}$ X $\xrightarrow{\text{distn.}}$ X
| | |
CO$_2$R   CH⟨OH⟩OR   CHO

[*1316*]   X = (CF$_2$)$_4$, R = CH$_3$ 96%   65%

Reduction of lactones leads to cyclic hemiacetals of aldehydes (lactols). With a stoichiometric amount of *lithium aluminum hydride* in tetrahydrofuran at –10 to –15 °C and using the inverse technique, γ-valerolactone is converted in a 58% yield to 2-hydroxy-5-methyltetrahydrofuran, and α-methyl-δ-caprolactone gives a 64.5–84% yield of 3,6-dimethyl-2-hydroxytetrahydropyran [1320]. Both compounds are in equilibrium with their open-chain forms, δ-hydroxy-valeraldehyde and α-methyl-δ-hydroxycapronaldehyde, respectively. Also *diisobutylaluminum hydride* in tetrahydrofuran solutions at subzero temperatures affords high yields of lactols from lactones [1315].

Reduction of lactones to aldehyde–hemiacetals is of the utmost importance in the realm of saccharides. The old method of converting aldonolactones to aldoses by means of sodium amalgam is superseded by the use of *sodium borohydride,* which usually gives good yields, is easy to handle, and is very suitable because of its solubility in water, the best solvent for sugars [1321]. However, the yields of sodium borohydride reductions are not always high [1322], and yields of *sodium amalgam* reductions can be greatly improved if the reductions are carried out at pH 3–3.5 with 2.5–3 g-atoms of sodium in the form of a 2.5% sodium amalgam with particle size of 4–8 mesh. Yields of 50–84% aldoses are thus obtained [1323].

The **reduction of esters to ethers** is of limited practical use. Good results are obtained only with esters of tertiary alcohols using 2 mol of *borane* generated from sodium borohydride and an enormous excess (30 mol) of boron trifluoride etherate in tetrahydrofuran and diglyme. Butyl, *sec*-butyl, and *tert*-butyl cholanates are converted to the corresponding ethers, 24-butoxy-, *sec*-butoxy-, and *tert*-butoxy-5β-cholanes in respective yields of 7%, 41%, and 76%. *tert*-Butyl 5α-pregnane-20S-carboxylate is reduced to 20S-*tert*-butoxymethylene-5α-pregnane in a 76% yield [1324] (equation **248**).

248   [1324]

Reduction of esters by *trichlorosilane* in tetrahydrofuran in the presence of *tert*-butyl peroxide and under ultraviolet irradiation gives predominantly ethers from esters of primary alcohols, whereas esters of tertiary alcohols are cleaved to acids and hydrocarbons. Esters of secondary alcohols give ethers and mixtures of acids and hydrocarbons in varying ratios. 1-Adamantyl trimethylacetate, for example, affords 50–100% yields of mixtures containing 2–42% 1-adamantyl neopentyl ether and 58–98% adamantane and trimethylacetic acid [1325].

A more useful way of reducing esters to ethers is a two-step procedure applied to the **reduction of lactones to cyclic ethers.** First the lactone is treated with *diisobutylaluminum hydride in* toluene at –78 °C, and the product, a lactol,

is subjected to the action of *triethylsilane* and boron trifluoride etherate at
–20 °C to –70 °C. γ-Phenyl-γ-butyrolactone is thus transformed to 2-phenyl-
tetrahydrofuran in a 75% yield, and the δ-lactone of 3-methyl-5-phenyl-5-
hydroxy-2-pentenoic acid is converted to 4-methyl-2-phenyl-2,3-dihydropyran
in a 72% yield [*1326*].

A direct one-step procedure for conversion of lactones to cyclic ethers is
carried out by adding a solution of the lactone in ether and *boron trifluoride
etherate* into a cooled suspension of *lithium aluminum hydride* in ether or
tetrahydrafuran and refluxing the reaction mixture for 2 h [*1327*]. *Lithium
aluminum hydride* alone reduces cyclic dilactones to crown ethers at 0 °C in
77–92% yields [*1328*].

Cleavage of esters to acids and hydrocarbons, already mentioned, is
achieved not only with hydrides but also by *catalytic hydrogenation* and
reduction with metals and metal compounds. For example, the acetate of
mandelic acid is converted to phenylacetic acid and acetic acid by hydrogenation
at 20 °C and 1 atm over palladium on barium sulfate in ethanol in the presence
of triethylamine in 10 min [*1329*] (equation **249**).

249

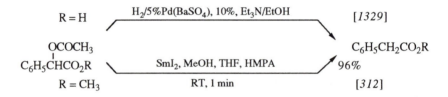

Cinnamyl acetate is cleaved to acetic acid and to (*E*)-1-phenylpropene on
heating at 70 °C in acetonitrile with 1-propyl-1,4-dihydronicotinamide in the
presence of tris(triphenylphosphine)rhodium chloride and lithium perchlorate
(97% yield) [*1495*]. *Tributylstannane* in the presence of tetrakis(triphenylphosphine)palladium in tetrahydrofuran cleaves the acetate of cinnamaldehyde
cyanohydrin to acetic acid and 4-phenyl-3-butenenitrile in a 99% yield without
affecting the other reducible functions [*1330*]. Similar cleavage is effected using
samarium diiodide in tetrahydrofuran in reasonable to high yields [*312*]
(equation **249**). α,α-Diphenylphthalide is reduced by refluxing for 5 h with *zinc*
in formic acid to α,α-diphenyl-*o*-toluic acid (92% yield) [*1331*]. Such reductions
are of immense importance in esters of benzyl-type alcohols in which the yields
of the acids are almost quantitative.

The most common *hydrogenolysis of benzyl esters* is accomplished by
catalytic hydrogenation over palladium catalysts. This reaction is very useful
because it removes benzylic protective groups under very gentle conditions
(room temperature and atmospheric pressure) without using media that could
damage sensitive functions in the molecule [*1332*]. Dibenzyl 4-methoxy-
carbonyl-3-indolylmethylacetamidomalonate is debenzylated to the free
dicarboxylic acid in 1.5 h in a 75% yield over palladium on charcoal in methanol
at 20 °C and 1 atm [*1333*] (equation **250**). Dibenzyl esters of acylmalonic acids
are hydrogenolyzed over 10% palladium on carbon or strontium carbonate at

30 °C. Under these conditions double decarboxylation takes place, giving ketones in 60–91% yields [*1334*] (equation **251**).

250 [*1333*]

$$\begin{array}{c} \text{CO}_2\text{CH}_3 \\ \text{CH}_2\text{C}(\text{CO}_2\text{CH}_2\text{C}_6\text{H}_5)_2 \\ | \\ \text{NHCOCH}_3 \end{array} \xrightarrow[\text{20 °C, 1 atm, 1.5 h}]{\text{H}_2/\text{Pd(C), MeOH}} \begin{array}{c} \text{CO}_2\text{CH}_3 \\ \text{CH}_2\text{C}(\text{CO}_2\text{H})_2 \\ | \\ \text{NHCOCH}_3 \quad 75\% \end{array}$$

251 [*1334*]

$$\begin{array}{c} \text{CH}_3(\text{CH}_2)_6\text{COCl} \\ + \\ \text{CH}_3(\text{CH}_2)_7\text{CNa}(\text{CO}_2\text{CH}_2\text{C}_6\text{H}_5)_2 \end{array} \longrightarrow \begin{array}{c} \text{CH}_3(\text{CH}_2)_6\text{CO} \\ | \\ \text{CH}_3(\text{CH}_2)_7\text{C}(\text{CO}_2\text{CH}_2\text{C}_6\text{H}_5)_2 \end{array} \xrightarrow[\text{EtOH, <30 °C}]{\text{H}_2/10\% \text{ Pd(C)}} \begin{array}{c} \text{CH}_3(\text{CH}_2)_6\text{CO} \\ | \\ \text{CH}_3(\text{CH}_2)_7\text{CH}_2 \\ 91\% \end{array}$$

Catalytic hydrogenolysis is frequently used for **cleavage of benzyl carbonates or carbamates.** The ester–acid or amide–acid resulting from the hydrogenolysis of alkylcarbonic or alkylcarbamic esters spontaneously decarboxylates to a hydroxy or amino compound (equation **252**).

252

$$\text{C}_6\text{H}_5\text{CH}_2\text{OCOOR} \xrightarrow{\text{H}_2/\text{Pd}} \text{C}_6\text{H}_5\text{CH}_3 + \text{HOCOOR} \longrightarrow \text{HOR} + \text{CO}_2$$

$$\text{C}_6\text{H}_5\text{CH}_2\text{OCONHR} \xrightarrow{\text{H}_2/\text{Pd}} \text{C}_6\text{H}_5\text{CH}_3 + \text{HOCONHR} \longrightarrow \text{H}_2\text{NR} + \text{CO}_2$$

Many examples are found in amino acids, in which benzyloxycarbonyl and *p*-nitrobenzyloxycarbonyl groups temporarily protect amino groups [*1335, 1336, 1337, 1338, 1339*]. Hydrogenolysis is very easy and takes place in preference to saturation of double bonds [*946*], prior to reduction of the nitro group [*1337, 1339*], and even in the presence of divalent sulfur in the molecule [*786, 946, 1332*]. However, the azido group is hydrogenated preferentially to the hydrogenolysis of the benzyl ester [*786*]. Cyclohexene and cyclohexadiene can be used as hydrogen donors in the catalytic hydrogen transfer over palladium on charcoal [*85, 1340*] (Procedure 9, p 299).

Cleavage of alkyl–oxygen bonds of other than benzyl-type esters takes place in esters of α-keto alcohols. 3β,17α-Diacetoxyallopregnan-20-one gives 3β-acetoxyallopregnan-20-one on refluxing with *zinc* in acetic acid (89% yield) [*1179*], and a lactone of *trans*-2,9-dioxo-1a-carboxymethyl-1-hydroxy-8-methyl-1,1a,2,3,4,4a-hexahydrofluorene is hydrogenolyzed to the free acid with *chromous chloride* in acetone at room temperature [*1341*] (equation **253**). Also, the vinylog of an ester of an α-keto alcohol, namely Δ⁴-cholesten-6β-ol-3-one acetate, undergoes alkyl–oxygen hydrogenolysis by refluxing with zinc dust in acetic acid and gives cholest-4-en-3-one in a 67% yield [*1342*].

253   [*1341*]

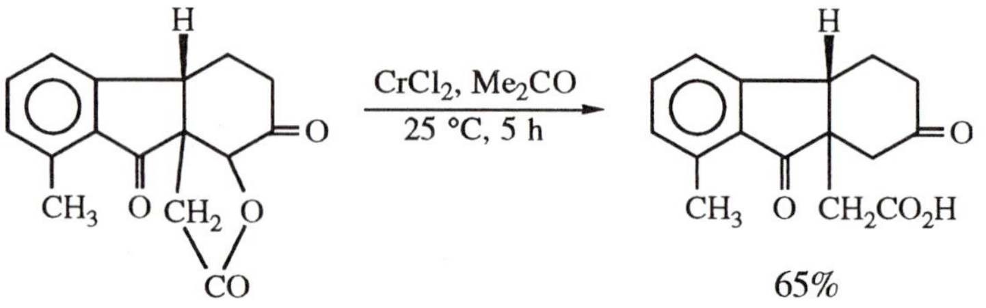

65%

Reactions reminiscent of pinacol reduction take place if esters are treated with sodium in aprotic solvents. The initially formed radical anion dimerizes and ultimately forms an α-hydroxy ketone, an acyloin. Such *acyloin condensation* of esters is especially useful with esters of α,ω-dicarboxylic acids of at least six carbons in the chain, because cyclic acyloins are formed in very good yields. The best yields are obtained if the reduction is carried out with an exactly stoichiometric amount of sodium (2 g-atoms per molecule of ester) under an inert gas in boiling toluene or xylene, which dissolves the reaction intermediates and keeps sodium in a molten state. Open-chain acyloins of 8–18 carbons are thus synthesized in 80–90% yields [*1343*], and cyclic acyloins of 9, 10, 12, 14, and 20 carbons in 30, 43, 65, 47, and 96% yields, respectively [*1344, 1345, 1346*] (equations **254** and **255**, Procedure 31, p 309).

254   [*1343*]

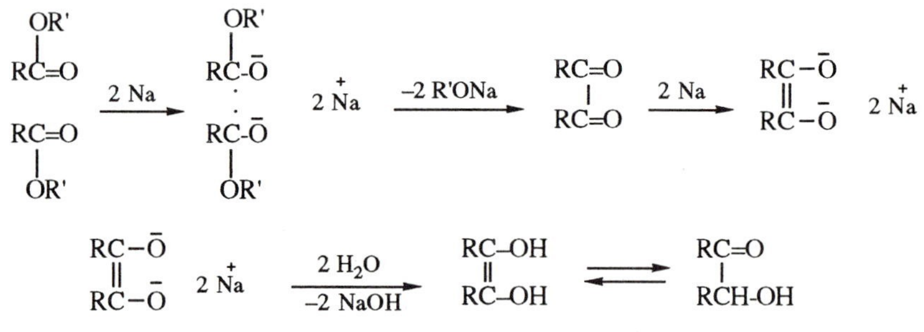

255   [*1345*]

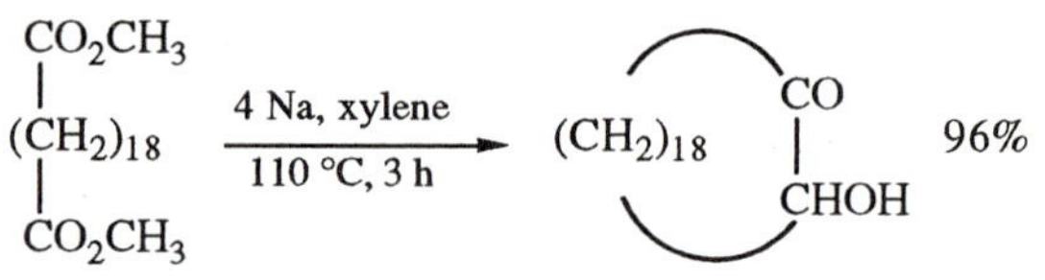

By far the most frequent **reduction of esters** is their *conversion to alcohols*. The reaction is important not only in the laboratory but also in industry, where it is used mainly for hydrogenolysis of fats to fatty alcohols and glycerol. *Lactones are reduced to diols.* The reduction of esters to alcohols has an interesting history. The oldest reduction, the *Bouveault–Blanc reduction* [*1347*], carried out by adding sodium into a solution of an ester in ethanol, did not give very high yields, possibly because of side reactions like Claisen condensation. With some modifications, good yields (65–75%) were obtained with esters of monocarboxylic acids [*1348*], dicarboxylic acids (73–75%) [*1349*], and even unsaturated acids containing nonconjugated double bonds (49–51%) [*1350*]. The main improvement was use of an inert solvent such as toluene or xylene and a secondary alcohol that is acidic enough to decompose sodium-containing intermediates but does not react too rapidly with sodium: the best of all is methylisobutylcarbinol (4-methyl-2-pentanol). Under these conditions the precise theoretical amounts of sodium and the alcohol may be used, and this precision prevents side reactions such as acyloin and Claisen condensations. The reaction is best carried out by adding a mixture of the ester, alcohol, and toluene or xylene into a stirred refluxing mixture of sodium and xylene at a rate sufficient to maintain reflux. Yields obtained in this modification are close to theoretical or at least 80% [*205*]. The amount of sodium and the alcohol is given by the stoichiometric equation **256** accounting for the individual reaction steps.

256

$$OC_2H_5$$
$$RC=O \xrightarrow{2\ Na} RC\text{-}\overset{+}{O}Na \xrightarrow[-EtONa]{+EtOH} RCH\text{-}ONa \xrightarrow{-EtONa}$$
$$\overset{-}{N}a$$

$$\xrightarrow{} RCH=O \xrightarrow{2\ Na} RCH\text{-}\overset{+}{O}Na \xrightarrow[-EtONa]{+EtOH} RCH_2ONa \xrightarrow[-NaOH]{+H_2O} RCH_2OH$$
$$\overset{+}{N}a$$

$$RCO_2C_2H_5 + 4\ Na + 2\ C_2H_5OH + H_2O = RCH_2OH + 3\ C_2H_5ONa + NaOH$$

The reduction as described is applicable to small-scale as well as large-scale operations provided safety precautions necessary for handling molten sodium and sodium dispersion are observed.

Catalytic hydrogenation of esters to alcohols is rather difficult; therefore, it is not surprising that feasible ways of hydrogenolyzing esters were developed. The main breakthrough was the discovery of special catalysts based on oxides of copper, zinc, chromium [*51, 52, 53*], and other metals [*55*]. Such catalysts are especially suited for hydrogenolysis of carbon–oxygen bonds. They are fairly resistant to catalytic poisons but require high temperatures (100–300 °C) and high pressures (140–350 atm) [*819, 1351, 1352, 1353*]. Yields of the alcohols are high to quantitative [*819, 1352*]. Side reactions include hydrogenolysis of carbon–oxygen bonds in diols resulting from hydrogenation of hydroxy or keto esters, hydrogenolysis of glycerol to propylene glycol during the hydrogenation of fats to fatty alcohols, and hydrogenolysis of some aromatic alcohols to hydrocarbons or basic heterocycles [*575*]. Unsaturated esters are usually converted to saturated alcohols on the *copper chromite* catalysts, but unsaturated alcohols are obtained in 37–65% yields if *zinc chromite is* used at 280–300 °C and 200 atm [*53*] (equation **257**).

257

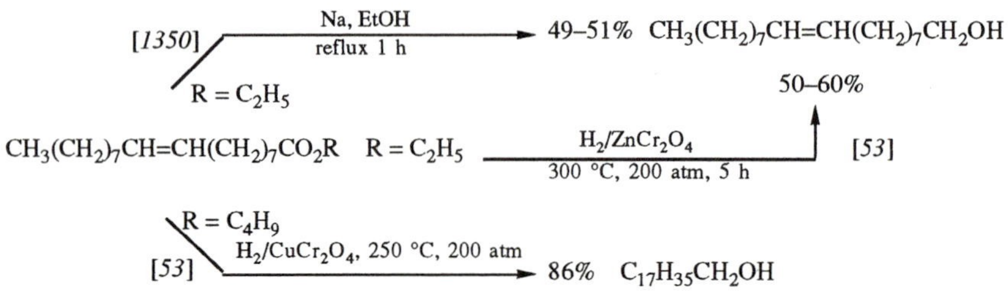

Benzene rings resist hydrogenation, but pyridine rings are hydrogenated to piperidine rings [*819*]. Also, pyrrole nuclei are often saturated to pyrrolidines with the concomitant hydrogenolysis of the intermediate alcohols. Esters of alkyl pyrrolecarboxylates are converted to alkylpyrroles and alkylpyrrolidines [*575*] (equation **258**).

258

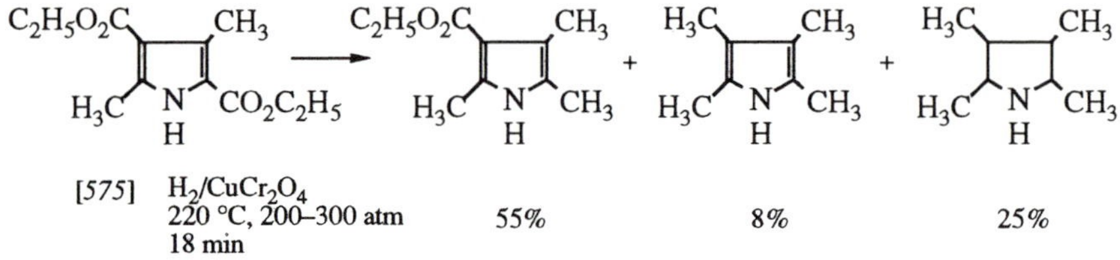

Hydrogenation temperatures can be lowered down to 125–150 °C when a large excess of the catalyst is used. Such a modification is especially useful in reductions of hydroxy and keto esters because the hydrogenolysis of diols or triols to alcohols is considerably lower, and yields of the products range from 60% to 80%. The amounts of catalysts used in such experiments are up to 1.5 times the weight of the ester under a pressure of 350 atm [*65*] (equation **259**).

259   [*65*]

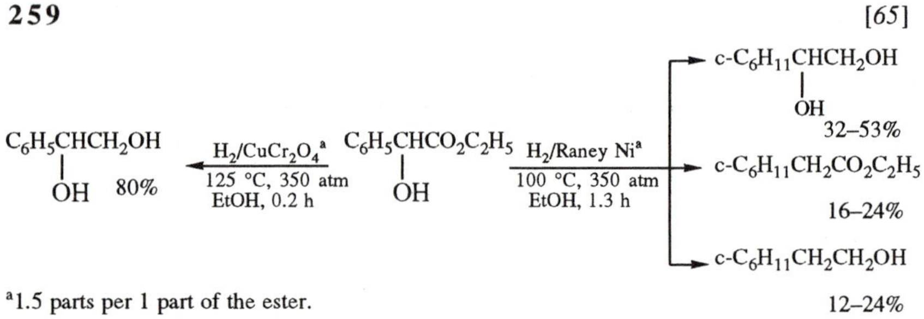

[a]1.5 parts per 1 part of the ester.   12–24%

With the same excess of catalysts, hydrogenation of the esters over Raney nickel could be carried out at temperatures as low as 25–125 °C at 350 atm with comparable results (80% yields). However, benzene rings are saturated under these conditions [*65*] (equation **259**). In addition to nickel and copper, zinc and chromium oxides and rhenium obtained by reduction of rhenium heptoxide also catalyze hydrogenation of esters to alcohols at 150–250 °C and 167–340 atm in 35–100% yields [*55*].

Both older methods for the reduction of esters to alcohols, catalytic hydrogenation and reduction with sodium, have given way to reductions with *hydrides* and *complex hydrides,* which have revolutionized the laboratory preparation of alcohols from esters.

Countless reductions of esters to alcohols have been accomplished using *lithium aluminum hydride.* Only 0.5 mol of this hydride is needed for reduction of 1 mol of the ester. Ester or its solution in ether is added to a solution of lithium aluminum hydride in ether. The heat of reaction brings the mixture to boiling. The reaction mixture is decomposed by ice–water and acidified with mineral acid to dissolve lithium and aluminum salts. Less frequently, sodium hydroxide is used for this purpose. Yields of alcohols are frequently quantitative [*110, 1354*]. Lactones afford glycols (diols) [*757*].

Esters are also reduced by *sodium aluminum hydride* (95–97% yields) [*115*] and by *lithium trimethoxyaluminum hydride* (2 mol per mol of the ester) [*123*] but not by *lithium tris(tert-butoxy)aluminum hydride* [*125*]. Another complex hydride, *sodium bis(2-methoxyethoxy)aluminum hydride,* reduces esters in benzene or toluene solutions (1.1–1.2 mol per ester group) at 80 °C in 15–90 min in 66–98% yields [*1249*]. *Magnesium aluminum hydride* (in the form of its tetrakistetrahydrofuranate) reduces methyl benzoate to benzyl alcohol in a 58% yield on refluxing for 2 h in tetrahydrofuran [*117*].

Alane, formed in situ from lithium aluminum hydride and aluminum chloride, reduces esters (1.3 mol per mol of ester) in tetrahydrofuran at 0 °C in 15–30 min in 83–100% yields [*1280*]. Because it does not reduce some other functions, it is well suited for selective reductions [*1279, 1280*].

Other reagents used for reduction are *boranes* and complex borohydrides. *Lithium borohydride,* whose reducing power lies between that of lithium aluminum hydride and that of sodium borohydride, reacts with esters sluggishly and requires refluxing for several hours in ether or tetrahydrofuran (in which it is more soluble) [*983*]. The reduction of esters with lithium borohydride is strongly catalyzed by boranes such as *B*-methoxy-9-borabicyclo[3.3.1]nonane and some other complex lithium borohydrides such as lithium triethylborohydride and lithium 9-borabicyclo[3.3.l]nonane. Addition of 10 mol% of such hydrides shortens the time necessary for complete reduction of esters in ether or tetrahydrofuran from 8 to 0.5–1 h [*1355*] and affects the selectivity for reductions of different other functions [*1491*].

Sodium borohydride reduces esters only under special circumstances. An excess of up to 16 mol of sodium borohydride refluxed with 1 mol of the ester in methanol for 1–2 h produces 72–93% alcohols from aliphatic, aromatic, heterocyclic, and α,β-unsaturated esters. The conjugated double bonds, especially if also conjugated with aromatic rings, are saturated in many instances [*1356*]. In polyethylene glycols, reductions of aliphatic, aromatic, and α,β-unsaturated esters are accomplished by heating at 65 °C for 10 h with 3 mol of sodium borohydride per mol of the ester [*1357*]. *Tetrabutylammonium borohydride* is unsuitable for the reduction of esters [*1004*]. On the other hand, *calcium borohydride* in tetrahydrofuran reduces esters at a rate higher than sodium borohydride and lower than lithium borohydride and in alcoholic solvents even higher than lithium borohydride. When esters are treated with stoichiometric amounts (0.5 mol) of *lithium borohydride* in ether or tetrahydrofuran or calcium borohydride in tetrahydrofuran, and the mixtures after addition of toluene are heated to 100 °C, while the more volatile solvents distill off, 70–96% yields of alcohols are usually obtained in 0.5–2 h [*1358*]. *Sodium trimethoxyborohydride* is also used for reduction of esters to alcohols at higher temperatures but in poor yields [*129*]. On the other hand, *sodium dimethylaminoborohydride* and *sodium* tert-*butylaminoborohydride* in tetra-hydrofuran at room temperature produce alcohols in 63–96% yields [*130*].

The efficient *lithium triethylborohydride* converts esters to alcohols in 94–100% yields in tetrahydrofuran at 25 °C in a few minutes when 2 mol of the hydride per mol of the ester is used [*133*].

Borane prepared by adding aluminum chloride to a solution of sodium borohydride in diethylene glycol dimethyl ether (diglyme) reduces aliphatic and aromatic esters to alcohols in quantitative yields in 3 h at 25 °C using a 100% excess, or in 1 h at 75 °C using a 25% excess over 2 mol of sodium borohydride per mol of the ester [*958*] (Procedure 22, p 305).

A solution of borane in tetrahydrofuran reduces esters at room temperature only slowly [*1252*]. Under such conditions, free carboxylic groups of acids are reduced preferentially: The monoethyl ester of adipic acid treated with 1 mol of borane in tetrahydrofuran at –18–25 °C gives an 88% yield of ethyl 6-hydroxyhexanoate [*1251*]. Borane–dimethyl sulfide in tetrahydrofuran is used for reduction of esters to alcohols on refluxing for 0.5–16 h and allowing the dimethyl sulfide to distill off (73–97% yields) [*1359*] (equation **260**).

260

$C_6H_5CO_2C_2H_5$ —————————————————→ $C_6H_5CH_2OH$

[*110*] LiAlH$_4$, Et$_2$O, reflux	90%
[*1249*] NaAlH$_2$(OCH$_2$CH$_2$OMe)$_2$, C$_6$H$_6$, 80 °C, 0.5 h	91%
[*1358*] LiBH$_4$, THF, PhMe, 100 °C, 1 h	87%
[*1355*] LiBH$_4$, MeOB⬡ , Et$_2$O, reflux 2 h	81%
[*1356*]10 NaBH$_4$, MeOH, reflux 1–2 h[a]	64%
[*130*] [a]NaBH$_3$NMe$_2$, THF, 25 °C, 15 h	90%
[*1358*] Ca(BH$_4$)$_2$, THF, PhMe, 100 °C, 1.5 h	87%
[*1359*] BH$_3$•Me$_2$S, THF, reflux 4 h	90%

[a]Methyl ester was reduced.

Other hydrides used for the conversion of esters to alcohols are *magnesium aluminum hydride* in tetrahydrofuran [*117, 758*] and *magnesium bromohydride* prepared by decomposition of ethylmagnesium bromide at 235 °C for 2.5 h at 0.5 mm [*1360*]. They do not offer special advantages.

UNSATURATED ESTERS

The reduction of unsaturated esters encompasses the reduction of esters containing double and triple bonds, usually in α,β-positions, sometimes conjugated with aromatic rings. Such esters may be converted to less saturated or completely saturated esters, unsaturated or less unsaturated alcohols, or saturated alcohols. The presence of aromatic rings in conjugation with the multiple bonds facilitates saturation of the bonds. Aromatic rings are hydrogenated only after the saturation of the double bonds.

Catalytic hydrogenation of **olefinic esters** under mild conditions affords *saturated esters* whether or not the double bonds are conjugated. Methyl 6-methoxy-1,2,3,4-tetrahydro-1-naphthylidenecrotonate absorbs 1 mol of hydrogen in 8.5 min on hydrogenation over Raney nickel in ethanol and acetic acid at room temperature and atmospheric pressure and gives methyl γ-(6-methoxy-1,2,3,4-tetrahydronaphthylidene)butyrate in 96% yield. Here the α,β-conjugated double bond is saturated in preference to the γ,δ bond [*1361*].

Hydrogenation over nickel catalyst at high temperatures and pressures *affects aromatic rings*. Over Urushibara nickel at 106–150 °C and 54 atm, ethyl benzoate gives ethyl hexahydrobenzoate in an 82% yield [*49*].

Methyl cinnamate is reduced quantitatively to methyl 3-phenylpropanoate by hydrogen over colloidal palladium [*1273*], and ethyl cinnamate is converted to ethyl 3-phenylpropanoate over tris(triphenylphosphine)rhodium chloride [*91*] and over copper chromite [*800*]. On the other hand, hydrogenation of ethyl cinnamate over Raney nickel gives ethyl 3-cyclohexylpropanoate [*1362*], and over copper chromite 3-phenylpropanol [*1352*] (equation **261**).

261

$$C_6H_5CH=CHCO_2R \longrightarrow C_6H_5CH_2CH_2CO_2R + C_6H_5CH=CHCH_2OH + C_6H_5CH_2CH_2CH_2OH$$

[*91*]	[a]H_2/(Ph$_3$P)$_3$RhCl, EtOH, 40–60 °C, 4–7 atm	93%		
[*800*]	[a]H_2/CuCr$_2$O$_4$, 150 °C, 175 atm	97%		
[*1362*]	H_2/Raney Ni, 250 °C, 200 atm	88%[c]		
[*1352*]	H_2/CuCr$_2$O$_4$, 250 °C, 220 atm			83%
[*1356*]	[b]10 NaBH$_4$ MeOH, reflux 1–2 h	29%	38%	33%
[*1366*]	[b]1 LiAlH$_4$ C$_6$H$_6$, 60 °C, 14.5 h		73%	
[*1371*]	[a]KBH$_4$, LiCl, THF, reflux 4–5 h		10%	66%
[*693*]	[a]1.1 NaAlH$_2$(OCH$_2$-CH$_2$OMe)$_2$, C$_6$H$_6$, 15–20 °C, 45 min		75.5%	
[*693*]	[a]2.4 NaAlH$_2$(OCH$_2$-CH$_2$OMe)$_2$, C$_6$H$_6$, 15–20 °C, 2 h			44.5%
[*1372*]	[a]Na, EtOH, reflux			d

[a]R = Me   [b]R = Et   [c]The product is ethyl 3-cyclohexylpropanoate   [d]The yield is not given.

Reduction of unsaturated esters can be done *stereoselectively* over chiral catalysts, usually atropoisomeric phosphine complexes of rhodium. Such reactions are very important for the preparation of optically active amino acids and their esters from α-acylamido-α,β-unsaturated acids or esters readily available by aldol-type condensations [96, 97, 99, 100, 101, 107] (equation **262**). The chiral hydrogenations are also applicable to dehydro dipeptides.

262

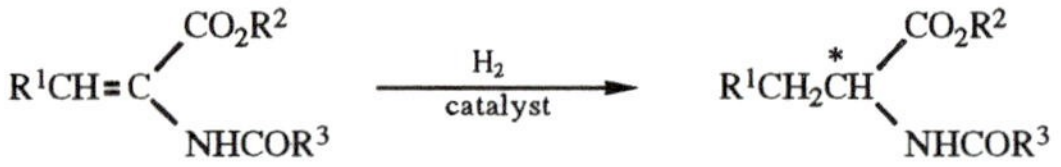

$$R^1CH=C\underset{NHCOR^3}{\overset{CO_2R^2}{\diagup}} \quad \xrightarrow[\text{catalyst}]{H_2} \quad R^1CH_2\overset{*}{CH}\underset{NHCOR^3}{\overset{CO_2R^2}{\diagup}}$$

				%	ee %	Config.
[96]	$R^1 = H, R^2 = H, R^3 = CH_3$	$(CH_3)_3COCH_2CHPPh_2$ Rh — CH_2PPh_2	EtOH, RT, 1 atm		92	S
[107]		$Rh[(S,S\text{-Chiraphos}]^a$	EtOH		91	R
[101]		$Rh\text{-Binap}^b$	EtOH, RT, 3–4 atm	97	67	R
[96]	$R^1 = C_6H_5, R^2 = H,$ $R^3 = CH_3$	$C_6H_5CH_2OCHPPh_2$ Rh — CH_2PPh_2	EtOH, RT, 1 atm		87	S
[107]		$Rh[(S,S\text{-Chiraphos}]^a$	EtOH		95	R
[101]		$Rh\text{-Binap}^b$	EtOH, RT, 3–4 atm	99	84	R
[100]		3,4-(R,R)-diphenyl-phosphinopyrrolidine Rh	MeOH, 22 °C, 50 atm		88	
[97]		$Rh\text{-Chiraphos}^a$	MeOH, 25 °C, 1 atm	100	87	R
[99]		$Rh\text{-Chiraphos}^{a,c}$	H_2O, C_6H_6 25 °C, 1 atm	100	88	
[99]		$Rh\text{-Chiraphos}^{a,c}$	H_2O, C_6H_6 25 °C, 1 atm	100	88	
[101]	$R^1 = C_6H_5, R^2 = H,$ $R^3 = C_6H_5$	$Rh\text{-Binap}^b$	EtOH, RT, 3–4 atm	96	96	R
[1499]	$R^1 = R^3 = CH_3$ $R^2 = $ 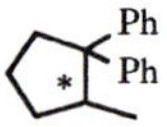 (norbornyl with Ph, Ph)	PtO_2	AcOH, AcOEt, Ac$_2$O RT, 3–5 atm, 3 h	89	96^d	

[a] See formula 7 (p 16).
[b] See formula 6 (p 16).
[c] Catalysts are water-soluble.
[d] Diastereomeric excess (de)

Sodium borohydride under normal conditions does not reduce ester groups but reduces (in 59–74% yields) double bonds in dialkyl alkylidenemalonates (and cyanoacetates), in which the double bonds are cross-conjugated with two carboxylic groups [1363, 1364]. A high yield (80%) is also obtained in the reduction of ethyl atropate (ethyl α-phenylacrylate), in which the double bond is conjugated with one carboxyl and one phenyl group [1364]. Ethyl acrylate gives, under the same conditions, only a 25% yield (possibly because of polymerization). The reduction is carried out by allowing equimolar quantities of sodium borohydride and the ester in ethanol or isopropyl alcohol to react first at 0–5 °C for 1 h and then at room temperature for 1–2 h [1364].

Conjugated double bonds are also reduced in methyl 3-methyl-2-butenoate with *tributylstannane* on irradiation with ultraviolet light at 70 °C (yield of methyl isovalerate is 90%) [*171*] and in diethyl maleate and diethyl fumarate, which afford diethyl succinate in respective yields of 95% and 88% on treatment with *chromous sulfate* in *N,N*-dimethylformamide at room temperature [*1259, 1365*]. Cyclohexyl crotonate is converted in an 82% yield to cyclohexyl butyrate by stirring with *samarium diiodide* in tetrahydrofuran and *tert*-butyl alcohol in the presence of 1,3-bis(dimethyldiamino)propane [*308*], and diethyl maleate gives 94% diethyl succinate by reduction with *diimide* [*376*].

Unsaturated alcohols can be obtained from unsaturated esters under special precautions. *Catalytic hydrogenation* over zinc chromite converts butyl oleate to oleyl alcohol (9-octadecenol) in a 63–65% yield even under very vigorous conditions (283–300 °C, 200 atm) [*53*].

Much more conveniently, even α,β-unsaturated esters can be transformed into α,β-unsaturated alcohols by very careful treatment with *lithium aluminum hydride* [*1366*], *sodium bis(2-methoxyethoxy)aluminum hydride* [*705*], or *diisobutylalane* [*1367*] (Procedure 20, p 305). An excess of the reducing agent must be avoided. Therefore, the inverse technique (addition of the hydride to the ester) is used, and the reaction is usually carried out at low temperature. In hydrocarbons as solvents, the reduction does not proceed further even at elevated temperatures. Methyl cinnamate is converted to cinnamyl alcohol in a 73% yield when an equimolar amount of the ester is added to a suspension of lithium aluminum hydride in benzene and the mixture is heated at 59–60 °C for 14.5 h [*1366*]. Ethyl cinnamate gives a 75.5% yield of cinnamyl alcohol on inverse treatment with 1.1 mol of sodium bis(2-methoxyethoxy)aluminum hydride at 15–20 °C for 45 min [*705*].

Complete reduction of unsaturated esters to **saturated alcohols** takes place when the esters are hydrogenated over Raney nickel at 50 °C and 150–200 atm [*45*] or over copper chromite at 250–300 °C and 300–350 atm [*53, 1351*]. In contrast to most examples in the literature, the reduction of ethyl oleate is achieved at atmospheric pressure and 270–280 °C over copper chromite, giving an 80–90% yield of octadecanol [*1368*]. α,β-Unsaturated lactones are reduced to saturated ethers or alcohols, depending on the reaction conditions. The results are demonstrated by the hydrogenation of coumarin, the δ-lactone of *o*-hydroxy-*cis*-cinnamic acid [*1369*] (equation **263**).

α,β-Unsaturated esters, especially those in which the double bond is also conjugated with an aromatic ring, are more comfortably converted to saturated alcohols on treatment with an excess of *lithium aluminum hydride* [*1370*], *potassium borohydride* in the presence of lithium chloride [*1371*], or *sodium bis(2-methoxyethoxy)aluminum hydride* [*705*].

Sodium borohydride, even in a very large excess, reduces methyl 2-nonenoate and methyl cinnamate incompletely to mixtures of saturated esters, unsaturated alcohols, and saturated alcohols [*1356*]. On the other hand, α,β-unsaturated esters of the pyridine and pyrimidine series are converted predominantly and even exclusively to saturated alcohols. Methyl 3-(γ-pyridyl)-acrylate gives, on refluxing for 1–2 h in methanol with 10 mol of sodium hydride per mol of the ester, 67% 3-(γ-pyridyl)propanol and 6% 3-(γ-pyridyl)-2-propenol; methyl 3-(6-pyrimidyl)acrylate gives 77% pure 3-(6-pyrimidyl)-propanol [*1356*].

263   [1369]

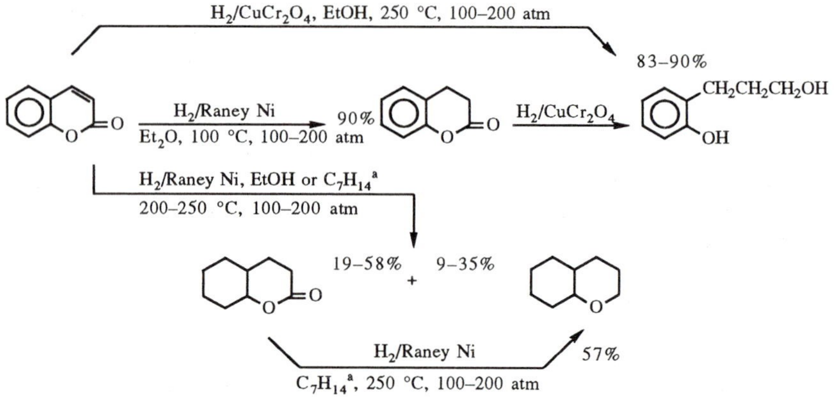

ᵃMethylcyclohexane

Complete reduction of α,β-unsaturated esters to saturated alcohols is also accomplished by *sodium* in alcohol [1372] (equation **261**).

Esters containing conjugated triple bonds may be reduced to *ethylenic esters* by controlled *catalytic hydrogenation* and to ***acetylenic or ethylenic alcohols*** *by lithium aluminum hydride.* Methyl phenylpropiolate treated with 0.5 mol of lithium aluminum hydride in ether at –78 °C for 5 min affords phenylpropargyl alcohol in a 96% yield, but addition of the same ester to 0.75 mol of lithium aluminum hydride in ether at 20 °C gives, after 15 min, an 83% yield of *trans*-cinnamyl alcohol [1373]. Treatment of dimethyl acetylenedicarboxylate with triphenylphosphine and deuterium oxide in tetrahydrofuran gives dimethyl dideuterofumarate in a 70% yield [1374].

SUBSTITUTION DERIVATIVES OF ESTERS

Reductions of halogen, nitro, diazo, and azido derivatives of esters resemble closely those of the corresponding derivatives of carboxylic acids (p 197).

Catalytic hydrogenation of **halogenated esters** may yield halogen-free esters. The tetraacetate of 6-iodo-6-deoxy-β-glucopyranose, on treatment with hydrogen and Raney nickel in methanol containing triethylamine, gives 52% 6-deoxy-β-glucose tetraacetate [1375].

In 2-bromo-2-deoxy-D-arabinono-1,4-lactone and 2-bromo-2-deoxy-D-xylono-1,4-lactone, hydrogenation over palladium on carbon in ethanol at room temperature hydrogenolyzes not only the bromine but also the neighboring hydroxyl to give 2,3-dideoxyaldono-1,4-lactone in a 76% yield [831] (equation **264**).

Use of *lithium aluminum hydride* in reducing halogenated esters to halogenated alcohols calls for certain precautions because halogens are hydrogenolyzed occasionally, and not even fluorine is immune [120]. More reliable hydrides that reduce the ester group without endangering the halogen are *alane* prepared in situ from lithium aluminum hydride and aluminum chloride [1279] or sulfuric acid [1280] and *calcium borohydride* obtained by adding

264 [*831*]

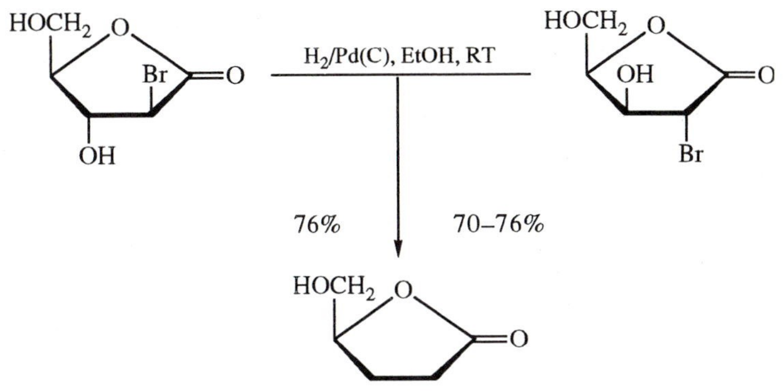

anhydrous calcium chloride to a mixture of sodium borohydride and tetrahydrofuran at 0 °C [*120*]. Alane reduces ethyl 3-bromopropanoate to 3-bromopropanol in better yields than lithium aluminum hydride, even when the inverse technique is used (72–90% as opposed to 4–52%) [*1279*]. Calcium borohydride converts ethyl 2-fluoro-3-hydroxypropanoate to 2-fluoro-1,3-propanediol in a 70% yield, whereas lithium borohydride gives only a 45% yield, and lithium aluminum hydride hydrogenolyzes fluorine [*120*].

4,4,4-Trichlorocrotonic acid is reduced with *zinc* in acetic acid to 50–60% 4,4-dichlorocrotonic acid, which is converted by *sodium amalgam* in an 80% yield to crotonic acid [*672*]. *Isopropyl alcohol* under ultraviolet irradiation substitutes hydrogen for α-fluorine in polyfluorinated esters [*426*] (equation **265**). In methyl 2,3-dichloro-2,3,3-trifluoropropionate, only the α-chlorine is hydrogenolyzed under similar conditions [*1376*] (equation **266**).

265 [*426*]

$$CH_3OCF_2CHFCO_2CH_3 \xrightarrow[\text{18–20 °C, 10 h}]{\text{Me}_2\text{CHOH, } h\nu} CH_3OCF_2CH_2CO_2CH_3$$

conversion 50%, yield 84%

266 [*1376*]

$$CClF_2CClFCO_2CH_3 \xrightarrow[\text{RT, 2 h}]{\text{Me}_2\text{CHOH, } h\nu} CClF_2CHFCO_2CH_3$$

86%

Reduction of nitro esters to amino esters is easy, because most of the reagents used for the reduction of a nitro group to an amino group do not attack the ester group. Ethyl 2-chloro-4-nitrobenzoate is quantitatively *hydrogenated* to either ethyl 2-chloro-4-aminobenzoate or 4-aminobenzoate over palladium on barium sulfate at room temperature after 3 h and 5 h, respectively [*1377*].

Complex hydrides may reduce both the nitro group and the ester. Thus, an excess of *lithium aluminum hydride* converts ethyl 2-nitrocinnamate to 3-phenyl-2-amino-1-propanol in a 72% yield. *Iron* and acetic or dilute hydrochloric acid can be safely used for the reduction of a nitro group to an amino group in nitro esters.

The problem arises when a **nitro ester** is to be *reduced to a nitro alcohol.* Nitro groups are not inert toward the best reagents for the reduction of esters to alcohols, complex hydrides. However the rate of reduction of a nitro group by lithium aluminum hydride is considerably lower than that of the ester group. Consequently, when a reduction is carried out at low enough temperatures (-30 to -60 °C) both aliphatic and aromatic nitro esters can be reduced to nitro alcohols [*1378*]. On treatment with 0.5 mol of *lithium aluminum hydride* in ether at -35 °C, methyl 4-nitropentanoate gives 61% 4-nitropentanol. Even better yields of nitro alcohols are obtained on treatment of nitro esters with *alane* or *borane.* Alane (2 mol) in tetrahydrofuran converts methyl 4-nitropentanoate to 4-nitropentanol at -10 °C in an 80% yield [*1280*] and ethyl *p*-nitrobenzoate to *p*-nitrobenzyl alcohol in a 68% yield [*1280*]. An even better reagent for this purpose is borane, prepared from sodium borohydride and aluminum chloride in diglyme; it reduces ethyl *p*-nitrobenzoate to *p*-nitrobenzyl alcohol at 50–75 °C in a 77% yield [*958*].

Diazo esters are reduced to ester hydrazones, ester hydrazines, or amino esters. Reduction of ethyl diazoacetate with aqueous *potassium sulfite* at 20–30 °C gives a derivative of ethyl glyoxylate hydrazone that is hydrolyzed to ethyl glyoxylate [*1379*]. *Sodium amalgam* reduces the same diazo ester to ethyl hydrazinoacetate in a 90% yield [*1380*] (equation **267**).

267

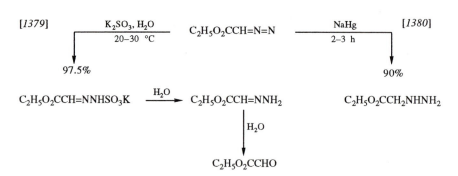

Catalytic hydrogenation over platinum oxide at room temperature converts ethyl diazoacetoacetate in ethanolic sulfuric acid to 2-amino-3-hydroxybutanoic acid (39% yield) by simultaneously reducing the diazo group to an amino group and the keto group to hydroxyl [*1167*].

Azido esters (triazo esters) are reduced to amino esters very easily. *tert*-Butyl azidoacetate *hydrogenated over* 5% palladium on carbon in methanol at room temperature and at atmospheric pressure affords a 75–85% yield of glycine *tert*-butyl ester [*1381*]. Unsaturated azido esters are reduced to unsaturated amino esters by *aluminum amalgam* in ether at room temperature in

71–86% yields [*223*]. Ethyl α-azidocrotonate gives a 75.6% yield of ethyl α-aminocrotonate, and ethyl α-azidocinnamate affords ethyl α-aminocinnamate in a 70.9% yield [*223*].

HYDROXY ESTERS, AMINO ESTERS, KETO ESTERS, OXIMINO ESTERS, AND ESTER ACIDS

Hydroxy esters are reduced at the ester group by intensive *catalytic hydrogenation* over nickel or copper chromite at high temperatures and high pressures [*800, 1382*]. As a consequence, secondary reactions ensue, such as the hydrogenolysis of hydroxyl groups and even carbon–carbon bond cleavage. Especially easy is the replacement by hydrogen of benzylic hydroxyl. Ethyl mandelate, for example, on hydrogenation over copper chromite at 250 °C and 100–170 atm gives a 70% yield of 2-phenylethanol and 13% ethylbenzene [*1382*]. But even nonactivated hydroxyls in aliphatic hydroxy esters are frequently hydrogenolyzed during the catalytic hydrogenation over nickel or copper chromite [*800*].

More reliable therefore is reduction with *lithium aluminum hydride, lithium borohydride*, or *calcium borohydride*, which do not hydrogenolyze hydroxy groups [*120*].

Reduction of amino esters to amino alcohols is accomplished by hydrogenation over Raney nickel at 50 °C and 150–200 atm with yields up to 98% [*45*] or by reduction with *lithium aluminum hydride* [*1383*].

In **keto esters** reduction can affect the *keto group* and convert it to an alcoholic group or to a methylene group, or else it can hydrogenolyze the *ester group* either to an acid or to an alcohol. The latter reduction may be accompanied by the reduction of the keto group at the same time.

Catalytic hydrogenation transforms keto esters to *hydroxy esters* under very gentle conditions. In cyclic keto esters, products of various configurations may result. Ethyl 2,2-dimethyl-6-oxocyclohexane-1-carboxylate, on hydrogenation over platinum oxide in acetic acid, gives 96.3% *cis-*, and over Raney nickel in methanol, 97% *trans-*ethyl 2,2-dimethyl-6-hydroxycyclohexane-1-carboxylate, both at room temperature and atmospheric pressure [*1091*].

Different stereoisomeric cyclohexanol carboxylates result also from reduction of ethyl 5-*tert*-butyl-1-methyl-2-oxocyclohexane-1-carboxylate with *sodium borohydride* (equatorial hydroxyl) or with *aluminum isopropoxide* (axial hydroxyl) [*1384*] (equation **268**).

268   [*1384*]

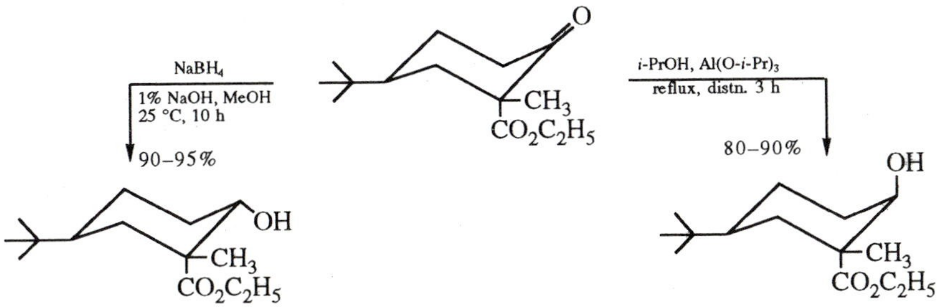

Metal amalgams may be used for reduction of the keto groups in keto esters provided the medium does not cause hydrolysis of the ester. Because of this circumstance *aluminum amalgam* in ether is preferable to *sodium amalgam* in aqueous solutions. Diethyl oxalacetate is reduced to diethyl malate by sodium amalgam in a 50% yield and with aluminum amalgam in an 80% yield [222].

Stereospecific reductions of α- and β-keto esters to optically pure hydroxy esters are achieved by *chiral reducing agents* or by *biochemical reduction*.

α-**Keto esters** are converted to α-*hydroxy esters* by treatment with *K-glucoride* in tetrahydrofuran at –78 °C in yields of 78–85% and enantiomeric excesses of 81–98% [1060]. These results are comparable to those obtained by reduction with baker's yeast (*Saccharomyces cerevisiae*) [1385, 1386] (Procedure 63, p 321) and chiral 1,4-dihydronicotinamide derivatives [1492, 1493, 1494] (equation **269**). β,γ-Unsaturated-α-keto esters are reduced to β,γ-unsaturated α-hydroxy esters with *lithium tris(sec-butyl)borohydride* in 90–96% yields and 93–97% enantiomeric excesses [1484].

269   RCOCO$_2$R' $\longrightarrow$ RCHCO$_2$R'
                                                              |
                                                             OH

	R = R ' = Me	Yield	ee	Config.
[1060]	K-Glucoride, THF, –78 °C, 6 h	80%	82–86%	S
[1386]	*Saccharomyces cerevisiae*, H$_2$O	79%	93%	S
	R = Ph, R' = Et			
[1492]	[a] L-Proline amide 1,4-dihydronicotinamide, Mg(ClO$_4$)$_2$, MeCN, 50 °C, 7 days	84%	83%	R
[1493]	[b] Bis(NADH) derivative, Mg(ClO$_4$)$_2$, MeCN, CHCl$_3$, RT, 1 h	67%	98%	R
[1494]	[c] Bridged bis(NADH) derivative Mg(ClO$_4$)$_2$, MeCN, CHCl$_3$, RT, 17 h	64%	96%	R

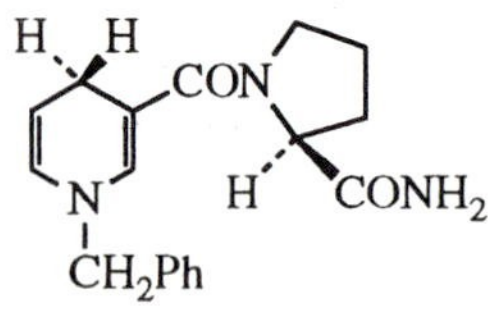

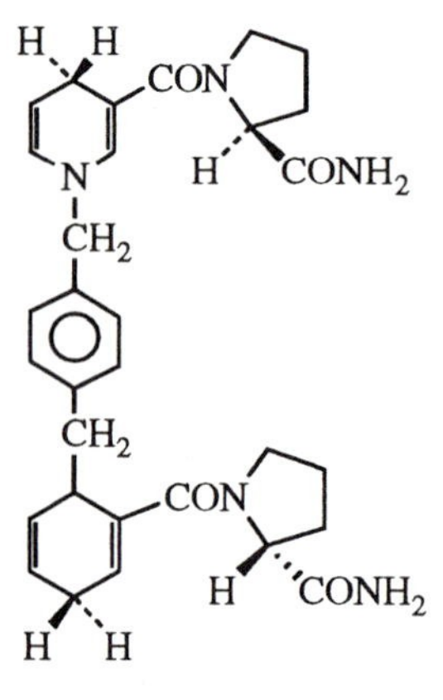

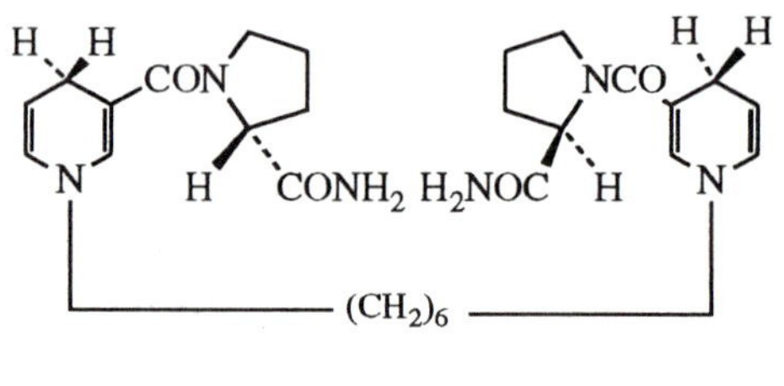

Best yields and diastereoselectivity in reduction of chiral β,γ-unsaturated α-keto esters to β,γ-unsaturated α-hydroxy esters is obtained with *L-Selectride (lithium tris(sec-butyl)borohydride)* in ether or tetrahydrofuran at -78 °C. 2-Phenylcyclohexyl 4-phenyl-2-oxo-3-butenoate gives the corresponding hydroxy ester in a 95% yield and 93% diastereomeric excess in tetrahydrofuran and a 96% yield and 96% diastereomeric excess in ether [*1484*].

Enantioselective reduction of **β-keto esters** to *β-hydroxy esters* is achieved by *homogeneous hydrogenation* using chiral catalysts such as Binap ruthenium diacetate [*103*] or dichloride [*105, 1387*] in almost quantitative chemical and optical yields. Reduction of ethyl acetoacetate with *baker's yeast* gave ethyl (+)-(S)-3-hydroxybutyrate in a 56.5% yield and 97% enantiomeric excess [*1486*].

Diastereoselectivity during reductions of chiral β-keto esters can be accomplished by *sodium borohydride* in the presence of manganese dichloride or by *tetrabutylammonium borohydride* [*1388*]. Various reagents give different ratios of *erythro* to *threo* isomers. The stereoselectivity is comparable with that experienced with *baker's yeast* [*1389*] (equation **270**).

270

$$\underset{\underset{CH_3}{|}}{RCOCHCO_2C_2H_5} \longrightarrow \underset{\underset{OH}{|}\ \underset{CH_3}{|}}{RCH\ CHCO_2C_2H_5} + \underset{\underset{CH_3}{|}}{\overset{\overset{OH}{|}}{RCHCHCO_2C_2H_5}}$$

		erythro	*threo*
[*1388*]	R = C_6H_5; $NaBH_4/MnCl_2$, MeOH, 0 °C, 30 min	72–86%	8–9%
[*1388*]	R = C_6H_5; $(C_4H_9)_4NBH_4$, MeOH, 0 °C, 10 min	9–12%	41–58%
[*1389*]	R = CH_3; *baker's yeast*, H_2O, 30 °C, 2 days	58%	13%

Very good diastereoselective reductions are accomplished by *dimethylphenylsilane*, giving predominantly *threo*-isomers from α-methylketo esters [*166*].

Biochemical reductions [*455*] are carried out by using microorganisms such as *Saccharomyces cerevisiae (baker's yeast)* [*1390*], which give yields of 44–86% and enantiomeric excesses of 73–99% [*478, 1385, 1390, 1391, 1392*], *Thermoanaerobicum brockii* [*478*], *Aspergillus niger* [*475*], *Geotrichum candidum* [*475*], and *Streptomyces* [*475*] (equation **271**).

Reductions of **keto esters to esters** are not very frequent. Both Clemmensen and Wolff–Kizhner reductions can hardly be used. The best way is *desulfurization of thioketals with Raney nickel*. Thus, ethyl acetoacetate is reduced to ethyl butyrate in a 70% yield, methyl benzoylformate (phenylglyoxylate) is reduced to methyl phenylacetate in a 79% yield, and other keto esters are reduced to esters in equally high yields (74–77%) [*1061*].

Another reagent capable of accomplishing the reduction of keto esters to esters is *sodium cyanoborohydride*, which converts isopropylidene acylmalonates to isopropylidene alkylmalonates in 50–85% yields [1393] (equation 272).

In an interesting reaction, β-santonin is reduced with *lithium* in liquid ammonia so that the *lactone is hydrogenolyzed to an acid* and one of the double bonds conjugated with the carbonyl is reduced. The other double bond as well as the keto group do not undergo reduction [1394] (equation 273).

271

$$CH_3COCH_2CO_2C_2H_5 \xrightarrow{\hspace{4cm}} CH_3\underset{\underset{OH}{|}}{C}HCH_2CO_2C_2H_5$$

			ee	
[1387]	H₂/RuCl₂[(R)-Binap], MeOH, 23–30 °C, 103 atm, 58 h	96%	99%	R
[478]	*Saccharomyces cerevisiae*, 30 °C		97%	S
[1390]	*Saccharomyces cerevisiae*, H₂O, MeOH	48%	96%	
[478]	*Thermoanaerobicum brockii*, 72 °C	80%	80%	S
[475]	*Aspergillus niger*, 30 °C, 2 days		80%	R
[475]	*Geotrichum candidum*, 30 °C, 3 days		89%	R
[475]	*Streptomyces sp.*, 1 PV 2645, 30 °C, 1 day		85%	R

272   [1393]

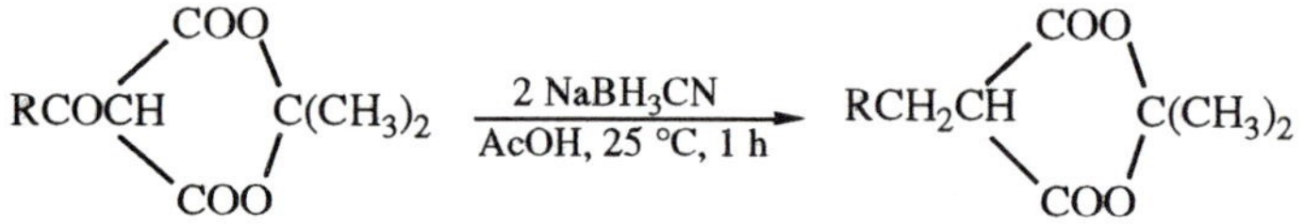

R = C₂H₅	80%
R = (CH₃)₂CH	50%
R = C₆H₅CH₂	85%

273   [1394]

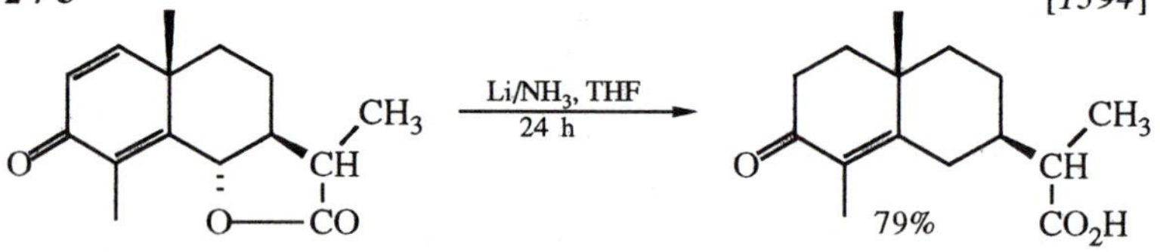

The reason why the carbonyl group in β-santonin remains intact may be that, after the reduction of the less hindered double bond, the ketone is enolized by lithium amide and is thus protected from further reduction. Indeed, treatment of ethyl 1-methyl-2-oxocyclopentane-1-carboxylate with lithium diisopropylamide in tetrahydrofuran at –78 °C enolizes the ketone and prevents its reduction with *lithium aluminum hydride* and *diisobutylalane* (Dibal). Reduction by these two reagents in tetrahydrofuran at –78 to –40 °C or –78 to –20 °C, respectively, affords *keto alcohols from several keto esters* in 46–95% yields. Ketones whose enols are unstable fail to give keto alcohols [1395].

Another way of avoiding reduction of the keto group in a keto ester is protection by acetalization. Ketals are evidently not reduced by lithium aluminum hydride in ether at 0 °C. Consequently, the diethyl ketal of ethyl *p*-acetylbenzoate (prepared in an 81% yield) is reduced with lithium aluminum hydride to the diethyl ketal of *p*-acetylbenzyl alcohol (93% yield). Hydrolysis with methanolic hydrochloric acid at 80 °C recovers the keto group in 95% yield (overall 71.5% based on the starting keto ester) [*1396*].

Reduction of oximino esters, that is, oximes of keto esters, is very useful for the preparation of amino esters. Reductions are very selective because the oximes are easily reduced by *catalytic hydrogenation* over 10% palladium on carbon in ethanol (78–82% yield) [*1397*], by *aluminum amalgam* in ether (52–87% yields) [*224, 1398*], or by *zinc* dust in acetic acid (77–78% yield) [*1399*]. None of these reagents attacks the ester group. The last-mentioned reaction gives an *N*-acetyl derivative [*1399*] (equation **274**).

274

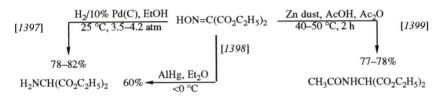

In ethyl 2-oximino-3-oxo-3-phenylpropanoate, *catalytic hydrogenation* over palladium on carbon reduces both the keto and the oximino group, giving a 74% yield of the ethyl ester of β-phenylserine (ethyl 2-amino-3-hydroxy-3-phenyl-propionate). The reduction is stereospecific, and only the *erythro* diastereomer is obtained, probably via a cyclic intermediate [*1400*]. Similarly, hydrogenation over Raney nickel at 25–30 °C and 1–3 atm converts ethyl α-oximinoaceto-acetate quantitatively to ethyl 2-amino-3-hydroxybutanoate [*46*].

In dicarboxylic acids with one free and one esterified carboxyl group, *only the free carboxyl* may be reduced with *hydrides* (e.g., alanes and boranes). The monoethyl ester of adipic acid is converted by an equimolar amount of *borane* in tetrahydrofuran at –18 to 25 °C to ethyl 6-hydroxyhexanoate in an 88% yield [*1251*] (Procedure 21, p 305), and monoethyl fumarate gives ethyl 4-hydroxycrotonate in a 67% yield [*1255*]. Borane is probably the acting compound when *sodium borohydride* is mixed with *iodine* to reduce ester acids to hydroxy esters. Monomethyl esters of sebacic and phthalic acids are thus reduced to hydroxy esters in the respective yields of 89% and 82% [*1253*].

Reduction with *sodium* in alcohols, on the other hand, **reduces only the ester group** and not the free carboxyl.

ORTHO ESTERS, THIO ESTERS, AND DITHIO ESTERS

Ortho esters are reduced to acetals. Refluxing with 0.25 mol of *lithium aluminum hydride* in ether–benzene solution for 4 h transforms 3-methyl-mercaptopropanoic acid trimethyl orthoester to 3-methylmercaptopropion-aldehyde dimethylacetal in a 97% yield [*1401*].

Thiol esters can be reduced to sulfides, aldehydes, or alcohols. *Alane* generated from lithium aluminum hydride and boron trifluoride etherate [*1402*] or aluminum chloride [*1403*] in ether reduces the carbonyl group to methylene and gives 37–93% yields of *sulfides*. Phenyl thiolbenzoate fails to give the sulfide, but methyl thiolbenzoate gives a 79% yield of benzyl methyl sulfide [*1403*] (equation **275**).

Desulfurization by refluxing with Raney nickel in 70% ethanol for 6 h converts thiol esters to *aldehydes* in 57–73% yields (an exception was a 22% yield) [*1404*] (equation **275**) (Procedure 7, p 299).

275

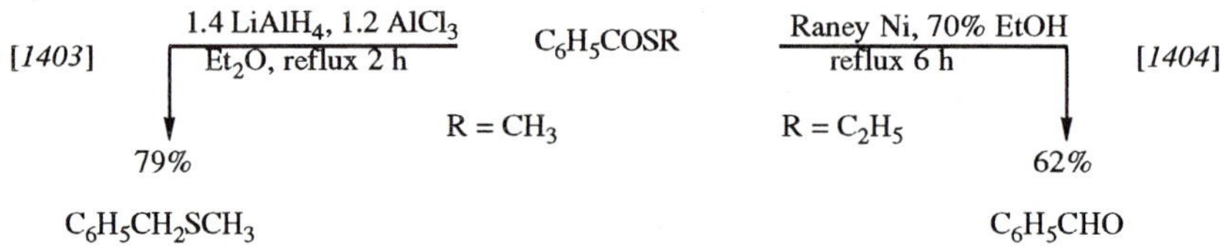

$$
\begin{array}{ccccc}
 & \dfrac{\text{1.4 LiAlH}_4,\ \text{1.2 AlCl}_3}{\text{Et}_2\text{O, reflux 2 h}} & C_6H_5COSR & \dfrac{\text{Raney Ni, 70\% EtOH}}{\text{reflux 6 h}} & \\
[1403] & & & & [1404] \\
 & \downarrow & R = CH_3 \qquad R = C_2H_5 & \downarrow & \\
 & 79\% & & 62\% & \\
 & C_6H_5CH_2SCH_3 & & C_6H_5CHO &
\end{array}
$$

Aldehydes are also obtained in 75–97% yields by a room-temperature reaction of thiol esters with *triethylsilane* in acetone in the presence of palladium on carbon [*162*]. **Reduction to alcohols** can be accomplished in 80–90% yields by treatment of the thiol esters with *sodium borohydride* in ethanol at room temperature [*1405*].

Desulfurization of a dithioester, methyl dithiophenylacetate, by refluxing with Raney nickel in 80% ethanol for 1 h, affords ethylbenzene in a 65% yield [*1406*].

CHAPTER 17

Reduction of Amides, Lactams, and Imides

Reduction of amides may yield aldehydes, alcohols, amines, and even hydrocarbons. Which of these is formed depends on the structure of the amide, the reducing agent, and to a certain extent reaction conditions.

A simplified scheme of reduction of the amide function shows that the first stage is formation of an intermediate with oxygen and nitrogen atoms linked to an sp^3 carbon. Such compounds tend to regenerate the original sp^2 system by elimination of ammonia or an amine. Thus an aldehyde is formed and may be isolated or reduced to an alcohol. Alternatively, the product is an amine resulting from direct hydrogenolysis of the sp^3 intermediate (equation **276**).

276

$$RCONR'_2 \xrightarrow{2\,H} \underset{\underset{OH}{|}}{RCH-NR'_2} \begin{array}{c} \xrightarrow{-HNR'_2} RCH{=}O \xrightarrow{2\,H} RCH_2OH \\ \\ \xrightarrow{2\,H} RCH_2NR'_2 + H_2O \end{array}$$

Reduction of amides to aldehydes is accomplished mainly by complex hydrides. Not every amide is suitable for reduction to aldehyde. Good yields are obtained only with some tertiary amides and *lithium aluminum hydride, lithium triethoxyaluminum hydride*, or *sodium bis(2-methoxyethoxy)aluminum hydride.* The nature of the substituents on nitrogen plays a key role. Amides derived from aromatic amines such as *N*-methylaniline [*1407*] and especially pyrrole, indole, and carbazole are most suitable for the preparation of aldehydes. By adding 0.25 mol of lithium aluminum hydride in ether to 1 mol of the amide in ethereal solution cooled to –10 to –15 °C, 37–60% yields of benzaldehyde are obtained from the benzoyl derivatives of the aforementioned heterocycles [*1408*], and a 68% yield is obtained from *N*-methylbenzanilide [*1407*]. Similarly 4,4,4-trifluorobutanal is prepared in an 83% yield by reduction of *N*-(4,4,4-trifluorobutanoyl)carbazole in ether at –10 °C [*1409*] (equation **277**).

Even better results than those reached by reduction of the aromatic heterocyclic amides are reported from reductions of *N*-acylethyleneimines with 0.25–0.5 mol of lithium aluminum hydride in ether at 0 °C: butanal, hexanal, 2,2-dimethylpropanal, 2-ethylhexanal, and cyclopropanecarboxaldehyde are prepared in 67–88% yields [*1410*].

Better reagents than lithium aluminum hydride alone are its alkoxy derivatives, especially *di-* and *triethoxyaluminum hydrides* prepared in situ from lithium aluminum hydride and ethanol in ethereal solutions. The best of all, *lithium triethoxyaluminum hydride*, gives higher yields than its trimethoxy and tris(*tert*-butoxy) analogs. When an equimolar quantity of this reagent is added to an ethereal solution of a tertiary amide derived from dimethylamine, diethylamine, *N*-methylaniline, piperidine, pyrrolidine, aziridine, or pyrrole, and the mixture is allowed to react at 0 °C for 1–1.5 h, aldehydes are isolated in 46–

92% yields [*124, 1411*]. The reaction is unsuccessful for the preparation of crotonaldehyde and cinnamaldehyde from the corresponding dimethylamides [*124*].

N-Acylsaccharins prepared by treatment of the sodium salt of saccharin with acyl chlorides are reduced by 0.5 molar amounts of *sodium bis(2-methoxyethoxy)aluminum hydride* in benzene at 0–5 °C to give 63–80% yields of aliphatic, aromatic, and unsaturated aldehydes [*1412*] (equation **277**). Fair yields (45–58%) of some aliphatic aldehydes are obtained by *electrolytic reduction* of tertiary and even secondary amides in undivided cells fitted with platinum electrodes and filled with solutions of lithium chloride in methylamine. However, many secondary and especially primary amides give 51–97% yields of alcohols under the same conditions [*192*].

277

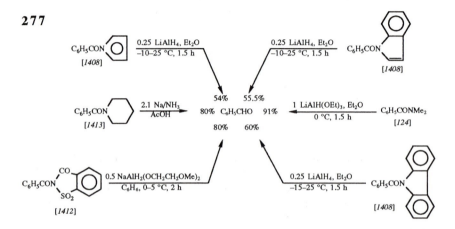

Some tertiary amides are reduced in low yields by *sodium* in liquid ammonia and ethanol. Better results are reached when acetic acid (ammonium acetate) is used as a proton source: benzaldehyde is isolated in yields up to 80% from the reduction of *N*-benzoylpiperidine [*1413*] (equation **277**).

Of the methods used for converting amides to aldehydes, the one using *lithium triethoxyaluminum hydride* is most universal, can be applied to many types of amides, and gives highest yields. In this way it parallels other methods for the preparation of aldehydes from acids or their derivatives (pp 191 and 207).

The reduction of amides to aldehydes competes successfully with other synthetic routes leading to aldehydes, but the **reduction of amides to alcohols** is only rarely used for preparative purposes. One such example is the conversion of trifluoroacetamide to trifluoroethanol in a 76.5% yield by *catalytic hydrogenation* over platinum oxide at 90 °C and 105 atm [*1414*].

Tertiary amides derived from pyrrole, indole, and carbazole are hydrogenolyzed to alcohols and amines by refluxing in ether with a 75% excess (0.88 mol) of *lithium aluminum hydride*. Benzoyl derivatives of these heterocycles afford 80–92.5% yields of benzyl alcohol and 86–90% yields of the amines [*1408*] (equation **278**).

278

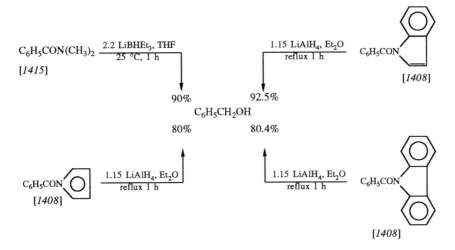

Good to excellent yields (62–90%) of alcohols are reported in reductions of dimethyl and diethyl amides of benzoic acid and aliphatic acids by *lithium triethylborohydride* (2.2 mol per mol of the amide) in tetrahydrofuran at room temperature [*133, 1415*], by *sodium dimethylaminoborohydride* (72–98% yields) [*130*], and by *sodium* in liquid ammonia (40–84% yields) [*1416*].

Usually alcohols accompany aldehydes in reductions with *lithium aluminum hydride* [*1408*] or *sodium bis(2-methoxyethoxy)aluminum hydride* [*705*] or in hydrogenolytic cleavage of trifluoroacetylated amines. Thus the *tert*-butyl ester of *N*-(*N*-trifluoroacetylprolyl)leucine is cleaved on treatment with *sodium borohydride* in ethanol to the *tert*-butyl ester of *N*-prolylleucine (92% yield) and trifluoroethanol [*1417*]. During catalytic hydrogenations over copper chromite, alcohols sometimes accompany amines that are the main products [*1418*].

The predominant application of the **reduction of amides** lies in their **conversion to amines**, realized most often by hydrides and complex hydrides. Primary, secondary, and tertiary amines can be prepared from the corresponding amides in high yields. A few other reductive methods are available for the same purpose.

Catalytic hydrogenation of amides does not take place with noble metal catalysts under gentle conditions but is accomplished over copper chromite at 210–250 °C and 100–300 atm using dioxane as a solvent [*1418, 1419*]. Because of the high temperatures needed for the reduction, the yields of the major products are diminished by side reactions, such as formation of secondary amines from primary amines produced from unsubstituted amides and hydrogenolysis of secondary and tertiary amines. Alcohols are also formed to a small extent, and aromatic hydrocarbons are isolated when the benzylic amines are the primary reduction products [*1418*]. Salicylamide gives *o*-cresol as the sole product, lauramide affords a 48% yield of dodecylamine and 49% didodecylamine, and lauranilide gives 37% *N*-dodecylaniline, 29% aniline, 14% dodecylamine, 5% diphenylamine, and 2% didodecylamine [*1418*].

Complications inherent in the high-temperature catalytic hydrogenation do not occur during reduction of amides to amines by *lithium aluminum hydride* and other hydrides and complex hydrides. Reduction of a tertiary amide requires 0.5 mol of lithium aluminum hydride, whereas secondary and primary amides require 0.75 and 1 mol of the hydride per mol of the amide, respectively. In practice a 25–30% excess over the stoichiometric amount is used for the disubstituted, and 200–250% excess for the monosubstituted amides. Refluxing for more than 1 h was found necessary, sometimes for up to 34 h [*1408*]. Yields of amines are around 90% or even higher [*1408, 1420, 1421*]. Amides insoluble in ether are placed in a Soxhlet apparatus and extracted and flushed into the reaction flask containing the hydride [*183*].

Isolation of amines is best accomplished by treating the reaction mixture with a calculated amount of water *(2n* milliliters of water for *n* grams of lithium aluminum hydride followed by *n* milliliters of 15% sodium hydroxide and 5*n* milliliters of water) [*1408*], or by quenching of the reaction mixture with triethanolamine that complexes the aluminum byproducts and allows for an easy isolation of the amines [*1247*] (Procedure 14, p 302). Other methods of decomposition can result in obstinate dispersions [*1408*].

High yields of amines are also obtained by reduction of amides with an excess of *magnesium aluminum hydride* (100% yield) [*758*], with *lithium trimethoxyaluminum hydride* at 25 °C (83% yield) [*123*], *sodium bis(2-methoxyethoxy)aluminum hydride* at 80 °C (84.5% yield) [*705*], *alane* in tetrahydrofuran at 0–25 °C (46–93% isolated yields) [*1280, 1422*], *sodium borohydride* and triethyloxonium fluoroborates at room temperature (81–94% yields) [*1423*], *sodium borohydride* in the presence of acetic or trifluoroacetic acid on refluxing (20–92.5% yields) [*1424*], *sodium dimethylamino- or tert-butylaminoborohydride* (it does not reduce secondary amides) [*130*], *borane* in tetrahydrofuran on refluxing (79–84% isolated yields) [*1425*], borane–dimethyl sulfide complex (5 mol) in tetrahydrofuran on refluxing (37–89% isolated yields) [*1359*], and by *electrolysis* in dilute sulfuric acid at 5 °C using a lead cathode (63–76% yields) [*1426*].

In the reductions with borohydrides and boranes, the isolation of products differs from that used in the reductions with lithium aluminum hydride, the solvent usually being removed by evaporation prior to the final workup [*1359, 1423, 1424, 1425*] (equation **279**).

279

$C_6H_5CONH_2$	⟶	$C_6H_5CH_2NH_2$

[*758*]	$Mg(AlH_4)_2$, THF	100%
[*1280*]	2 AlH_3, THF, 25 °C	82%
[*1424*]	5 $NaBH_4$, 5 AcOH, dioxane, reflux 2 h	76%
[*1425*]	1.7 BH_3, THF, reflux 1 h	87%
[*1268*]	3 Na/NH_3, EtOH	74%

Sodium and sodium amalgam may be used for reduction of amides, but the yields of amines are generally very low. Primary aromatic amides (benzamides) are reduced at the carbonyl function with 3.3 equivalents of *sodium* in liquid ammonia and ethanol, whereas in *tert*-butyl alcohol reduction takes place in the aromatic ring, giving 1,4-dihydrobenzamides [*1268*]. Occasionally **primary amides** can be reduced to **hydrocarbons** by hydrogenation over a large amount of a specially prepared platinum catalyst at 250 °C. Benzamide gives a mixture of toluene and methylcyclohexane (49%) and benzene and cyclohexane (48%), and phthalimide affords 54% *o*-xylene, 35% toluene and methylcyclohexane, and 11% benzene and cyclohexane [*735*].

Reduction of lactams to amines resembles closely the reduction of amides, except that *catalytic hydrogenation* is much easier and is accomplished even under mild conditions. α-Norlupinone (1-azabicyclo[4.4.0]-2-oxodecane) is converted quantitatively to norlupinane (1-azabicyclo[4.4.0]decane) over platinum oxide in 1.25% aqueous hydrochloric acid at room temperature and atmospheric pressure after 16 h [*1427*]. Reduction of the same compound by *electrolysis* in 50% sulfuric acid over a lead cathode gives a 70% yield [*1427*].

Reduction of 5,5-dimethyl-2-pyrrolidone with 3 mol of *lithium aluminum hydride* by refluxing for 8 h in tetrahydrofuran gives 2,2-dimethylpyrrolidine in 67–79% yields [*1428*]. Reduction of ε-caprolactam is accomplished by heating with *sodium bis(2-methoxyethoxy)aluminum hydride* [*705*], by successive treatment with triethyloxonium fluoroborate and *sodium borohydride* [*1423*], and by refluxing with *borane–dimethyl sulfide* complex [*1359*] (equation **280**).

280

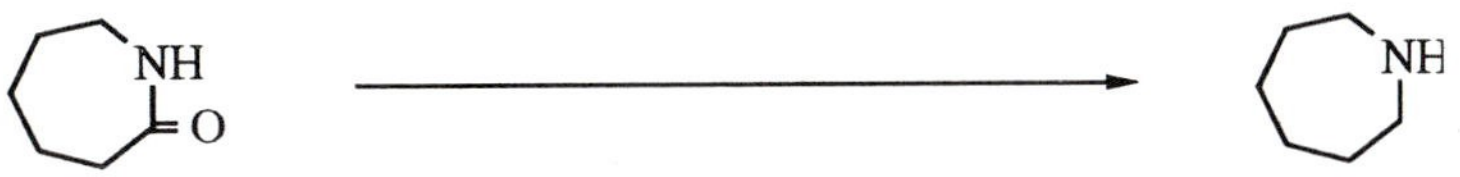

[*705*]   1.7 NaAlH$_2$(OCH$_2$CH$_2$OMe)$_2$, PhMe, 100 °C, 1 h   81.5%
[*1423*] 1.1 Et$_3$O$^+$BF$_4^-$, CH$_2$Cl$_2$, 25 °C; 2.4 NaBH$_4$, EtOH, 25 °C   92%
[*1359*] BH$_3$•Me$_2$S, THF, reflux   74%

On treatment with hydrides and complex hydrides, α- and β-lactams may suffer ring cleavage and give amino alcohols.

Imides are reduced to lactams or amines. Succinimide is reduced *electrolytically* to pyrrolidone [*1429*] but is hydrogenolyzed by *sodium borohydride* in water to γ-hydroxybutyramide [*1430*] (equation **281**).

281

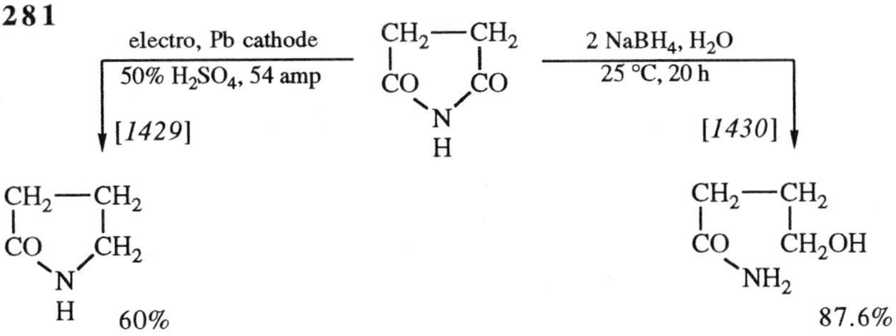

N-Methylsuccinimide is converted to *N*-methylpyrrolidone (92% yield) on heating for 18 min at 100 °C with 3 equivalents of *sodium bis(2-methoxy-ethoxy)aluminum hydride* [*705*]. *N*-Phenylsuccinimide is transformed into *N*-phenylpyrrolidone (67% yield) by *electroreduction* on a lead cathode in dilute sulfuric acid [*1426*].

Borane generated from sodium borohydride and boron trifluoride etherate in diglyme reduces *N*-aryl substituted succinimides to *N*-arylpyrrolidones in 42–72% yields [*1431*].

Glutarimides disubstituted in α or β positions are hydrogenolyzed by *sodium borohydride* in aqueous methanol mainly to disubstituted δ-hydroxy-valeramides [*1430*], but *N*-methylglutarimide is reduced *electrolytically* in sulfuric acid to a mixture of *N*-methylpiperidone (15–68%) and *N*-methylpiper-idine (7–62%) [*1432*], and by *lithium aluminum hydride* in ether to *N*-methyl-piperidine in an 85% yield [*1433*].

Phthalimide is *hydrogenated catalytically* at 60–80 °C over palladium on barium sulfate in acetic acid containing an equimolar quantity of sulfuric or perchloric acid to phthalimidine [*1434*]. The same compound is produced in a 76–80% yield by hydrogenation over nickel at 200 °C and 200–250 atm [*44*] and in a 75% yield over copper chromite at 250 °C and 190 atm [*1435*]. Reduction with *lithium aluminum hydride,* on the other hand, reduces both carbonyls and gives isoindoline (5% yield) [*1435*], also obtained by *electroreduction* on a lead cathode in sulfuric acid (72% yield) [*1435*] (equation **282**).

282

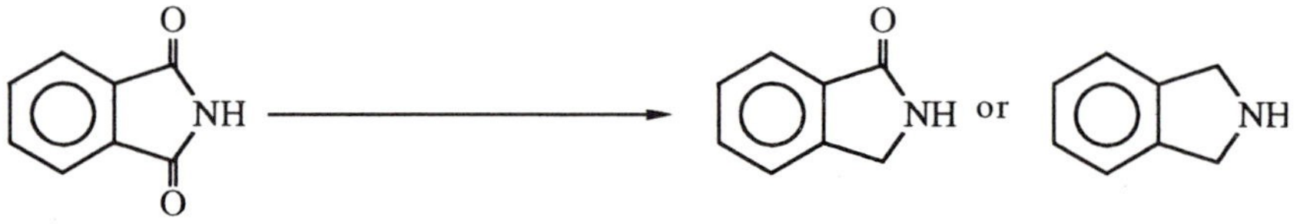

[*44*]	H$_2$/Ni, C$_6$H$_{11}$Me, 200 °C, 200–250 atm 76–80%	
[*1435*]	H$_2$/CuCr$_2$O$_4$, dioxane, 250 °C, 190 atm 75%	
[*1435*]	LiAlH$_4$, Et$_2$O	5%
[*1435*]	electrolysis, Pb cathode, 4.5 amp, 47 °C	72%

Boiling with *zinc* in alkaline medium converts phthalimide to phthalide in a 95.6% yield [*1483*].

AMIDES AND LACTAMS CONTAINING DOUBLE BONDS AND REDUCIBLE FUNCTIONAL GROUPS

Because of great differences in rates of hydrogenation, amides and lactams containing conjugated double bonds (*enamides*) are hydrogenated easily to saturated amides or lactams. α-Acetamidocinnamic acid on *hydrogenation* over platinum oxide in acetic acid at room temperature and 3 atm gives an 85–86% yield of α-acetamidohydrocinnamic acid [*1436*]. α,β-Unsaturated amides having chiral auxiliary groups bonded to nitrogen give saturated amides with very high stereoselectivity, both by *catalytic hydrogenation* and by reduction with *lithium aluminum hydride* with cobalt dichloride [*68*]. Such hydrogenations are especially important for the preparation of acylamido acids or esters from enamide derivatives obtained by aldol-type condensations. The use of chiral catalysts leads to optically active amino acids (equation **262**). Unsaturated *N,N*-dibenzylamides undergo reduction of the double bond by *samarium diiodide* in 80–99% yields [*308*].

Reduction of α,β-unsaturated lactams, 5,6-dihydro-2-pyridones, with *lithium aluminum hydride, lithium alkoxyaluminum hydrides*, and *alane* leads to the corresponding piperidines. 5-Methyl-5,6-dihydro-2-pyridone (with no substituent on nitrogen) gives on reduction with lithium aluminum hydride in tetrahydrofuran only a 9% yield of 2-methylpiperidine, but 1,6-dimethyl-5,6-dihydro-2-pyridone and 6-methyl-1-phenyl-5,6-dihydro-2-pyridone afford 1,2-dimethylpiperidine and 2-methyl-1-phenylpiperidine in respective yields of 47% and 65% with an excess of lithium aluminum hydride, and 91% and 92% with *alane* generated from lithium aluminum hydride and aluminum chloride in ether. *Lithium mono-, di-, and triethoxyaluminum hydrides* also give satisfactory yields (45–84%) [*1437*].

Hydrogenation of *N*-methyl-2-pyridone over platinum oxide in acetic acid at 26 °C and 2–3.5 atm gives *N*-methyl-2-piperidone in an 87.5% yield [*582*].

Reductions of alkylpyridones with *lithium aluminum hydride* or *alane* are very complex, and their results depend on the position of the substituents and on the reducing reagent. The pyridones can be viewed as doubly unsaturated lactams with α,β- and γ,δ-conjugated double bonds, and so the products result from all possible additions of hydride ion: 1,2, 1,4, or 1,6. Consequently, the products of reduction are alkylpiperidines and alkylpiperideines with double bonds in 3,4 or 4,5 positions [*600, 1438, 1501*] (equation **283**).

283   [*600,1438*]

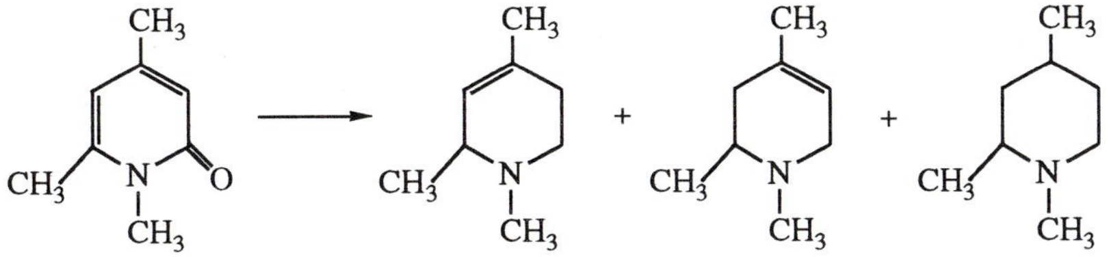

| [*1438*] | 2 LiAlH$_4$, THF, reflux 5 h | 13.5% | 2.5% | 3.8% *cis* 5.6% *trans* |
| [*1438*] | 1 LiAlH$_4$, 1 AlCl$_3$, Et$_2$O, reflux 5 h | 7.7% | 11.8% | 2% *cis* |

In 5-bromouracil (5-bromo-2,4-dioxopyrimidine), bromine is replaced by hydrogen in a 53% yield using 1-benzyl-1,4-dihydronicotinamide at 180 °C [*453*].

Amides containing nitro groups are reduced to diamino compounds with *alane*. *N,N*-Dimethyl-*p*-nitrobenzamide on reduction with *lithium alumi-*

num hydride in the presence of sulfuric acid in tetrahydrofuran gives a 98% yield of *N,N*-dimethyl-*p*-aminobenzylamine [1422].

Amides of keto acids are reduced to amides of hydroxy acids biochemically using *Saccharomyces cerevisiae* to give optically pure products [1385]. Surprisingly, stereoselective reduction of the keto group results from the treatment of α,β-keto amides with *zinc borohydride* [1439] or with *dimethylphenylsilane*. Both *erythro* or *threo* hydroxyamides are obtained, depending on whether the catalyst of the reaction is proton or a fluoride ion [1440] (equation **284**).

284

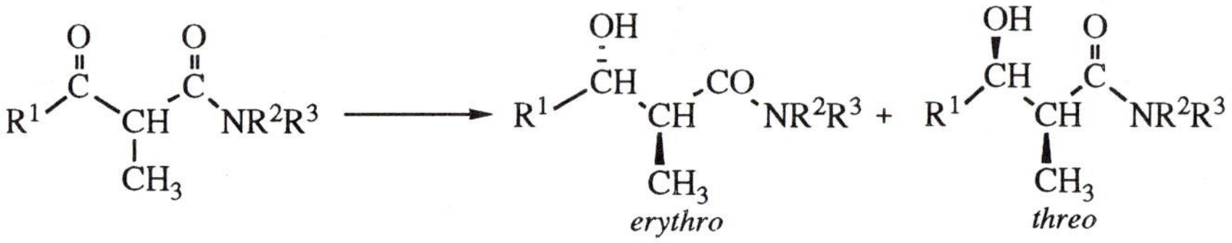

	erythro	threo
[1439] $R^1 = CH_3$, $R^2 = H$, $R^3 = C_4H_9$		
$Zn(BH_4)_2$, Et_2O, -78 °C, 15 min	89%	2%
[1440] $R^1 = C_6H_5$, $R^2 = R^3 = C_2H_5$		
$SiHPhMe_2$, CF_3CO_2H, DMPU, 0 °C, 6 h	97%	2%
$SiHPhMe_2$, TASF, DMPU, 0 °C, 12 h	2%	97%

DMPU = 1,3-dimethyl-3,4,5,6-tetrahydro-2-(1*H*)-pyrimidinone
TASF = tris(diethylamino)sulfonium difluorotrimethylsilicate

Refluxing with *lithium aluminum hydride* in ether for 6 h reduces both the ketonic and the amidic carbonyl in *N*-methyl-5-phenyl-5-oxopentanamide and gives an 82% yield of 5-methylamino-1-phenylpentanol [1441].

The **amidic group** in methyl *N*-acetyl-*p*-aminobenzoate is reduced *in preference to an ester group* with *borane* in tetrahydrofuran (1.5–1.8 mol per mol of the amide), giving a 66% yield of methyl *p*-*N*-ethylaminobenzoate. Similarly, 1-benzyl-3-methoxycarbonyl-5-pyrrolidone affords methyl 1-benzyl-3-pyrrolidinecarboxylate in a 54% yield, and 1,2-diethyl-5-ethoxycarbonyl-3-pyrazolidone gives ethyl 1,2-diethylpyrazolidine-3-carboxylate in a 60% yield. However, methyl hippurate (*N*-benzoylaminoacetate) under comparable conditions gives only 11% methyl 2-benzylaminoacetate and 85% 2-benzylaminoethanol [1442].

REDUCTIVE AMINATION WITH CARBOXYLIC ACIDS

When a compound containing a primary or secondary amino group (or a compound that can generate such on reduction) is reduced in the presence of acids such as formic acid, acetic acid, trifluoroacetic acid, propionic acid, and butyric acid, the product is a **secondary or tertiary amine** containing one alkyl group with the same number of carbons as the acid used. When indole is heated with 4 mol of *sodium borohydride* in acetic acid at 50–60 °C the product is *N*-ethylindoline resulting from alkylation at the nitrogen atom. The same compound is obtained on treatment of indoline with sodium borohydride in acetic acid (86–88% yields) [*607*]. Similar treatment of quinoline with 7 mol of sodium borohydride and acetic acid gives a 65% yield of 1-ethyl-1,2,3,4-tetra-hydroquinoline, and heating isoquinoline at 50–60 °C with 7 mol of sodium borohydride and formic acid affords a 78% yield of 2-methyl-1,2,3,4-tetrahydroisoquinoline [*620*].

The reaction has broad applications, and a large number of secondary and especially tertiary amines were prepared in isolated yields ranging from 60% to 84% [*1443*]. Although the mechanism of this reaction is not clear, most likely the key step is reduction of the acid by borane, generated in situ from sodium borohydride and the acid, to an aldehyde that reacts with the amine, as described in the section on reductive amination (p 187) (equation **285**).

285   [*1443*]

$$C_6H_5CH_2NHCH_3 \xrightarrow[\text{3 NaBH}_4,\ 50-55\ °C]{\text{Me}_2CHCO_2H} \underset{(CH_3)_2CHCHOH}{C_6H_5CH_2NCH_3} \longrightarrow \underset{(CH_3)_2CHCH_2}{C_6H_5CH_2NCH_3}\ 77\%$$

AMIDINES, THIOAMIDES, IMIDOYL CHLORIDES, AND HYDRAZIDES

Several derivatives of carboxylic acids yield *aldehydes* on reduction and are frequently prepared just for this purpose.

Benzamidine, *N*-ethylbenzamidine, and *N*-phenylbenzamidine afford benzaldehyde on reduction with *sodium* in liquid ammonia with or without ethanol in yields of 54–100% [*1413*].

Thioamides are converted to aldehydes by cautious *desulfurization* with *Raney nickel* [*1444*] or by treatment with *iron* and acetic acid [*249*]. More intensive desulfurization with Raney nickel [*1445*]; *electroreduction* [*249*]; and reduction with *lithium aluminum hydride* [*1446*], *sodium borohydride* [*1447*], or *sodium cyanoborohydride* [*1447*] give *amines* in good to excellent yields.

Imidoyl chlorides substituted at nitrogen (usually with an aryl group) are prepared from anilides and phosphorus pentachloride and are *reduced to aldimines* (*Schiff bases*) *or amines.* Homogeneous *hydrogenation* over bis(triphenylphosphine)palladium dichloride gives aldimines, whereas hydrogenation over palladium on barium sulfate affords secondary amines, both in high yields [*95*] (equation **286**).

286   [*95*]

$$C_6H_5CH=NC_6H_4CH_3\text{-}p \xleftarrow[\substack{C_6H_6,\ 120\ °C,\ 50\ atm,\ 4\ h}]{H_2/(Ph_3P)_2PdCl_2,\ Et_3N} C_6H_5CCl=NC_6H_4CH_3\text{-}p \xrightarrow[\substack{C_6H_6,\ 120\ °C,\ 50\ atm,\ 4\ h}]{H_2/Pd(BaSO_4),\ Et_3N}$$

$$C_6H_5CH=NC_6H_4CH_3\text{-}p\quad 87\% \qquad\qquad 78\%\quad C_6H_5CH_2NHC_6H_4CH_3\text{-}p$$

Aldimines are also obtained by the reduction with *lithium tris(*tert-*butoxy)aluminum hydride*. N-Cyclohexylimidoyl chlorides of propanoic, octanoic, and benzoic acids are converted to imines on treatment with a 120–300% excess of lithium tris(*tert*-butoxy)aluminum hydride in tetrahydrofuran at –78 to 25 °C (60–85% yields) [*1448*]. **Reduction of imidoyl chlorides to aldehydes** (equation **287**) is accomplished with *magnesium* [*1449*], but better yields are obtained with *stannous chloride (Sonn–Müller reduction)* [*259, 1450*] and *chromous chloride* [*1451*]. Chromous chloride is especially useful in the preparation of α,β-unsaturated aldehydes. Cinnamaldehyde is obtained from the imidoyl chloride of cinnamic acid in an 80% yield [*1451*].

The **hydrazide** of 2,2-diphenyl-3-hydroxypropanoic acid is reduced with *lithium aluminum hydride* in N-ethylmorpholine at 100 °C to 3-amino-2,2-diphenylpropanol in a 72.5% yield [*1452*]. Much more useful is **reduction** of **N-arenesulfonylhydrazides of acids to aldehydes** *(* the *McFadyen–Stevens reduction)* [*388, 389*] based on an alkali-catalyzed thermal decomposition according to equation **287**.

287

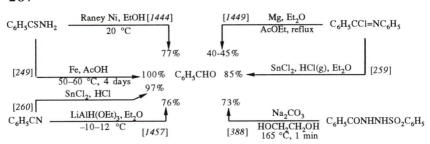

The original method called for addition of an excess of solid sodium carbonate to a solution of a benzenesulfonylhydrazide in ethylene glycol. The optimum temperature of decomposition was around 160 °C, and the optimum time was around 1 min. Yields of aromatic aldehydes ranged from 42% to 87%. The method failed with *p*-nitrobenzaldehyde, cinnamaldehyde, and aliphatic aldehydes [*388*]. Later studies revealed that essential features of the reaction are the presence of a solid material in the reaction mixture and a short time of heating. A modification based on the addition of ground glass to the mixture gives higher yields of some aldehydes. Even *p*-nitrobenzaldehyde is obtained, albeit in only a 15% yield [*1453*]. The reaction is generally unsuitable for the synthesis of aliphatic aldehydes. Only a few exceptions were recorded such as preparation of cyclopropane carboxaldehyde (16% yield) [*1454*], pivalaldehyde (2,2-dimethylpropanal) (40% yield after heating for only 30 s) [*1455*], and apocamphane-l-carboxaldehyde (60% yield) [*1455*].

Reduction of *p*-toluenesulfonylhydrazides by *complex hydrides* yields **hydrocarbons**. The N-tosylhydrazide of stearic acid gives a 50–60% yield of octadecane on reduction with *lithium aluminum hydride* [*1048*].

CHAPTER 18

Reduction of Nitriles

Nitriles of carboxylic acids are an important source of primary amines that are produced by many reducing agents. However, a few reagents effect reduction to other compounds.

Aromatic nitriles are converted to aldehydes in 50–95% yields on treatment with 1.3–1.7 mol of *sodium triethoxyaluminum hydride* in tetrahydrofuran at 20–65 °C for 0.5–3.5 h [*1456*]. More universal reducing agents are *lithium trialkoxyaluminum hydrides,* which also reduce aliphatic nitriles. They are generated in situ from lithium aluminum hydride and an excess of ethyl acetate or butanol, are used in equimolar quantities in ethereal solutions at –10 to 12 °C, and produce aldehydes in isolated yields ranging from 55% to 84% [*1457*] (equation **287**). Reduction of nitriles is also accomplished with *diisobutylalane* but in a very low yield [*1367*], with *potassium 9-(sec-amyl)-9-borabicyclo[3.3.1]nonane* in tetrahydrofuran at room temperature in yields of 65–89% [*137*], and with *nickel–aluminum alloy* in formic acid (63–69% yields) [*73*].

The complex hydride reductions of nitriles to aldehydes compare favorably with the classic *Stephen reduction,* which consists of treatment of aliphatic and aromatic nitriles with anhydrous *stannous chloride* and gaseous hydrogen chloride in ether or diethylene glycol [*260, 389, 1458*]. An advantage of the Stephen method is its applicability to polyfunctional compounds containing reducible groups, such as carbonyls, that are reduced by hydrides but not by stannous chloride [*1459*].

Alternative methods for the conversion of aromatic nitriles to aldehydes are heating with *Raney nickel and sodium hypophosphite* in water–acetic acid–pyridine (1:1:2) solutions at 40–45 °C for 1–1.5 h (40–90% yields) [*1460*], heating with nickel aluminum alloy and formic acid at 75–80 °C for 30 min (35–83% yields) [*1461*], or even refluxing for 30 min with Raney nickel alloy in 75% aqueous formic acid (44–100% yields) [*1462*]. Reduction with *zinc* and nickel dichloride transforms **nitriles into alcohols,** albeit in poor yields [*241*]. A very rare **conversion of nitriles to hydrocarbons** with the same number of carbon atoms is accomplished by their *hydrogenation* over a specially activated platinum on silica gel at 150–225 °C in yields of 95–99% [*735*]. The same reaction can be accomplished in 40–100% yields by treatment of the nitriles with ammonium formate and 10% palladium on carbon in methanol solution at room temperature [*75*] (equation **288**).

Catalytic hydrogenation converts **nitriles to amines** in good to excellent yields. The primary product, primary amines, may react with the intermediate imines to form secondary and even tertiary amines whose amounts depend on the reaction conditions. Elevated temperatures favor formation of secondary and tertiary amines by condensation accompanied by elimination of ammonia. Such side reactions can be suppressed by carrying out the hydrogenations in the presence of ammonia, which affects the equilibrium of the condensation reaction. Another way of preventing the formation of secondary and tertiary

288

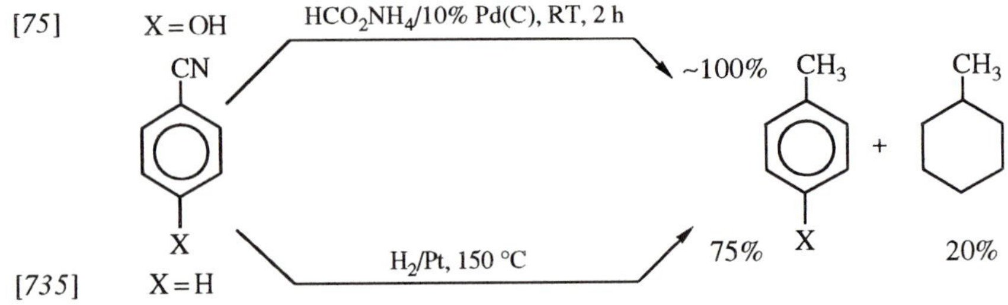

[75]   X = OH   HCO$_2$NH$_4$/10% Pd(C), RT, 2 h

~100% +

[735]   X = H   H$_2$/Pt, 150 °C   75%   20%

amines is to carry out the hydrogenation in acylating solvents such as acetic acid or acetic anhydride.

Excellent yields of primary amines are obtained by hydrogenation of nitriles over 5% rhodium on alumina in the presence of ammonia. The reaction is carried out at room temperature and at a pressure of 2–3 atm and is finished in 2 h, giving 63–92.5% yields of primary amines. Under such conditions no hydrogenolysis of benzyl residues is noticed [1463]. Hydrogenation over Raney nickel at room temperature and atmospheric [1464] or slightly elevated pressures [46] also gives high yields of primary amines, especially in the presence of ammonia [1464]. Comparable results are achieved using nickel boride catalyst [1465]. In addition, high yields of primary amines are obtained in hydrogenations over Raney nickel at temperatures of 90–130 °C and pressures of 30–270 atm, especially in the presence of ammonia [1466, 1467, 1468]. Under such conditions Raney nickel gives even better yields of primary amines than rhodium or platinum on charcoal [1468]. Hydrogenation over nickel on infusorial earth at 115–150 °C and 100–175 atm affords high proportions of secondary amines [44].

A reducing agent of choice for conversion of nitriles to primary amines without producing secondary or tertiary amines is *lithium aluminum hydride*. It is used in an equimolar ratio in ether at up to refluxing temperature and gives yields of 40–90% [184, 757, 1023] (Procedure 14, p 302). Modified lithium aluminum hydrides such as *lithium trimethoxyaluminum hydride* also give high yields of primary amines [123].

Even better yields are obtained with *alane* produced in situ from lithium aluminum hydride and 0.5 mol of 100% sulfuric acid in tetrahydrofuran [1280] or 1 mol of aluminum chloride in ether [1023] (Procedure 19, p 304). The reduction is carried out at room temperature, and 1 or 1.3 mol of the alane is used per mole of the nitrile. Comparative experiments show somewhat higher yields of amines than those obtained by lithium aluminum hydride alone [1023] (equation **289**).

Isolation of the products is accomplished by decomposition with sodium potassium tartrate [757], with water and dilute sodium hydroxide [184], or by successive treatment with 6 N sulfuric acid followed by alkalinization with potassium hydroxide [1023].

In contrast to lithium aluminum hydride, borohydrides are not suitable for the reduction of nitriles. Lithium triethylborohydride gives erratic results [133], sodium trimethoxyborohydride reacts slowly even at higher temperature, and sodium borohydride reduces nitriles to amines only rarely. However, in the presence of 0.33 mol of aluminum chloride in diglyme, *borane* is produced and gives very good yields of primary amines when used in a 25% excess at temperatures not exceeding 50 °C. Capronitrile affords a 65% yield of hexylamine, and *p*-tolunitrile gives an 85% yield of *p*-methylbenzylamine [*958*]. Equally good results are obtained with borane made in diglyme from sodium borohydride and boron trifluoride etherate [*1252*] (equation **289**).

289

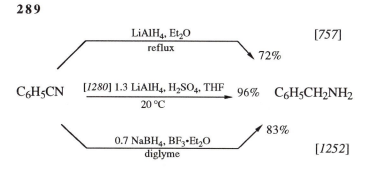

$$C_6H_5CN \xrightarrow[20\ °C]{[1280]\ 1.3\ LiAlH_4,\ H_2SO_4,\ THF} 96\% \quad C_6H_5CH_2NH_2$$

A long time before the hydrides were known, nitriles were reduced to primary amines in good yields by adding *sodium* into solutions of nitriles in boiling alcohols. Seven gram-atoms of sodium and at least 3 mol of butyl alcohol were used per mole of nitrile, giving an 86% yield of pentylamine and a 78% yield of hexylamine from the corresponding nitriles [*1469*]. The amount of sodium can be cut down to 4 gram-atoms if toluene or xylene is used as a diluent [*205*]. *Zinc* and nickel dichloride reduce nitriles to a mixture of alcohols and amines, both in poor yields [*241*].

NITRILES CONTAINING DOUBLE BONDS, TRIPLE BONDS, AND REDUCIBLE FUNCTIONS

A nitrile group undergoes reductions very readily. It is therefore important to choose selective reagents for the reduction of nitriles containing reducible functions.

A **β,γ-unsaturated nitrile is converted** to the corresponding ***unsaturated amine*** by *catalytic hydrogenation*. Hydrogenation over Raney cobalt at 60 °C and 95 atm gives a 90% yield of 2-(1-cyclohexenyl)ethylamine from 1-cyclohexenylacetonitrile. The reduction of the nitrile group precedes even the saturation of the double bond [*1471*]. Reduction with *lithium aluminum hydride* in ether gives a 74% yield of the same product [*1471*]. Catalytic hydrogenation over other catalysts usually reduces the carbon–carbon double bond in α,β-**unsaturated nitriles** preferentially. Thus ethyl 2-ethoxycarbonylmethylcyclopentylidenecyanoacetate is reduced in 95% yields to ethyl 2-ethoxycarbonylmethylcyclopentylcyanoacetate over platinum oxide *(cis* isomer) and over palladium on alumina *(trans* isomer) [*1472*]. In 1-cyclohexene-1-carbonitrile, lithium aluminum hydride reduces the double bond; at –15 °C, it produces cyclohexanecarboxaldehyde, and at room temperature, cyclohexylmethylamine [*178*] (equation **290**).

290   [*178*]

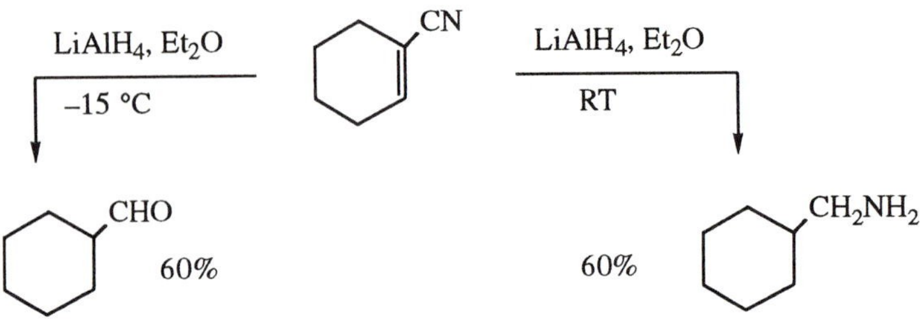

LiAlH₄, Et₂O
―――――――――
–15 °C

LiAlH₄, Et₂O
―――――――――
RT

CHO
60%

CH₂NH₂

60%

Sodium borohydride usually does not reduce the nitrile function, so it may be used for *selective reductions of conjugated double bonds in α,β-unsaturated nitriles* in fair to good yields [*1363, 1364*]. In addition some special reagents are effective for reducing carbon–carbon double bonds preferentially: *copper hydride* prepared from cuprous bromide and *sodium bis(2-methoxyethoxy)-aluminum hydride* [*1473*], *magnesium* in methanol [*1474*], *zinc* and zinc chloride in ethanol or isopropyl alcohol [*1475*], and *triethylammonium formate* with *N,N*-dimethylformamide [*441*].

Triple bonds can be selectively reduced to double bonds by *catalytic hydrogenation* over Raney nickel in dioxane at 60 °C and 30 atm. 4,4'-Bis(cyanophenyl)acetylene is thus converted to *cis*-4,4'-dicyanostilbene in an 87% yield [*505*].

In nitriles containing a nitro group, the nitro group is reduced preferentially by *iron* or *stannous chloride*. 2-Chloro-6-nitrobenzonitrile is reduced to 2-chloro-6-aminobenzonitrile with iron in methanol and hydrochloric acid in an 89% yield [*1476*], and *o*-nitrobenzonitrile is converted to *o*-aminobenzonitrile with stannous chloride and hydrochloric acid in an 80% yield [*256*].

Nitriles of keto acids are reduced with *lithium aluminum hydride* at both functions. Benzoyl cyanide affords 2-amino-1-phenylethanol in an 86% yield [*1477*]. Selective reduction of the nitrile group in an 87% yield without the reduction of the carbonyl is achieved by *stannous chloride* [*1459*]. **Oximes of keto nitriles** are reduced preferentially at the oximino group by *catalytic hydrogenation* in acetic anhydride over 5% platinum on carbon (85% yield) [*1478*] or by *aluminum amalgam* (78–82% yield) [*1479*].

SUMMARY

Reduction of *carboxylic acids* to alcohols is achieved by high-temperature, high-pressure hydrogenation over special catalysts (industrial applications) and by complex hydrides. Unsaturated acids are saturated by a low-temperature, low-pressure catalytic hydrogenation.

Esters of carboxylic acids are reduced to aldehydes by complex hydrides under special reaction conditions. High-temperature, high-pressure hydrogenation converts esters to alcohols (industrial process). The old reduction with sodium in alcohol has been completely replaced by reduction with complex hydrides. Unsaturated esters are reduced to saturated esters by mild catalytic hydrogenation and to unsaturated or saturated alcohols by complex hydrides.

Amides are mostly reduced to amines by complex hydrides. Reduction to aldehydes is rare.

Nitriles can be reduced to aldehydes by special complex hydrides and by stannous chloride. Their reduction to amines is effected by catalytic hydrogenation, complex hydrides, and sodium in alcohols.

CHAPTER 19

Reduction of Organometallic Compounds

Reduction of organometallic compounds is usually very easy. Hydrogenolysis of carbon–halogen bonds in organolithium compounds takes place even without a catalyst, just by treatment of alkyllithium compounds with *hydrogen* at room temperature and 0.2–7 atm. The ease of the formation of a hydrocarbon increases in the series R_2Ca < RLi < RNa < RK < RRb < RCs [*5*]. Organometallic compounds of magnesium, zinc, and lead require catalysts *(nickel or copper chromite),* higher temperatures (160–200 °C), and higher pressures (125 atm) [*1480*].

Complex hydrides are used for reductions of organometallic compounds with good results. Trimethyllead chloride is reduced with *lithium aluminum hydride* in dimethyl ether at –78 °C to trimethylplumbane in 95% yield [*1481*], and 2-methoxycyclohexylmercury chloride is reduced with *sodium borohydride* in 0.5 N sodium hydroxide to methyl cyclohexyl ether in 86% yield [*1482*].

Reduction with *sodium dithionite* has been used for conversion of aryl *arsonic* and aryl *stibonic acids to arsenobenzenes* [*359*] and *stibinobenzenes* in good yields [*360*].

CORRELATION TABLES

The correlation tables show how the main types of compounds are reduced to various products by different reducing agents. The tables list *only the reactions contained in this book*. If a reagent for a certain conversion is not listed, that does not necessarily mean that it cannot accomplish a particular reduction. Reagents listed in *parentheses* apply only under special conditions or are used rarely.

Because the equations are not always placed in the text describing the reductions, pages contiguous to those shown in the tables should also be examined.

Table 1. Reduction of Alkanes, Alkenes, Alkadienes, and Alkynes

Starting compound	Product	Reducing agent	Page
$\diagup_C\hspace{-6pt}\diagdown\!-\!C\diagup\!\diagdown$	$-CH\ +\ HC-$	H_2/Pt, H_2/Ni R_3SiH/CF_3CO_2H	53
$\diagdown_C\!=\!C\diagup$	$\diagdown_{CH-CH}\diagup$	H_2/Pt, Pd, Rh H_2/Ru, Ni, Raney Ni R_3SiH/CF_3CO_2H $LiAlH_4$/$TiCl_3$/$CoCl_2$ or $NiCl_2$ B_2H_6/$EtCO_2H$ Electroreduction Na, Li, Zn/$NiCl_2$ N_2H_2	53–55
$\diagdown_C\!=\!C\!-\!C\!=\!C\diagup$	$\diagdown_{CH-C=C-CH}\diagup$	H_2/Pt, Ni_2B, *P-1* Ni Na, Li, K/NH_3 N_2H_2	56–57
	$\diagdown_{CHCHCHCH}\diagup$	H_2/Pt	56
$-C\equiv C-$	$\overset{\displaystyle\diagdown}{\underset{H}{}}C\!=\!C\overset{\displaystyle\diagup}{\underset{H}{}}$	H_2/Pd deact. H_2/Raney Ni, H_2/Ni_2B $NH_4PH_2O_2$/Pd $Et_3N{\cdot}HCO_2H$/Pd $LiAlH_4$/$NiCl_2$ BH_3, AlH(i-Bu)$_2$ BH(CHMeCHMe$_2$)$_2$ MgH_2 Electroreduction (Ni)	58–59
	$\overset{\displaystyle\diagdown}{\underset{H}{}}C\!=\!C\overset{\displaystyle H}{\underset{\diagdown}{}}$	$LiAlH_4$/$NiCl_2$ LiAlHMe(i-Bu)$_2$ Electroreduction Na, Li, K/NH_3[a] Li or Ca/(CH_2NH_2)$_2$	59–60
	$-CH_2\!-\!CH_2-$	H_2/PtO_2, Pd, Ni Na/t-BuOH	61 59

[a]Mixtures of both stereoisomers are obtained from macrocyclic alkynes.

Table 2. Reduction of Aromatic Hydrocarbons

Starting compound	Product	Reducing agent	Page
		Electroreduction	62–
		Na, Li, K, Ca/NH$_3$, ROH	63
		Ca/(CH$_2$NH$_2$)$_2$	
		H$_2$/PtO$_2$, RhO$_2$, Rh	61–
		H$_2$/Ni, Raney Ni	62
		N$_2$H$_4$(N$_2$H$_2$)	
		H$_2$/Pt, Pd, Ni; C$_6$H$_8$/Pd	63–
		Electroreduction	64
		LiAlH$_4$	
		Na/NH$_3$[a],	
		AlHg[b], Li/MeNH$_2$	
		Li/MeNH$_2$[c]	
	cis	H$_2$/Raney Ni	64–
		LiAlH$_4$/FeCl$_2$, CoCl$_2$ or	65
	trans	NiCl$_2$, N$_2$H$_4$	
		Electroreduction	
		Na/MeOH, Zn	
		CrSO$_4$, SmI$_2$	
		H$_2$/Pd, Re$_2$S$_7$	64–
		Electroreduction	65
		Na/EtOH	
		N$_2$H$_2$	
		Na/NH$_3$, ROH	65–
		Li/MeNH$_2$, (CH$_2$NH$_2$)$_2$	67
		H$_2$/CuCr$_2$O$_4$	65–
		Na/NH$_3$, ROH	67
		Li/MeNH$_2$, (CH$_2$NH$_2$)$_2$	66–
		K/MeNH$_2$,	67
		Ca/(CH$_2$NH$_2$)$_2$	
		H$_2$/PtO$_2$, Pt, Ni	65–
			66
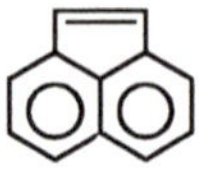	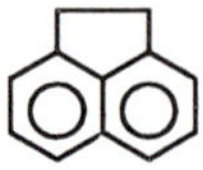	LiAlH$_4$	66

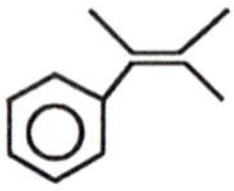

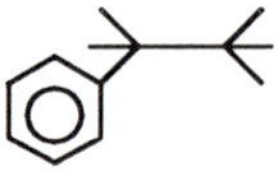

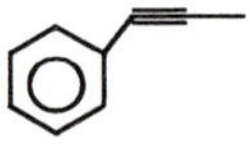

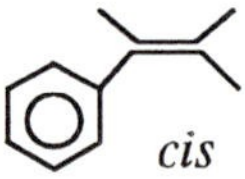

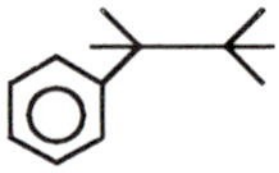

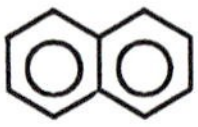

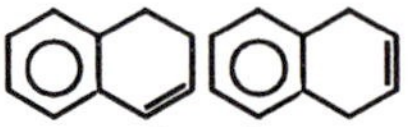

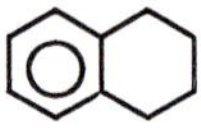

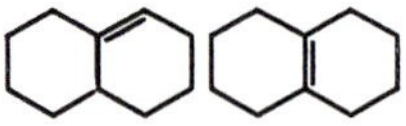

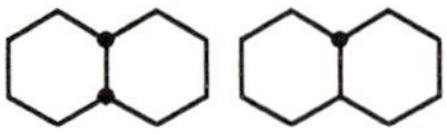

Table 2. Continued

Starting compound	Product	Reducing agent	Page
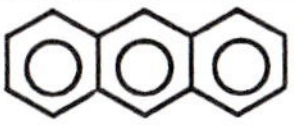 [d]	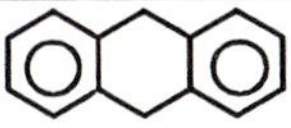	Na/EtOH	67
	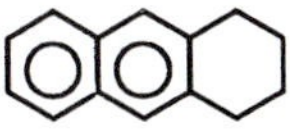	H$_2$/Pd	67
	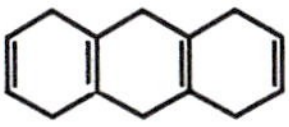	Ca/(CH$_2$NH$_2$)$_2$	67
	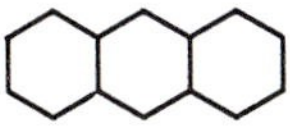	H$_2$/catalysts	67
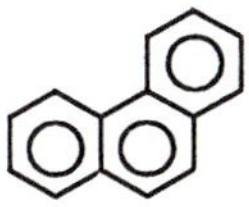 [d]	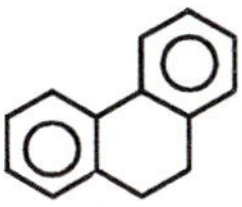 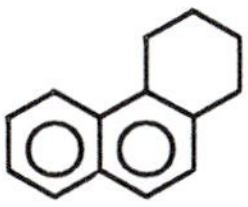	H$_2$/CuCr$_2$O$_4$, Raney Ni Na/ROH	68
	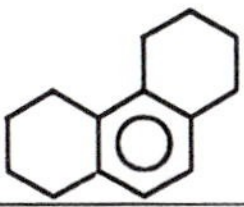 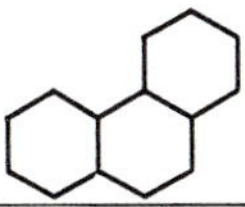	H$_2$/Raney Ni H$_2$/CuCr$_2$O$_4$	68

[a]Partial reduction of the ring may occur.

[b]The reaction is not general.

[c]Ethylcyclohexene is also formed from styrene.

[d]More detailed partial reduction is shown in equations 60–62.

Table 3. Reduction of Heterocyclic Aromatic Compounds

Starting compound	Product	Reducing agent	Page
		H$_2$/Pd, Raney Ni, Ni	66–68
		H$_2$/Re$_2$S$_7$, Re$_2$Se$_7$ H$_2$/CoS$_n$	68
		Zn[a] PH$_4$I[a]	69
		H$_2$/Pt, Raney Ni, Ni H$_2$/CuCr$_2$O$_4$ Zn[a]	68–69
	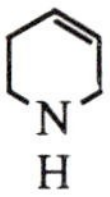	Complex hydrides[b] Alane[b] Electroreduction Na/ROH (HCO$_2$H)	70–71
	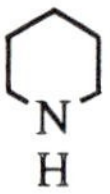	H$_2$/PtO$_2$, Pd, Rh, RuO$_2$ H$_2$/Raney Ni, CuCr$_2$O$_4$ Complex hydrides[b] Na/ROH, (HCO$_2$H) Ni-Al, KOH	69–71
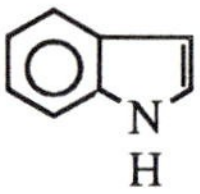	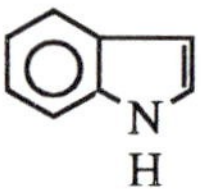	H$_2$/PtO$_2$, H$_2$/CuCr$_2$O$_4$ NaBH$_4$, NaBH$_3$CN BH$_3$·NR$_3$ Na, Li/NH$_3$, Zn	72
	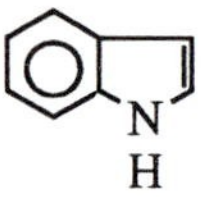	H$_2$/Raney Ni	73
	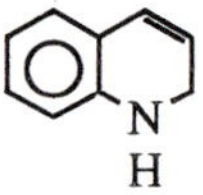	Na/NH$_3$	73,74
	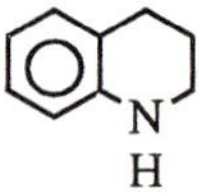	H$_2$/Pt, Raney Ni H$_2$/Ni, CuCr$_2$O$_4$ BH$_3$·NR$_3$, NaBH$_3$CN Ni-Al, KOH, Zn/NiCl$_2$	73,74
	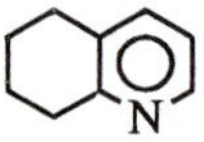	H$_2$/PtO$_2$, CF$_3$CO$_2$H H$_2$/Pd, Rh, CF$_3$CO$_2$H	73
	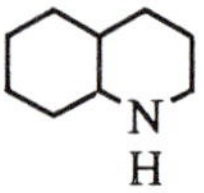	H$_2$/Pt, PtO$_2$, AcOH	73, 74

Table 3. Continued

Starting compound	Product	Reducing agent	Page
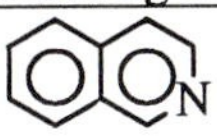	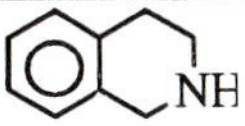 NH	H_2/PtO_2, AcOH Na/NH_3, EtOH Ni-Al, KOH	74
	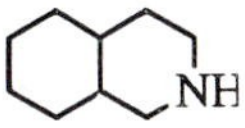 NH	H_2/PtO_2, AcOH, H_2SO_4	74
Acridine Phenanthridine Benzoquinolines	Partial and total reduction	H_2/PtO_2, Raney Ni $CuCr_2O_4$, Na/ROH	74,75
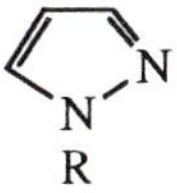 R	R = H 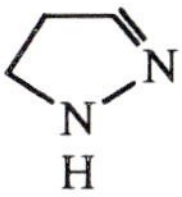	$H_2/Pd(BaSO_4)$, AcOH	76
	R = Ph 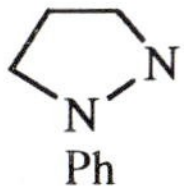 Ph	$H_2/Pd (BaSO_4)$, AcOH	76
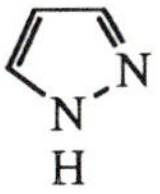 H	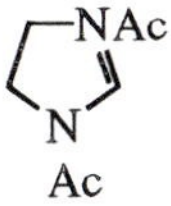 -NAc N Ac	H_2/PtO_2, Ac_2O	76
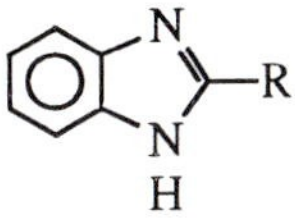 -R H	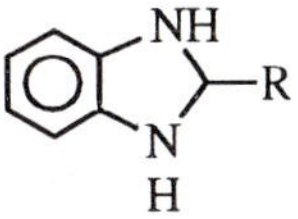 NH -R N H	$LiAlH_4$	76
	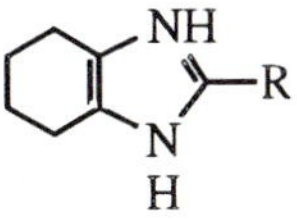 NH -R N H	H_2/PtO_2, AcOH	76
 N[c] N	 N N	$H_2/Pd(C)$, HCl	76
 N N	 N H_4 N H	Electroreduction	76
	 N N H	$H_2/Pd(C)$	76

Table 3. Continued

Starting compound	Product	Reducing agent	Page
		$H_2/Pd/AcOH$	76
		Electroreduction	76,77
	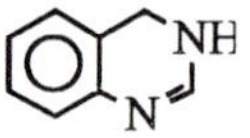	H_2/PtO_2	76,77
	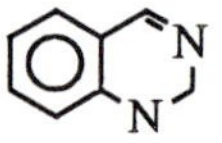	$NaBH_4/CF_3CO_2H$	76,77
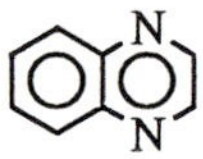	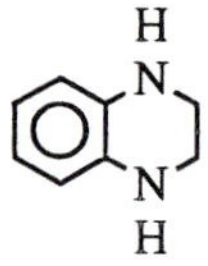	H_2/PtO_2, Raney Ni $LiAlH_4$ $NaBH_4/CF_3CO_2H$	76,77
	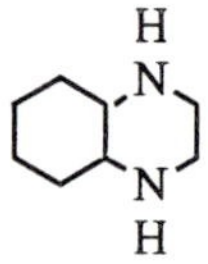	H_2/PtO_2, Pd(C), Rh H_2/Raney Ni	77
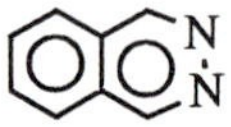	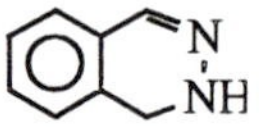	Electroreduction, KOH	76,77
	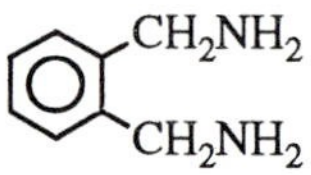	$H_2/Pd(C)$, Raney Ni Electroreduction, HCl	76,77 76,78
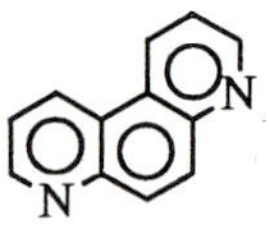	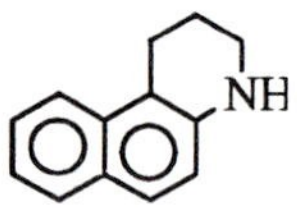	H_2/Raney Ni, 20 °C	78
	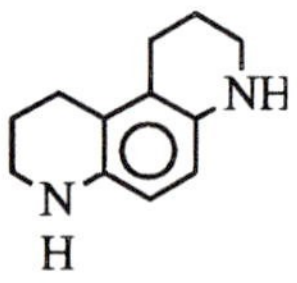	H_2/Raney Ni, 100 °C	78

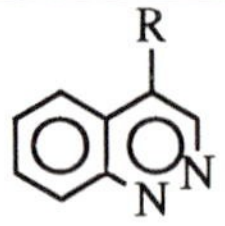

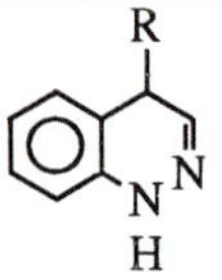

[a]Reductions occur in pyrrole derivatives.

[b]More detailed reduction of pyridine and its derivatives is given in equations 66 and 67.

[c]1,2-Diazine (pyridazine) resists hydrogenation over palladium at room temperature.

Table 4. Reduction of Halogen Compounds

Starting compound (X = Cl, Br, I, rarely F	Product	Reducing agent	Page
R–C̣–X	R–C̣–X	[a]H_2/Pt, Rh, Raney Ni/KOH $LiAlH_4$, $NaBH_4$, $NaAlH_4$ Et_3SiH, R_3SnH, R_2SnH_2 $NaBH_3CN$/$SnCl_2$ $Zn(BH_4)_2$[b] $LiAlH(OR)_3$, $LiBHEt_3$ Mg^c, Zn^c, $CrSO_4{}^c$	81, 82 81, 82

R–C̣–X
   X
	R–C̣–H
	   X
		$LiAlH_4$,
		$NaAlH_2(OCH_2CH_2OMe)_2$
		Bu_3SnH, electroreduction
		Na/NH_3, AlHg, $CrSO_4$, SmI_2
		Fe, Na_3AsO_3
		NaH_2PO_2, Na_2SO_3
		$(EtO)_2PHO$, i-PrOH/$h\upsilon$
			83, 84

	R–C̣–H H	H_2/PtO_2, Pd, Raney Ni $LiAlH_4$ $NaAlH_2(OCH_2CH_2OMe)_2$ Bu_3SnH Electroreduction, Na/ROH $CrSO_4$	82–84 83
—C̣X—C̣X—	—C̣=C̣—	R_3SnH, Mg, Zn	84
	—C̣H—C̣H—	H_2/Raney Ni, Li/t-BuOH	84
⟩C=C̣–C̣–X[d]	⟩CH–C̣H–C̣–X[d]	H_2/Pt, Pd, Rh	85
	⟩C=C̣–C̣–H	$LiAlH_4$, $CrSO_4$ Li, Na/t-BuOH, Zn	85, 86
	⟩CH–C̣H–C̣–H	H_2/Raney Ni, KOH	85, 86

⟩C=C⟨
     X
	⟩C=C⟨
	     H
		$LiAlH_4$, $NaBH_4$, Bu_3SnH
		Li, Na/t-BuOH, Zn
			86, 87

Table 4. Continued

Starting compound (X = Cl, Br, I, rarely F	Product	Reducing agent	Page
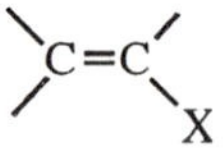 $\diagup\!\!\diagup C = C\diagdown\!\!\diagdown$ with X	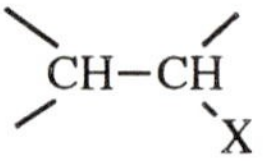 $\diagdown CH - CH \diagup$ with X	H_2/Pt, Pd, Rh	85
	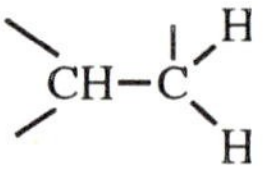 $\diagdown CH - C \diagdown$ with H, H	H_2/Raney Ni/KOH	85
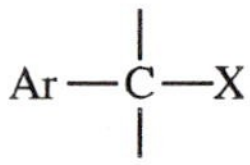 $Ar - \overset{\mid}{\underset{\mid}{C}} - X$	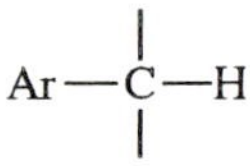 $Ar - \overset{\mid}{\underset{\mid}{C}} - H$	H_2/Pd, KOH, H_2/Raney Ni, KOH $LiAlH_4$, BH_3	87, 88
$Ar - X$	$Ar - H$	H_2/Pd, KOH, H_2/Raney Ni, KOH; HCO_2H/Pd $Et_3N\text{-}HCO_2H$/Pd, $LiAlH_4$ $NaAlH_2(OCH_2CH_2OMe)_2$ $NaAlH_4$, $NaBH_4$ $NaBH_3NMe_2$ Et_3SiH, R_3SnH, Na/NH_3 Mg, Zn, $H_2Fe(CO)_4$ $EtSH/AlCl_3$, benzyl-dihydronicotinamide	87–90

[a]Hydrogenolysis of fluorine and chlorine in haloalkanes is possible only under unusually vigorous conditions, and it is practically unfeasible.
[b]Reduction takes place only with tertiary chlorides.
[c]Only for X = Br or I.
[d]Similar reagents may be applied to acetylenic compounds.

Table 5. Reduction of Nitro Compounds

Starting compound	Product	Reducing agent	Page
$\diagdown$CH—NO$_2$	$\diagdown$CHNHOH	B$_2$H$_6$, SmI$_2$	92
	$\diagdown$C=NOH	H$_2$/catal., SnCl$_2$, TiCl$_3$ CrCl$_2$, Na$_2$S$_2$O$_3$	92
	$\diagdown$CHNH$_2$	H$_2$/PtO$_2$, AlHg, Fe, SmI$_2$	91
	$\diagdown$CH$_2$	H$_2$/Pt(SiO$_2$)	93
—CH=CHNO$_2$	—CH$_2$CH$_2$NO$_2$	H$_2$/Pd, H$_2$/(Ph$_3$P)$_3$RhCl Et$_3$N·HCO$_2$H/Pd LiAlH$_4$, NaBH$_4$ LiBH (s-Bu)$_3$, Zn(BH$_4$)$_2$ Bu$_3$SnH, 1,4-dihydropyridine derivative	93
	—CH$_2$CH=NOH	H$_2$/Pd LiAlH$_4$, Zn(BH$_4$)$_2$, Fe, Na$_2$SnO$_2$, CrCl$_2$	93,94
	—CH$_2$CH$_2$NHOH	LiAlH$_4$, BH$_3$·THF	94
	—CH$_2$CH$_2$NH$_2$	H$_2$/Pd, LiAlH$_4$, Fe	93,94
ArNO$_2$	ArNHOH	H$_2$/Pd, N$_2$H$_4$/Raney Ni Electroreduction, AlHg Zn/NH$_4$Cl	95
	ArN(O)=NAr	H$_2$/Pd NaAlH$_2$(OCH$_2$CH$_2$OMe)$_2$ Na$_3$AsO$_3$, sugars	96 95 95,96
	ArN=NAr	LiAlH$_4$, Mg(AlH$_4$)$_2$ NaAlH$_2$(OCH$_2$CH$_2$OMe)$_2$ Si/NaOH, Zn/NaOH	95,96
	ArNHNHAr	H$_2$/Pd, Zn/KOH	95,96

Table 5. Continued

Starting compound	Product	Reducing agent	Page
$ArNO_2$	$ArNH_2$	H_2/PtO_2, RhO_2–PtO_2, Pd	95–
		H_2/Raney Ni, $CuCr_2O_4$, Re_2S_7	97
		HCO_2H/Cu, HCO_2NH_4/Pd	
		N_2H_4/Ni, $Et_3N\cdot HCO_2H$	
		KBH_4/Cu_2Cl_2	96
		Zn, Fe, $SnCl_2$, $TiCl_3$	96,97
		$Zn/NiCl_2$, $FeSO_4$, Na_2S, H_2S	96
		$Na_2S_2O_4$	96
	Hydroaromatic amines	H_2/PtO_2, RhO_2–PtO_2	97
		HCO_2H/Ni	96,97
$HalC_6H_4NO_2$	$ArNO_2$	$Et_3N\cdot HCO_2H/Pd$	98
		NaI/HCO_2H	
		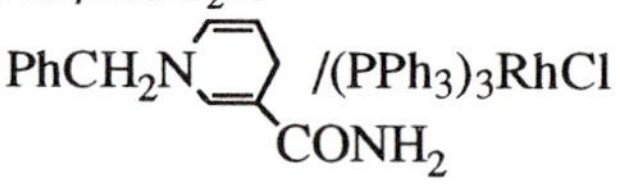 $PhCH_2N\underset{CONH_2}{\diagup\!\!\!\diagdown}$ /$(PPh_3)_3RhCl$	
	$HalC_6H_4NH_2$	N_2H_4/Raney Ni, Fe, $Na_2S_2O_4$	99
	$ArNH_2$	N_2H_4/Pd, Zn, $SnCl_2$, NH_4SH	99

Table 6. Reduction of Nitroso, Diazonium, and Azido Compounds[a]

Starting compound	Product	Reducing agent	Page
ArNO	ArNHOH	Ph_3P, $(EtO)_3P$	99
	ArN=NAr	$LiAlH_4$	99
	$ArNH_2$	KBH_4/Cu_2Cl_2	95
		NH_4SH, $Na_2S_2O_4$	
$Ar\overset{+}{N}_2\overset{-}{X}$	$ArNHNH_2$	Zn, $SnCl_2$, Na_2SO_3	100
	ArH	$NaBH_4$, Sn	100
		Na_2SnO_2	
		H_3PO_2, $(Me_2N)_3P$	
		EtOH, $HCONH_2$	
		N-benzyl-1,4-dihydronicotinamide	
RN_3	RNH_2	H_2/PtO_2, H_2/Pd, HCO_2NH_4/Pd	100
		$LiAlH_4$, $NaBH_4$, Mg, Ca	
		AlHg, $TiCl_3$, VCl_2	
		H_2S, $Na_2S_2O_4$	
		HBr, $HS(CH_2)_3SH$	
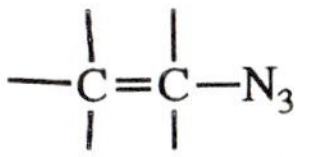 $-\overset{\mid}{C}=\overset{\mid}{C}-N_3$	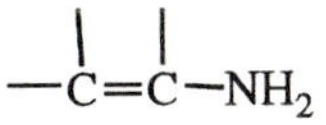 $-\overset{\mid}{C}=\overset{\mid}{C}-NH_2$	AlHg, $HS(CH_2)_3SH$	100
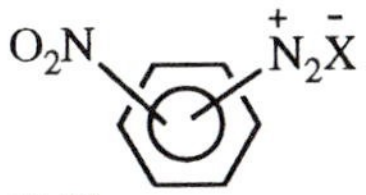 O_2N ⬡ $\overset{+}{N}_2\overset{-}{X}$	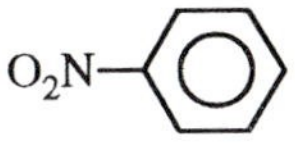 O_2N—⬡	$NaBH_4$	100
		Na_2SnO_2	
		H_3PO_2, EtOH	
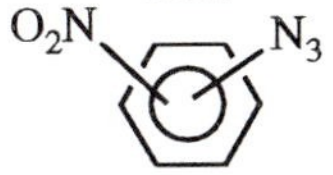 O_2N ⬡ N_3	 O_2N ⬡ NH_2	H_2S, HBr	100
		$HS(CH_2)_3SH$	
	 H_2N ⬡ NH_2	H_2/catalysts	
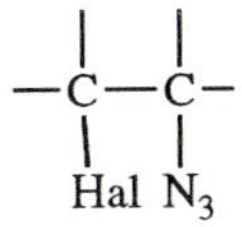 $-\overset{\mid}{\underset{Hal}{C}}-\overset{\mid}{\underset{N_3}{C}}-$	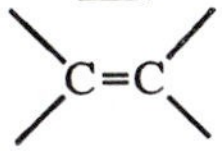 $\diagdown C=C \diagup$	$LiAlH_4$, $NaBH_4$	101
	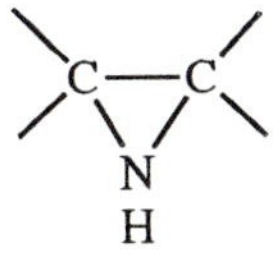 $\diagdown C \overset{}{\underset{N}{\frown}} C \diagup$ (N–H)	B_2H_6, $LiAlHCl_3$	101

[a]Reductions of other nitrogen compounds such as hydroxylamino, azo, and hydrazo compounds are reviewed in Table 11. Aliphatic diazo compounds are reviewed in Tables 16 and 24.

Table 7. Reduction of Alcohols and Phenols

Starting compound	Product	Reducing agent	Page
ROH	RH	H_2/V_2O_5, MoS_2,WS_2 Pd/C, Raney Ni $LiAlH_4/AlCl_3$ R_3SiH/CF_3CO_2H[a] Raney Ni and $H_2/Pd(C)$[a] + RN=C=NR, H_2/Pd + CS_2, MeONa, BH_3, Et_3SiH Bu_3SnH o-$Me_2NCH_2C_6H_4SnMe_2H$ RSH, K/crown ether + $ArSO_2Cl$, NaI, Bu_3SnH	103–104
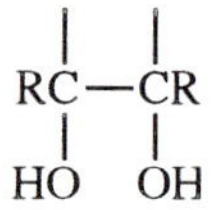 $\underset{\underset{OH}{\mid}}{RC}\!-\!\underset{\underset{OH}{\mid}}{CR}$	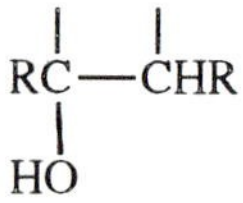 $\underset{\underset{HO}{\mid}}{RC}\!-\!\overset{\mid}{C}HR$	$H_2/CuCr_2O_4$	104
	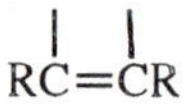 $RC\!=\!CR$	$TiCl_3/K$, $(RO)_3P$	104
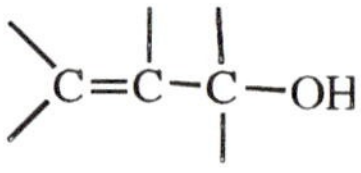 $\underset{}{}C\!=\!C\!-\!\overset{\mid}{C}\!-\!OH$	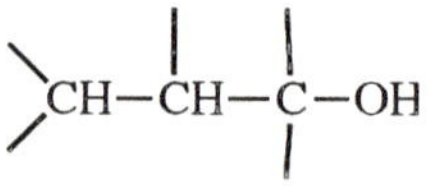 $CH\!-\!CH\!-\!\overset{\mid}{C}\!-\!OH$	H_2/catalysts, Na/NH_3[b], $Li/EtNH_2$[b], N_2H_2	104–105
	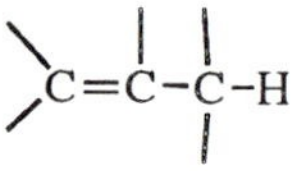 $C\!=\!C\!-\!\overset{\mid}{C}\!-\!H$	$LiAlH_4/AlCl_3$, Na/NH_3, Ph_3PI_2, Ph_3PHI, $C_5H_5N\cdot SO_3$ + $LiAlH_4$	105
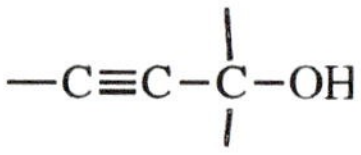 $-C\!\equiv\!C\!-\!\overset{\mid}{C}\!-\!OH$	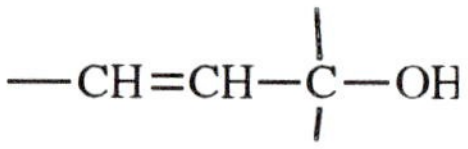 $-CH\!=\!CH\!-\!\overset{\mid}{C}\!-\!OH$	H_2/Pd deactivated $LiAlH_4$ $LiAlH_2(OCH_2CH_2OMe)_2$ Zn, $CrSO_4$	105
	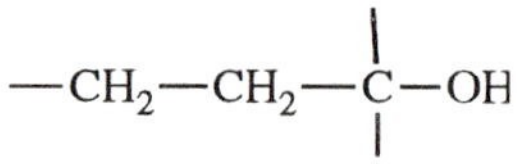 $-CH_2\!-\!CH_2\!-\!\overset{\mid}{C}\!-\!OH$	H_2/catalyst	
ArOH	Hydroaromatic alcohols[c]	H_2/PtO_2, RhO_2–PtO_2, Raney Ni, Ni Li, Na/NH_3[d]	108– 109 109

Table 7. Continued

Starting compound	Product	Reducing agent	Page
ArOH	ArH	+ RN=C=NR, H_2/Pd	106–
		+ BrCN + Et_2NH, H_2/Pd	107
		+ RSO_2F, H_2/Pd	
		$Et_3N \cdot HCO_2H$/Pd	
		$NaBH_4$/Pd	
		$LiAlH_4$ distn., Zn distn.	
		distn. HI/AcOH[d]	
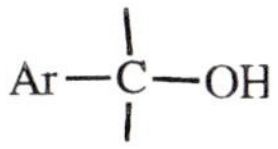 Ar—C—OH (with bonds above and below C)	Ring-hydrogenated alcohols[c]	H_2/PtO_2	108– 109
	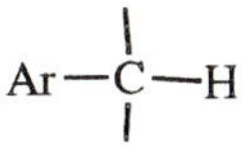 Ar—C—H (with bonds above and below C)	H_2/Pd, Raney Ni, Ni $CuCr_2O_4$ $AlHCl_2$ $NaBH_4$, BH_3, Zn, HI, P_2I_4	107 108
	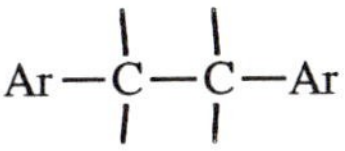 Ar—C—C—Ar (with bonds)	$TiCl_3$/$LiAlH_4$	107– 108
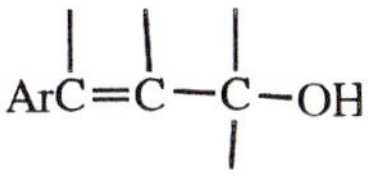 ArC=C—C—OH (with bonds)	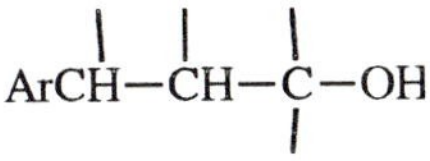 ArCH—CH—C—OH (with bonds)	H_2/PtO_2, RhO_2–PtO_2, PdO $LiAlH_4$	108
	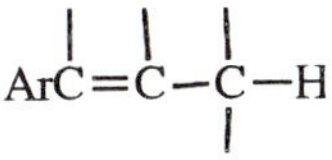 ArC=C—C—H (with bonds)	H_2/PtO_2/AcOH	108
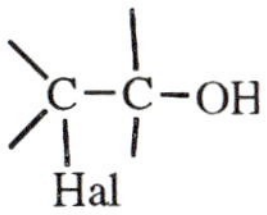 C—C—OH with Hal (structure)	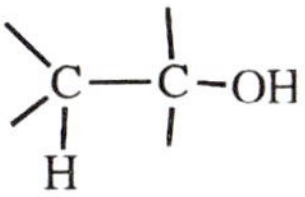 C—C—OH with H (structure)	H_2/Raney Ni; Raney Ni + H_2/Pd(C)[a], *i*-PrOH, *h*υ	109– 110
	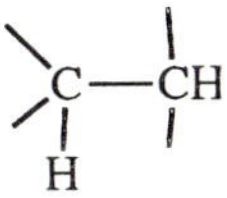 C—CH with H (structure)	H_2/Pd(C)	109– 110
	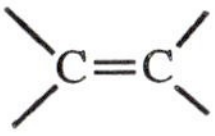 C=C (structure)	$TiCl_3$/$LiAlH_4$[e]	110

Table 7. Continued

Starting compound	Product	Reducing agent	Page
Halogenated phenols	Phenols	$HCO_2NH_4/Pd(C)$ $EtSH/AlCl_3$	109
$\overset{\displaystyle |}{\underset{\displaystyle NO_2}{>C-\overset{\displaystyle |}{C}}-OH}$	$\overset{\displaystyle |}{\underset{\displaystyle NH_2}{>C-\overset{\displaystyle |}{C}}-OH}$	$H_2/Pt; Fe, Sn, Na_2S$ $Na_2S_2O_4$	110

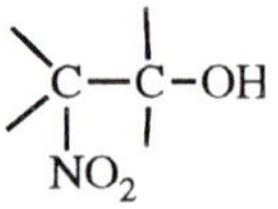

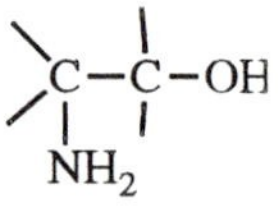

[a]Reduction applies only to tertiary alcohols.

[b]Reduction applies only to nonallylic unsaturated alcohols.

[c]Phenols give cyclohexanols; naphthols give decahydronaphthols on reduction over platinum oxide, tetrahydronaphthols on reduction over Raney nickel, and dihydro derivatives by reduction with sodium.

[d]Reduction of naphthols gives dihydro derivatives.

[e]Reduction applies to bromohydrins only.

Table 8. Reduction of Ethers

Starting compound	Product	Reducing agent	Page
		(H$_2$/Raney Ni) (H$_2$/CuCr$_2$O$_4$) (LiAlH$_4$/AlCl$_3$)	110– 111
$\underset{/}{\overset{\backslash}{C}}=\overset{|}{C}-OR$	$\underset{/}{\overset{\backslash}{C}}=\overset{|}{C}H$	Na, Li, K/NH$_3$ Me$_2$CHMgBr	111
$\underset{/}{\overset{\backslash}{C}}=\overset{|}{C}-\overset{|}{C}-OR$	$\underset{/}{\overset{\backslash}{C}}=\overset{|}{C}-\overset{|}{C}-H$	LiAlH$_4$, LiAlH$_2$(i-Bu)$_2$ LiBHEt$_3$, Li/EtNH$_2$	111
ArOR	ArH	(Na/EtOH)	112
	Dihydro- and tetrahydro-aromatic ethers and hydrocarbons	H$_2$/Raney Ni Electroreduction Li, Na/NH$_3$, ROH Na/EtOH	112 113
$Ar-\overset{|}{\underset{|}{C}}-OR$	$Ar-\overset{|}{\underset{|}{C}}-H$	H$_2$/Pd c-C$_6$H$_{10}$/Pd Na/NH$_3$, Ca/NH$_3$ Na/BuOH	111, 112
Hal-ArOR	ArOR	Zn/NiCl$_2$, NaI/Ph$_3$P	113

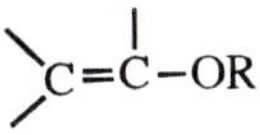

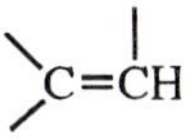

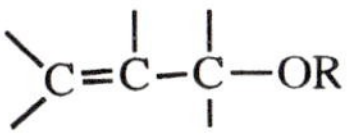

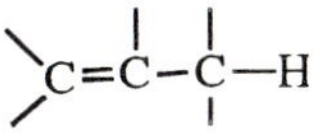

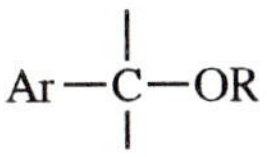

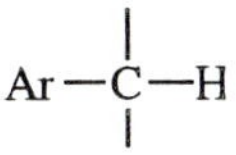

Table 9. Reduction of Epoxides, Peroxides, and Ozonides

Starting compound	Product	Reducing agent	Page
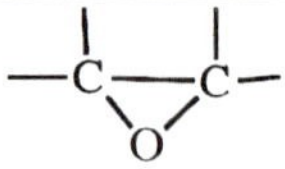 —C—C— (epoxide, with O bridge)	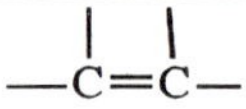 —C=C—	Zn, Zn/Cu TiCl$_3$/LiAlH$_4$, CrCl$_2$, titanocene FeCl$_3$/BuLi, SmI$_2$ (EtO)$_2$PONa/Te (C$_5$H$_5$)$_2$TiCl$_2$/Mg	113
	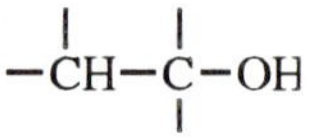 —CH—C—OH	H$_2$/PtO$_2$, LiAlH$_4$, NaAlH$_2$(OCH$_2$CH$_2$OMe)$_2$ LiAlH$_4$/AlCl$_3$, NaBH$_3$CN/ BH$_3$·BF$_3$, AlH$_3$ Li/(CH$_2$NH$_2$)$_2$ NaH, C$_5$H$_{11}$OH, NiCl$_2$	113– 114
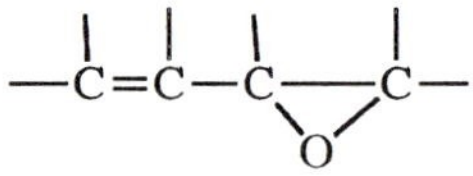 —C=C—C—C— (with epoxide O)	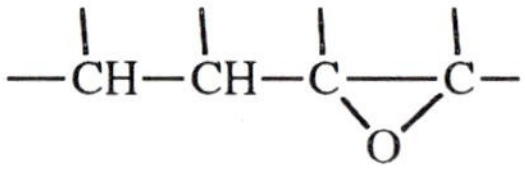 —CH—CH—C—C— (with epoxide O)	H$_2$/Pd(C)	114
	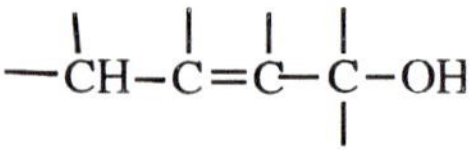 —CH—C=C—C—OH	LiAlH$_4$/CuCN, BH$_3$·THF	114
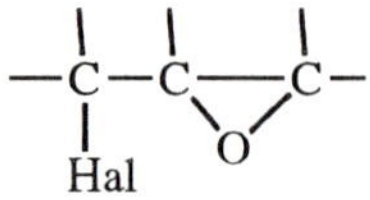 —C—C—C— (with Hal and epoxide O)	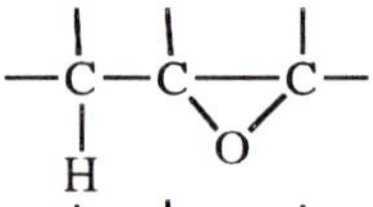 —C—C—C— (with H and epoxide O)	Electroreduction	115
	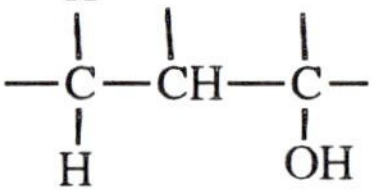 —C—CH—C— (with H and OH)	LiAlH$_4$, LiBHEt$_3$, AlH(i-Bu)$_2$	115
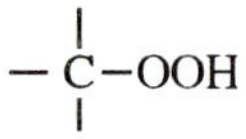 —C—OOH	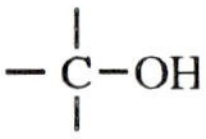 —C—OH	H$_2$/PtO$_2$, Pd, Raney Ni Zn, Na$_2$SO$_3$ R$_3$P, (RO)$_3$P	115– 116
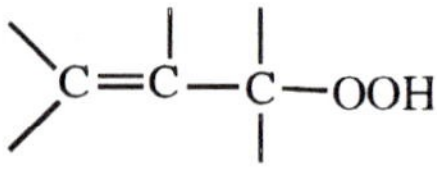 C=C—C—OOH	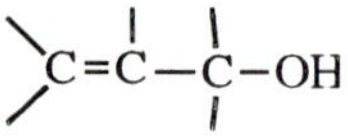 C=C—C—OH	H$_2$/Pd, Zn, Na$_2$SO$_3$, Ph$_3$P	116
	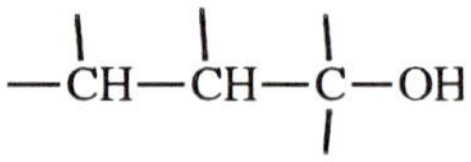 —CH—CH—C—OH	H$_2$/Pd, Zn	115– 116
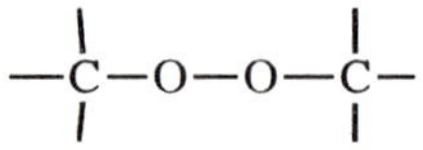 —C—O—O—C—	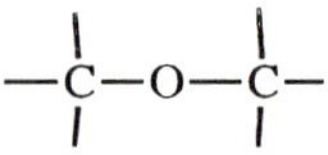 —C—O—C—	Et$_3$P, Ph$_3$P	115– 116
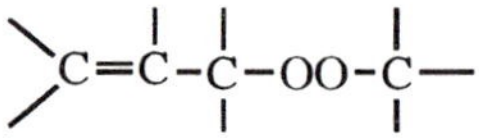 C=C—C—OO—C—	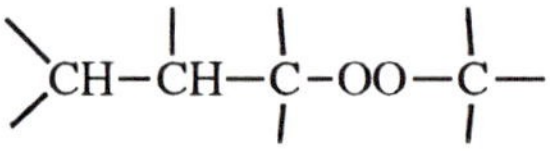 CH—CH—C—OO—C—	H$_2$/Pd	116

Table 9. Continued

Starting compound	Product	Reducing agent	Page
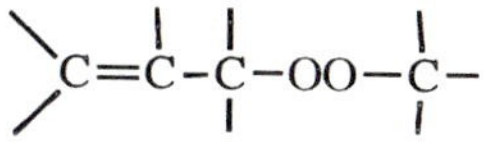 $\rangle$C=C-C-OO-C-	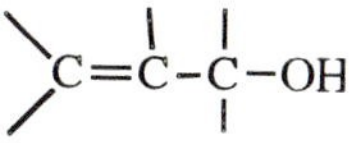 $\rangle$C=C-C-OH	H_2/Pd	116
	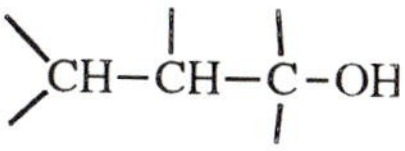 $\rangle$CH-CH-C-OH	H_2/Pd	116
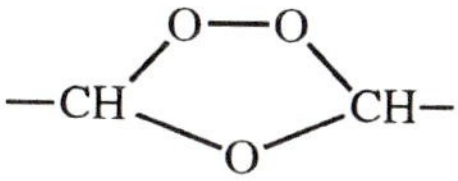 $-CH\overset{O-O}{\underset{O}{\diagdown\quad\diagup}}CH-$	—CHO + OHC—	H_2/Pd Zn/AcOH Me$_2$S	116, 117

Table 10. Reduction of Sulfur Compounds (Except Sulfur Derivatives of Aldehydes, Ketones, and Acids)

Starting compound	Product	Reducing agent	Page
RSH	RH	Raney Ni, Ni$_2$B, Fe, (EtO)$_3$P	119, 120
RSR[a]	RH	Raney Ni, Ni$_2$B	119, 120
		LiAlH$_4$/TiCl$_3$	
		LiAlH$_4$/CuCl$_2$	
		Electroreduction	122
	RSH	Na/NH$_3$	119, 120
$-\overset{\displaystyle}{\underset{\displaystyle|}{C}}\overset{\displaystyle S}{\diagup\diagdown}\overset{\displaystyle}{\underset{\displaystyle|}{C}}-$	$-\overset{\displaystyle|}{C}=\overset{\displaystyle|}{C}-$	Ph$_3$P, (EtO)$_3$P	119, 120
$-\overset{\displaystyle|}{C}=\overset{\displaystyle|}{C}-\overset{\displaystyle|}{C}SR$	$-\overset{\displaystyle|}{C}H-\overset{\displaystyle|}{C}H-\overset{\displaystyle|}{C}SR$	H$_2$/(Ph$_3$P)$_3$RhCl	119
RSSR	RH	Raney Ni, Ni$_2$B	120,
	RSH	LiAlH$_4$, LiBHEt$_3$, NaBH$_4$,	121
		Al/NH$_3$, N$_2$H$_4$	
	RSR	(Me$_2$N)$_3$P, (Et$_2$N)$_3$P	121
$\underset{\displaystyle O}{\overset{\displaystyle \|}{R\!S\!R}}$	RH	Raney Ni, Ni$_2$B, t-BuLi	120, 122
		AlHg	
	RSR	BBr$_3$, Me$_2$BBr, 9-BBN-Br	121, 122
		H$_2$/Pd, Cl$_3$SiH/NaI	
		Electroreduction, AlHg	
		SnCl$_2$, TiCl$_3$, MoCl$_3$	
		VCl$_2$, CrCl$_2$, SmI$_2$	
		NaI/(CF$_3$CO)$_2$O	
$\underset{\displaystyle O}{\overset{\displaystyle O \atop \|}{\underset{\displaystyle \|}{R\!S\!R'}}}$	RH	Raney Ni	120
	RSR'	LiAlH$_4$, AlH(i-Bu)$_2$	123
		Zn/AcOH	
	RSO$_2$H + R'H	LiAlH$_4$/CuCl$_2$, Na/NH$_3$	123, 124
		NaHg, AlHg, SmI$_2$	
		Na$_2$S$_2$O$_4$	
RSCN	RSH	LiAlH$_4$	121
RSCl	RSSR	LiAlH$_4$	123, 124
RSO$_2$H	RSSR	LiAlH$_4$, NaH$_2$PO$_2$[b]	123, 124
		EtOPH$_2$(O)[b]	
RSOCl	RSSR	LiAlH$_4$	124

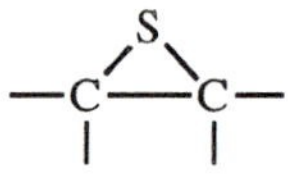

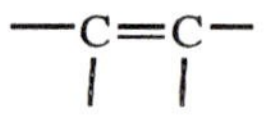

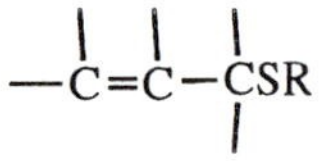

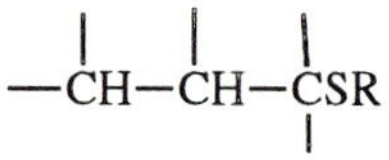

Table 10. Continued

Starting compound	Product	Reducing agent	Page
RSO_2Cl	RSH	$LiAlH_4$, $HI(P + I_2)$	124, 125
	RSO_2H	$LiAlH_4/-20\ °C$, Zn, Na_2S Na_2SO_3	124, 125
RSO_2OR'	RH	H_2/Raney Ni, NaI/Zn	125
	RSH	$LiAlH_4$	125
	RSO_2H	H_2/Raney Ni, $C_{10}H_8 \cdot Na_2$	126
	R'H	H_2/Raney Ni $R_3N \cdot HCO_2H/Pd$, $LiAlH_4$ $LiBHEt_3$, $NaBH_4$	125, 126
	R'OH	H_2/Raney Ni, $LiAlH_4$ $C_{10}H_8 \cdot Na_2$, NaHg	125, 126
RSO_2NHR'	RH	$C_{10}H_8 \cdot Na_2$, Na/ROH	126, 127
	RSH	Zn, $SnCl_2$, HI/PH_4I	
	RSO_2H	$C_{10}H_8 \cdot Na_2$, Na/ROH	126, 127
	$R'NH_2$	$C_{10}H_8 \cdot Na_2$, Na/ROH Zn, $SnCl_2$, HI/PH_4I	126, 127

[a]Benzyl sulfides are not desulfurized but reduced to mercaptans and toluene.
[b]Reduction is carried out with alkali salts.

Table 11.   Reduction of Amines and Their Derivatives

Starting compound	Product	Reducing agent	Page
RNH_2	RH	$H_2/PtO_2(SiO_2)$, H_2/Ni^a H_2NOSO_3H	129
$RCH=CHCH_2NH_2$	$R(CH_2)_3NH_2$	H_2/catalysts Na/NH_3, MeOH	129
$RCH=CHNR'_2$	$RCH_2CH_2NH_2$	$AlHCl_2$, $NaBH_4$, $NaBH_3CN$ $Zn(BH_3CN)_2$, HCO_2H	129
	$RCH=CH_2 + NHR'_2$	AlH_3	129
$ArNH_2$, ArNHR, ArNRR'	Dihydro and tetrahydro aromatic amines	Na/NH_3 + ROH; Na/ROH	129
	Hexahydro amines	H_2/PtO_2, H_2/Ni, Li/$EtNH_2$	129
$ArCH_2NR_2$	$ArCH_3$	H_2/PdO, $H_2/Pd(C)$, $LiAlH_4$	130
Hal-$ArNH_2$	$ArNH_2$	$LiAlH_4$	130
$ArCH_2\overset{+}{N}R_3\overset{-}{X}$	$ArCH_3$	NaHg	130
$Ar\overset{+}{N}R_3\overset{-}{X}$	ArH	Electroreduction	130
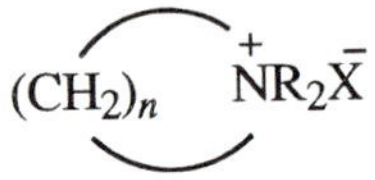 $(CH_2)_n$ $\overset{+}{N}R_2\overset{-}{X}$	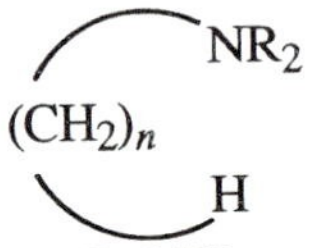 $(CH_2)_n$ NR_2 H	H_2/Raney Ni, $NaAlH_4$	130
R_2NNO	R_2NH	$LiAlH_4$	131
	R_2NNH_2	$LiAlH_4$, Zn/AcOH, $TiCl_2$	131
R_2NNO_2	R_2NNH_2	Electroreduction	131
R_3NO	R_3N	H_2/Pd^b, HCO_2NH_4/Pd NaH_2PO_2, BH_3, Fe/AcOH $TiCl_2$, $TiCl_3$, $CrCl_2$, SO_2 $P(OMe)_3$, $P(OEt)_3$, CS_2 HCO_2COCH_3	131
		$AlH_3{}^c$, $XAlH_4{}^c$, electro-reduction^c, Na^c	131
RNHOH	RNH_2	Zn/AcOH, biochemical reduction	132, 133
RNHNH$_2$	RNH_2	H_2/Raney Ni, Na/NH_3 Zn, AlHg	133

Table 11.   Continued

Starting compound	Product	Reducing agent	Page
RNHNHR	RNH$_2$	H$_2$/Pd[d]	133
RN=NR	RNHNHR	H$_2$/Pd, Bu$_3$SnH, N$_2$H$_2$ NaHg, AlHg, Zn/NH$_3$, H$_2$S	133, 134
	RNH$_2$	H$_2$/Pd, KBH$_4$/Cu$_2$Cl$_2$, AlHg Zn/AcOH, Na$_2$S$_2$O$_4$ (EtO)$_2$P(S)SH	134
RN=NR | O	RNH$_2$	KBH$_4$/Cu$_2$Cl$_2$	134
RN=C=S	RNHCH=S	Bu$_3$SnH	134

[a]This reduction is very unusual.

[b]Nitroamine oxide is reduced selectively to nitroamine.

[c]Pyridine ring is also reduced in pyridine oxides.

[d]All reagents used to reduce azo compounds to amines are likely to cleave hydrazo compounds to amines..

Table 12. Reduction of Phosphorus Compounds

Starting compound	Product	Reducing agent	Page
$R_3P=O$	R_3P	$PhSiH_3$, SmI_2	134
$(RO)_2PCl=O$	$(RO)_2POH$	$NaBH_4$	134, 135
RPCl=O | OR'	RPOH | OR'	$NaBH_4$	134, 135

Table 13. Reduction of Aldehydes

Starting compound	Product	Reducing agent	Page
RCHO	RCH$_2$OH	H$_2$/PtO$_2$, H$_2$/Raney Ni	137,
		LiAlH$_4$, LiAlH(OR)$_3$	138
		Na$_2$AlH$_2$(OCH$_2$CH$_2$OMe)$_2$	
		LiBH$_4$, NaBH$_4$, Zn(BH$_4$)$_2$	
		Bu$_4$NBH$_3$CN, KBHPh$_3$	
		Bu$_4$NBH(OAc)$_3$	
		R$_3$SnH, LiH/FeCl$_2$	
		LiH, NaH/FeCl$_2$	
		Zn/NiCl$_2$, Fe/AcOH	
		Na$_2$S$_2$O$_4$	
		i-PrOH/Al(O-*i*-Pr)$_3$	
	RCH$_3$	ZnHg/HCl, N$_2$H$_4$/KOH	137,
			138
	RCH=CHR	TiCl$_3$/K or Mg	137,
			138
RCH=CHCHO	RCH$_2$CH$_2$CHO	H$_2$/Pd, H$_2$/Ni, H$_2$/HCo(CO)$_4$	138,
		Et$_3$N•HCO$_2$H/Pd	139
	RCH=CHCH$_2$OH	H$_2$/PtO$_2$, H$_2$/Os, LiAlH$_4$	139
		Mg(AlH$_4$)$_2$, LiBH$_4$, NaBH$_4$	
		NaBH(OMe)$_3$, 9-BBN,	
		Ph$_2$SnH$_2$	
		i-PrOH/Al(O-*i*-Pr)$_3$	
	RCH$_2$CH$_2$CH$_2$OH	H$_2$/Ni, electroreduction	139
ArCHO	ArCH$_2$OH	H$_2$/Pt, H$_2$/Pd, H$_2$/Ni,	139–
		H$_2$/Raney Ni	141
		H$_2$/CuCr$_2$O$_4$, HCO$_2$H/Cu	
		Ni–Al/NaOH, LiAlH$_4$, AlH$_3$	
		LiBH$_4$, NaBH$_4$, Zn(BH$_4$)$_2$	
		NaBH(OMe)$_3$, Bu$_4$NBH$_4$	
		NaBH$_3$NMe$_2$	
		Bu$_4$NBH$_3$CN, 9-BBN	
		Bu$_3$SnH, Ph$_2$SnH$_2$,	
		NaH/FeCl$_3$, Zn/NiCl$_2$,	
		Na$_2$S$_2$O$_4$,	
		i-PrOH/Al(O-*i*-Pr)$_3$, CH$_2$O	

Table 13. Continued

Starting compound	Product	Reducing agent	Page
ArCHO	ArCH$_3$	H$_2$/Pt, H$_2$/Pd, H$_2$/Ni	141,
		H$_2$/CuCr$_2$O$_4$, AlH$_3$	142
		Et$_3$SiH/BF$_3$, electroreduction	
		Li/NH$_3$, ZnHg/HCl	
		N$_2$H$_4$/KOH	
		1. TosNHNH$_2$; 2. LiAlH$_4$	
	ArCH=CHAr	TiCl$_3$/Li	142
	Hydroaromatic methyl derivative	H$_2$/Pt/AcOH	142
ArCH=CHCHO	ArCH$_2$CH$_2$CHO	H$_2$/Pd, H$_2$/Ni	142,
		H$_2$/(Ph$_3$P)$_3$RhCl	143
	ArCH=CHCH$_2$OH	H$_2$/Os, H$_2$/PtO$_2$, LiAlH$_4$	143
		AlH$_3$	
		NaBH$_4$, NaBH(OMe)$_3$	
		Zn(BH$_4$)$_2$, NaBH$_3$CN	
		9-BBN, Bu$_4$NBH$_4$	
		BH$_3$·Me$_2$S	
		Bu$_3$SnH, Ph$_2$SnH$_2$	
		i-PrOH/Al(O-i-Pr)$_3$	
	ArCH=CHCH$_3$	NaBH$_4$/Pd(C)[a]	144
	ArCH$_2$CH=CH$_2$	1. TosNHNH$_2$, 2.	144
		NaBH(OAc)$_3$, BH(OB$_2$)$_2$	
	ArCH$_2$CH$_2$CH$_2$OH	H$_2$/Raney Ni	143
		Electroreduction	
		LiAlH$_4$	
	ArCH$_2$CH$_2$CH$_3$	ZnHg/HCl	144
ArC≡CCHO	RCH=CHCHO	Et$_3$N·HCO$_2$H/Pd(C)	143

[a]This reduction is very rare.

Table 14. Reduction of Derivatives of Aldehydes

Starting compound	Product	Reducing agent	Page
RCHCl～～CHO	RCHCl～～CH$_2$OH[a]	H$_2$/catalysts, Na, Zn i-PrOH/Al(O-i-Pr)$_3$	144
RCCl=CHCHO	RCH$_2$CH$_2$CHO	H$_2$/Pd(C)	144
O$_2$N—C$_6$H$_4$—CHO	O$_2$N—C$_6$H$_4$—CH$_2$OH	AlH$_3$, Bu$_3$SnH Na$_2$S•9H$_2$O i-PrOH/Al(O-i-Pr)$_3$	144, 145
	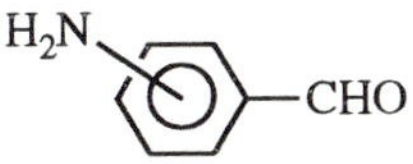 H$_2$N—C$_6$H$_4$—CHO	SnCl$_2$, TiCl$_3$, FeSO$_4$ Na$_2$S	144, 145
	H$_2$N—C$_6$H$_4$—CH$_2$OH	H$_2$/catalysts	144
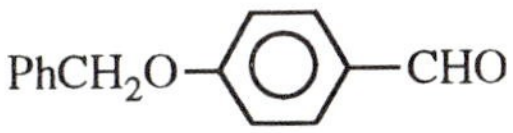 PhCH$_2$O—C$_6$H$_4$—CHO	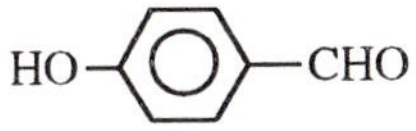 HO—C$_6$H$_4$—CHO	H$_2$/Pd	144
RCH(OR')$_2$	RCH$_2$OR'	AlH$_3$	144–146
RCH(OCOR')$_2$	RCH$_2$OCOR'	Fe/AcOH	146
RCH(O—CH$_2$CH$_2$—S)	RCH$_2$OCH$_2$CH$_2$SH	AlH$_3$, Ca/NH$_3$	145, 146
RCH(S—CH$_2$CH$_2$—S)	RCH$_2$SCH$_2$CH$_2$SH RCH$_3$	Ca/NH$_3$ Raney Ni, Bu$_3$SnH Na/NH$_3$[b], Ca/NH$_3$[b]	145 145–147
RCH=NR'	RCH$_2$NHR'	H$_2$/Pt, H$_2$/Ni, LiAlH$_4$ NaAlH$_4$, LiBH$_4$, NaBH$_4$ NaHg, K/C	146, 147
ArCH=NR'	ArCH$_3$	H$_2$/Pd	147
RCH=NOH	RCH$_2$NH$_2$	H$_2$/Raney Ni, LiAlH$_4$ Na/ROH	147
RCH=NNHSO$_2$Ar	RCH$_3$	LiAlH$_4$, NaBH$_4$ BH$_3$•THF/BzOH NaBH$_3$CN	148

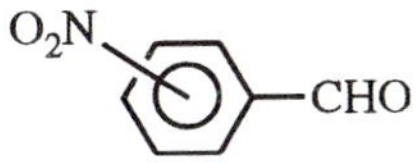

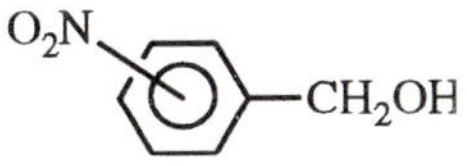

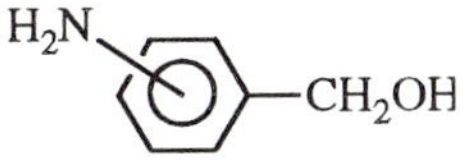

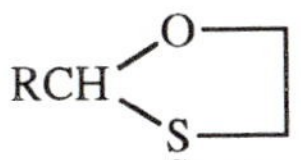

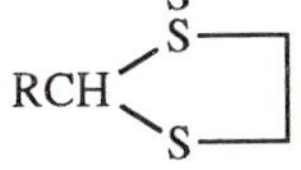

[a]Halogen may or may not survive.
[b]Reduction occurs only if R = Ar.

Table 15. Reduction of Ketones

Starting compound	Product	Reducing agent	Page
RCOR' (R = aliphatic or aromatic residue)	RCH(OH)R'[a]	H_2/PtO_2, $H_2/RhO_2 \cdot PtO_2$	149–151
		H_2/Ru, $H_2/CuCr_2O_4$[b]	157–163
		H_2/Ni, $H_2/Raney\ Ni$[b]	
		$HCO_2NH_4/Pd(C)$[b], or Raney Ni[b]	
		$Ni\text{-}Al/NaOH$[b], $LiAlH_4$[b], $Mg(AlH_4)_2$[b],	
		$LiAlH\ (i\text{-}Bu)_2Bu$,	
		$LiAlH(OMe)_3$[b],	
		$LiAlH(OEt)_3$[b]	
		$LiAlH(OCMe_3)_3$, $AlHCl_2$[b]	
		$LiBH_4$[b], $LiBHEt_3$[b], $NaBH_4$[b],	
		$NaAlH_2(OCH_2CH_2OMe)$[b],	
		BH_3, $BH_3 \cdot t\text{-}BuNH_2$	
		$BH(s\text{-}C_5H_{11})_2$, $BH(C_6H_{11}\text{-}c)_2$	
		$NaBH_3NMe_2$[b], $NaBH(OMe)_3$,	
		Bu_4NBH_4, $KBHPh_3$	
		Bu_4NBH_3CN[b], $PhMe_2SiH$[b],	
		$Zn(BH_3CN)_2$[b], $Zn(BH_4)_2$[b],	
		$PhMe_2SiH$[b], Bu_3SnH[b],	
		R_2SnH_2[b]	
		Electroreduction	
		$Na/EtOH$, Li, K, Ca/NH_3,	
		$Zn/NiCl_2$[b]	
		$Na_2S_2O_4$[b], $i\text{-}PrOH/Al(O\text{-}i\text{-}Pr)_3$	
		(CH_2O)[c], $EtMeCHCH_2OH/$	
		$Al(OCH_2CHMeEt)_3$[b,c]	
		Binal-H,[b,c]	
		Diphenyloxaza-borolidine $\cdot BH_3$[b,c]	
		Oxazaphospholidine-BH_3[b,c]	
		K-Glucoride[b,c]	
		Diisopinocampheylborane[b,c]	
		Biochemical reduction[b,c]	

Table 15. Continued

Starting compound	Product	Reducing agent	Page
RCOR' (R = aliphatic or aromatic residue)	RCH$_2$R'	H$_2$/Pd[d], H$_2$/PtO$_2$[d], H$_2$/Raney Ni[d], H$_2$/CuCr$_2$O$_4$[d], H$_2$/MoS$_2$[d] AlH$_3$, AlHCl$_2$[d] i-Bu$_2$AlH/AlBr$_3$[d] LiAlH$_4$/P$_2$I$_4$[d], NaBH$_4$[d] NaBH$_3$CN/ZnI$_2$[d], Et$_3$SiH Zn/HCl, N$_2$H$_4$/KOH NaBH$_4$/CF$_3$CO$_2$H[c], LiBHEt$_3$[c] 1. (CH$_2$SH)$_2$, TosH 2. Raney Ni 1. TosNHNH$_2$, 2. NaBH$_4$ or NaBH$_3$CN or BH$_3$/BzOH[d] HI/P[d] Electroreduction, Na/EtOH	151 155– 157 163, 164
	RC(OH)R' $\vert$ RC(OH)R'	Na, Mg, AlHg, (TiCl$_3$/Mg) i-PrOH/$h\nu$[d]	151, 155, 156, 164
	RCR' $\Vert$ RCR'	TiCl$_3$/LiAlH$_4$, TiCl$_3$/Li, K, Mg TiCl$_3$/Zn–Cu	151, 155, 156
RCH=CHCOR' (R = aliphatic or aromatic residue)	RCH$_2$CH$_2$COR'	H$_2$/PtO$_2$, H$_2$/Pt, H$_2$/Pd, H$_2$/Rh H$_2$/Ph$_3$PCuH, (Ph$_3$P)$_3$RhCl, Ni–Al/NaOH, Ni/Zn, AlH(OR)$_2$ AlH(NR$_2$)$_2$, NaAlH$_2$(OCH$_2$CH$_2$OMe)$_2$ Ph$_3$SnH, CuH, NaHFe$_2$(CO)$_8$ Electroreduction, Li/PrNH$_2$ C$_8$K, ZnHg/AcOH, Na$_2$S$_2$O$_4$ Biochemical reduction	165– 167
	RCH=CHCHR' $\vert$ OH	LiAlH$_4$, LiAlH(OMe)$_3$, AlH$_3$ AlH(i-Bu)$_2$, LiBH$_3$Bu, NaBH$_4$ NaBH$_3$CN, Zn(BH$_4$)$_2$, 9-BBN i-PrOH/Al(O-i-Pr)$_3$	166, 167
	RCH$_2$CH$_2$CHR' $\vert$ OH	H$_2$/Ni, H$_2$/CuCr$_2$O Ni–Al/NaOH, LiAlH$_4$/THF LiAlH(OCMe$_3$)$_3$, NaBH$_4$	167, 168

Table 15. Continued

Starting compound	Product	Reducing agent	Page
	RCH=CHCH$_2$R'	(H$_2$/Pd) 1. TosNHNH$_2$, 2. NaBH$_3$CN or BH$_3$·BF$_3$ or catechol borane	167, 168
RCH = CHCOR'	RCH$_2$CH$_2$CH$_2$R'	ZnHg/HCl, N$_2$H$_4$/KOH	168
	RCHCH$_2$COR' | RCHCH$_2$COR'	(NaHg)	168
RC≡CCOR'	RCH=CHCOR'	H$_2$/Pd[e], CrCl$_2$, CrSO$_4$	168
	RCH$_2$CH$_2$COR'	H$_2$/Pd[e]	168
	RC≡CCH(OH)R'	NaBH$_3$CN, Bu$_4$NBH$_3$CN LiBHEt$_3$, chiral boranes[c] o-Me$_2$NCH$_2$C$_6$H$_4$SnMe$_2$H$_2$	168, 169

[a]Palladium oxide is ineffective for reduction of aliphatic ketones but is the catalyst of choice for aromatic ketones; platinum oxide favors hydrogenolysis of benzylic hydroxyl, and both platinum and rhodium tend to hydrogenate the ring.

[b]References are also on pp 152–156.

[c]Where applicable, reduction yields chiral alcohols.

[d]Applies only to aromatic ketones.

[e]Applies to nonconjugated systems.

Table 16. Reduction of Substitution Derivatives of Ketones

Starting compound	Product	Reducing agent	Page
RCOCHR'X[a,b]	RCOCH$_2$R[c]	H$_2$/Pd, Me$_3$SiSiMe$_3$/Pd	169–
		Zn, ZnHg, CrCl$_2$, NaI/HCl	171
		VCl$_2$, Ph$_3$PHI, Na$_2$S$_2$O$_4$	
		HOCH$_2$OSONa, HBr, HI	
		(EtO)$_2$PHO	
	RCH(OH)CHR'X	LiAlH$_4$, NaBH$_4$	169,
		o-Me$_2$NCH$_2$C$_6$H$_4$SnMe$_2$H	170
		Biochemical reduction	
		i-PrOH/Al(O-i-Pr)$_3$	
	RCH(OH)CH$_2$R'	Raney alloys, NaOH	170,
	RCH=CHR'	i-PrOH/Al(O-i-Pr)$_3$[d]	171
		N$_2$H$_4$/AcOK	170,
			171
RCOCHR'NO$_2$[b]	RCOCH$_2$R´	Et$_3$SiH/Na$_2$S$_2$O$_4$	172
	RCOCHR'NH$_2$	Zn/NiCl$_2$, Fe[e], SnCl$_2$[e]	172
	RCHCHR'NO$_2$	LiAlH$_4$, NaBH$_4$, Ca(BH$_4$)$_2$	172
	|	i-PrOH/Al(O-i-Pr)$_3$	
	OH	Baker's yeast, glucose[f]	
	RCH$_2$CHR'NH$_2$	H$_2$/Pd[g]	172,
			173
RCOCR'N$_2$	RCOCH$_2$R'	H$_2$/PdO + CuO or HCl	173
		HI, EtOH[h], H$_3$PO$_2$[h]	
	RCOCHR'NH$_2$	H$_2$/PdO	173
	RCHCHR'NH$_2$	LiAlH$_4$	173
	|		
	OH		
	RCOCR=NNH$_2$	H$_2$/PdO	173
RCOCHR'N$_3$	RCHCHR'NH$_2$	LiAlH$_4$	173
	|		
	OH		

[a]X = Cl, Br, or I.

[b]The substituent is usually, but not necessarily, in a position α to the carbonyl group.

[c]Reduction takes place only with very reactive halogen.

[d]Reduction is not general.

[e]Reduction applies only to the nitro group in an aromatic ring.

[f]Nitro group is reduced to azo group.

[g]Reduction applies only to the keto group adjacent to and nitro group in the aromatic ring.

[h]Reduction applies to diazonium group in aromatic ring.

Table 17. Reduction of Hydroxy Ketones, Diketones, and Quinones

Starting compound	Product	Reducing agent	Page
RCOCH(OH)R'[a]	RCHCH(OH)R' | OH	H_2/$CuCr_2O_4$ Biochemical reduction	174
	RCOCH$_2$R'	Me$_3$SiI, Zn, ZnHg, Sn, Ph$_2$PLi HI	174
	RCH$_2$CH$_2$R'	ZnHg/HCl	174, 175
RCO~~~CH$_2$OH	RCO~~~CH$_2$OH	LiAlH(OR)$_3$, Bu$_4$NBH(OAc)$_3$ BH$_3$·NH$_3$, BH$_3$-t-BuNH$_2$ Bu$_3$SnH, i-Pr$_2$CHOH/Al$_2$O$_3$ TiCl$_3$/Zn, KBHPh$_3$	176
RCOCOR'[a]	RCOCH(OH)R'	H$_2$/Ru[b], H$_2$/Ru-Binap Zn, TiCl$_3$, VCl$_2$,H$_2$S P(OEt)$_3$, Ph$_2$C(OH)C(OH)Ph$_2$ Biochemical reduction	177, 180
	RCHCH(OH)R' | OH	H$_2$/Ru/Binap	179, 180
	RCOCH$_2$R'	ZnHg/HCl[c], HI, H$_2$S N$_2$H$_4$/KOH	177, 178
	RCH$_2$CH$_2$R'	ZnHg/HCl[c], 2N$_2$H$_4$/KOH	177, 178
	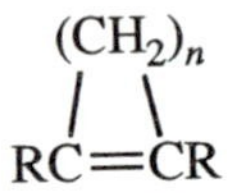 (CH$_2$)$_n$ / \\ RC=CR	TiCl$_3$/ZnCu[d]	179
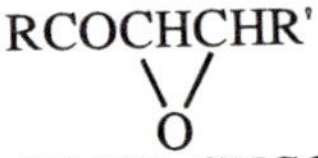 RCOCHCHR' \\ / O	RCOCH=CHR'	Zn/AcOH, CrCl$_2$	175
COCH=CHCO | | R R'	COCH$_2$CH$_2$CO | | R R'	Sn, CrCl$_2$	180
 O=⟨O⟩=O	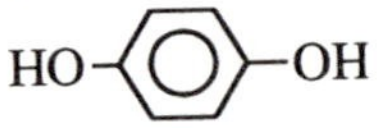 HO-⟨O⟩-OH	H$_2$/Pt[e], H$_2$/Pd, LiAlH$_4$, SnCl$_2$ TiCl$_3$, VCl$_2$, CrCl$_2$, H$_2$S, SO$_2$	180, 181
	 O=⟨⟩ H[f] / H	Sn	181
	 HO ⟨⟩ H / H OH	LiAlH$_4$, LiBHEt$_3$, NaBH$_4$ 9-BBN	181
	 ⟨O⟩	Al, Zn/ZnCl$_2$, SnCl$_2$, HI	181

[a]Both functions are not necessarily vicinal.
[b]Reduction applies only to γ-diketone.
[c]Clemmensen reduction gives rearranged ketones and hydrocarbons.
[d]Reduction occurs only if $n \geq 2$.
[e]In mineral acid cyclohexanol is formed.
[f]Reduction applies to anthrahydroquinone.

Table 18. Reduction of Ketals, Thioketals, Ketimines, Ketoximes, and Hydrazones

Starting compound	Product	Reducing agent	Page
$RR'C(OR'')_2^a$	$RR'CHOR''$	H_2/Pd, H_2/Rh, H_2/Ru AlH_3, i-Bu_2Al, BH_2Cl	181, 182
$RR'C(OR'')(SR'')^a$	$RCOR'$	Raney Ni	182
	$RR'CHOR''$	Ca/NH_3	182
	$RR'CHSR''$	AlH_3	182
$RR'C(SR'')_2^a$	$RR'CHSR''$	Ni, BH_3, PhSH/NaH	183
	$RR'CH_2$	Ni, Raney Ni Bu_3SnH, N_2H_4	183
$RR'C=NH$	$RR'CHNH_2^b$	H_2/catalysts, $NaBH_4$, KBH_4 $NaAlH_2(OCH_2CH_2OMe)_2$ i-PrOH (Raney Ni)	184
$RR'C=NOH$	$RR'C$ H NHOH	BH_3, $NaBH_4$, $NaBH_3CN$	184
	$RR'CHNH_2$	H_2/Pd, Pt, Rh; H_2/Raney Ni N_2H_4/Raney Ni, $LiAlH_4$, NaHg, Na/ROH, SmI_2, $SnCl_2$	185, 186
$RCOCR'=NOH$	$RCOCHR'NH_2$	H_2/PtO_2	186
$RC=NOH$ ⎸ NO_2	$RC=NOH$ ⎸ NO_2	$(NH_4)_2S$	186
$RR'C=NNHR''$	$RR'CHNHNHR''$	H_2/Pt, electroreduction	186
	$RR'CHNH_2$	H_2/PtO_2, AlHg	187
	$RR'CO$	$TiCl_3$	187
	$RR'CH_2$	KOH, heat; $LiAlH_4^c$, $NaBH_4^c$, $NaBH_3CN^c$	187 151, 156, 164

[a]Included are cyclic ketals and thioketals with five- or six-membered rings:

[b]See also p 184 and pp 187–189 (**Reductive Alkylation**).
[c]Reduction applies to tosylhydrazones only.

Table 19. Reduction of Carboxylic Acids

Starting compound	Product	Reducing agent	Page
RCO$_2$H	RCHO	AlH(NR$_2$)$_2$ Me$_2$CHCMe$_2$BHX(X = Cl, Br) Electroreduction Li/MeNH$_2$, NaHg	191, 192
	RCH$_2$OH	H$_2$/RuO$_2$, H$_2$Ru(C), Re$_2$O$_7$ H$_2$/Re$_2$S$_7$, H$_2$/CuCr$_2$O$_4$ LiAlH$_4$, NaBH$_4$/BF$_3$ NaAlH$_2$(OCH$_2$CH$_2$OMe)$_2$ Zn(BH$_4$)$_2$/CF$_3$CO$_2$H	192
RCH=CHCO$_2$H	RCH$_2$CH$_2$CO$_2$H	H$_2$/catalysts, H$_2$/(Ph$_3$)$_3$RhCl NaHg, CrSO$_4$	192, 193
	RCH=CHCH$_2$OH	LiAlH$_4$, BH$_3$	193
RC≡CCO$_2$H	RCH=CHCO$_2$H	H$_2$/Ni (*cis*) Na (*trans*) CrSO$_4$ (*trans*)	193, 194
	RCH=CHCH$_2$OH	LiAlH$_4$ (*trans*)	193, 194
ArCO$_2$H	RCO$_2$H[a]	H$_2$/Rh[b], Li/NH$_3$/EtOH[c] Na/NH$_3$/EtOH[c], NaHg[d] Na/C$_5$H$_{11}$OH	195, 196
	ArCHO	AlH(NR$_2$)$_2$ Me$_2$CHCMe$_2$BHX (X = Cl, Br) NaHg + RNH$_2$	194
	ArCH$_2$OH	LiAlH$_4$, NaAlH$_4$ NaAlH$_2$(OCH$_2$CH$_2$OMe)$_2$ NaBH$_4$•BF$_3$, NaBH$_4$/I$_2$ Electroreduction	194, 195
	ArCH$_3$	SiHCl$_3$/Pr$_3$N NaAlH$_2$(OCH$_2$CH$_2$OMe)$_2$[e]	195

Table 19. Continued

Starting compound	Product	Reducing agent	Page
$ArCH=CHCO_2H$	$ArCH_2CH_2CO_2H$	H_2/Pd, $H_2/Raney\ Ni$, $H_2/CuCr_2O_4$ $H_2/(Ph_3P)_3RhCl$ NiAl/NaOH, NaHg, $CrSO_4$ Electroreduction	196
	$ArCH=CHCH_2OH$	$NaBH_4/I_2$	
	$ArCH_2CH_2CH_2OH$	$LiAlH_4$	197
$ArC\equiv CCO_2H$	$ArCH=CHCO_2H$	H_2/Pt (*cis*), $CrSO_4$ (*trans*)	197
	$ArCH_2CH_2CO_2H$	H_2/Pt	197

[a]R = hydrogenated aromatic ring.
[b]Total hydrogenation of the ring.
[c]Partial or total hydrogenation of the ring results.
[d]Applied only in anthracene series.
[e]Occurs only with OH or NH_2 in *ortho* or *para* positions.

Table 20. Reduction of Derivatives of Carboxylic Acids

Starting compound	Product	Reducing agent	Page
$RCHXCO_2H^{a,b}$	$RCHXCH_2OH$	BH_3, AlH_3, $LiAlH_4$ (inverse technique)	197
$XC_6H_4CO_2H^b$	$C_6H_5CO_2H$	H_2/Raney Ni, NaOH; H_2/Pd[a]	197
	$XC_6H_4CH_2OH$	AlH_3, BH_3	197
	$XC_6H_4CH_3$	$SiHCl_3$/Pr_3N	197
$RCX=CHCO_2H^{a,b}$	$RCH=CHCO_2H$	H_2/Pd($BaSO_4$)[c], ZnNaHg	197, 198
	$RCHXCH_2CO_2H$	H_2/Ru-Binap	
	$RCH_2CH_2CO_2H$	H_2/Pd(C)[c,d], Zn	197, 198
		Biochemical reduction	
$O_2NC_6H_4CO_2H$	$H_2NC_6H_4CO_2H$	$(NH_4)_2S$	198, 199
	$O_2NC_6H_4CH_2OH$	BH_3	
$\overset{-\;+}{X}N_2C_6H_4\,CO_2H$	$ArCO_2H$	$NaBH_4$, EtOH/hv	198
$RSSC_6H_4CO_2H$	$HSC_6H_4CO_2H$	Zn/AcOH	199
$RCHCHCO_2H$ $\diagdown\!\diagup$ O	$RCHCH_2CO_2H$ $|$ OH	$NaBH_4$	199
	RCH_2CHCO_2H $|$ OH	$NaBH_4$	199
$RCH(NH_2)CO_2H$	$RCH(NH_2)CH_2OH$	$NaBH_4$/I_2	199
$RCOCO_2H^e$	$RCOCH_2OH$	$BH_3\cdot THF$	199
	$RCH(OH)CO_2H$	H_2/Raney Ni, H_2/Ru-Binap	199, 200
		$SiHEt_3$/CF_3CO_2H	
		$NaBH_4$, NaHg, Zn	
		L-, K-Selectride	200, 201
		Biochemical reduction	
		$LiAlH(OCEt_3)_3$	
	RCH_2CO_2H	$SiHEt_3$/CF_3CO_2H; ZnHg N_2H_4, HI/P	201

[a]The substituent is not necessarily in the α-position or in the β-position.
[b]X = Cl, Br, or seldom F.
[c]Reduction is not general.
[d]Reduction applies to X = F.
[e]The keto group is not necessarily in the α-position.

Table 21. Reduction of Acyl Chlorides and Acid Anhydrides

Starting compound	Product	Reducing agent	Page
RCOCl	RCHO	$H_2/Pd(BaSO_4)+S$, Et_3SiH/Pd $H_2/PtO_2(S)$, H_2/Ni $LiAlH(OCMe_3)_3$, $NaBH_4$ CuH, $(EtO)_3P$, $NaBH_4$	203, 204
	RCH_2OH	$LiAlH_4$, $NaAlH_4$ $NaAlH_2(OCH_2CH_2OMe)_2$ AlH_3, $NaBH_4$, $NaBH(OMe)_3$ Bu_4NBH_4, $Zn(BH_3CN)_2$ t-BuMgCl	204, 205
RCOOCOR'	$RCH_2OH+R'CO_2H$	$NaBH_4$ $H_2/CuCr_2O_4$, $H_2/(Ph_3)_3RuCl_2$ $LiAlH_4/-55\ °C$, $NaBH_4$, $LiBH_4$ $LiBHEt_3$, $LiBH(CHMeEt)_3$, $KBH(CHMeEt)_3$ $NaHg$, Zn	205 205
	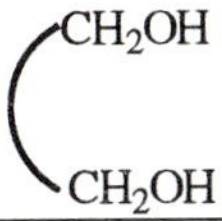	$LiAlH_4$ $NaAlH_2(OCH_2CH_2OMe)_2$	206

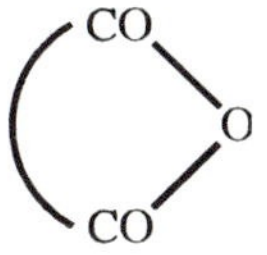

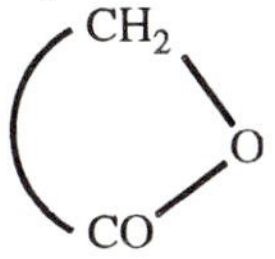

Table 22. Reduction of Esters of Carboxylic Acids

Starting compound	Product	Reducing agent	Page
RCO_2R'	RCHO	$(LiAlH_4)$, $KAlH_4$	207–
		$LiAlH_2 (NEt_2)_2$	209
		$NaAlH_4$, $NaAlH_2(CHMeEt)_2$	
		$AlH(CHMeEt)_2$, $LiAlH(OCMe_3)_3$	
		$NaAlH_2(OCH_2CH_2OMe)_2$	
		$NaAlHEt_2NC_5H_{10}$, $NaBH_4$	
		$AlH(N{\diagup}{\diagdown}NMe)_2$, NaHg	
	RCH_2OR	$NaBH_4/BF_3 \cdot Et_2O$, $SiHCl_3$	209
	$RCO_2H + R'H$	H_2/Pd^a, Bu_3SnH, Zn^a, $CrCl_2$	210–
		SmI_2	212
	$RCH_2OH +$	$H_2/Raney\ Ni$, $H_2/CuCr_2O_4$	213–
	$R'OH$	$H_2/ZnCr_2O_4$, H_2/Re_2O_7, $LiAlH_4$,	215
		$LiAlH(OMe)_3$, $NaAlH_4$	
		$NaAlH_2(OCH_2CH_2OMe)_2$	
		$LiAlH_4/AlCl_3$, $MgHBr$	
		$Mg(AlH_4)_2$, $LiBH_4$, $NaBH_4$	
		$Ca(BH_4)_2$, $NaBH(OMe)_3$	
		$LiBHEt_3$, $NaBH_4/AlCl_3$	
		$BH_3 \cdot Me_2S$	
		$NaBH_3NEt_2$, $NaBH_3NHBu$-t	
	$RCH_3 + R'OH$	$H_2/CuCr_2O_4{}^b$	214
	RCO	Na/ROH	212
	$\mid$	Na/xylene	
	RCHOH		
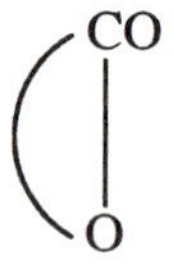 $\underset{O}{\overset{CO}{\big(}}$	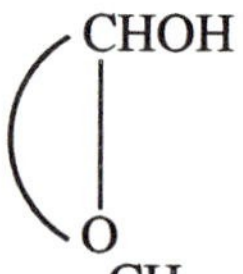 $\underset{O}{\overset{CHOH}{\big(}}$	$AlH(i\text{-}Bu)_2$, NaHg, $NaBH_4$	209
	$\underset{O}{\overset{CH_2}{\big(}}$	$LiAlH_4$, AlH_3	209,2
		$AlH(i\text{-}Bu)_2$	10
		$Et_3SiH/BF_3 \cdot Et_2O$	

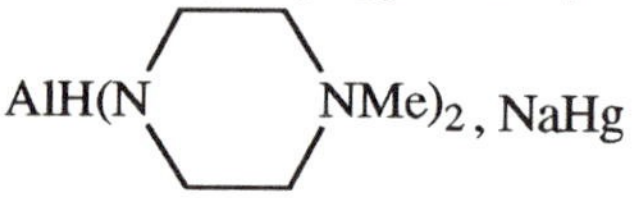

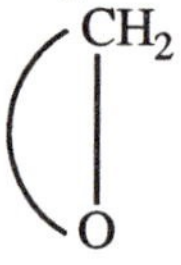

^{a}Reduction applies to benzyl-type only.
^{b}Reduction applies to some aromatic esters.

Table 23. Reduction of Esters of Unsaturated Carboxylic Acids

Starting compound	Product	Reducing agent	Page
$RCH=CHCO_2R'$[a]	$RCH_2CH_2CO_2R'$	H_2/Pd, $H_2/Raney$ Ni[b], $H_2/CuCr_2O_4$, $H_2/(Ph_3P)_3RhCl$ Stereoselective reduction $(NaBH_4)$, Bu_3SnH N_2H_2, SmI_2, $CrSO_4$	217–220 218 219
	$RCH=CHCH_2OH$	$H_2/ZnCr_2O_4$, $LiAlH_4$, $(NaBH_4)$ $NaAlH_2(OCH_2CH_2OMe)_2$ $AlH(t\text{-}Bu)_2$	219
	$RCH_2CH_2CH_2OH$	$H_2/Raney$ Ni, $H_2/CuCr_2O_4$ $LiAlH_4$ $NaAlH_2(OCH_2CH_2OMe)_2$ $NaBH_4$[c], $KBH_4/LiCl$, Na	219
$RC≡CCO_2R'$	$RCH=CHCO_2R'$	H_2/Pd (deact.), Ph_3P	220
	$RC≡CCH_2OH$	$LiAlH_4$	220
	$RCH=CHCH_2OH$	$LiAlH_4$ (excess)	220

[a]R = alkyl or aryl.
[b]Hydrogenation of the aromatic ring occurs at high temperatures.
[c]Reduction occurs only if R = pyridyl or pyrimidyl.

Table 24. Reduction of Substitution Derivatives of Esters

Starting compound	Product	Reducing agent	Page
$RCHXCO_2R'^{a,b}$	RCH_2CO_2R	H_2/Pd, H_2/Raney Ni, Zn NaHg, i-PrOH/$h\nu$	220, 221
	$RCHXCH_2OH$	$(LiAlH_4)$; AlH_3, $LiBH_4$ $Ca(BH_4)_2$	220, 221
$RCH(NO_2)CO_2R'^b$	$RCH(NH_2)CO_2R'$	H_2/Pd, Fe	221, 222
	$RCH(NO_2)CH_2OH$	$LiAlH_4$, AlH_3, BH_3	222
O_2NArCO_2R'	H_2NArCO_2R'	H_2/Pd, Fe	
RCN_2CO_2R'	$RC(:NNH_2)CO_2R'$	K_2SO_3	222
	$RCH(NH)CO_2R'$ | NH_2	NaHg	222
	$RCH(NH_2)CO_2R'$	H_2/PtO_2	222
$RCH(N_3)CO_2R'^b$	$RCH(NH_2)CO_2R$	H_2/Pd, AlHg	222, 223

[a] X = F, Cl, or Br.

[b] The substituent is not necessarily in the α-position and may be in an aromatic ring.

Table 25. Reduction of Functionalized Esters

Starting compound	Product	Reducing agent	Page
ArCH(OH)CO$_2$R	ArCH(OH)CH$_2$OH	LiAlH$_4$, LiBH$_4$, Ca(BH$_4$)$_2$	223
	ArCH$_2$CH$_2$OH	H$_2$/Ni, H$_2$/CuCr$_2$O$_4$	223
RCH(NH$_2$)CO$_2$R'	RCH(NH$_2$)CH$_2$OH	H$_2$/Raney Ni, LiAlH$_4$	223
RCOCO$_2$R'	RCH(OH)CO$_2$R'	H$_2$/PtO$_2$, Raney Ni	224,
		H$_2$/Ru-Binap, NaBH$_4$	225
		Bu$_4$NBH$_4$, K-Glucoride	
		NaHg, AlHg, PhSiMe$_2$H	
		i-PrOH/Al(O-i-Pr)$_3$	
		LiAlH(t-Bu)$_3$	
		LiBH(i-Bu)$_3$	
		Biochemical reduction	
	RCH$_2$CO$_2$R'	2R"SH; Raney Ni	225,
		desulfurization, NaBH$_3$CN	226
	RCOCH$_2$OH	1. LiN(i-Pr)$_2$, 2. LiAlH$_4$	226,
		1. HC(OR)$_3$, 2. LiAlH$_4$	227
RC(:NOH)CO$_2$R'	RCH(NH$_2$)CO$_2$R'	H$_2$/Pd, AlHg, Zn	227
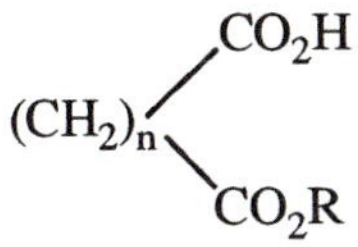 (CH$_2$)$_n$ ⟨CO$_2$H / CO$_2$R	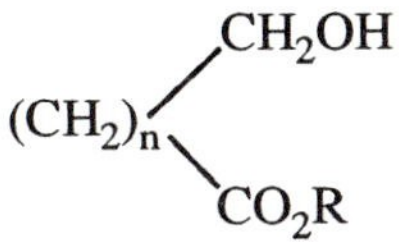 (CH$_2$)$_n$ ⟨CH$_2$OH / CO$_2$R	AlH$_3$, BH$_3$, NaBH$_4$/I$_2$	227
	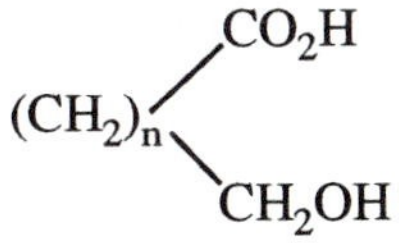 (CH$_2$)$_n$ ⟨CO$_2$H / CH$_2$OH	Na/ROH	227
RC(OR')$_3$	RCH(OR')$_2$	LiAlH$_4$	227
RCOSR'	RCH$_2$SR'	AlH$_3$	227, 228
	RCHO	Raney Ni, Et$_3$SiH/Pd	228
	RCH$_2$OH	NaBH$_4$	228
RC(S)SR'	RCH$_3$	Raney Ni	228

Table 26. Reduction of Amides, Lactams, and Imides

Starting compound	Product	Reducing agent	Page
RCONR'R"	RCHO[a]	LiAlH$_4$, LiBH(OEt)$_3$	229,
		LiAlH(OCMe$_3$)$_3$	230
		NaAlH$_2$(OCH$_2$CH$_2$OMe)$_2$	
		Electroreduction, Na/NH$_3$	
	RCH$_2$OH	(H$_2$/PtO$_2$), LiAlH$_4$, LiBHEt$_3$	230,
		NaAlH$_2$(OCH$_2$CH$_2$OMe)$_2$	231
		NaBH$_4$, NaBH$_3$NMe$_2$	
		Na/NH$_3$	
	RCH$_2$NR'R"	H$_2$/CuCr$_2$O$_4$, LiAlH$_4$	231–
		LiAlH(OMe)$_3$, Mg(AlH$_4$)$_2$	233
		NaAlH$_2$(OCH$_2$CH$_2$OMe)$_2$	
		AlH$_3$, BH$_3$, NaBH$_3$NMe$_2$	
		Electroreduction, Na, NaHg	233
		NaBH$_3$NHC(CH$_3$)$_3$	
	RCH$_3$[b]	H$_2$/Pt	

Lactam reductions:

Starting compound	Product	Reducing agent	Page
(ring) CO / NH	(ring) CH$_2$ / NH	H$_2$/PtO$_2$, electroreduction	233
		LiAlH$_4$, NaBH$_4$, BH$_3$·Me$_2$S	
		NaAlH$_2$(OCH$_2$CH$_2$OMe)$_2$	
(ring) CO / NH \ CO	(ring) CH$_2$ / NH \ CO	H$_2$/Pd, Ni, CuCr$_2$O$_4$	233,
		LiAlH$_4$	234
		NaAlH$_2$(OCH$_2$CH$_2$OMe)$_2$	
		BH$_3$, electroreduction	
	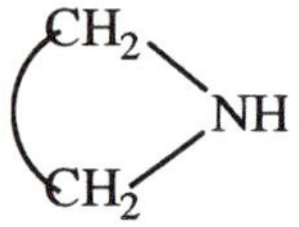 (ring) CH$_2$ / NH \ CH$_2$	LiAlH$_4$, electroreduction	234
RCH=CNHCOR' | CO$_2$H	RCH$_2$CHNHCOR' | CO$_2$H	H$_2$/PtO$_2$	234
		H$_2$/homogeneous chiral catalysts	218
RCH=CHCNHR'[c] ‖ O	RCH$_2$CH$_2$CONHR'	SmI$_2$	235
	RCH$_2$CH$_2$CH$_2$NHR'	LiAlH$_4$, LiAlH(OR)$_3$ AlH$_3$	235
O$_2$N—CONR$_2$	H$_2$N—CH$_2$NR$_2$	AlH$_3$	235
RCOCONHR'	RCH(OH)CONHR'	Zn(BH$_4$)$_2$, SiHPhMe$_2$	236
		Biochemical reduction	
	RCH(OH)CH$_2$NHR'	LiAlH$_4$	236
RO$_2$C—CONH$_2$	RO$_2$C—CH$_2$NH$_2$	BH$_3$	236

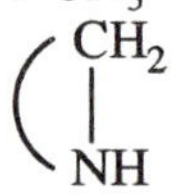

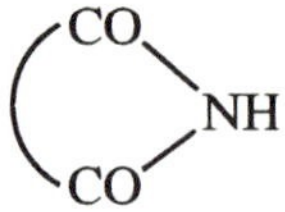

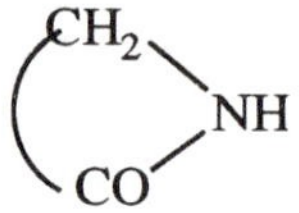

[a]Reduction applies only with R' = alkyl or aryl, and R" = aryl.
[b]Hydrogenolysis occurs only if R' = R" = H.
[c]Reduction applies to α,β-unsaturated lactams.

Table 27. Reduction of Amidines, Thioamides, Imidoyl Chlorides, and Hydrazides

Starting compound	Product	Reducing agent	Page
RC(:NH)NH$_2$	RCHO	Na/NH$_3$	237
RCSNH$_2$	RCHO	Raney Ni, Fe	237
	RCH$_2$NH$_2$	Raney Ni (excess), LiAlH$_4$	237
		NaBH$_4$, NaBH$_3$CN	
		Electroreduction	
RCCl=NR'	RCH=NR'	H$_2$/(Ph$_3$P)$_2$PdCl$_2$	237,
		LiAlH(OCMe$_3$)$_3$	238
		Mg, SnCl$_2$, CrCl$_2$	
	RCH$_2$NHR'	H$_2$/Pd(BaSO$_4$)	237,
			238
RCONHNH$_2$	RCH$_2$NH$_2$	LiAlH$_4$	237
RCONHNHSO$_2$Ar	RCHO	Na$_2$CO$_3$	238
	RCH$_3$	LiAlH$_4$	238

Table 28. Reduction of Nitriles

Starting compound	Product	Reducing agent	Page
RCN	RCHO	Raney Ni/NaH$_2$PO$_2$ Ni–Al/HCO$_2$H NaAlH(OEt)$_3$ LiAlH(OR)$_3$, AlH(i-Bu)$_2$ SnCl$_2$	239
	RCH$_2$OH	Zn/NiCl$_2$	239
	RCH$_3$	H$_2$/Pd, HCO$_2$NH$_4$/Pd	239, 240
	RCH$_2$NH$_2$	H$_2$/Rh, H$_2$/Pt, H$_2$/Raney Ni H$_2$/NiB$_2$, LiAlH$_4$ LiAlH(OMe)$_3$, AlH$_3$, BH$_3$ Na/ROH, (Zn/NiCl$_2$)	240, 241
RCH=CHCN	RCH$_2$CH$_2$CN	H$_2$/PtO$_2$, H$_2$/Pd Et$_3$N·HCO$_2$H NaAlH$_2$(OCH$_2$CH$_2$OMe)$_2$ NaBH$_4$, CuH, Mg/ROH, Zn	242
	RCH=CHCH$_2$NH$_2$	[a]H$_2$/Raney Co, (LiAlH$_4$)	241
	RCH$_2$CH$_2$CH$_2$NH$_2$	LiAlH$_4$	242
RC≡C—CN	RCH=CH—CN	H$_2$/Raney Ni	242
O$_2$N〜CN	H$_2$N〜CN	Fe, SnCl$_2$	242
RCOCN	RCOCH$_2$NH$_2$	SnCl$_2$	242
	RCH(OH)CH$_2$NH$_2$	LiAlH$_4$	242
RC(=NOH)CN	RCH(NH$_2$)CN	H$_2$/Pt(C), AlHg	242

[a]The nitrile contained β,γ-double bond.

PROCEDURES

Because some of the procedures involve work with highly flammable and even pyrophoric compounds, reading of the safety regulations on p xxiii prior to carrying out the experiment is in order.

1. Catalytic Hydrogenation Under Atmospheric Pressure

Procedure

The apparatus for this procedure is shown in Figure 6. A 25–250-mL ground-glass flask, A, fitted with a magnetic stirring bar is charged with the catalyst, the compound to be hydrogenated, and the solvent (if necessary.) The amount of catalyst is relative to the amount of the hydrogenated compound, and the reaction conditions are summarized in Table 1 (p 8), which gives general guidelines for hydrogenations of most common types of compounds. The solvent is chosen on the basis of the solubility of the compound to be hydrogenated. The ground-glass joint of the hydrogenating flask is lightly greased or sealed with a Teflon sleeve, and the flask is attached to the hydrogenation apparatus by means of steel springs or rubber bands. The three-way stopcock, B, which connects the hydrogenating flask to a 50–500-mL graduated tube, C, filled completely with a liquid from the reservoir, D, is set so as to cut off the graduated tube and to connect the hydrogenating flask to a glass tube fitted with another three-way stopcock, E, and a manometer, F. Stopcock E is opened to outlet G connected to an aspirator or a vacuum pump, and the apparatus is evacuated. If a low-boiling solvent is used, flask A must be cooled to prevent effervescence under reduced pressure. When the pressure in the apparatus has dropped to 5–20 mm, stopcock E is opened to outlet H, through which hydrogen is cautiously introduced into the apparatus using a hydrogen regulator on a hydrogen tank.

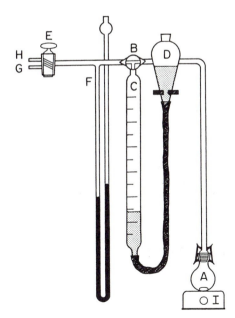

Figure 6. Apparatus for hydrogenation at atmospheric pressure. The text explains the alphabet letters.

When the pressure in the manometer has risen to about 1 atm, stopcock E is switched to the vacuum pump and the apparatus is evacuated again. Such flushing of the apparatus with hydrogen is necessary to displace air and may be repeated once more.

Then stopcock E is set to connect the apparatus to the source of hydrogen again, hydrogen is introduced until its pressure reaches 1 atm or slightly more, and stopcock B is opened to allow the hydrogen to enter graduated tube C. Hydrogen displaces the liquid into reservoir D. By lowering reservoir D, the height of the liquid is adjusted so that it is level in both graduated tube C and reservoir D. At this moment stopcock E is turned to cut off the hydrogenating assembly A–D from the source of hydrogen, the volume of hydrogen at exactly atmospheric pressure is read, and reservoir D is placed in its original position as shown in Figure 6.

Magnetic stirrer I is started, and the hydrogenation begins. Its progress is followed by the decrease in the volume of hydrogen enclosed in tube C. When the consumption of hydrogen stops, the final reading is taken after equalizing the liquid levels in both graduated tube C and reservoir D. If the hydrogenation has not been finished after the volume of hydrogen has been used up, additional hydrogen is supplied as described for the initial filling.

After the hydrogenation has come to an end, the exact final volume of hydrogen used for the hydrogenation is calculated from the volume read at atmospheric pressure and adjusted to normal conditions, that is, 760 mm and 25 °C (298.15 K).

The excess hydrogen (if any) is bled through valve E, the hydrogenating flask is disconnected, the catalyst is removed by centrifugation or filtration under the necessary precautions (see p xxiii), and the filtrate is worked up depending on the nature of the products.

Notes

All ground-glass joints and stopcocks must be gently greased and held in position by springs or rubber bands. The manometer need not be sealed on but may be connected to the apparatus by a rubber or plastic hose. Also, the hydrogenating flask may be attached to the apparatus by means of rubber or plastic tubing. This arrangement is preferable in hydrogenations involving Raney nickel, for which shaking of the flask is better than magnetic stirring because Raney nickel sticks to the magnetic stirring bar. In this case solvents that cause swelling of the plastic tubing must be avoided because they may extract organic material and contaminate the product of hydrogenation [67]. The manometric liquid used for an apparatus of any size is usually water. For precise work in a small apparatus (50–100-mL volume), mercury is preferred. For hydrogenations at slightly elevated pressures (1–4 atm), more complicated apparatuses have been designed and described in the literature.

2. Catalytic Hydrogenation with Hydrogen Generated from Sodium Borohydride [4]

Hydrogenation Ex Situ

The special apparatus (Figure 7) consists of two 500-mL Erlenmeyer flasks, A and B, having slightly rounded bottoms and fitted with Teflon-coated magnetic stirring bars and injection ports; a mercury-filled bubbler; a back-up prevention valve, C; and a mercury-filled automatic dispenser, D, with the delivery tube of a buret, E, inserted to be air-tight.

Flask B (reactor) connected to bubbler C is charged with 100 mL of ethanol, 5 mL of a 0.2 M solution of chloroplatinic acid in ethanol, and 5 g of activated charcoal. Flask B is attached to flask A (generator), and the whole apparatus is flushed with nitrogen. Buret E is filled with 150 mL of 1 M sodium borohydride solution in 1 M sodium hydroxide. A 1 M solution of sodium borohydride in ethanol (20 mL) is injected into reaction flask B with vigorous stirring to prepare the catalyst. Concentrated hydrochloric acid (5 mL) or glacial acetic acid (4 mL) is injected into generator A followed by 30 mL of 1 M sodium borohydride solution (to flush the apparatus with hydrogen). A solution of the compound to be hydrogenated (0.5 mol of double bond or its equivalent) is injected into reactor B. As soon as hydrogen starts reacting, the pressure in the apparatus decreases and the solution of sodium borohydride is sucked in continuously from buret E through automatic dispenser D until the hydrogenation ends.

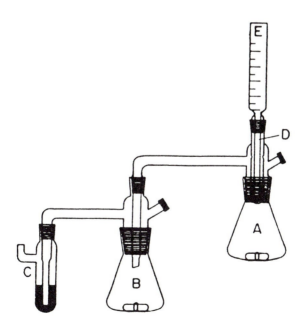

Figure 7. Apparatus for hydrogenation with hydrogen generated from sodium borohydride. The alphabet letters are explained in the text.

Hydrogenation In Situ

Hydrogenation of a compound may be carried out in the same vessel in which hydrogen is generated by decomposition of sodium borohydride. For this purpose the apparatus shown in Figure 7 is simplified by leaving out flask B.

Flask A connected to bubbler C is charged with 100 mL of ethanol, 5 mL of 0.2 M solution of chloroplatinic acid in ethanol, and 5 g of activated charcoal, and is flushed with nitrogen. Buret E is filled with 150 mL of a 1 M ethanolic solution of sodium borohydride. A 1 M ethanolic solution (20 mL) of sodium borohydride is injected with vigorous stirring to prepare the catalyst followed by 20 mL of concentrated hydrochloric acid (or 10 mL of glacial acetic acid). Finally a solution of 0.5 mol of a compound to be hydrogenated is added through the injection port. The stock solution of sodium borohydride is drained continuously from automatic dispenser D as soon as the absorption of hydrogen has started and ends automatically when the reaction has finished.

The advantage of hydrogenation with hydrogen generated from sodium borohydride is easy handling of relatively large amounts of compounds to be reduced in a relatively small apparatus (much larger equipment would be needed for regular catalytic hydrogenation at atmospheric pressure). The disadvantages include the more expensive source of hydrogen and the more complicated isolation of products, which cannot, as in regular catalytic hydrogenation, be obtained just by filtration and evaporation of the solvent but require extraction from the reaction mixture.

3. Catalytic Hydrogenation Under Elevated Pressure

A medium-pressure apparatus is assembled from a stainless steel bomb, standard parts, valves, unions, and copper or stainless steel tubing according to Figure 8. Stainless steel bomb A (30–500-mL, tested for 125 atm) is fitted with a magnetic stirring bar and charged with the catalyst, the compound to be hydrogenated, and the solvent (if necessary). It is connected tightly to a tree carrying two valves and a pressure gauge by means of a union using Teflon tape around the threads. After closing valve B, the apparatus is connected at valve C by means of copper or stainless steel tubing to a regulator of a hydrogen tank. To test the apparatus for leakage, it is pressurized with hydrogen until the pressure gauge for 100 atm, D reads 50–100 atm. At this moment, valve C is closed. If the pressure of hydrogen drops immediately and continues dropping within a few minutes, some of the connections are leaking and the leak must be stopped. If tightening of the connections does not prevent the leakage, the apparatus must be disassembled and the connections cleaned or replaced. If the pressure does not show any decrease within 5 min, or if a small initial drop of pressure due to absorption of hydrogen by the catalyst does not continue, valve B is connected to an aspirator or a vacuum pump and slowly opened, and the apparatus is evacuated to remove the air originally present. This flushing with hydrogen may be repeated once more.

Thereafter valve B is closed, and through valve C hydrogen is introduced until pressure gauge D indicates the desired pressure. The pressure should be well below the maximum pressure for which the gauge and the bomb are

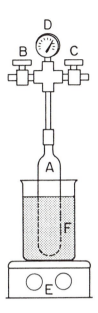

Figure 8. Apparatus for hydrogenation at medium pressure. Alphabet letters are explained in the text.

designed. This precaution is necessary because, during exothermic hydrogenations, the temperature of the reaction may raise the pressure considerably, especially if the reaction mixture occupies a large portion of the container.

After both valves B and C have been closed and the pressure reading has been recorded, magnetic stirring is started. Progress of the reaction is followed by the pressure drop. When the pressure no longer falls, the hydrogenation is finished or has stopped at an intermediate stage (for a multifunctional substrate). If hydrogenation is finished, the hydrogen pressure is released by opening valve B, the bomb is disconnected, and the contents are worked up.

If hydrogenation is unfinished, it may be tried at a higher temperature by applying an oil bath and by heating the reaction mixture with continued stirring. The hydrogen pressure rises because of the rise in temperature. After the temperature has stabilized, any pressure drop indicates that further hydrogenation is taking place. Stirring and heating may be continued until no more absorption of hydrogen takes place at the same temperature. After cooling, the remaining hydrogen is released through valve B and the apparatus is disassembled.

If, during the hydrogenation at room temperature or at an elevated temperature, the pressure of hydrogen has dropped to zero, more hydrogen may be needed for completing the reduction. In this case the apparatus is refilled with hydrogen by opening valve C connected to the source of hydrogen. After repressurizing the apparatus to the desired pressure, valve C is closed and the hydrogenation is continued.

After the hydrogenation has definitely finished, the final reading of the hydrogen pressure is recorded and the excess of hydrogen is carefully bled off by opening valve B. Only then is the bomb disconnected. The contents are filtered, the bomb is rinsed once or twice with the solvent, the washings are used to wash the catalyst on the filter, and the filtrate is worked up depending on the nature of the products.

Notes

As soon as the apparatus is pressurized for the hydrogenation and valve C is closed, the master valve on the hydrogen cylinder should be closed for safety reasons. While removing the catalyst, precautions must be taken during the filtration of pyrophoric catalysts (see p xxiii).

4. Preparation of Palladium Catalyst [*34*]

Powdered sodium borohydride (0.19 g, 5 mmol) is added portionwise at room temperature over 5–10 min to a stirred suspension of 0.443 g (2.5 mmol) of powdered palladium dichloride in 40 mL of absolute methanol. Stirring is continued until the evolution of hydrogen has ceased (about 20 min). The solvent is decanted from the black settled catalyst, which is washed twice or three times with a solvent to be used in hydrogenations.

5. Preparation of the Lindlar Catalyst [*39*]

To a suspension of 50 g of purest precipitated calcium carbonate in 400 mL of distilled water is added 50 mL of a 5% solution of palladous chloride ($PdCl_2$), and the mixture is stirred for 5 min at room temperature and for 10 min at 80 °C. The hot suspension is shaken with hydrogen until no more absorption takes place and then filtered with suction. The cake on the filter is washed thoroughly with distilled water. The cake is then stirred vigorously in 500 mL of distilled water; the suspension is treated with a solution of 5 g of lead acetate in 100 mL of water, stirred for 10 min at 20 °C, and finally stirred for 40 min in a boiling water bath. The catalyst is filtered with suction, washed thoroughly with distilled water, and dried at 40–50 °C in vacuo.

The Lindlar catalyst is commercially available.

6. Reduction with Raney Nickel

Hydrogenolysis of Halogens [697]

A halogen derivative (0.1 mol) dissolved in 30 mL of methanol is dissolved or dispersed in 100 mL of 1 N methanolic potassium hydroxide. After addition of a suspension of 19 g of methanol-wet and hydrogen-saturated Raney nickel (commercially available) in 70 mL of methanol, hydrogenation is carried out at room temperature and atmospheric pressure until the volume of hydrogen corresponding to the hydrogenolyzed halogen has been absorbed (25–500 min). The reaction mixture is filtered with suction, taking precautions necessary for handling pyrophoric catalysts (see p xxiii); the residue on the filter is washed several times with methanol, and the filtrate is worked up depending on the properties of the product.

With polyhalogen derivatives, the amount of potassium hydroxide must be equivalent to the number of halogens to be replaced by hydrogen.

7. Desulfurization with Raney Nickel

Preparation of Aldehydes from Thiol Esters [1404]

A solution of 10 g (0.06 mol) of ethyl thiolbenzoate in 200 mL of 70% ethanol is refluxed for 6 h with 50 g of Raney nickel. After the removal of the catalyst, the filtrate is distilled and the aldehyde is isolated from the distillate by treatment with a saturated (40%) solution of sodium bisulfite. The yield of the benzaldehyde–sodium bisulfite addition complex is 8.0 g (62%).

Deactivation of Raney Nickel [47]

For desulfurization of compounds containing reducible functions, the Raney nickel (20 g) is deactivated prior to the desulfurization by stirring and refluxing with 60 mL of acetone for 2 h. This removes the hydrogen that is adsorbed in Raney nickel.

8. Preparation of Nickel Catalyst [13]

To a stirred suspension of 1.24 g (5 mmol) of powdered nickel acetate in 50 mL of 95% ethanol in a 250-mL flask is added 5 mL of 1 M sodium borohydride in 95% ethanol at room temperature. Stirring is continued until the evolution of gas has ceased, usually within 30 min. The flask with the colloidal material is used directly in the hydrogenation.

9. Catalytic Transfer of Hydrogen

General [85]

In a flask fitted with an efficient stirrer and a nitrogen inlet, 0.001 mol of a compound to be hydrogenated dissolved in 4 mL of ethanol or, preferably, glacial acetic acid is treated with an equal weight of 10% palladium on charcoal and 0.94 mL (0.01 mol) of cyclohexadiene. The flask is kept at 25 °C, and a slow stream of nitrogen is passed through the mixture. After at least 2 h the mixture is filtered through infusorial earth (Celite), the solid is washed with the solvent, and the filtrate is evaporated in vacuo to give a 90–100% yield of the hydrogenated product.

Reduction of a Cyano Group to a Methyl Group [75]

1-Naphthonitrile (0.77 g, 5 mmol) is dissolved in 12 mL of methanol, the flask with the stirred solution is evacuated (behind a safety screen) to remove air, a slow stream of nitrogen is passed through the solution, and 0.77 g of 10% palladium on carbon is added, followed immediately by 1.26 g (20 mmol) of finely powdered ammonium formate. The reaction mixture is stirred at room temperature for 16 h. The mixture is brought to a boil under vacuum, the catalyst is filtered off, the filtrate is evaporated, and the residue is shaken with 20 mL of ether and 10 mL of water. The ether layer is separated, washed with 20 mL of a saturated solution of sodium chloride, dried with anhydrous magnesium sulfate, and evaporated. Distillation of the residue gives 0.53 g (75%) of 1-methylnaphthalene, bp 239–241 °C at 760 mmHg.

Deoxygenation of Phenols via Trifluoroacetates [814]

To a mixture of 0.276 g (1 mmol) of 2-naphthyl trifluoroacetate, 0.42 mL (3 mmol) of triethylamine, 4 mg (0.02 mmol) of palladium acetate, and 10 mg (0.04 mmol) of triphenylphosphine in 2 mL of *N,N*-dimethylformamide is added 0.075 (2 mmol) of 99% formic acid. The mixture is stirred at 60 °C for 1 h under nitrogen, diluted with brine, and extracted with ether. The ether extract is washed twice with brine, dried with anhydrous sodium sulfate, and evaporated. Chromatography of the residue (0.150 g) on 5 g of silica gel gives on elution with hexane 0.116 g (91%) of naphthalene.

10. Homogeneous Hydrogenation

Preparation of the Catalyst Tris(triphenylphosphine)rhodium Chloride [92]

Rhodium chloride trihydrate (1 g, 5.2 mmol) and triphenylphosphine (6 g, 23 mmol) are refluxed in 120 mL of ethanol. The solid tris(triphenylphosphine)-rhodium chloride is filtered with suction and washed with ethanol and ether. The yield is 3.5 g (73%). The catalyst is commercially available.

Reduction of Unsaturated Aldehydes to Saturated Aldehydes [93]

In a flask of a hydrogenation apparatus, 0.2 g (0.00216 mol) of tris(triphenyl-phosphine)rhodium chloride is dissolved in 75 mL of de-aerated benzene under hydrogen at 25 °C and 700 mm of pressure. After the addition of 1.5 mL (0.0183 mol) of crotonaldehyde (2-butenal), hydrogenation is carried out with vigorous stirring for 16 h. The solution is washed with three 10-mL portions of a saturated solution of sodium hydrogen sulfite containing some solid metabisulfite, each portion being left in contact with the benzene solution for about 1 h. The combined aqueous layers and the solid are neutralized with sodium hydrogen carbonate, the solution is extracted with three 10-mL portions of ethyl ether, and the combined ether solutions are dried over sodium sulfate and evaporated to give 1.0 mL (61.2%) of butanal.

Enantioselective Reduction of Double Bonds [101]

A 100-mL pressure bottle is fitted with a magnetic stirring bar, flushed with argon, and evacuated. A solution of 7.4 mg (0.009 mmol) of (–)-(*S*)-2,2'-bis(diphenylphosphine)-1,1'-binaphthylrhodium perchlorate and 216 mg (0.810

mmol) of (Z)-α-benzamidocinnamic acid in 30 mL of ethanol is placed in the bottle. The bottle is filled with hydrogen under 3 atm, and the mixture is stirred at room temperature for 48 h. It is then transferred into a 100-mL flask, and the solvent is evaporated. The residue is quickly dissolved in 30 mL of 0.5 N sodium hydroxide, a small amount of a yellow–brown solid is filtered off through a pad of infusorial earth (Celite), and the flask and the pad are washed with two 20-mL portions of cold water. The combined filtrate and washings are extracted with two 40-mL portions of ether, and the aqueous layer is acidified with 1 N hydrochloric acid. The solution is extracted with three 50-mL portions of ether, the combined extracts are dried with anhydrous sodium sulfate, and the solution is evaporated to give, after drying at 80 °C at 0.1 mmHg, 209 mg (96%) of (+)-*N*-benzoyl-(*R*)-alanine, $[\alpha]_D^{21}$ +40° (*c* 1.0, MeOH); ee 96%.

11. Preparation of Alane (Aluminum Hydride) [*177*]

A 300-mL flask fitted with a magnetic stirring bar, an inlet port closed by a septum, and a reflux condenser connected to a calcium chloride tube is charged with 100 mL of a 1.2 M solution of lithium aluminum hydride in tetrahydrofuran and 100 mL of anhydrous tetrahydrofuran. By means of a syringe, 5.88 g (0.060 mol) of 100% sulfuric acid is added slowly with vigorous stirring. The solution is stirred for 1 h and then allowed to stand at room temperature to permit the precipitated lithium sulfate to settle. The clear supernatant solution is transferred by a syringe to a bottle and stored under nitrogen. For precise work, the titer of the solution is found by volumetric analysis for hydrogen.

12. Preparation of Lithium Tris(*tert*-butoxy)aluminum Hydride [*1300*]

A 1-L flask fitted with a stirrer, a reflux condenser, and a separatory funnel is charged with 7.6 g (0.2 mol) of lithium aluminum hydride and 500 mL of anhydrous ether. A solution of 44.4 g (0.6 mol) of anhydrous *tert*-butyl alcohol in 250 mL of ether is added slowly from the separatory funnel to the stirred contents of the flask. (The hydrogen evolved is vented to a hood.) During the addition of the last third of the alcohol, a white precipitate is formed. The solvent is decanted, and the flask is evacuated with heating on the steam bath to remove the residual ether and *tert*-butyl alcohol. The solid residue, lithium tris-(*tert*-butoxy)aluminum hydride, is stored in bottles protected from atmospheric moisture. Solutions 0.2 M in reagent are prepared by dissolving the solid in diglyme. The reagent is commercially available.

13. Analysis of Hydrides and Complex Hydrides [*67, 109*]

Of many analytical methods described in the literature, the most simple and sufficiently accurate one is decomposition of lithium aluminum hydride suspended in dioxane by a dropwise addition of water and by measurement of the evolved hydrogen gas.

For this purpose the apparatus shown in Figure 6 could be used, provided that the flask is replaced by a flask with a side arm closed by a septum. Then, with the stopcock B closed to connect the flask only to the graduated tube, lithium aluminum hydride is weighed into the flask and covered with a large amount of dioxane. Alternatively, an exact aliquot of a solution of lithium aluminum hydride to be analyzed is pipetted into the flask. The loaded flask containing a magnetic stirring bar is connected to the apparatus and, with stirring, water is added dropwise slowly through the septum with a hypodermic syringe. The volume of the evolved hydrogen is measured after it does not change any more. Sufficient time must be allowed because the decomposition of the lithium aluminum hydride is exothermic and warms up the apparatus [67]. Reading of the volume is to be done only when the level of the water in the graduated tube is stable. For accurate measurement, correction for the partial vapor pressure of dioxane and water must be considered.

14. Reduction with Lithium Aluminum Hydride

Acidic Quenching: Reduction of Aldehydes and Ketones [110]

In a 2-L, three-necked flask equipped with a dropping funnel, a mechanical stirrer, and an efficient reflux condenser protected from moisture by a calcium chloride tube, 19 g (0.5 mol) of lithium aluminum hydride is dissolved in 600 mL of anhydrous ether. From the dropping funnel, 200 g (1.75 mol) of heptanal is introduced at a rate producing gentle reflux. Ten minutes after the last addition and with continued stirring, water is added dropwise and cautiously. The flask is cooled if, during the exothermic decomposition of excess hydride, refluxing becomes too vigorous. The mixture is poured into 200 mL of ice–water and treated with 1 L of 10% sulfuric acid. The ether layer is separated, the aqueous layer is extracted with two 100-mL portions of ether, the combined ether solutions are dried and evaporated, and the residue is fractionated to give an 86% yield of heptanol, bp 175–175.5 °C at 750 mmHg.

Alkaline Quenching: Reduction of Nitriles [184]

A solution of 12.5 g (0.10 mol) of caprylonitrile (octanenitrile) in 20 mL of anhydrous ethyl ether is slowly added to a stirred and ice-cooled solution of 3.8 g (0.10 mol, 100% excess) of lithium aluminum hydride in 200 mL of anhydrous ethyl ether. With continued cooling and vigorous stirring, 4 mL of water, 3 mL of a 20% solution of sodium hydroxide, and 14 mL of water are added in succession. The ether solution is decanted from the white granular residue, the residue is washed twice with ether, all the ether solutions are combined, the ether is distilled off, and the product is distilled at 53–54 °C at 6 mm to yield 11.5–11.9 g (89–92%) of octylamine.

Quenching with Triethanolamine [1247]

N-(3-Hydroxypropyl)formamide (64.1 g, 0.062 mol) is added dropwise to a stirred mixture of 42.3 g (1.1 mol) of lithium aluminum hydride in 2000 mL of anhydrous ether over a period of 2.5 h. After stirring the mixture for an additional 12 h at room temperature, 160 mL (1.2 mol) of triethanolamine is added over 1 h. The mixture is stirred for an additional hour. Distilled water (40 mL, 2.2 mol) is added over 1 h, and the mixture is stirred for 12 h, during which time the hydrolyzed residues have formed a solid grey–white mass at the bottom of the reaction flask. The solid is broken to a "sand" consistency using a glass rod and filtered off by gravity filtration. The residue is extracted with six 500-mL portions of ether, the combined extracts are filtered and evaporated, and the residue is distilled under reduced pressure to give 46.6 g (84%) of 3-hydroxypropylmethylamine, bp 94–100 °C at 82 mmHg or 170–175 °C at 760 mmHg.

15. Reduction with Lithium Triethoxyaluminum Hydride

Reduction of Amides to Aldehydes [124]

Into a magnetically stirred solution of 1.15 g (0.01 mol) of *N,N*-dimethylbutyramide in 10 mL of diethyl ether is added, over 30 min, 40 mL of 0.25 M lithium triethoxyaluminum hydride (0.01 mol) in diethyl ether under cooling with an ice bath. After stirring the mixture for an additional hour at the same temperature, 20 mL of diethyl ether is added. The product, butanal, is isolated as 2,4-dinitrophenylhydrazone in an 89% yield.

16. Reduction with Lithium Tris-*tert*-butoxyaluminum Hydride

Reduction of Acyl Chlorides to Aldehydes [1300]

A 500-mL flask fitted with a stirrer, separatory funnel, low-temperature thermometer, and nitrogen inlet and outlet is charged with a solution of 37.1 g (0.20 mol) of *p*-nitrobenzoyl chloride in 100 mL of diglyme. The flask is flushed with dry nitrogen and cooled to approximately –78 °C with a dry-ice–acetone bath. A solution (200 mL) of 50.8 g (0.20 mol) of tris-(*tert*-butoxy)aluminum hydride in diglyme is added over 1 h with stirring, avoiding any major rise in temperature. The cooling bath is removed, and the flask is allowed to warm up to room temperature over approximately 1 h. The contents are then poured onto crushed ice. The precipitated *p*-nitrobenzaldehyde (with some unreacted acyl chloride) is filtered with suction until dry and then extracted several times with 95% ethanol. The combined extracts are filtered, and the filtrate is evaporated to give 24.5 g (81%) of *p*-nitrobenzaldehyde, mp 103–104 °C, which after recrystallization from water or aqueous ethanol affords 20.3 g (67%) of pure product, mp 104–105 °C.

17. Reduction with Lithium Aluminum Hydride and Diethylamine

Reduction of Esters to Aldehydes [1318]

An oven-dried 25-mL flask fitted with a side arm and a bent adapter connected to a mercury bubbler is charged with 5 mL of 1 M lithium aluminum hydride in tetrahydrofuran (5 mmol). The tetrahydrofuran is evaporated under reduced pressure, and 5 mL of pentane and 0.73 g (10 mmol) of diethylamine are injected into the flask. To the slurry thus formed is added 0.72 g (5 mmol) of ethyl caproate, and the mixture is stirred for 1 h at room temperature. The reaction mixture is cooled to 0 °C, and the precipitate is filtered and washed with several portions of pentane. The separated liquid portion is stirred with 5 mL of

2 N hydrochloric acid for 6 h at room temperature. The separated organic layer is distilled to give 0.42 g (84%) of capronaldehyde, bp 130–131 °C.

18. Reduction with Lithium (1,1'-Binaphthyl-2,2'-dioxy)ethoxy-aluminum Hydride

Enantioselective Reduction of Butyrophenone to (–)-(S)-1-Phenylbutanol [156]

Into a flame-dried, long-necked flask equipped with a rubber septum and filled with argon, 5.1 mL (8.31 mmol) of 1.63 M lithium aluminum hydride in tetrahydrofuran and 4.2 mL (8.4 mmol) of 2.0 M ethanol in tetrahydrofuran are introduced via a syringe, followed by a solution of 2.41 g (8.43 mmol) of (–)-(S)-2,2'-dihydroxy-1,1'-binaphthyl (Binal-H) in 13 mL of tetrahydrofuran. After stirring for 30 min at room temperature, the flask is cooled to –100 °C in a liquid-nitrogen–methanol bath. A solution of 0.370 g (2.5 mmol) of butyrophenone in 2.5 mL of tetrahydrofuran is added dropwise over 8 min at –100 °C. The mixture is stirred for an additional 3 h at –100 °C and then at –78 °C (dry-ice bath) for 16 h. After the addition of 1 mL of methanol at –78 °C, the mixture is warmed to room temperature, treated with 20 mL of 2 N hydrochloric acid, and extracted with ether. The organic extract is dried and concentrated by a bulb-to-bulb distillation at 150–170 °C at 19 mmHg. The colorless oil (0.375 g) consists of a mixture of the unreacted butyrophenone and (–)-(S)-1-phenylbutanol (92% yield by GC). Preparative GC at 160 °C affords 0.291 g (78%) of (–)-(S)-1-phenylbutanol, mp 46–47 °C after recrystallization from benzene; $[\alpha]_D^{22}$ –45.2° (c 4.81, benzene, ee 100%). The distillation residue is recovered (–)-(S)-Binal-H, $[\alpha]_D^{25}$ –34.5° (c 1.80, THF).

The reagent Binal-H is commercially available. Its synthesis and resolution are described in reference 101.

19. Reduction with Alane (Aluminum Hydride) In Situ

Reduction of Nitriles to Amines [1023]

A 1-L three-necked flask equipped with a gas-tight mechanical stirrer, a dropping funnel, and an efficient reflux condenser connected to a dry-ice trap is charged with 3.8 g (0.1 mol) of lithium aluminum hydride and 100 mL of anhydrous ether. Through the dropping funnel, a solution of 13.3 g (0.1 mol) of a solution of anhydrous aluminum chloride in 150 mL of ether is added rapidly. Five minutes after the last addition of the halide, a solution of 19.3 g (0.1 mol) of diphenylacetonitrile in 200 mL of ether is added dropwise to the well-stirred solution. One hour after the last addition of the nitrile, decomposition of the reaction mixture and excess halide is carried out by dropwise addition of water followed by 140 mL of 6 N sulfuric acid diluted with 100 mL of water. The clear mixture is transferred to a separatory funnel, the ether layer is separated, and the aqueous layer is extracted with four 100-mL portions of ether. After cooling by ice–water, the aqueous layer is alkalinized with solid potassium hydroxide to pH 11, diluted with 600 mL of water, and extracted with four 100-mL portions of ether. The combined ether extracts are stirred with anhydrous calcium sulfate (Drierite) and filtered, and the filtrate is evaporated under reduced pressure. The residue (21.5 g) is distilled to give a 91% yield of 2,2-diphenylethylamine, bp 184 °C at 17 mmHg, mp 44–45 °C.

20. Reduction with Diisobutylalane

Reduction of α,β-Unsaturated Esters to Unsaturated Alcohols [1367]

Diisobutylalane is available as a 1 M solution in hexane, Dibal-H (Aldrich). Diisobutylalane (diisobutylaluminum hydride) (0.85 mol) is added under nitrogen over 10 h to a stirred solution of 34.4 g (0.2 mol) of diethyl fumarate in 160 mL of benzene. The temperature rises to 50 °C. After standing overnight at room temperature, the reaction mixture is decomposed by the addition of 76.8 g (2.4 mol) of methanol in 150 mL of benzene followed by 45 g (2.5 mol) of water. The aluminum salts are filtered with suction and washed several times with methanol. The filtrate is distilled to give 12.3 g (70%) of *trans*-2-butene-1,4-diol, bp 86–88 °C at 0.5 mmHg.

21. Reduction with Borane

Reduction of Carboxylic Acids to Alcohols [1251]

An oven-dried 100-mL flask with a side arm closed with a septum is fitted with a magnetic stirring bar and a reflux condenser connected to a mercury bubbler. The flask is cooled to room temperature under nitrogen, charged with 4.36 g (0.025 mol) of adipic acid monoethyl ester followed by 12.5 mL of anhydrous tetrahydrofuran, and cooled to –18 °C by immersion in an ice–salt bath. Then 10.5 mL of 2.39 M (or 25 mL of 1 M) solution of borane in tetrahydrofuran (0.025 mol) (standard solution, commercially available) is slowly added dropwise over 19 min. The resulting clear reaction mixture is stirred well, and the ice–salt bath is allowed to warm slowly to room temperature over 16 h. The mixture is hydrolyzed with 15 mL of water at 0 °C. The aqueous phase is treated with 6 g of potassium carbonate (to decrease the solubility of the alcohol–ester in water), the tetrahydrofuran layer is separated, and the aqueous layer is extracted three times with a total of 150 mL of ether. The combined ether extracts are washed with 30 mL of a saturated solution of sodium chloride, dried over anhydrous magnesium sulfate, and evaporated in vacuo to give 3.5 g (88%) of a colorless liquid, which on distillation yields 2.98 g (75%) of ethyl 6-hydroxyhexanoate, bp 79 °C at 0.7 mmHg.

22. Reduction with Borane In Situ

Reduction of Esters to Alcohols [958]

In a 1-L three-necked flask equipped with a mechanical stirrer, a dropping funnel, a thermometer, and a reflux condenser, 8.5 g (0.225 mol) of sodium borohydride is dissolved in 250 mL of diglyme with stirring. After addition of 125 g (0.4 mol) of ethyl stearate, a solution of 0.084 mol of aluminum chloride

(42 mL of a 2 M solution in diglyme) is added through the dropping funnel while the mixture is stirred vigorously. The rate of addition is adjusted so that the temperature inside the flask does not rise above 50 °C. After all of the aluminum chloride has been added, the reaction mixture is stirred for 1 h at room temperature followed by heating on a steam cone for 0.5–1 h.

After cooling to room temperature, the mixture is poured onto a mixture of 500 g of crushed ice and 50 mL of concentrated hydrochloric acid. The precipitate is filtered with suction, washed with ice–water, pressed, and dried in vacuo. Recrystallization from aqueous alcohol affords 98.2 g (91%) of 1-octadecanol, mp 58–59 °C.

23. Reduction with Thexylchloroborane–Dimethyl Sulfide

Preparation of Thexylchloroborane–Dimethyl Sulfide [143, 144]

2,3-Dimethyl-2-butene (1.26 g, 15 mmol) is added to 15 mmol of 2.55 M boron–dimethyl sulfide in tetrahydrofuran while the temperature is maintained below 10 °C by means of an ice bath. After the solution has been stirred for 1 h at 25 °C, an equimolar amount of dry hydrogen chloride in ether is added at 0–5 °C. This addition is accompanied by a rapid and quantitative evolution of hydrogen. The solution contains thexylchloroborane.

Reduction of Carboxylic Acids to Aldehydes [144]

An oven-dried 100–mL flask, fitted with a side arm and a bent adapter connected to a mercury bubbler, is flushed with nitrogen and charged with 10.53 g (54 mmol) of 6-bromohexanoic acid in 16 mL of dichloromethane. The flask is immersed in an ice–water bath, and a precooled solution of 118.8 mmol (10% excess) (39.6 mL of a 3 M solution) of thexylchloroborane–dimethyl sulfide in dichloromethane is added dropwise with vigorous stirring. After the complete evolution of hydrogen, the ice–water bath is removed and the reaction mixture is stirred for 15 min at room temperature. The solvents are evaporated, and the residue is treated with a saturated solution of sodium bisulfite. The adduct is treated with formaldehyde to give 6-bromohexanal in an 86% yield.

24. Reduction with Sodium Borohydride

Reduction of Ketones to Alcohols [67]

A 125-mL flask fitted with a magnetic stirring bar, a reflux condenser, a thermometer, and a separatory funnel is charged with 7.2 g (9 mL, 0.1 mol) of 2-butanone (methyl ethyl ketone). From the separatory funnel, a solution of 1.5 g (0.04 mol, 60% excess) of sodium borohydride in 15 mL of water is added dropwise with stirring at such a rate as to raise the temperature of the reaction mixture to 40 °C and maintain it at 40–50 °C. If the temperature rises above 50 °C, the reaction mixture can be cooled with a water bath. After the addition has been completed (approximately 30 min), the mixture is stirred until the temperature drops to 30 °C. It is then transferred to a separatory funnel and saturated with sodium chloride. The aqueous layer is drained, and the organic layer is dried with anhydrous potassium carbonate. Distillation affords 5.5–6.0 g (73–81%) of 2-butanol, bp 90–95 °C.

25. Reduction with Zinc Borohydride

Preparation of Zinc Borohydride [*121*]

Freshly fused zinc chloride (3.4 g, 0.025 mol) is added to a solution of 1.95 g (0.052 mol) of sodium borohydride in 50 mL of redistilled 1,2-dimethoxyethane, and the mixture is stirred overnight at 0–5 °C. Filtration under nitrogen gives a clear 0.5 M solution of zinc borohydride. The solution can be stored in a refrigerator for several weeks.

Selective Reduction of Ketones to Alcohols [*1005*]

A solution of 1 mmol of zinc borohydride in 1,2-dimethoxyethane is added to a solution of 0.084 g (1 mmol) of cyclopentanone in 2 mL of dimethoxyethane. The mixture is stirred and cooled to –15 °C with an ice–salt bath. After 30 min the reaction is quenched by a dropwise addition of 0.5 N hydrochloric acid. The organic layer is separated, and the aqueous layer is extracted with three 10-mL portions of ether. The combined organic solutions are washed twice with 10-mL portions of water, dried with anhydrous magnesium sulfate, and filtered. The filtrate is evaporated to give cyclopentanol in a quantitative yield according to NMR assessment.

Zinc borohydride does not reduce conjugated ketones but reduces conjugated aldehydes.

26. Reduction with Sodium Cyanoborohydride

Reduction of Enamines to Amines [*138*]

To a solution of 0.4 g (0.002 mol) of ethyl 3-(*N*-morpholino)crotonate in 4 mL of methanol at 25 °C is added a trace of bromocresol green (pH 3.8–5.4) followed by the addition of 2 N hydrogen chloride in methanol until the color turns yellow. Then 0.13 g (0.002 mol) of sodium cyanoborohydride is added with stirring, and the methanolic hydrogen chloride is added dropwise to maintain the yellow color. After stirring at 25 °C for 1 h, the solution is poured into 5 mL of 0.1 N sodium hydroxide, saturated with sodium chloride, and extracted with three 10–mL portions of ether. The combined extracts are evaporated in vacuo, and the residue (0.315 g) is dissolved in 10 mL of 1 N hydrochloric acid, and the solution is extracted with two 10-mL portions of ether. The aqueous layer is alkalinized to pH > 9 with 6 N potassium hydroxide, saturated with sodium chloride, and extracted with three 10-mL portions of ether. The combined extracts are dried over anhydrous magnesium sulfate and evaporated in vacuo to give 0.260 g (65%) of chromatographically pure ethyl 3-morpholinobutyrate.

27. Reduction with Triethylsilane

Ionic Reduction of Alkenes Capable of Forming Carbonium Ions [181]

To a mixture of 1.16 g (0.01 mol) of triethylsilane and 2.28 g (0.02 mol) of trifluoroacetic acid is slowly added 0.96 g (0.01 mol) of 1-methylcyclohexene. The mixture is maintained at 50 °C for 10 h, and then poured into water. The hydrocarbon layer is separated, and the aqueous layer is extracted with ether. The combined organic layers are neutralized, washed with water, dried, and distilled to give a 67% yield of methylcyclohexane. Triethylsilane is commercially available.

28. Reduction with Stannanes

Hydrogenolysis of Alkyl and Aryl Halides [659]

In a 50-mL flask fitted with a reflux condenser, 5.5 g (0.0189 mol) of tributylstannane (tributyltin hydride, commercially available) is added to 3.5 g (0.018 mol) of octyl bromide. An exothermic reaction ensues. The reaction mixture is cooled if the temperature rises above 50 °C. After 1 h the mixture is distilled, giving an 80% yield of octane and a 90% yield of tributyltin bromide.

1-Bromo-2-phenylethane requires heating with tributylstannane at 100 °C for 4 h to give an 85% yield of ethylbenzene and 87% tributyltin bromide, and benzyl chloride requires heating with tributylstannane at 150 °C for 20 min to give a 78% yield of toluene and an 80% yield of tributyltin chloride.

29. Electrolytic Reduction

Partial Reduction of Aromatic Rings [194]

In an electrolytic cell (Figure 9) consisting of platinum electrodes (2 cm x 5 cm in area) and cathode and anode compartments separated by an asbestos divider, each compartment is charged with 17 g (0.4 mol) of lithium chloride and 450 mL of anhydrous methylamine. Isopropylbenzene (12 g, 0.1 mol) is placed in the cathode compartment, and a total of 50,000 coulombs (2.0 A, 90 V) is passed through the solution in 7 h. After evaporation of the solvent, the mixture is hydrolyzed by the slow addition of water and extracted with ether; the ether extracts are dried and evaporated to give 9.0 g (75%) of product boiling at 149–153 °C and consisting of 89% of a mixture of isomeric isopropylcyclohexenes and 11% recovered isopropylbenzene.

Similar electrolysis without the divider affords an 82% yield of a product containing 78% 2,5-dihydroisopropylbenzene, 6% isopropylcyclohexene, and 13% recovered isopropylbenzene.

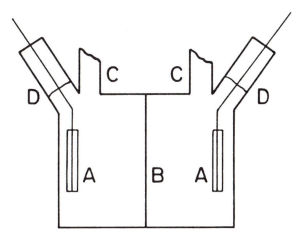

Figure 9. Electrolytic cell. Key: A, electrodes; B, diaphragm; C, reflux condensers; and D, glass seals.

30. Reduction with Sodium (Birch Reduction)

Partial Reduction of Aromatic Rings [547]

An apparatus is assembled from a 250-mL three-necked flask fitted with a magnetic stirring bar, an inlet tube reaching almost to the bottom, and an efficient dry-ice reflux condenser connected to a calcium chloride tube. The remaining opening for addition of solids is stoppered, the reflux condenser is charged with dry ice and acetone, and the flask is immersed in a dry-ice–acetone bath. From a tank of ammonia, 100 mL of liquid ammonia is introduced. α-Naphthol (10 g, 0.07 mol) and 2.7 g (0.068 mol) of powdered sodamide are added followed by 12.5 g of *tert*-amyl alcohol and finally by 3.2 g (0.13 g-atom) of sodium in small pieces. After the blue color has disappeared, the cooling bath and the reflux condenser are removed and ammonia is allowed to evaporate. Water (100 mL) is added, and then the solution is extracted several times with ether and acidified with dilute hydrochloric acid. The precipitated oil solidifies and gives, after recrystallization from petroleum ether (bp 80–100 °C), 8.5 g (83.5%) of 5,8-dihydro-α-naphthol.

31. Reduction with Sodium

Acyloin Condensation of Esters [1343]

A 5-L three-necked flask fitted with a high-speed stirrer (2000–2500 rpm), a reflux condenser, and a separatory funnel is charged with 115 g (5 g-atoms) of sodium and 3 L of xylene and heated in an oil bath (or a heating mantle) to

105 °C. The air in the flask is displaced by nitrogen. When the sodium has melted, stirring is started and the sodium is dispersed in a finely divided state in xylene. From the separatory funnel, 535 g (2.5 mol) of methyl laurate is added at such a rate as to maintain the temperature below 110 °C. This step requires approximately 1 h. Stirring is continued for an additional 30 min. Small particles of unreacted sodium are then destroyed by the addition of 1–2 mol of methanol (40–80 mL). After cooling to 80 °C, 0.5–1.0 L of water is added cautiously. The organic layer is separated, washed once or twice with water, neutralized with a slight excess of a mineral acid, and washed free of acid with a solution of sodium bicarbonate. The xylene is removed by steam distillation, and the oily residue is poured into a suitable vessel to solidify. The crude product contains 80–90% acyloin and is purified by crystallization from 95% ethanol. (Precautions necessary for work with sodium and flammable liquids must be observed.)

32. Reduction with Sodium Naphthalene

Cleavage of Sulfonamides to Amines [922]

In an Erlenmeyer flask capped with a rubber septum, a mixture of naphthalene, 1,2-dimethoxyethane, and enough sodium to yield an 0.5–1.0 M solution of anion radical is stirred with a glass-covered stirring bar for 1–1.5 h, by which time the anion radical will have formed and scavenged the oxygen inside the flask. A solution of one-sixth to one-third of an equivalent of the sulfonamide in dimethoxyethane is injected by syringe, and the mixture is stirred at room temperature for approximately 1 h. Quenching with water produces amines in 68–94% isolated yields.

33. Reduction with Sodium Amalgam

Preparation of Sodium Amalgam [67]

This experiment must be carried out in an efficient hood. If a cube of sodium is not entirely submerged, its dissolution may be accompanied by a flash of fire.

Clean sodium (5 g, 0.217 g-atom) is cut into small cubes of 3–5-mm sides. Each individual cube is placed on the surface of 245 g of mercury in a mortar, pressed by a pestle under the surface of the mercury, and held there until it dissolves. Further cubes are treated in the same way without any delay. The dissolution is accompanied by a hissing sound and sometimes by slight effervescence. It is strongly exothermic and warms up the mercury well above room temperature. Warmer mercury dissolves sodium more rapidly. The addition of sodium must therefore be accelerated in order not to let the mercury cool. When all the sodium has dissolved, the liquid is poured into another dish and allowed to cool, whereafter it is crushed to pieces and stored in a closed bottle.

Reduction of α,β-Unsaturated Acids [67]

In a 250-mL thick-walled bottle with a ground-glass stopper, 7.4 g (0.05 mol) of cinnamic acid is dissolved in a solution of 2 g (0.05 mol) of sodium hydroxide in

65 mL of water until the reaction is alkaline to phenolphthalein. A total of 150 g of sodium amalgam containing 3 g (0.13 g-atom) of sodium is added in pieces. After each addition the bottle is shaken vigorously until the solid amalgam changes to liquid mercury. Only then is the next piece of the amalgam added. When all the amalgam has reacted, the bottle is warmed for 10 min at 60–70 °C, the contents of the bottle are transferred into a separatory funnel, and the mercury layer is drained. The aqueous layer is transferred into an Erlenmeyer flask and acidified with concentrated hydrochloric acid to Congo Red. Hydrocinnamic acid precipitates as an oil that soon crystallizes. Crystallization from 45 °C warm water, slightly acidified with hydrochloric acid, gives 5.5–6.0 g (73–80%) of hydrocinnamic acid, mp 48–49 °C.

34. Reduction with Aluminum Amalgam

Reduction of Aliphatic–Aromatic Ketones to Pinacols [218]

In a 1–L flask fitted with a reflux condenser, 31.8 g (0.265 mol) of acetophenone is dissolved in 130 mL of absolute ethanol and 130 mL of dry sulfur-free benzene, and 0.5 g of mercuric chloride and 8 g (0.296 g-atom) of aluminum foil are added. Heating of the mixture initiates a vigorous reaction that is allowed to proceed without external heating until it moderates. Then the flask is heated to maintain reflux until all of the aluminum has dissolved (2 h). After cooling, the reaction mixture is treated with dilute hydrochloric acid, and the product is extracted with benzene. The combined organic layers are washed with dilute acid, a solution of sodium carbonate, and a saturated solution of sodium chloride. They are then dried with anhydrous sodium sulfate, and the solvent is evaporated under reduced pressure. A rapid vacuum distillation affords a fraction at 160–170 °C at 0.5 mmHg, which, after recrystallization from petroleum ether (bp 65–110 °C), gives 18.0 g (56%) of 2,3-diphenyl-2,3-butanediol, mp 100–123 °C.

35. Reduction with Nickel–Aluminum Alloy

Partial Reduction of Aromatic Heterocycles [230]

To a stirred mixture of 5.05 g (39 mmol) of quinoline, 100 mL of methanol, and 100 mL of 1 M potassium hydroxide in a 500-mL round-bottom flask fitted with a reflux condenser, 17 g of nickel–aluminum alloy is added over 37 min. After stirring for an additional 16 min, the mixture is filtered through a pad of infusorial earth (Celite) and the residue on the filter is washed with 400 mL of dichloromethane. The aqueous layer is separated and extracted three times with the organic layer, and the extract is dried with anhydrous sodium sulfate and evaporated. Distillation of the residue gives 4.39 g (84%) of 1,2,3,4-tetrahydroquinoline as a yellow oil.

36. Reduction with Zinc (Clemmensen Reduction)

Reduction of Ketones to Hydrocarbons [*67, 991*]

In a 1-L flask fitted with a reflux condenser, 400 g (5.3 mol, 3.2 equivalent) of mossy zinc is treated with 800 mL of a 5% aqueous solution of mercuric chloride for 1 h. The solution is decanted; then 100 g (0.835 mol) of acetophenone is added followed by as much hydrochloric acid diluted with the same volume of water as is needed to cover all the zinc. The mixture is refluxed for 6 h during which time additional dilute hydrochloric acid is added in small portions. After cooling, the upper layer is separated, washed free of acid, dried, and distilled to give 70 g (79%) of ethylbenzene, bp 135–136 °C.

37. Reduction with Zinc in Alkaline Solution

Reduction of Nitro Compounds to Hydrazo Compounds

To a mixture of 50 g (0.4 mol) of nitrobenzene, 180 mL of 30% sodium hydroxide, 20 mL of water, and 50 mL of ethanol, 100–125 g (1. 5–1.9 g-atom) of zinc dust is added portionwise with efficient mechanical stirring until the red liquid turns light yellow. After stirring for an additional 15 min, 1 L of cold water is added, the mixture is filtered with suction, the solids on the filter are washed with water, and the hydrazobenzene is extracted from the solids by boiling with 750 mL of ethanol. The mixture is filtered while hot, the filtrate is cooled, the precipitated crystalline hydrazobenzene is filtered with suction, and the mother liquor is used for repeated extraction of the zinc residue. Recrystallization of the crude product from an alcohol–ether mixture gives hydrazobenzene of mp 126–127 °C. This is E. Fischer's procedure, as given in Beilstein's *Handbuch der organischen Chemie,* 4th ed.; Berlin, Germany, 1932; Hauptwerk, Vol. XV, p 68.

38. Reduction with Zinc and Sodium Iodide

Reductive Cleavage of Sulfonates to Hydrocarbons [*916*]

A mixture of a solution of 0.300 g of an alkyl methanesulfonate or *p*-toluene-sulfonate in 3–6 mL of 1,2-dimethoxyethane, 0.300 g of sodium iodide, 0.300 g of zinc dust, and 0.3 mL of water is stirred and refluxed for 4–5 h. After dilution with ether, the mixture is filtered; the solution is washed with water, 5% aqueous hydrochloric acid, 5% aqueous potassium hydrogen carbonate, 5% aqueous sodium thiosulfate, and water. After drying with anhydrous sodium sulfate, the solution is evaporated and the residue worked up, giving a 26–84% yield of alkane.

39. Reduction with Iron

Partial Selective Reduction of Dinitro Compounds [*242]*

A 500-mL three-necked, round-bottomed flask fitted with a robust mechanical stirrer, a reflux condenser, and an inlet is charged with 50 g (0.3 mol) of *m*-

dinitrobenzene, 14 g of 80-mesh iron filings, 45 mL of water, and 7.5 mL of a 33% solution of ferric chloride (or 50 mL of water and 3.7 g of sodium chloride). The flask is immersed in a preheated water bath, stirring is started at 35 rpm and additional iron filings are added to the mixture in two 13-g portions after 30 min and 1 h. After 2.5 h of stirring and heating, the reaction mixture is cooled and extracted with ether, giving an essentially quantitative yield of *m*-nitroaniline.

40. Reduction with Tin

Preparation of Tin Amalgam [251]

To a solution of 15 g of mercuric chloride in 100 mL of water is added 100 g of 30-mesh tin metal with vigorous shaking. After a few minutes, when all the tin has acquired a shiny coating of mercury, the liquid is decanted and the tin amalgam is washed repeatedly with water until the washings are clear. It is stored under distilled water.

Reduction of Quinones to Hydroquinones [251]

To 5.7 g (0.0527 mol) of *p*-benzoquinone is added 10 g (approximately 0.072 mol) of tin amalgam and 50 mL of glacial acetic acid. The mixture is heated on a steam bath. After 3 min green crystals of quinhydrone precipitate but soon dissolve to give a light-yellow solution. After 0.5 h the solution is filtered, the solvent is removed in vacuo, and the residue is recrystallized from benzene–acetone to give 5.0 g (88%) of hydroquinone, mp 169–190 °C.

41. Reduction with Stannous Chloride

Deoxygenation of Sulfoxides [263]

A solution of 0.005 mol of a sulfoxide, 2.26 g (0.01 mol) of stannous chloride dihydrate, and 2.0 mL of concentrated hydrochloric acid in 10 mL of methanol is refluxed for 2–22 h. The mixture is cooled, diluted with 20 mL of water, and extracted with two 25-mL portions of benzene. The dried extracts are evaporated in vacuo, leaving almost pure sulfide that can be purified further by filtration of its solution through a short column of alumina and by a vacuum distillation or sublimation. Yields are 62–93%.

42. Reduction with Chromous Chloride

Preparation of Chromous Chloride [265]

Zinc dust (10 g) is shaken vigorously with a solution of 0.8 g of mercuric chloride and 0.5 mL of concentrated hydrochloric acid in 10 mL of water for 5 min. The supernatant liquid is decanted, 20 mL of water and 2 mL of concentrated hydrochloric acid are added to the residue, and 5 g of chromic chloride is added portionwise with swirling in a current of carbon dioxide. The dark-blue solution is kept under carbon dioxide until used.

Reduction of Iodo Ketones to Ketones [265]

A solution of 1 g of 2-iodo-Δ^4-3-ketosteroid in 50–100 mL of acetone is treated under carbon dioxide portionwise with 20 mL of the solution of chromous chloride. After 10–30 min water is added, and the product is filtered or extracted with ether and recrystallized. The yield of pure Δ^4-3-ketosteroid is 60–63%.

43. Reduction with Titanium Trichloride

Reduction of 2,4-Dinitrobenzaldehyde to 2-Amino-4-nitro-benzaldehyde [772]

To a solution of 9.25 g (0.06 mol) of titanium trichloride in 57 mL of boiled water is added 66 g of concentrated hydrochloric acid, and the solution is diluted to 1 L with boiled water. The mixture is heated to boiling, a current of carbon dioxide is passed through it, and a hot solution of 1.96 g (0.01 mol) of 2,4-dinitrobenzaldehyde in ethanol is added. The dark-blue color fades almost instantaneously. Addition of an excess of sodium acetate precipitates 2-amino-4-nitrobenzaldehyde, which crystallizes on cooling. Repeated extraction with benzene and evaporation of the extract gives 0.8 g (50%) of 2-amino-4-nitro-benzaldehyde.

44. Reduction with Low-Valence Titanium

Intermolecular Reductive Coupling of Carbonyl Compounds [286]

A mixture of 0.328 g (0.0473 mol) of lithium and 2.405 g (0.0156 mol) of titanium trichloride in 40 mL of dry 1,2-dimethoxyethane is stirred and refluxed under argon for 1 h. After cooling, a solution of 0.004 mol of ketone in 10 mL of 1,2-dimethoxyethane is added, and the refluxing is continued for 16 h. The mixture is cooled to room temperature, diluted with petroleum ether, and filtered through activated magnesium silicate (Florisil) on a sintered glass filter. The residue is cautiously quenched by a slow addition of methanol, and the filtrate is concentrated under reduced pressure to give up to a 97% yield of crude alkylidenealkane.

A modified procedure suitable for intramolecular reductive coupling is carried out with low-valence titanium prepared by reduction of titanium trichloride with a zinc–copper couple followed by the extremely slow addition of ketone to the refluxing reaction mixture (0.0003 mol over 9 h by use of a motor-driven syringe pump) [560].

45. Reduction with Vanadous Chloride

Preparation of Vanadous Chloride [293]

Vanadium pentoxide (60 g, 0.33 mol) is covered with 250 mL of concentrated hydrochloric acid and 250 mL of water in a 1-L flask equipped with a gas outlet tube dipping under water. Amalgamated zinc is prepared by treating 90 g of 20-mesh zinc for 10 min with 1 L of 1 N hydrochloric (or sulfuric) acid containing 0.1 mol of mercuric chloride; decanting the solution; and washing the zinc thoroughly with water, 1 N sulfuric acid, and finally water. The zinc is added to the mixture of vanadium pentoxide and hydrochloric acid, and the mixture is allowed to stand for 12 h to form a brown solution. An additional 200 mL of concentrated hydrochloric acid is added giving, after 24 h, a clear purple solution of vanadous chloride. The solution is approximately 1 M and is stored over amalgamated zinc or, after filtration, in an atmosphere of nitrogen. In time the purple solution turns brown, but the purple color is restored by addition of a small amount of concentrated hydrochloric acid.

Reduction of Azides to Amines [297]

A 1 M aqueous solution of vanadous chloride (10 mL) is added to a stirred solution of 1 g of an aryl azide in 10 mL of tetrahydrofuran. After the evolution of nitrogen has subsided, the mixture is poured into 30 mL of aqueous ammonia. The resultant slurry is mixed with benzene and filtered with suction, and the filter cake is washed with benzene. The benzene layer of the filtrate is separated, the aqueous layer is extracted with benzene, the combined benzene solutions are dried, the solvent is evaporated, and the residue is distilled to give 70–95% of the amine.

46. Reduction with Samarium Diiodide

Preparation of Samarium Diiodide [304]

Samarium powder (3 g, 0.02 mol) is placed in a 500-mL flask fitted with a 250-mL dropping funnel containing 2.82 g (0.01 mol) of 1,2-diiodoethane in 250 mL of dry tetrahydrofuran. The solution is slowly added to the contents of the flask with magnetic stirring. No induction period is observed if the solvent is very pure. The intense blue–green solution is thus obtained in 0.04 M concentration.

Deoxygenation of Epoxides [304]

To a solution of 2 mmol of samarium diiodide in 50 mL of tetrahydrofuran in a 100-mL flask is added a solution of 0.134 (1 mmol) g of 1-phenyl-2,3-epoxy-propane in 5 mL of tetrahydrofuran. If necessary, 2 mmol of methanol or *tert*-butyl alcohols is added. After 1 day at room temperature under nitrogen, the initial blue–green color turns yellow, indicating the end of the reaction. Sometimes a yellow precipitate is formed. The mixture is treated with 0.1 N hydrochloric acid to dissolve the SmI_3 species. Organic products are extracted twice with ether. The organic layer is washed with water, sodium thiosulfate, water again, and brine. The ether solution is dried with anhydrous magnesium sulfate and evaporated to remove ether and tetrahydrofuran. The yield of 3-phenylpropane is 82% as found by gas–liquid chromatography. The 18% byproducts are 3-phenyl-2-propanone and 3-phenyl-2-propanol.

47. Reduction with Hydriodic Acid

Reduction of Acyloins to Ketones [1178]

To a mixture of 0.372 g (0.012 mol) of red phosphorus and 4.57 g (0.036 mol) of iodine in 30 mL of carbon disulfide stirred for 10 min is added 2.12 g (0.01 mol) of benzoin (alone or dissolved in 10 mL of benzene) followed by 0.8 g (0.01 mol) of pyridine a few minutes later. The dark reaction mixture is allowed to stand for 3.5 h at room temperature, then poured into aqueous thiosulfate. The organic layer is separated, and the aqueous layer is extracted with 25 mL of benzene. The combined organic solutions are dried with anhydrous magnesium sulfate and evaporated to give a crude product that, after passing through a short column of alumina followed by crystallization, affords 1.76 g (90%) of deoxybenzoin, mp 58–60 °C.

48. Reduction with Hydrogen Sulfide [328]

Reduction of α-Diketones to Acyloins

Hydrogen sulfide is introduced into an ice-cooled solution of 0.2 mol of a 1,2-diketone and 0.02 mol of piperidine in 30 mL of *N,N*–dimethylformamide for 1–4 h. Elemental sulfur is precipitated during the introduction of hydrogen sulfide. The mixture is acidified with dilute hydrochloric acid, and the sediment is filtered with suction and dissolved in warm methanol. The undissolved sulfur is separated, and the product is isolated by evaporation of the methanol and purified by crystallization. If the product after the acidification is liquid, it is isolated by ether extraction and distillation after drying the ether extract with anhydrous sodium sulfate. The yield of benzoin from benzil after 1 h of treatment with hydrogen sulfide is quantitative.

Reduction of α-Diketones to Ketones

If hydrogen sulfide is introduced for 4 h into a solution of 0.2 mol of an α-diketone in methanol and 0.02 mol of pyridine, a monoketone is obtained. The yield of deoxybenzoin from benzoin is quantitative.

    The experiments are done in a fume hood, because hydrogen sulfide is extremely toxic (comparable to hydrogen cyanide).

49. Reduction with Sulfur Dioxide

Deoxygenation of Amine Oxides [961]

A slow stream of sulfur dioxide is passed through a solution of 9.5 g (0.1 mol) of pyridine oxide in 100 mL of dioxane for 3 h. The solution is cooled and evaporated under reduced pressure of an aspirator. The residue is alkalinized with a 20% potassium carbonate solution, and the product is extracted with ether. The extract is dried with anhydrous potassium carbonate and evaporated, and the crude product is distilled to give 5.2 g (66%) of pyridine.

50. Reduction with Sodium Sulfite

Partial Reduction of Geminal Polyhalides [348]

In a 250-mL flask fitted with a magnetic stirring bar, a separatory funnel, and a column surmounted by a reflux condenser with an adapter for regulation of the reflux ratio is placed 29 g (0.105 mol) of 1,1-dibromo-1-chlorotrifluoroethane. The flask is heated at 65–72 °C, and 100 g of a solution containing 14.7% sodium sulfite (0.117 mol) and 6.3% sodium hydroxide (0.157 mol) is added. The rate of addition and the intensity of heating are regulated so that the product boiling at 50 °C condenses in the reflux condenser without flooding the column. The reflux ratio is set so as to keep the temperature in the column head at 50 °C. The distillation is continued until all the product has been removed. The yield of 1-bromo-1-chloro-2,2,2-trifluoroethane is 80–90% after drying.

51. Reduction with Sodium Hydrosulfite (Dithionite)

Reduction of Nitro Compounds to Amino Compounds [354]

A suspension of 1 g (0.005 mol) of 2-chloro-6-nitronaphthalene in 60 mL of hot ethanol is treated gradually with a solution of 3.5g (0.02 mol) of sodium hydrosulfite (dithionite) ($Na_2S_2O_4$) in 16 mL of water. The mixture is refluxed for 1 h, cooled, and filtered from most of the inorganic matter. The filtrate is diluted with 2–3 volumes of water. The precipitate is redissolved by heating, the hot solution is filtered, and the almost pure halogenoamine is separated after cooling and recrystallized from 40% aqueous pyridine to give an almost quantitative yield of 2-amino-6-chloronaphthalene, mp 123 °C.

52. Reduction with Sodium Dithionite in the Presence of a Phase-Transfer Catalyst

Conjugate Reduction of Double Bond in Enones [351]

A solution of 0.385 g (1 mmol) of cholest-4-en-3-one in toluene and a solution of 1.57 g (9 mmol) of sodium dithionite and 0.76 g (9 mmol) of sodium bicarbonate in water in the presence of methyltrialkylammonium chloride (Adogen-336) as a phase-transfer catalyst is stirred and refluxed for 2 h. The aqueous layer is separated, the organic layer is concentrated under reduced pressure, and the crude product is purified by crystallization to give 0.34 g (88%) of 5-α-cholestan-3-one, mp 125–127 °C (Lit. mp is 129 °C).

53. Reduction with Hydrazine

The Wolff–Kizhner Reduction of Ketones to Hydrocarbons: Huang Minlon Modification [385]

In a flask fitted with a take-off adapter surmounted by a reflux condenser and a thermometer reaching to the bottom of the flask, a mixture of 40.2 g (0.30 mol) of propiophenone, 40 g of potassium hydroxide, 300 mL of triethylene glycol, and 30 mL of 85% hydrazine hydrate (0.51 mol) is refluxed for 1 h. The aqueous liquor is removed by means of the take-off adapter until the temperature of the liquid rises to 175–178 °C, and refluxing is continued for 3 h. The reaction mixture is cooled, combined with the aqueous distillate, and extracted with ether. The ether extract is washed free of alkali, dried, and distilled over sodium to give 29.6 g (82.2%) of propylbenzene, bp 160–163 °C.

54. Reduction with Hypophosphorous Acid

Replacement of Aromatic Primary Amino Groups by Hydrogen [779]

To a solution of 3.65 g (0.03 mol) of p-aminobenzylamine in 60 g (0.45 mol) of 50% aqueous hypophosphorous acid placed in a 200-mL flask, 50 mL of water is added, and the mixture is cooled to 5 °C. With stirring, a solution of 2.3 g (0.033 mol) of sodium nitrite in 10 mL of water is added dropwise over 10–15 min. The mixture is stirred for a half hour at 5 °C and allowed to warm to room temperature and stand for about 4 h. After strong alkalinization with sodium hydroxide, the product is obtained by continuous extraction with ether, drying of the extract with potassium hydroxide, and distillation. The yield of benzylamine is 2.67 g (84%), bp 181–183 °C.

55. Reduction with Ammonium Hypophosphite

Preparation of Ammonium Hypophosphite Hydrate [397]

Gaseous ammonia is slowly passed into 100 mL of 50% aqueous hypophosphorous acid until white crystals start to precipitate. The reaction mixture is kept at 10–15 °C for 5 h. The crystals are filtered to give 52 g (56%) of ammonium hypophosphite monohydrate $H_2PO_2NH_4 \cdot H_2O$. The salt is stored in tightly closed bottles. It is soluble in water and N,N-dimethylacetamide; slightly soluble in methanol and ethanol; and insoluble in ether, tetrahydrofuran, and benzene.

Semireduction of Triple Bond [397]

A solution of 3.68 g (36.4 mmol) of ammonium hypophosphite monohydrate in 30 mL of degassed water is added to a solution of 2.3 mL (12.64 mmol) of diphenylacetylene and 0.05 mL (0.43 mmol) of quinoline in 20 mL of benzene. The mixture is stirred in a stream of argon for 5 min. After the addition of 0.202 g (19 mmol) of 10% palladium on carbon, the progress of the reaction is monitored by gas–liquid chromatography. After 12 h, 20 mL of ether is added, and the catalyst is filtered off. The filtrate is washed with water, dried with anhydrous magnesium sulfate, and evaporated to give 2.2 g (96.5%) of a mixture of 88% cis-stilbene, 8% trans-stilbene, and 0.5% 1,2-diphenylethane.

56. Reduction with Trialkyl Phosphites

Desulfurization of Mercaptans to Hydrocarbons [883]

A mixture of 83 g (0.5 mol) of triethyl phosphite (previously distilled from sodium) and 73 g (0.5 mol) of octyl mercaptan (octanethiol) in a Pyrex flask fitted with an efficient column is irradiated with a 100-W General Electric S-4 bulb at a distance of 12.5 cm from the flask. After 6.25 h of irradiation, the mixture is distilled to give 50.3 g (88%) of octane, bp 122–124.5 °C.

57. Reduction with Methyltrichlorosilane and Sodium Iodide

Reductive Cleavage of Ethers [419]

Methyltrichlorosilane (1.8 g, 12 mmol) and a dialkyl ether (10 mmol) are added successively under a nitrogen atmosphere to a stirred solution of 1.8 g (12 mmol) of sodium iodide in dry acetonitrile. The reaction is monitored by thin-layer chromatography in hexane and by ^{1}H NMR spectroscopy, and quenched with water. The mixture is extracted with ether. The extract is washed successively with aqueous sodium thiosulfate, water, and brine; dried with anhydrous sodium sulfate; and evaporated to give 55–100% yields of alcohols.

58. Reduction with Alcohols

The Meerwein–Ponndorf–Verley Reduction [430]

A mixture of 20 g (0.1 mol) of aluminum isopropoxide, 0.1 mol of an aldehyde or a ketone, and 100 mL of dry isopropyl alcohol is placed in a 250-mL flask surmounted by an efficient column fitted with a column head providing for variable reflux. The mixture is heated in an oil bath or by a heating mantle until the byproduct of the reaction, acetone, starts distilling. The reflux ratio is adjusted so that the temperature in the column head is kept at about 55 °C (the boiling point of acetone), and acetone only is collected while the rest of the condensate, mainly isopropyl alcohol (bp 82 °C), flows down to the reaction flask. When no more acetone is noticeable in the condensate based on the test for acetone by 2,4-dinitrophenylhydrazine, the reflux regulating stopcock is opened and most of the isopropyl alcohol is distilled off through the column. The residue in the distilling flask is cooled, treated with 200 mL of 7% hydrochloric acid, and extracted with benzene; the benzene extract is washed with water, dried, and either distilled if the product of the reduction is volatile or evaporated in vacuo if the product is nonvolatile or solid. Yields of the alcohols are 80–90%.

Reduction with Alcohol on Alumina [990]

Neutral aluminum oxide (Woelm W-200) is heated in a quartz vessel at 400 °C at 0.06 mmHg for 24 h. A glass-wool plug placed in the neck of the quartz flask prevents alumina from escaping from the vessel. About 5 g of the dehydrated alumina is transferred inside a nitrogen-filled glove bag to an oven-dried, tared 25-mL round-bottom flask containing a magnetic stirring bar. The flask is stoppered and removed from the glove bag. Approximately 5 mL of an inert solvent such as diethyl ether, chloroform, carbon tetrachloride, or hexane containing 0.5 g of isopropyl alcohol is added. After stirring for 0.5 h at 25 °C, 0.001 mol of an aldehyde or ketone in about 1 mL of a solvent is added and stirring is continued for 2 h at 25 °C. Methanol (5 mL) is added, and the mixture is stirred for 15 min and then suction-filtered through infusorial earth (Celite). The Celite is washed with 35 mL of methanol, and the combined filtrates are evaporated at reduced pressure.

This method reduces aldehydo group preferentially to a keto group in yields up to 88% and with 95% selectivity.

59. Reductive Amination (Leuckart Reaction)

Preparation of Secondary Amines from Ketones [446]

In a distilling flask fitted with a thermometer reaching nearly to the bottom is placed 4 mol of ethylammonium formate prepared by adding ethylamine to cold formic acid. The mixture is heated to 180–190 °C and maintained at this temperature as long as water distils. After cooling, 1 mol of acetophenone is added and the mixture is heated to boiling. The water and amine distil and are collected in concentrated hydrochloric acid. If any ketone codistils, it is returned to the flask at intervals. The temperature is raised to 190–230 °C and maintained there for 4–8 h. After cooling, the mixture is diluted with 3–4 volumes of water, which precipitates an oil. The aqueous layer is mixed with the hydrochloric acid used for collecting the distilled amine, and the mixture is evaporated to give the hydrochloride of the excess amine used. The oil is refluxed for several hours with 150–200 mL of hydrochloric acid. After the hydrolysis is complete, the solution is filtered through moist paper. The filtrate, if not clear, is extracted once or twice with ether, and the amine is liberated by basifying with ammonia and isolated by extraction with ether and distillation. The yield of ethyl-α-phenethylamine is 70%.

60. Reduction with 1,4-Dihydropyridine Derivatives

Conjugate Reduction of Nitro Compounds [1470]

A mixture of 0.149 g (1 mmol) of β-nitrostyrene, 0.278 g (1.1 mmol) of diethyl 1,4-dihydro-2,6-dimethylpyridine-3,5-dicarboxylate (the "Hantzsch ester"), and 100 mg of silica gel in 4 mL of benzene is heated to 60 °C. After 20 h, the benzene is evaporated and the residue is chromatographed on silica gel. Elution with hexane gives 0.146 g (96%) of 2-phenyl-1-nitroethane.

61. Biochemical Reduction

Reduction of Diketones to Keto Alcohols [474]

In a 2-L rotating flask, 0.154 g (0.94 mmol) of 1,2,3,4,5,6,7,8-octahydro-naphthalene-1,5-dione is treated with 440 mL of phosphate buffer at pH 7, 4 g of saccharose, and 20 g of wet centrifuged mycelium of *Curvularia falcata* for 20 h. Extraction with ether and evaporation of the extract affords 0.194 g of a dark-brown resinous material, which, after chromatography on 11.6 g of aluminum oxide activity III, gives on elution with benzene and benzene–ether (9:1) 0.120 g (77%) of 5-(*S*)-5-hydroxy-1-oxo-1,2,3,4,5,6,7,8-octahydronaphthalene, mp 93–94 °C (ether–heptane), $[\alpha]_D$ –55° (*c* 0.76, 1.05). The first benzene eluates contain 3% of 5-(*S*), 9-(*R*)-5-hydroxy-*trans*-decalone.

Reduction of 1,2-Diketones to 1,2-Diols [1075]

A 1–L Erlenmeyer flask containing 250 mL of a sterile solution of 6% glucose, 4% peptone, 4% yeast extract, and 4% malt extract is inoculated with a culture of *Cryptococcus macerans* and is shaken at 30 °C for 2 days. After addition of 0.100 g (0.545 mmol) of di-(2-pyridyl)ethanedione, the shaking is continued for 7 days. The suspension is then alkalinized with 10% potassium hydroxide and extracted three times with 250-mL portions of ethyl acetate. The combined extracts are dried over anhydrous sodium sulfate and evaporated in vacuo to give an 80% yield of *threo*- and 5% of *erythro*-di(2-pyridyl)ethane-1,2-diol. The mixture is separated by thick-layer chromatography over silica gel by elution with a 1:1 mixture of ethyl acetate and hexane to give 0.072 g (70%) of *threo*-di(2-pyridyl)ethane-1,2-diol, mp 92–93 °C (after crystallization from 50% aqueous ethanol), $[\alpha]_D^{25}$–51.7° (*c* 2.44, EtOH).

62. Enantioselective Reduction with Baker's Yeast

Conjugate Reduction of Unsaturated Nitro Compounds [456]

A mixture of 10 g of dry yeast, 5 g of glucose, and 50 mL of tap water is stirred at room temperature for 10 min. After addition of 0.1 g (0.6 mmol) of 1-nitro-2-phenylpropene, the stirring is continued at room temperature for 2 days. The reaction mixture is extracted with ethyl acetate, the extract is evaporated, and the residue is chromatographed on silica gel. Elution with hexane–diethyl ether gives 0.5 g (50%) of 1-nitro-2-phenylpropane in 97.9% enantiomeric excess.

63. Microbial Reductions

Enantioselective Reduction of a Keto Ester to a Hydroxy Ester by Baker's Yeast in Benzene [1386]

Ethyl 3-methyl-2-oxobutanoate (144 mg, 1 mmol) is added to suspension of 10 g of dry *baker's yeast* in 50 mL of benzene. An aqueous buffer solution (8 mL, pH 5) is added, and the mixture is stirred at 30 °C for 24 h. After filtration, the filtrate is evaporated under reduced pressure, and the residue is chromatographed to give 72 mg (49%) of ethyl (*R*)-2-hydroxy-3-methyl butanoate, ee 90%.

Enantioselective Reduction of a Diketone Using Horse Liver Alcohol Dehydrogenase [470]

2,6-Dioxo-*trans*-decalin (1 g, 6.8 mmol) and 700 mg of nicotinamide diphosphate (NAD+) are dissolved at 25 °C in 400 mL of a 0.1 M phosphate buffer, pH 6.5, containing 6 mL of ethanol to give a solution 15 mM in the diketone, 2.6 mM in NAD$^+$, and 35 mM in ethanol. To the mixture is added 150–300 units of *horse liver alcohol dehydrogenase*, and the mixture is kept in the dark at 25 °C for 5–8 days while the progress of the reaction is monitored by gas–liquid chromatography. After the reaction has been completed, the mixture is saturated with sodium chloride and extracted with four 50-mL portions of chloroform. The extracts are dried with anhydrous magnesium sulfate, filtered, and evaporated under reduced pressure. The residue is chromatographed on silica gel, and the product is eluted with chloroform, hexanes, and methanol (80:19:1) to give on evaporation 0.64 g (64%) of (+)-(1R,3S,6S)-6-oxo-*trans*-2-decalol, mp 77–79 °C (pentane), $[\alpha]_D^{25}$ + 24.5 ° (c 0.7); ee > 98%.

Enantioselective Reduction of a Ketone to an Alcohol Using Lactobacillus kefir [468]

Wet cells (1 g) of *Lactobacillus kefir*, previously stored at –80 °C, are suspended in 3 mL of a 0.1 M phosphate buffer (pH 8.5) containing 5 mmol of dithioerythritol, added to an equal volume of 0.1-mm glass beads, and disintegrated in a bead beater three times for 3 min at 0 °C. Total volume of the mixture is 50 mL. Cell debris is removed by centrifugation for 25 min at 15,000 rpm. A portion of 8 mL of the supernatant is added to 0.435 g (2.5 mmol) of ω,ω,ω-trifluoroacetophenone, 15 mg of NADPH, and 1 mL of 2-propanol in 8 mL of 50 mM phosphate–2 mM magnesium chloride buffer (pH 7.1). When problems with the solubility of the ketones are encountered, 10 mL of hexane is added to the reaction mixture. After the reaction has been completed as determined by lack of product formation (12–36 h), the aqueous layer is extracted with three 15-mL portions of ether. The combined organic solutions are dried with sodium sulfate, the filtrate is evaporated, and the residue is purified by silica gel chromatography. (S)-1-Phenyl-2,2,2-trifluoroethanol (0.313 g, 71%) is obtained in higher than 99% enantiomeric excess.

REFERENCES

The numbers in square brackets at the end of the references refer to the page of the book where the reference is quoted.

[1]   von Wilde, M. P. *Chem. Ber.* **1874**, *7*, 352 [3].
[2]   Sabatier, P.; Senderens, J. B. *C. R. Hebd. Seances Acad. Sci.* **1897**, *124*, 1358 [3].
[3]   Ipatiew, W. *Chem. Ber.* **1907**, *40*, 1270, 1281 [3].
[4]   Brown, C. A.; Brown, H. C. *J. Am. Chem. Soc.* **1962**, *84*, 2829; *J. Org. Chem.* **1966**, *31*, 3989 [3, 10, 295].
[5]   Gilman, H.; Jacoby, A. L.; Ludeman, H. *J. Am. Chem. Soc.* **1938**, *60*, 2336 [6, 245].
[6]   Mann, R. S.; Lien, T. R. *J. Catal.* **1969**, *15*, 1 [6].
[7]   Horiuti, J.; Polanyi, M. *Trans. Faraday Soc.* **1934**, *30*, 1164 [6].
[8]   Baker, R. H.; Schuetz, R. D. *J. Am. Chem. Soc.* **1947**, *69*, 1250 [6, 7, 9, 13, 61, 62, 65, 66, 108, 109, 129].
[9]   Schuetz, R. D.; Caswell, L. R. *J. Org. Chem.* **1962**, *27*, 486 [6, 62].
[10]   Siegel, S.; Smith, G. V.; Dmuchovsky, B.; Dubbell, D.; Halpern, W. *J. Am. Chem. Soc.* **1962**, *84*, 3132 [6, 13, 62].
[11]   Siegel, S.; Dunkel, M.; Smith, G. V.; Halpern, W.; Cozort, J. *J Org. Chem.* **1966**, *31*, 2802 [6].
[12]   Siegel, S.; Dmuchovsky, B. *J. Am. Chem. Soc.* **1962**, *84*, 3132 [6].
[13]   Brown, C. A. *J. Org. Chem.* **1970**, *35*, 1900 [6, 7, 10, 53, 54, 56].
[14]   Lozovoi, A. V.; Dyakova, M. K. *Zh. Obshch. Khim.* **1940**, *10*, 1; *Chem. Abstr.* **1940**, *34*, 4728 [6, 53, 61, 63].
[15]   Beamer, R. L.; Belding, R. H.; Fickling, C. S. *J. Pharm. Sci.* **1969**, *58*, 1142 [7].
[16]   Baker, G. L.; Fritschel, S. J.; Stille, J. R.; Stille, J. K. *J. Org. Chem.* **1981**, *46*, 2954 [7].
[17]   Kagan, H. B.; Dang, T.-P. *J. Am. Chem. Soc.* **1972**, *94*, 6429 [7].
[18]   Morrison, J. D.; Masler, W. F. *J. Org. Chem.* **1974**, *39*, 270 [7].
[19]   Valentine, D., Jr.; Sun, R. C.; Toth, K. *J. Org. Chem.* **1980**, *45*, 3703 [7, 192].
[20]   Valentine, D., Jr.; Johnson, K. K.; Priester, W.; Sun, R. C.; Toth, K.; Saucy, G. *J. Org. Chem.* **1980**, *45*, 3698 [7, 16].
[21]   Helpern, J. *Science (Washington, D.C.)* **1982**, *217*, 401 [7, 15, 16, 17].
[22]   Knowles, W. S. *Acc. Chem. Res.* **1983**, *16*, 106 [7, 15, 16, 17].
[23]   Valentine, D., Jr.; Scott, J. W. *Synthesis* **1978**, 329 [7, 16].
[24]   Augustine, R. L.; Migliorini, D. C.; Foscante, R. E.; Sodano, C. S.; Sisbarro, M. J. *J. Org. Chem.* **1969**, *34*, 1075 [7, 166].
[25]   Skita, A.; Schneck, A. *Chem. Ber.* **1922**, *55*, 144 [7, 13].
[26]   Adams, R.; Shriner, R. L. *J. Am. Chem. Soc.* **1923**, *45*, 2171 [7].
[27]   Adams, R.; Voorhees, V.; Shriner, R. L. *Org. Synth., Collect. Vol.* **1932**, *1*, 463 [7].
[28]   Maxted, E. B.; Akhtar, S. *J. Chem. Soc.* **1959**, 3130 [7, 11].
[29]   Baltzly, R. *J. Org. Chem.* **1976**, *41*, 920, 928, 933 [7, 12].
[30]   Brown, H. C.; Brown, C. A. *J. Am. Chem. Soc.* **1962**, *84*, 1493, 1494, 1495 [8, 9].
[31]   Baltzly, R. *J. Am. Chem. Soc.* **1952**, *74*, 4586 [8, 11, 12, 107].
[32]   Starr, D.; Hixon, R. M. *Org. Synth., Collect. Vol.* **1943**, *2*, 566 [9, 66, 68].
[33]   Shriner, R. L.; Adams, R. *J. Am. Chem. Soc.* **1924**, *46*, 1683 [9, 140].

[34] Russell, T. W.; Duncan, D. M.; Hansen, S. C. *J. Org. Chem.* **1977**, *42*, 551 [9, 138, 139, 142, 143, 298].

[35] Mozingo, R. *Org. Synth., Collect. Vol.* **1955**, *3*, 685 [9].

[36] Woodward, R. B.; Sondheimer, F.; Taub, D.; Heusler, K.; McLamore, W. M. *J. Am. Chem. Soc.* **1952**, *74*, 4223 [9].

[37] Weygand, C.; Meusel, W. *Chem. Ber.* **1943**, *76*, 498 [9, 165].

[38] Mosettig, E.; Mozingo, R. *Org. React. (N.Y.)* **1948**, *4*, 362 [9, 12, 203].

[39] Lindlar, H. *Helv. Chim. Acta* **1952**, *35*, 446 [9, 11, 58, 60, 105, 106, 298].

[40] Freifelder, M.; Yew Hay, Ng; Helgren, P. F. *J. Org. Chem.* **1965**, *30*, 2485 [9].

[41] Nishimura, S. *Bull. Chem. Soc. Jpn.* **1961**, *34*, 32 [9, 62, 95, 96, 97, 108, 149, 152, 157].

[42] Nishimura, S.; Onoda, T.; Nakamura, A. *Bull. Chem. Soc. Jpn.* **1960**, *33*, 1356 [9, 108].

[43] Ham, G. E.; Coker, W. P. *J. Org. Chem.* **1964**, *29*, 194 [9, 81, 85].

[44] Adkins, H.; Cramer, H. I. *J. Am. Chem. Soc.* **1930**, *52*, 4349 [10, 62, 68, 72, 73, 107, 108, 129, 140, 141, 149, 150, 152, 153, 234, 240].

[45] Adkins, H.; Pavlic, A. A. *J. Am. Chem. Soc.* **1946**, *68*, 1471; **1947**, *69*, 3039 [10, 219, 223].

[46] Adkins, H.; Billica, H. R. *J. Am. Chem. Soc.* **1948**, *70*, 695; *Org. Synth., Collect. Vol.* **1955**, *3*, 176 [10, 62, 95, 108, 109, 137, 138, 140, 143, 150, 152, 153, 185, 186, 196, 227, 240].

[47] Spero, G. B.; McIntosh, A. V., Jr.; Levin, R. H. *J. Am. Chem. Soc.* **1948**, *70*, 1907 [10, 145, 183, 299].

[48] Brown, C. A.; Ahuja, V. K. *Chem. Commun.* **1973**, 553 [10].

[49] Motoyama, I. *Bull. Chem. Soc. Jpn.* **1960**, *33*, 232 [10, 62, 108, 217].

[50] Brunet, J.-J.; Gallois, P.; Caubère, P. *J. Org. Chem.* **1980**, *45*, 1937, 1946 [11, 53, 58, 59, 60, 138, 140, 149, 152, 153, 162, 163, 167].

[51] Adkins, H.; Connor, R. *J. Am. Chem. Soc.* **1931**, *53*, 1091 [11, 69, 70, 73, 95, 97, 140, 149, 150, 152, 153, 167, 168, 196, 213].

[52] Connor, R.; Folkers, K.; Adkins, H. *J. Am. Chem. Soc.* **1932**, *54*, 1138 [11, 213].

[53] Sauer, J.; Adkins, H. *J. Am. Chem. Soc.* **1937**, *59*, 1 [11, 213, 214, 219].

[54] Carnahan, J. E.; Ford, T. A.; Gresham, W. F.; Grigsby, W. E.; Hager, G. F. *J. Am. Chem. Soc.* **1955**, *77*, 3766 [11, 192].

[55] Broadbent, H. S.; Campbell, G. C.; Bartley, W. J.; Johnson, J. H. *J. Org. Chem.* **1959**, *24*, 1847 [11, 192, 213, 215].

[56] Broadbent, H. S.; Slaugh, L. H.; Jarvis, N. L. *J. Am. Chem. Soc.* **1954**, *76*, 1519 [11, 68, 95, 97, 192].

[57] Broadbent, H. S.; Whittle, C. W. *J. Am. Chem. Soc.* **1959**, *81*, 3587 [11, 64, 68].

[58] Landa, S.; Macák, J. *Collect. Czech. Chem. Commun.* **1958**, *23*, 1322 [11, 156].

[59] Adams, R.; Garvey, B. S. *J. Am. Chem. Soc.* **1926**, *48*, 477 [11, 139].

[60] Weygand, C.; Meusel, W. *Chem. Ber.* **1943**, *76*, 503 [12].

[61] Adkins, H.; Billica, H. R. *J. Am. Chem. Soc.* **1948**, *70*, 3118 [12, 13].

[62] Busch, M.; Stöve, H. *Chem. Ber.* **1916**, *49*, 1063 [12, 83, 87].

[63] Horner, L.; Schläfer, L.; Kämmerer, H. *Chem. Ber.* **1959**, *92*, 1700 [12, 81, 82, 83, 84, 85, 87, 89, 109].

[64] Ubbelohde, L.; Svanoe, T. *Angew. Chem.* **1919**, *32*, 257 [12, 13].

[65] Adkins, H.; Billica, H. R. *J. Am. Chem. Soc.* **1948**, *70*, 3121 [12, 214, 215].

[66]. Hudlický, M. *J. Fluorine Chem.* **1979**, *14*, 189 [13, 85].

[67] Hudlický, M. Virginia Polytechnic Institute and State University, unpublished results [13, 14, 28, 96, 196, 197, 198, 208, 294, 301, 302, 306, 312].

[68] Oppolzer, W.; Mills, R. J.; Réglier, M. *Tetrahedron Lett.* **1986**, *27*, 183 [14, 16, 50].

[69] Leggether, B. E.; Brown, R. K. *Can. J. Chem.* **1960**, *38*, 2363 [15].

[70]   Kuhn, L. P. *J. Am. Chem. Soc.* **1951**, *73*, 1510 [15].
[71]   Ayyangar, N. R.; Brahme, K. C.; Kalkote, U. R.; Srinivasan, K. V. *Synthesis* **1984**, 938 [15, 44, 95, 96].
[72]   Davies, R. R.; Hodgson, H. H. *J. Chem. Soc.* **1943**, 281 [15, 46, 95, 96, 97, 140].
[73]   van Es, T.; Staskun, B. *Org. Synth., Collect. Vol.* **1988**, *6*, 631 [15, 239].
[74]   Pandey, P. N.; Purkayastha, M. L. *Synthesis* **1982**, 876 [15, 46, 88, 89].
[75]   Brown, G. R.; Foubister, A. J. *Synthesis* **1982**, 1036 [15, 46, 239, 240, 299].
[76]   Ram, S.; Spicer, L. D. *Synth. Commun.* **1992**, *22*, 2673 [15, 152, 153].
[77]   Chen, F.-E.; Zhang, H.; Yuan, W.; Zhang, W. W. *Synth. Commun.* **1991**, *21*, 107 [15, 46, 152, 153].
[78]   Ram, S.; Ehrenkaufer, R. E. *Synthesis* **1988**, 91 [15, 46, 95, 97, 100, 143].
[79]   Weiz, J. R.; Patel, B. A.; Heck, R. F. *J. Org. Chem.* **1980**, *45*, 4926 [15, 46, 58, 60].
[80]   Terpko, M. O.; Heck, R. F. *J. Org. Chem.* **1980**, *45*, 4992 [15, 46, 95, 97, 98].
[81]   Chen, Q.-Y.; Ya-bo He, H. *Synthesis* **1988**, 896 [15, 106].
[82]   Olah, G. A.; Prakash, G. K. S. *Synthesis* **1978**, 397 [15, 64].
[83]   Rebeller, M.; Clement, G. *Bull. Soc. Chim. Fr.* **1964**, 1302 [15, 115, 116].
[84]   Eberhardt, M. K. *Tetrahedron* **1967**, *23*, 3029 [15, 64].
[85]   Felix, A. M.; Heimer, E. P.; Lambros, T. J.; Tzougraki, C.; Meyerhofer, J. *J. Org. Chem.* **1978**, *43*, 4194 [15, 211, 299].
[86]   Orchin, M. *J. Am. Chem. Soc.* **1944**, *66*, 535 [15, 66, 67].
[87]   Nishiguchi, T.; Tachi, K.; Fukuzume, K. *J. Org. Chem.* **1975**, *40*, 237 [15].
[88]   Citron, J. D. *J. Org. Chem.* **1969**, *34*, 1977 [15, 203, 204].
[89]   Yasui, S.; Nakamura, K.; Ohno, A. *Chem. Lett.* **1984**, 377 [15, 88, 98].
[90]   Brieger, G.; Nestrick, T. J. *Chem. Rev.* **1974**, *74*, 567 [15].
[91]   Harmon, R. E.; Parsons, J. L.; Cooke, D. W.; Gupta, S. K.; Schoolenberg, J. *J. Org. Chem.* **1969**, *34*, 3684 [15, 93, 143, 165, 167, 196, 217].
[92]   Birch, A. J.; Walker, K. A. M. *J. Chem. Soc. C* **1966**, 1894 [15, 300].
[93]   Jardine, F. H.; Wilkinson, G. *J. Chem. Soc. C* **1967**, 270 [15, 300].
[94]   Ireland, R. E.; Bey, P. *Org. Synth., Collect. Vol.* **1988**, *6*, 459 [15, 165].
[95]   Tanaka, M.; Kobayashi, T.-a. *Synthesis* **1985**, 967 [15, 237, 238].
[96]   Padmakumari Amma, J.; Stille, J. K. *J. Org. Chem.* **1982**, *47*, 468 [16, 218].
[97]   Toth, I.; Hanson, B. E.; Davis, M. E. *Tetrahedron: Asymmetry* **1990**, *1*, 913 [16, 218].
[98]   Takaya, H.; Agutawa, S.; Noyori, R. *Org. Synth.* **1988**, *67*, 20 [16].
[99]   Amrani, Y.; Lecompte, L.; Sinou, D.; Bakos, J.; Toth, I.; Heil, B. *Organometallics* **1989**, *8*, 542 [16, 218].
[100]   Nagel, U.; Kinzel, E. *Chem. Commun.* **1986**, 1098; *Chem. Ber.* **1986**, *119*, 1731 [16, 218].
[101]   Miyashita, A.; Takaya, H.; Souchi, T.; Noyori, R. *Tetrahedron* **1984**, *40*, 1245 [16, 218, 300, 304].
[102]   Vineyard, B. D.; Knowles, W. S.; Sabacky, M. J.; Bachman, G. L.; Weinkauff, D. J. *J. Am. Chem. Soc.* **1977**, *99*, 5946 [16].
[103]   Taber, D. F.; Silverberg, L. J. *Tetrahedron Lett.* **1991**, *32*, 4227 [16, 225].
[104]   Kawano, H.; Ishii, Y.; Saburi, M.; Uchida, Y. *Chem. Commun.* **1988**, 87 [16, 179, 180].
[105]   Kitamura, M.; Tokunaga, M.; Ohkuma, T.; Noyori, R. *Org. Synth.* **1992**, *71*, 1 [16, 50, 225].
[106]   Hayashi, T.; Kawamura, N.; Ito, Y. *J. Am. Chem. Soc.* **1987**, *109*, 7876 [16, 196].
[107]   Fryzuk, M. D.; Bosnich, B. *J. Am. Chem. Soc.* **1977**, *99*, 6262 [16, 50, 218].
[108]   Ojima, I.; Kogure, T.; Yoda, Y. *Org. Synth.* **1984**, *63*, 18 [16, 200].

[109] Finholt, A. E.; Bond, A. C.; Schlesinger, H. I. *J. Am. Chem. Soc.* **1947**, *69*, 1199 [19, 301].

[110] Nystrom, R. F.; Brown, W. G. *J. Am. Chem. Soc.* **1947**, *69*, 1197 [19, 29, 137, 138, 139, 140, 149, 150, 153, 204, 205, 206, 215, 216, 302].

[111] Schlesinger, H. I.; Brown, H. C.; Finholt, A. E. *J. Am. Chem. Soc.* **1953**, *75*, 205 [19].

[112] Vít, J.; Procházka, V.; Donnerová, Z.; Čásenský, B.; Hrozinka, I. Czech Patent 120,100, 1966; *Chem. Abstr.* **1967**, *67*, 23634 [19].

[113] Zweifel, G.; Brown, H. C. *Org. React. (N.Y.)* **1963**, *13*, 1 [19].

[114] Bruce, M. I. *Chem. Br.* **1975**, *11*, 237 [20].

[115] Piřha, J.; Heřmánek, S.; Vít, J. *Collect. Czech. Chem. Commun.* **1960**, *25*, 736 [20, 194, 204, 215].

[116] Zakharin, L. I.; Gavrilenko, V. V.; Maslin, D. N.; Khorlina, I. M. *Tetrahedron Lett.* **1963**, 2087 [20, 207, 208].

[117] Plešek, J.; Heřmánek, S. *Collect. Czech. Chem. Commun.* **1966**, *31*, 3060 [20, 149, 215, 216].

[118] Schlesinger, H. I.; Brown, H. C. *J. Am. Chem. Soc.* **1940**, *62*, 3429 [20].

[119] Banus, M. D.; Bragdon, R. W.; Hinckley, A. A. *J. Am. Chem. Soc.* **1954**, *76*, 3848 [20].

[120] Tolman, V.; Vereš, K. *Collect. Czech. Chem. Commun.* **1972**, *37*, 2962 [20, 220, 221, 223].

[121] Gensler, W. J.; Johnson, F.; Sloan, A. D. B. *J. Am. Chem. Soc.* **1960**, *82*, 6074, 6078 [20, 149, 307].

[122] Brändström, A.; Junggren, U.; Lamm, B. *Tetrahedron Lett.* **1972**, 3173 [20, 21].

[123] Brown, H. C.; Weissman, P. M. *J. Am. Chem. Soc.* **1965**, *87*, 5614 [20, 26, 215, 232, 240].

[124] Brown, H. C.; Tsukamoto, A. *J. Am. Chem. Soc.* **1964**, *86*, 1089 [20, 230, 303].

[125] Brown, H. C.; McFarlin, R. F. *J. Am. Chem. Soc.* **1958**, *80*, 5372 [20, 203, 215].

[126] Xiang, Y. B.; Snow, K.; Belley, M. *J. Org. Chem.* **1993**, *58*, 993 [20, 201].

[127] Vít, J.; Čásenský, B.; Macháček, J. French Patent 1,515,582, 1967; *Chem. Abstr.* **1969**, *70*, 115009 [20].

[128] Čásenský, B.; Macháček, J.; Abrham, K. *Collect. Czech. Chem. Commun.* **1972**, *37*, 1178 [20].

[129] Brown, H. C.; Mead, E. J. *J. Am. Chem. Soc.* **1953**, *75*, 6263 [20, 139, 140, 143, 149, 153, 204, 205, 216].

[130] Hutchins, R. O.; Learn, K.; El-Tebany, F.; Stercho, Y. P. *J. Org. Chem.* **1984**, *49*, 2438 [20, 87, 140, 141, 149, 153, 216, 231, 232].

[131] Yoon, N. M.; Ahn, J. H.; An, D. K.; Shon, Y. S. *J. Org. Chem.* **1993**, *58*, 1941 [20, 208].

[132] Nutaitis, C. F.; Gribble, G. W. *Tetrahedron Lett.* **1983**, *24*, 4287 [20, 137, 176].

[133] Brown, H. C.; Kim, S. C.; Krishnamurthy, S. *J. Org. Chem.* **1980**, *45*, 1 [21, 121, 149, 153, 181, 205, 206, 216, 231, 240].

[134] Brown, H. C.; Krishnamurthy, S. *J. Am. Chem. Soc.* **1972**, *94*, 7159 [21].

[135] Brown, C. A. *J. Am. Chem. Soc.* **1973**, *95*, 4100; *J. Org. Chem.* **1974**, *39*, 3913 [21].

[136] Yoon, N. M.; Kim, K. E.; Kang, J. *J. Org. Chem.* **1986**, *51*, 226 [21, 137, 150, 151, 157, 176].

[137] Cha, J. S.; Yoon, M. S. *Tetrahedron Lett.* **1989**, *30*, 3677 [21, 239].

[138] Borch, R. F.; Bernstein, M. D.; Durst, H. D. *J. Am. Chem. Soc.* **1971**, *93*, 2897 [21, 129, 187, 190, 307].

[*139*] Hutchins, R. O.; Kandasamy, D. *J. Am. Chem. Soc.* **1973**, *95*, 6131 [21, 137, 138, 141, 149, 150, 153, 162, 163].

[*140*] Whitesides, G. M.; San Filipo, J. *J. Am. Chem. Soc.* **1970**, *92*, 6611 [21].

[*141*] Brown, H. C.; Zweifel, G. *J. Am. Chem. Soc.* **1959**, *81*, 1512; **1961**, *83*, 3834 [21, 58, 59].

[*142*] Brown, H. C.; Zweifel, G. *J. Am. Chem. Soc.* **1961**, *83*, 1241 [21].

[*143*] Brown, H. C.; Mandal, A. K.; Yoon, N. M.; Singaram, B.; Schwier, J. R.; Jadhav, P. K. *J. Org. Chem.* **1982**, *47*, 5069 [21, 306].

[*144*] Brown, H. C.; Cha, J. S.; Nazer, B.; Yoon, N. M. *J. Am. Chem. Soc.* **1984**, *106*, 8001 [21, 191, 194, 198, 306].

[*145*] Cha, J. S.; Kim, J. E.; Oh, S. Y.; Lee, J. C.; Lee, K. W. *Tetrahedron Lett.* **1987**, *28*, 2389 [21, 191, 194, 198].

[*146*] Brown, H. C.; Knights, E. F.; Scouten, C. G. *J. Am. Chem. Soc.* **1974**, *96*, 7765 [21].

[*147*] Brown, H. C.; Jadhav, P. K.; Mandal, A. K. *J. Org. Chem.* **1982**, *47*, 5074 [21].

[*148*] Brown, H. C.; Desai, M. C.; Jadhav, P. K. *J. Org. Chem.* **1982**, *47*, 5065 [21, 22, 154].

[*149*] Chandrasekharan, J.; Ramachandran, P. V.; Brown, H. C. *J. Org. Chem.* **1985**, *50*, 5446 [21, 22, 45, 154].

[*150*] Midland, M. M.; Greer, S.; Tramontano, A.; Zderic, S. A. *J. Am. Chem. Soc.* **1979**, *101*, 2352 [21, 22, 137, 140, 141, 154].

[*151*] Midland, M. M.; Kazubski, A. *J. Org. Chem.* **1982**, *47*, 2814 [22, 168, 169].

[*152*] Krishnamurthy, S.; Vogel, F.; Brown, H. C. *J. Org. Chem.* **1977**, *42*, 2534 [22].

[*153*] Brunel, J. M.; Pardigon, O.; Faure, B.; Buono, G. *Chem. Commun.* **1992**, 287 [22, 150, 154].

[*154*] Corey, E. J.; Bakshi, R. K.; Shibata, S. *J. Am. Chem. Soc.* **1987**, *109*, 5551 [22, 150, 154].

[*155*] Brown, H. C.; Cho, B. T.; Park, W. S. *J. Org. Chem.* **1988**, *53*, 1231 [22, 23].

[*156*] Noyori, R.; Tomino, I.; Tanimoto, Y.; Nishizawa, M. *J. Am. Chem. Soc.* **1984**, *106*, 6709 [22, 23, 154, 304].

[*157*] Deloux, L.; Srebnik, M. *Chem. Rev.* **1993**, *93*, 763 [22].

[*158*] Ashby, E. C.; Lin, J. J. *J. Org. Chem.* **1978**, *43*, 2567 [22, 55, 58, 65].

[*159*] Ashby, E. C.; Lin, J. J.; Goel, A. B. *J. Org. Chem.* **1978**, *43*, 757 [23, 59].

[*160*] Citron, J. D.; Lyons, J. E.; Sommer, L. H. *J. Org. Chem.* **1969**, *34*, 638 [23, 88].

[*161*] Cole, S. J.; Kirwan, N. J.; Roberts, B. P.; Willis, C. R. *J. Chem. Soc., Perkin Trans. 1* **1991**, 103 [23, 28, 81, 82, 104].

[*162*] Fukuyama, T.; Lin, S.-C.; Li, L. *J. Am. Chem. Soc.* **1990**, *112*, 7050 [23, 228].

[*163*] Benkeser, R. A.; Landesman, H.; Foster, D. J. *J. Am. Chem. Soc.* **1952**, *74*, 648 [23].

[*164*] Marsi, K. L. *J. Org. Chem.* **1974**, *39*, 265; *J. Am. Chem. Soc.* **1969**, *91*, 4724 [23, 134].

[*165*] Benkeser, R. A.; Foster, D. J. *J. Am. Chem. Soc.* **1952**, *74*, 5314 [23].

[*166*] Fujita, M.; Hiyama, T. *J. Am. Chem. Soc.* **1984**, *106*, 4629 [23, 153, 155, 225].

[*167*] Benkeser, R. A.; Smith, W. E. *J. Am. Chem. Soc.* **1968**, *90*, 5307; *J. Org. Chem.* **1977**, *42*, 2534 [23, 46].

[*168*] Li, G. S.; Ehler, D. F.; Benkeser, R. A. *Org. Synth., Collect. Vol.* **1988**, *6*, 747 [23, 195].

[*169*] Neumann, W. P.; Niermann, H. *Justus Liebig's Ann. Chem.* **1962**, *653*, 164 [23, 24].

[*170*] Kuivila, H. G.; Beumel, O. F., Jr. *J. Am. Chem. Soc.* **1958**, *80*, 3798; **1961**, *83*, 1246 [23, 24, 135, 140, 141, 143, 149, 153, 159, 167].

[*171*] Pereyre, M.; Colin, G.; Valade, J. *Tetrahedron Lett.* **1967**, 4805 [23, 219].

[*172*]   Adams, C. M.; Schemenaur, J. E. *Synth. Commun.* **1990**, *20*, 2359 [23, 137, 138, 140, 141, 176].
[*173*]   van Koten, G.; Schaap, C. A.; Noltes, J. G. *J. Organomet. Chem.* **1975**, *99*, 157 [23, 24].
[*174*]   Vedejs, E.; Duncan, S. M.; Haight, A. R. *J. Org. Chem.* **1993**, *58*, 3046 [23, 24, 26, 28, 82, 88, 104, 168, 169].
[*175*]   Sorrell, T. N.; Pearlman, P. S. *J. Org. Chem.* **1980**, *45*, 3449 [23, 24, 203, 204].
[*176*]   Mahoney, W. S.; Brestensky, D. M.; Stryker, M. *J. Am. Chem. Soc.* **1988**, *110*, 291 [23, 165].
[*177*]   Brown, H. C.; Yoon, N. M. *J. Am. Chem. Soc.* **1966**, *87*, 1464 [26, 301].
[*178*]   Mousseron, M.; Jacquier, R.; Mousseron-Canet, M.; Zagdoun, R. *Bull. Soc. Chim. Fr.* **1952**, 1042; *C. R. Hebd. Seances Acad. Sci.* **1952**, *235*, 177 [26, 242].
[*179*]   Brown, W. G. *Org. React. (N.Y.)* **1951**, *6*, 469 [26, 28].
[*180*]   Lane, C. F. *Chem. Rev.* **1976**, *76*, 773 [26].
[*181*]   Kursanov, D. N.; Parnes, Z. N.; Bassova, G. I.; Loim, N. M.; Zdanovich, V. I. *Tetrahedron* **1967**, *23*, 2235 [28, 55, 156, 308].
[*182*]   LeNoble, W. J. *Chem. Eng. News* **1983**, *61*(19), 2 [28].
[*183*]   Uffer, A.; Schlitter, E. *Helv. Chim. Acta* **1948**, *31*, 1397 [29, 232].
[*184*]   Amundsen, L. H.; Nelson, L. S. *J. Am. Chem. Soc.* **1951**, *73*, 242 [29, 30, 240, 302].
[*185*]   Caine, D. *Org. React. (N.Y.)* **1976**, *23*, 1 [32].
[*186*]   Hardegger, E. P.; Plattner, P. A.; Blank, F. *Helv. Chim. Acta* **1944**, *27*, 793 [32].
[*187*]   Stork, G.; White, W. N. *J. Am. Chem. Soc.* **1956**, *78*, 4604 [32, 35].
[*188*]   Sicher, J.; Svoboda, M.; Závada, J. *Collect. Czech. Chem. Commun.* **1965**, *30*, 421 [32, 60].
[*189*]   Campbell, K. N.; Young, E. E. *J. Am. Chem. Soc.* **1943**, *65*, 965 [32, 59].
[*190*]   Horner, L.; Röder, H. *Chem. Ber.* **1968**, *101*, 4179 [33].
[*191*]   Benkeser, R. A.; Kaiser, E. M.; Lambert, R. F. *J. Am. Chem. Soc.* **1964**, *86*, 5272 [33, 63].
[192]   Benkeser, R. A.; Watanabe, H.; Mels, S. J.; Sabol, M. A. *J. Org. Chem.* **1970**, *35*, 1210 [33, 230].
[*193*]   Tafel, J. *Chem. Ber.* **1900**, *33*, 2209 [33].
[*194*]   Benkeser, R. A.; Kaiser, E. M. *J. Am. Chem. Soc.* **1963**, *85*, 2858 [33, 63, 308].
[*195*]   Junghans, K. *Chem. Ber.* **1973**, *106*, 3465; **1974**, *107*, 3191 [33, 35, 55, 64].
[*196*]   Ferles, M.; Havel, M.; Tesařová, A. *Collect. Czech. Chem. Commun.* **1966**, *31*, 4121 [33, 70].
[*197*]   Bhuvaneswari, N.; Venkatachalam, C. S.; Balasubramanian, K. K. *Tetrahedron Lett.* **1992**, *33*, 1499 [33, 115].
[*198*]   Kunugi, A.; Abe, K. *Chem. Lett.* **1984**, 159 [33, 122].
[*199*]   Law, H. D. *J. Chem. Soc.* **1912**, *101*, 1016 [33, 139, 143, 166, 167].
[*200*]   Ruff, O.; Geisel, E. *Chem. Ber.* **1906**, *39*, 828 [34].
[*201*]   Slaugh, L. H.; Raley, J. H. *J. Org. Chem.* **1967**, *32*, 2861 [34, 63, 66].
[*202*]   Benkeser, R. A.; Robinson, R. E.; Sauve, D. M.; Thomas, O. H. *J. Am. Chem. Soc.* **1955**, *77*, 3230 [34, 63].
[*203*]   Benkeser, R. A.; Schroll, G.; Sauve, D. M. *J. Am. Chem. Soc.* **1955**, *77*, 3378 [34, 60].
[*204*]   Reggel, L.; Friedel, R. A.; Wender, I. *J. Org. Chem.* **1957**, *22*, 891 [34, 63, 66, 67].
[*205*]   Hansley, V. L. *Ind. Eng. Chem.* **1947**, *39*, 55; **1951**, *43*, 1759 [35, 213, 241].
[*206*]   Ferguson, L. N.; Reid, J. C.; Calvin, M. *J. Am. Chem. Soc.* **1946**, *68*, 2502 [35, 192, 194].
[*207*]   Cason, J.; Way, R. L. *J. Org. Chem.* **1949**, *14*, 31 [35, 82].

References   329

[208]   Levina, R. Y.; Kulikov, S. G. *Zh. Obshch. Khim.* **1946**, *16*, 117; *Chem. Abstr.* **1947**, *41*, 115 [35].

[209]   Bryce-Smith, D.; Wakefield, B. J. *Org. Synth., Collect. Vol.* **1973**, *5*, 998 [35].

[210]   Adams, R.; Adams, E. W. *Org. Synth., Collect. Vol.* **1932**, *1*, 459 [35, 151].

[211]   Hwu, J. R.; Chua, V.; Schroeder, J. E.; Barrans, R. E., Jr.; Khoudary, K. P.; Wang, N.; Wetzel, J. M. *J. Org. Chem.* **1986**, *51*, 4731 [35, 112].

[212]   Benkeser, R. A.; Belmonte, F. G. *J. Org. Chem.* **1984**, *49*, 1662 [35, 60].

[213]   Benkeser, R. A.; Laugal, J. A.; Rappa, A. *Tetrahedron Lett.* **1984**, *25*, 2089 [35, 63, 64, 66, 67].

[214]   Maiti, S. N.; Spevak, P.; Reddy, A. V. N. *Synth. Commun.* **1988**, *18*, 1201 [35, 100, 101].

[215]   Corey, E. J.; Chaykovsky, M. *J. Am. Chem. Soc.* **1965**, *87*, 1345, 1351 [36, 122, 123, 173, 174, 175].

[216]   Schreibmann, P. *Tetrahedron Lett.* **1970,** 4271 [36].

[217]   Kuhn, R.; Fischer, H. *Chem. Ber.* **1961**, *94*, 3060 [36, 64].

[218]   Newman, M. S. *J. Org. Chem.* **1961**, *26*, 582 [36, 155, 310].

[219]   Inoi, T.; Gericke, P.; Horton, W. J. *J. Org. Chem.* **1962**, *27*, 4597 [36, 83].

[220]   Calder, A.; Hepburn, S. P. *Org. Synth.* **1972**, *52*, 77 [36, 91].

[221]   Wislicenus, H. *Chem. Ber.* **1896**, *29*, 494 [36, 95, 96].

[222]   Wislicenus, H. *J. Prakt. Chem.* **1896**, [2], *54*, 18, 54, 65 [36, 133, 134, 224].

[223]   Shin, C.; Yonezawa, Y.; Yoshimura, J. *Chem. Lett.* **1976**, 1095 [36, 100, 223].

[224]   Drinkwater, D. J.; Smith, P. W. G. *J. Chem. Soc. C* **1971**, 1305 [36, 227].

[225]   Braun, W.; Mecke, R. *Chem. Ber.* **1966**, *99*, 1991 [36, 181].

[226]   Pascali, V.; Umani-Ronchi, A. *Chem. Commun.* **1973**, 351 [36, 123, 124].

[227]   Schroeck, C. W.; Johnson, C. R. *J. Am. Chem. Soc.* **1971**, *93*, 5305 [36, 123, 124].

[228]   Fischer, E.; Groh, R. *Justus Liebigs Ann. Chem.* **1911**, *383*, 363 [36, 187].

[229]   Sato, R.; Akaishi, R.; Goto, T.; Saito, M. *Bull. Chem. Soc. Jpn.* **1987**, *60*, 773 [36, 121].

[230]   Lunn, G.; Sansone, E. G. *J. Org. Chem.* **1986**, *51*, 513 [36, 70, 72, 73, 74, 311].

[231]   Yamamura, S.; Hirata, Y. *J. Chem. Soc. C* **1968**, 2887 [36, 37, 164].

[232]   Dekker, J.; Martins, F. J. C.; Kruger, J. A. *Tetrahedron Lett.* **1975**, 2489 [36, 178].

[233]   Rieke, R. D.; Uhm, S. J.; Hudnall, P. M. *Chem. Commun.* **1973**, 269 [36, 89].

[234]   Gardner, J. H.; Naylor, C. A. *Org. Synth., Collect. Vol.* **1943**, 2, 526 [36].

[235]   Leonard, N. J.; Barthel, E., Jr. *J. Am. Chem. Soc.* **1950**, *72*, 3632 [36, 164, 175].

[236]   Martin, E. L. *Org. React. (N.Y.)* **1942**, *1*, 155 [36, 137, 144, 151, 156, 163, 168].

[237]   Risinger, G. E.; Mach, E. E.; Barnett, K. W. *Chem. Ind. (London)* **1965**, 679 [36, 156].

[238]   Burdon, J.; Price, R. C. University of Birmingham, England, private communication [36] .

[239]   Vedejs, E. *Org. React. (N.Y.)* **1975**, *22*, 401 [36].

[240]   Kreiser, W. *Justus Liebigs Ann. Chem.* **1971**, *745*, 164 [37, 177, 178].

[241]   Nose, A.; Kudo, T. *Chem. Pharm. Bull.* **1990**, *38*, 2097 [37, 40, 56, 73, 74, 96, 97, 137, 138, 140, 141, 153, 172, 239, 241].

[242]   Lyons, R. E.; Smith, L. T. *Chem. Ber.* **1927**, *60*, 173 [37, 96, 97, 312].

[243]   Hodgson, H. H.; Whitehurst, J. S. *J. Chem. Soc.* **1945**, 202 [37, 96, 97].

[244]   Clarke, H. T.; Dreger, E. E. *Org. Synth., Collect. Vol.* **1932**, *1*, 304 [37].

[245]   Matsumura, K. *J. Am. Chem. Soc.* **1930**, *52*, 4433 [37].

[246]   Rakoff, H.; Miles, B. H. *J. Org. Chem.* **1961**, *26*, 2581 [37].

[247]   den Hertog, J.; Overhoff, J. *Recl. Trav. Chim. Pays-Bas* **1950**, *69*, 468 [37, 132].

[248] Blomquist, A. T.; Dinguid, L. I. *J. Org. Chem.* **1947**, *12*, 718, 723 [37, 119].
[249] Kindler, K. *Justus Liebigs Ann. Chem.* **1923**, *431*, 187 [37, 237, 238].
[250] Carter, P. H.; Craig, J. G.; Lack, R. E.; Moyle, M. *Org. Synth., Collect. Vol.* **1973**, *5*, 339 [38, 174].
[251] Schaefer, J. P. *J. Org. Chem.* **1960**, *25*, 2027 [38, 180, 313].
[252] Meyer, K. H. *Org. Synth., Collect. Vol.* **1932**, *1*, 60 [38, 181].
[253] Hartman, W. W.; Dickey, J. B.; Stampfli, J. G. *Org. Synth., Collect. Vol.* **1943**, *2*, 175 [38, 110].
[254] Braun, V. J.; Sobecki, W. *Chem. Ber.* **1911**, *44*, 2526, 2533 [39, 92].
[255] Oelschläger, H.; Schreiber, O. *Justus Liebigs Ann. Chem.* **1961**, *641*, 81 [39, 172].
[256] Reissert, A.; Grube, F. *Chem. Ber.* **1909**, *42*, 3710 [39, 242].
[257] Russig, F. *J. Prakt. Chem.* **1900**, [2], *62*, 30 [39, 180].
[258] Badger, G. M.; Gibb, A. R. M. *J. Chem. Soc.* **1949**, 799 [39, 181].
[259] Sonn, A.; Müller, E. *Chem. Ber.* **1919**, *52*, 1927 [39, 238].
[260] Stephen, H. *J. Chem. Soc.* **1925**, *127*, 1874 [39, 238, 239].
[261] Bischler, A. *Chem. Ber.* **1889**, *22*, 2801 [39, 100].
[262] Pschorr, R. *Chem. Ber.* **1902**, *35*, 2729, 2738 [39, 185].
[263] Ho, T.-L.; Wong, C. M. *Synthesis* **1973**, 206 [39, 121, 122, 313].
[264] Hanson, J. R.; Mehta, S. *J. Chem. Soc. C* **1969**, 2349 [39, 180, 181].
[265] Rosenkranz, G.; Mancer, O.; Gatica, J.; Djerassi, C. *J. Am. Chem. Soc.* **1950**, *72*, 4077 [39, 170, 313, 314].
[266] Castro, C. E. *J. Am. Chem. Soc.* **1961**, *83*, 3262 [39].
[267] Williamson, K. L.; Hsu, Y.; Young, E. I. *Tetrahedron* **1968**, *24*, 6007 [39, 86].
[268] Akita, Y.; Inaba, M.; Uchida, H.; Ohta, A. *Synthesis* **1977**, 792 [39, 92, 121, 122, 130].
[269] Akita, Y.; Misu, K.; Watanabe, T.; Ohta, A. *Chem. Pharm. Bull.* **1976**, *24*, 1839 [39, 132].
[270] Allen, W. S.; Bernstein, S.; Feldman, L. I.; Weiss, M. J. *J. Am. Chem. Soc.* **1960**, *82*, 3696 [39, 114, 175].
[271] Castro, C. E.; Kray, W. C., Jr. *J. Am. Chem. Soc.* **1963**, *85*, 2768 [39, 82, 83, 85].
[272] Hanson, J. R.; Premuzic, E. *Tetrahedron* **1967**, *23*, 4105 [39, 92].
[273] Castro, C. E.; Stephens, R. D. *J. Am. Chem. Soc.* **1964**, *86*, 4358 [39, 65, 105, 106, 193, 197].
[274] Smith, A. B., III; Levenberg, P. A.; Suits, J. Z. *Synthesis* **1986**, 184 [39, 168].
[275] Hanson, J. R.; Premuzic, E. *J. Chem. Soc. C* **1969**, 1201 [39, 180].
[276]   Nozaki, H.; Arantani, T.; Noyori, T. *Tetrahedron* **1967**, *23*, 3645 [39, 83].
[277] Williamson, K. L.; Hsu, Y.; Young, E. I. *Tetrahedron* **1968**, *24*, 6007 [39, 86].
[278] McCall, J. M.; Ten Brink, R. E. *Synthesis* **1975**, 335 [39, 132].
[279] Timms, G. H.; Wildsmith, E. *Tetrahedron Lett.* **1971**, 195 [39, 185].
[280] Ho, T.-L.; Wong, C. M. *Synthesis* **1974**, 45 [39, 96].
[281] Entwistle, I. D.; Johnstone, R. A. W.; Wilby, A. H. *Tetrahedron* **1982**, *38*, 419 [39, 131].
[282] McMurry, J. E.; Silvestri, M. *J. Org. Chem.* **1975**, *40*, 1502 [39, 187].
[283] Ho, T.-L.; Wong, C. M. *Synth. Commun.* **1973**, *3*, 37 [39, 121, 122].
[284] McMurry, J. E.; Silvestri, M. G.; Fleming, M. P.; Hoz, T.; Grayston, M. W. *J. Org. Chem.* **1978**, *43*, 3249 [39, 107, 108, 110, 113].
[285] Fleming, M. P.; McMurry, J. E. *J. Am. Chem. Soc.* **1974**, *96*, 4708 [39, 151, 157].
[286] McMurry, J. E.; Fleming, M. P.; Kees, K. L.; Krepski, L. R. *J. Org. Chem.* **1978**, *43*, 3255 [39, 104, 137, 142, 151, 155, 156, 157, 164, 179, 314].
[287] Tyrlik, S.; Wolochowicz, I. *Bull. Soc. Chim. Fr.* **1973**, 2147 [39, 138, 151, 155, 157, 164].

[288] Tyrlik, S.; Wolochowicz, I. *Chem. Commun.* **1975**, 781 [39].
[289] Castedo, L.; Saa, J. M.; Suan, R.; Tojo, G. *J. Org. Chem.* **1981**, *46*, 4292 [39, 142, 151].
[290] McMurry, J. E.; Kees, K. L. *J. Org. Chem.* **1977**, *42*, 2655 [39, 179].
[291] McMurry, J. E. *Acc. Chem. Res.* **1974**, *7*, 281 [39].
[292] Ho, T.-L. *Synth. Commun.* **1979**, *9*, 241 [39].
[293] Slaugh, L. H.; Raley, J. H. *Tetrahedron* **1964**, *20*, 1005 [40, 314].
[294] Ho, T.-L.; Olah, G. A. *Synthesis* **1976**, 807 [40, 171].
[295] Ho, T.-L.; Olah, G. A. *Synthesis* **1976**, 815 [40, 177, 178, 181].
[296] Olah, G. A.; Prakash, S. G.; Ho, T.-L. *Synthesis* **1976**, 810 [40, 121, 122].
[297] Ho, T.-L.; Henniger, M.; Olah, G. A. *Synthesis* **1976**, 815 [40, 100, 315].
[298] Smith, L. I.; Opie, J. W. *Org. Synth., Collect. Vol.* **1955**, *3*, 56 [40, 96, 144, 145].
[299] Fieser, L. F.; Heyman, H. *J. Am. Chem. Soc.* **1942**, *64*, 376, 380 [40, 100].
[300] Adams, R.; Marvel, C. S. *Org. Synth., Collect. Vol.* **1932**, *1*, 358 [40, 83].
[301] Bigelow, H. E.; Palmer, A. *Org. Synth., Collect. Vol.* **1943**, *2*, 5 [40, 95, 96].
[302] Gemal, A. L.; Luche, J. L. *Tetrahedron Lett.* **1980**, *21*, 3195 [40, 170].
[303] Olah, G. A.; Arvanaghi, M.; Vankar, Y. D. *J. Org. Chem.* **1980**, *45*, 3531 [40].
[304] Girard, P.; Namy, J. L.; Kagan, H. B. *J. Am. Chem. Soc.* **1980**, *102*, 2693 [40, 82, 113, 121, 137, 315].
[305] Molander, G. A.; Kenny, C. *J. Org. Chem.* **1991**, *56*, 1439 [40].
[306] Inanaga, J.; Yokoyama, Y.; Baba, Y.; Yamaguchi, M. *Tetrahedron Lett.* **1991**, *32*, 5559 [40, 65].
[307] Walborsky, H. M.; Topolski, M. *J. Org. Chem.* **1992**, *57*, 370 [40, 82, 83].
[308] Inanaga, J.; Sakai, S.; Handa, Y.; Yamaguchi, M.; Yokoyama, Y. *Chem. Lett.* **1991**, 2117 [40, 219, 235].
[309] Handa, Y.; Inanaga, J.; Yamaguchi, M. *Chem. Commun.* **1989**, 298 [40, 134].
[310] Kende, A. S.; Mendoza, J. S. *Tetrahedron Lett.* **1991**, *32*, 1699 [40, 92].
[311] Künzer, H.; Stahnke, M.; Sauer, G.; Wiechert, R. *Tetrahedron Lett.* **1991**, *32*, 1949 [40, 123].
[312] Kusuda, K.; Inanaga, J.; Yamaguchi, M. *Tetrahedron Lett.* **1989**, *30*, 2945 [40, 210].
[313] Renaud, R. N.; Stephens, J. C. *Can. J. Chem.* **1974**, *52*, 1229 [41, 181].
[314] Blatt, A. H.; Tristram, E. W. *J . Am. Chem. Soc.* **1952**, *74*, 6273 [41, 98].
[315] Bock, K.; Lundt, I.; Pedersen, C. *Carbohydr. Res.* **1979**, *68*, 313; **1981**, *90*, 7, 17 [41, 110].
[316] Gunther, F. A.; Blinn, R. C. *J. Am. Chem. Soc.* **1950**, *72*, 4282 [41, 107].
[317] Konieczny, M.; Harvey, R. G. *J. Org. Chem.* **1979**, *44*, 4813 [41, 107].
[318] Bradsher, C. K.; Vingiello, F. A. *J. Org. Chem.* **1948**, *13*, 786 [41, 156].
[319] Platt, K. L.; Oesch, F. *J. Org. Chem.* **1981**, *46*, 2601 [41, 201].
[320] Bickel, C. L.; Morris, R. *J. Am. Chem. Soc.* **1951**, *73*, 1786 [41, 170].
[321] Wagner, A. W. *Chem. Ber.* **1966**, *99*, 375 [41, 125].
[322] Pojer, P. M.; Ritchie, E.; Taylor, W. C. *Austr. J. Chem.* **1968**, *21*, 1375 [41, 173].
[323] Smith, P. A. S.; Brown, B. B. *J. Am. Chem. Soc.* **1951**, *73*, 2438 [41, 100, 101].
[324] Bastian, J. M.; Ebnöter, A.; Jucker, E.; Rissi, E.; Stoll, A. P. *Helv. Chim. Acta* **1971**, *54*, 277 [41].
[325] Newman, M. S. *J. Am. Chem. Soc.* **1951**, *73*, 4994 [41].
[326] Kellogg, R. M.; Schaap, A. P.; Harper, E. T.; Wynberg, H. *J. Org. Chem.* **1968**, *33*, 2902 [41].
[327] Brady, O. L.; Day, J. N. D.; Reynolds, C. V. *J. Chem. Soc.* **1929**, 2264 [42, 96, 97, 98].
[328] Mayer, R.; Hiller, G.; Nitzschke, M.; Jentzsch, J. *Angew. Chem.* **1963**, *75*, 1011 [42, 177, 178, 316].

[329]   Robertson, R. *Org. Synth., Collect. Vol.* **1932**, *1*, 52 [42, 96, 198, 199].
[330]   Boon, W. R. *J. Chem. Soc.* **1949**, S 230 [42, 186].
[331]   Hartman, W. W.; Silloway, H. L. *Org. Synth., Collect. Vol.* **1955**, *3*, 82 [42, 97, 98, 110].
[332]   Huber, D.; Andermann, G.; Leclerc, G. *Tetrahedron Lett.* **1988**, *29*, 635 [42, 98].
[333]   Ueno, K. *J. Am. Chem. Soc.* **1952**, *74*, 4508 [42, 96, 134].
[334]   Beard, H. G.; Hodgson, H. H. *J. Chem. Soc.* **1944**, 4 [42].
[335]   Kober, E. *J. Org. Chem.* **1961**, *26*, 2270 [42].
[336]   Node, M.; Kawabata, T.; Ohta, K.; Fujimoto, M.; Fujita, E.; Fuji, K. *J. Org. Chem.* **1984**, *49*, 3641 [42, 109].
[337]   Massiot, G.; Mulamba, T.; Lévy, J. *Bull. Soc. Chim. Fr.* **1982**, 241 [42, 183].
[338]   Pappas, J. J.; Keaveney, W. P.; Gancher, E.; Berger, M. *Tetrahedron Lett.* **1966**, 4273 [42, 117].
[339]   Kremers, E.; Wakeman, N.; Hixon, R. M. *Org. Synth., Collect. Vol.* **1932**, *1*, 511 [42, 99].
[340]   Weil, H.; Traun, M.; Marcel, S. *Chem. Ber.* **1922**, *55*, 2664, 2671 [42, 134].
[341]   Lieber, E.; Sherman, E.; Henry, R. A.; Cohen, J. *J. Am. Chem. Soc.* **1951**, *73*, 2327 [42, 100, 101].
[342]   Schmidt, J.; Kampf, A. *Chem. Ber.* **1902**, *35*, 3124 [42, 180, 181].
[343]   Kozlov, V. V.; Smolin, D. D. *Zh. Obshch. Khim.* **1949**, *19*, 740; *Chem. Abstr.* **1950**, *44*, 3479 [42, 124].
[344]   Dodgson, J. W. *J. Chem. Soc.* **1914**, 2435 [42, 180, 181].
[345]   Hock, H.; Lang, S. *Chem. Ber.* **1943**, *76*, 169; **1944**, *77*, 257, 262 [42, 115].
[346]   Fuller, A. T.; Tonkin, I. M.; Walker, J. *J. Chem. Soc.* **1945**, 633 [42, 124, 125].
[347]   Bamberger, E.; Kraus, E. *Chem. Ber.* **1896**, *29*, 1829, 1834 [42, 100].
[348]   Madai, H.; Müller, R. *J. Prakt. Chem.* **1963**, [4], *19*, 83 [42, 83, 84, 317].
[349]   Chung, S.-K.; Hu, Q.-Y. *Synth. Commun.* **1982**, *12*, 261 [42, 170].
[350]   Louis-Andre, O.; Gelbard, G. *Tetrahedron Lett.* **1985**, *26*, 831 [42, 165, 166].
[351]   Akamanchi, K. G.; Patel, H. S.; Meenakshi, R. *Synth. Commun.* **1992**, *22*, 1655 [42, 166, 317].
[352]   Conant, J. B.; Corson, B. B. *Org. Synth., Collect. Vol.* **1943**, *2*, 33 [42, 99, 110].
[353]   Prelog, V.; Wiesner, K. *Helv. Chim. Acta* **1948**, *31*, 870 [42, 96, 110].
[354]   Hodgson, H. H.; Ward, E. R. *J. Chem. Soc.* **1947**, 327 [42, 96, 99, 317].
[355]   Kamimura, A.; Kurata, K.; Ono, N. *Tetrahedron Lett.* **1989**, *30*, 4819 [42, 172].
[356]   Fieser, L. F. *Org. Synth., Collect. Vol.* **1943**, *2*, 35, 39 [42, 134].
[357]   Adams, R.; Blomstrom, D. C. *J. Am. Chem. Soc.* **1953**, *75*, 3405 [42, 100].
[358]   Park, K. K.; Lee, C. W.; Choi, S. Y. *J. Chem. Soc., Perkin Trans. 1* **1992**, 601 [43, 123].
[359]   Ehrlich, P.; Bertheim, A. *Chem. Ber.* **1912**, *45*, 756 [43, 245].
[360]   Schmidt, H. *Justus Liebigs Ann. Chem.* **1920**, *421*, 221 [43, 245].
[361]   de Vries, J. G.; Kellogg, R. M. *J. Org. Chem.* **1980**, *45*, 4126 [43, 137, 138, 140, 141, 149, 150, 153, 157].
[362]   Mager, H. I. X.; Berends, W. *Recl. Trav. Chim. Pays-Bas* **1960**, *79*, 282 [43].
[363]   Harris, A. R. *Synth. Commun.* **1987**, *17*, 1587 [43, 170].
[364]   Hamersma, J. W.; Snyder, E. I. *J. Org. Chem.* **1965**, *30*, 3985 [43, 133, 134].
[365]   Thiele, J. *Justus Liebigs Ann. Chem.* **1892**, *271*, 127 [43].
[366]   Wade, P. A.; Amin, N. V. *Synth. Commun.* **1982**, *12*, 287 [43, 105].
[367]   Corey, E. J.; Mock, W. L. *J. Am. Chem. Soc.* **1962**, *84*, 685 [43].
[368]   Diels, O.; Schmidt, S.; Witte, W. *Chem. Ber.* **1938**, *71*, 1186 [43].
[369]   Wolinsky, J.; Schultz, T. *J. Org. Chem.* **1965**, *30*, 3980 [43, 56].
[370]   Ohno, M.; Okamoto, M. *Org. Synth., Collect. Vol.* **1973**, *5*, 281 [43, 56].
[371]   van Tamelen, E. E.; Dewey, R. S. *J. Am. Chem. Soc.* **1961**, *83*, 3729 [43].

[372] Hünig, S.; Müller, H. R.; Thier, W. *Angew. Chem., Int. Ed. Engl.* **1965**, *4*, 271 [43].
[373] Corey, E. J.; Pasto, D. J.; Mock, W. L. *J. Am. Chem. Soc.* **1961**, *83*, 2957 [43].
[374] Corey, E. J.; Mock, W. L.; Pasto, D. J. *Tetrahedron Lett.* **1961**, 347 [43, 65].
[375] Corey, E. J.; Mock, W. L. *J. Am. Chem. Soc.* **1962**, *84*, 685 [43].
[376] Moriarty, R. M.; Vaid, R. K.; Duncan, M. P. *Synth. Commun.* **1987**, *17*, 703 [43, 65, 219].
[377] Hudlický, T.; Boros, C. H.; Boros, E. E. *Synthesis* **1992**, 174 [43, 57].
[378] Hudlický, T.; Butura, G.; Fernby, S. unpublished results [43, 57].
[379] Hanuš, J.; Vořišek, J. *Collect. Czech. Chem. Commun.* **1929**, *1*, 223 [44].
[380] Müller, E.; Kraemer-Willenberg, H. *Chem. Ber.* **1924**, *57*, 575 [44, 63].
[381] Fletcher, T. L.; Namkung, M. J. *J. Org. Chem.* **1958**, *23*, 680 [44, 95, 97, 99].
[382] Mosby, W. L. *Chem. Ind. (London)* **1959**, 1348; *J. Org. Chem.* **1959**, *24*, 421 [44, 95, 99].
[383] Furst, A.; Berlo, R. C.; Hooton, S. *Chem. Rev.* **1965**, *65*, 51 [44].
[384] Todd, D. *Org. React. (N.Y.)* **1948**, *4*, 378 [44, 137, 144, 151, 157, 163].
[385] Huang-Minlon *J. Am. Chem. Soc.* **1946**, *68*, 2487 [44, 151, 156, 157, 163, 164, 201, 317].
[386] Soffer, M. D.; Soffer, M. B.; Sherk, K. W. *J. Am. Chem. Soc.* **1945**, *67*, 1435 [44, 157, 163].
[387] Buu-Hoi, N. P.; Hoan, N.; Xuong, N. D. *Recl. Trav. Chim. Pays-Bas* **1952**, *71*, 285 [44].
[388] McFadyen, J. S.; Stevens, T. S. *J. Chem. Soc.* **1936**, 584 [44, 238].
[389] Mosettig, E. *Org. React. (N.Y.)* **1954**, *8*, 232 [44, 238, 239].
[390] Maiti, S. N.; Spevak, P.; Singh, M. P.; Micetich, R. G. *Synth. Commun.* **1988**, *18*, 575 [44, 121].
[391] Work, J. B. *Inorg. Synth.* **1946**, *2*, 141 [44].
[392] Robertson, A. V.; Witkop, B. *J. Am. Chem. Soc.* **1962**, *84*, 1697 [44, 69].
[393] Suzuki, H.; Tani, H.; Kubota, H.; Sato, N.; Tsuji, J.; Osuka, A. *Chem. Lett.* **1983**, 247 [44, 107].
[394] Lutz, R. E.; Allison, R. K. et al. *J. Org. Chem.* **1947**, *12*, 617, 681 [44, 100, 173].
[395] Kornblum, N. *Org. React. (N.Y.)* **1944**, *2*, 262 [44, 46, 100].
[396] Chang-Ming, Hu; Ming-Hu, Tu *J. Fluorine Chem.* **1991**, *55*, 101 [44, 84].
[397] Khai, B. T.; Arcelli, A. *Chem. Ber.* **1993**, *126*, 2265 [45, 58, 318].
[398] Leone-Bay, A. *J. Org. Chem.* **1986**, *51*, 2378 [45, 174].
[399] Horner, L.; Jurgeleit, W. *Justus Liebigs Ann. Chem.* **1955**, *591*, 138 [45, 115, 116].
[400] Davis, R. E. *J. Org. Chem.* **1958**, *23*,1767 [45, 119, 120].
[401] Horner, L.; Hoffman, H. *Angew. Chem.* **1956**, *68*, 473 [45].
[402] Kamiya, N.; Tanmatu, H.; Ishii, Y. *Chem. Lett.* **1992**, 293 [45].
[403] Schuetz, R. D.; Jacobs, R. L. *J. Org. Chem.* **1958**, *23*, 1799; **1961**, *26*, 3467 [45, 119].
[404] Brink, M. *Synthesis* **1975**, 807 [45].
[405] Mori, T.; Nakahara, T.; Nozaki, H. *Can. J. Chem.* **1969**, *47*, 3266 [45, 177].
[406] Saegusa, T.; Kobayashi, S.; Kimura, Y.; Yokoyama, T. *J. Org. Chem.* **1977**, *42*, 2797 [45].
[407] Bunyan, P. J.; Cadogan, J. I. G. *J. Chem. Soc.* **1963**, 42 [45, 99].
[408] Dirlam, J. P.; McFarland, J. W. *J. Org. Chem.* **1977**, *42*, 1360 [45, 132].
[409] Hirao, T.; Masunaga, T.; Ohshiro, Y.; Agawa, T. *J. Org. Chem.* **1981**, *46*, 3745 [45, 83].
[410] Clive, D. L. J.; Beaulieu, P. L. *J. Org. Chem.* **1982**, *47*, 1124 [45].
[411] Clive, D. L. J.; Menchen, S. M. *J. Org. Chem.* **1980**, *45*, 2347 [45, 113].

[*412*]   Pinnick, H. W.; Reynolds, M. A.; McDonald, R. T., Jr.; Brewster, W. D. *J. Org. Chem.* **1980**, *45*, 930 [45, 123, 124].

[*413*]   Newman, M. S.; Blum, S. *J. Am. Chem. Soc.* **1964**, *86*, 5598 [45, 144].

[*414*]   Harpp, D. N.; Gleason, J. G.; Snyder, J. P. *J. Am. Chem. Soc.* **1968**, *90*, 4181 [45, 121].

[*415*]   Harpp, D. N.; Smith, R. A. *Org. Synth., Collect. Vol.* **1988**, *6*, 130 [45, 121].

[*416*]   Guindon, Y.; Atkinson, J. G.; Morton, H. E. *J. Org. Chem.* **1984**, *49*, 4538 [45, 121, 122].

[*417*]   Urata, H.; Suzuki, H.; Moro-Oka, Y.; Ikawa, T. *J. Organomet. Chem.* **1982**, *234*, 367 [45, 169, 170].

[*418*]   Numazawa, M.; Nagaoka, M.; Kunitama, Y. *Chem. Commun.* **1984**, 31 [46, 174].

[*419*]   Olah, G. A.; Husain, A.; Singh, B. P.; Mehrotra, A. *J. Org. Chem.* **1983**, *48*, 3667 [46, 121, 122, 319].

[*420*]   Barclay, L. B.; Farrar, M. W.; Knowles, W. S.; Raffelson, H. *J. Am. Chem. Soc.* **1954**, *76*, 5017 [46].

[*421*]   Hodgson, H. H.; Turner, H. S. *J. Chem. Soc.* **1942**, 748; **1943**, 86 [46, 100].

[*422*]   Kato, M.; Tamano, T.; Miwa, T. *Bull. Chem. Soc. Jpn.* **1975**, *48*, 291 [46, 100, 198].

[*423*]   van Leusen, A. M.; Smid, P. M.; Strating, J. *Tetrahedron Lett.* **1967**, 1165 [46, 173].

[*424*]   Portella, C.; Iznaden, M. *Tetrahedron* **1989**, *45*, 6467 [46].

[*425*]   Liška, F.; Dĕdek, V.; Holík, M. *Collect. Czech. Chem. Commun.* **1971**, *36*, 2846 [46, 81, 84, 109, 110].

[*426*]   Šendrik, V. P.; Paleta, O.; Dĕdek, V. *Collect. Czech. Chem. Commun.* **1976**, *41*, 874 [46, 221].

[*427*]   Pošta, A.; Paleta, O.; Voves, J. *Collect. Czech. Chem. Commun.* **1974**, *39*, 2801 [46, 81, 84].

[*428*]   Reinholdt, K.; Margaretha, P. *Helv. Chim. Acta* **1983**, *66*, 2534 [46, 170].

[*429*]   Bachmann, W. E. *Org. Synth., Collect. Vol.* **1943**, *2*, 71 [46, 155, 156].

[*430*]   Wilds, A. L. *Org. React. (N.Y.)* **1944**, *2*, 178 [46, 137, 144, 149, 153, 166, 319].

[*431*]   Pinkey, D. T.; Rigby, R. D. G. *Tetrahedron Lett.* **1969**, 1267 [46, 89].

[*432*]   Winstein, S. *J. Am. Chem. Soc.* **1939**, *61*, 1610 [46, 170].

[*433*]   Werner, E. A. *J. Chem. Soc.* **1917**, 844 [46, 189].

[*434*]   Davidson, D.; Weiss, M. *Org. Synth., Collect. Vol.* **1943**, *2*, 590 [46, 141].

[*435*]   Mole, T. *J. Chem. Soc.* **1960**, 2132 [46, 153].

[*436*]   Johns, R. B.; Markham, K. R. *J. Chem. Soc.* **1962**, 3712 [46, 172].

[*437*]   Newbold, B. T.; LeBlanc, R. P. *J. Chem. Soc.* **1965**, 1547 [46, 95].

[*438*]   Anwer, M. K.; Spatola, A. F. *Tetrahedron Lett.* **1985**, *26*, 1381 [46, 87, 109].

[*439*]   Cortese, N. A.; Heck, R. F. *J. Org. Chem.* **1977**, *42*, 3491; **1978**, *43*, 3985 [46, 87, 138, 139].

[*440*]   Threadgill, M. D.; Gledhill, A. P. *J. Chem. Soc., Perkin Trans. 1* **1986**, 873 [47, 100].

[*441*]   Nanjo, K.; Suzuki, K.; Sekiya, M. *Chem. Lett.* **1976**, 1169 [47, 242].

[*442*]   Lukeš, R.; Pliml, J. *Collect. Czech. Chem. Commun.* **1950**, *15*, 464 [47, 71].

[*443*]   Kocián, O.; Ferles, M. *Collect. Czech. Chem. Commun.* **1979**, *44*, 1167 [47, 74, 157].

[*444*]   de Benneville, P. L.; Macartney, J. H. *J. Am. Chem. Soc.* **1950**, *72*, 3073 [47, 129].

[*445*]   Clarke, H. T.; Gillespie, H. B.; Weisshaus, S. Z. *J. Am. Chem. Soc.* **1933**, *55*, 4571 [47, 189].

[*446*]   Novelli, A. *J. Am. Chem. Soc.* **1939**, *61*, 520 [47, 189, 190, 320].

[447] Moore, M. L. *Org. React. (N.Y.)* **1949**, *5*, 301 [47, 187, 189].
[448] Blicke, F. F.; Powers, L. D. *J. Am. Chem. Soc.* **1929**, *51*, 3378 [47, 153].
[449] Greenwood, F. L.; Whitmore, F. C.; Crooks, H. M. *J. Am. Chem. Soc.* **1938**, *60*, 2028 [47, 204].
[450] Respess, W. L.; Tamborski, C. *J. Organomet. Chem.* **1969**, *18*, 263 [47].
[451] Ducoux, J.-P.; Le Ménez, P.; Kunesh, N.; Wenkert, E. *J. Org. Chem.* **1993**, *58*, 1290 [47, 111].
[452] Meerwein, H.; Hinz, G.; Majert, H.; Sönke, H. *J. Prakt. Chem.* **1936**, [2], *147*, 226 [47].
[453] Sako, M.; Hirota, K.; Maki, Y. *Tetrahedron* **1983**, *39*, 3919 [47, 235].
[454] Bucciarelli, M.; Forni, A.; Moretti, I.; Torre, G. *Synthesis* **1983**, 897 [48, 150, 154, 169].
[455] Sih, C. J.; Chen, C.-S. *Angew. Chem.* **1984**, *96*, 556; *Int. Ed. Engl.* **1984**, *23*, 570 [48, 49, 200, 225].
[456] Ohta, H.; Ozaki, K.; Tsuchihashi, G.-i. *Chem. Lett.* **1987**, 191 [48, 49, 321].
[457] Högberg, H.-E.; Hedenström, E.; Fägerhag, J. *J. Org. Chem.* **1992**, *57*, 2052 [48, 49].
[458] Brooks, D. W.; Mazdiyasni, H.; Grothans, P. G. *J. Org. Chem.* **1987**, *52*, 3223 [48, 49].
[459] de Carvalho, M.; Okamota, M. T.; Moran, P. S. S.; Rodrigues, J. A. R. *Tetrahedron* **1991**, *47*, 2073 [48, 49].
[460] Manzocchi, A.; Casati, R.; Fiecchi, A.; Santaniello, E. *J. Chem. Soc., Perkin Trans. 1* **1987**, 2753 [48, 49].
[461] Seebach, D.; Sutter, M. A.; Weber, R. H. *Org. Synth.* **1984**, *63*, 1 [48, 49].
[462] Naoshima, Y.; Maeda, J.; Munakata, Y.; Nishiyama, T.; Kamezawa, M.; Tachibana, H. *Chem. Commun.* **1990**, 964 [48, 49].
[463] Gramatica, P.; Manitto, P.; Monti, D.; Speranza, G. *Tetrahedron* **1988**, *44*, 1299 [48, 49].
[464] Simon, H.; Bader, J.; Gunther, H.; Newmann, S.; Thanos, J. *Angew. Chem., Int. Ed. Engl.* **1985**, *24*, 539 [49, 50].
[465] Hashimoto, H.; Simon, H. *Angew. Chem., Int. Ed. Engl.* **1975**, *14*, 106 [49, 50, 198].
[466] Kitazume, T.; Ishikawa, N. *Chem. Lett.* **1983**, 237 [49, 50].
[467] Osborn, J. A.; Jardine, F. H.; Young, J. F.; Wilkinson, G. *J. Chem. Soc. A* **1966**, 1711 [15].
[468] Bradshaw, C. W.; Hummel, W.; Wong, C.-H. *J. Org. Chem.* **1992**, *57*, 1532 [49, 50, 154, 155, 322].
[469] Bradshaw, C. W.; Fu, H.; Shen, G.-J.; Wong, C.-H. *J. Org. Chem.* **1992**, *57*, 1526 [49, 50, 154, 155].
[470] Dodds, D. R.; Jones, J. B. *J. Am. Chem. Soc.* **1988**, *110*, 577 [49, 50, 150, 180, 321].
[471] Casy, G.; Lee, T. V.; Levell, H. *Tetrahedron Lett.* **1992**, *33*, 817 [49, 50, 200, 201].
[472] Kim, M.-J.; Whitesides, G. M. *J. Am. Chem. Soc.* **1988**, *110*, 2959 [49, 50, 200, 201].
[473] Ohta, H.; Konishi, J.; Tsuchihashi, G. *Chem. Lett.* **1983**, 1895 [49, 166].
[474] Baumann, P.; Prelog, V. *Helv. Chim. Acta* **1959**, *42*, 736 [50, 320].
[475] Bernardi, R.; Cardillo, R.; Ghiringhelli, D.; de Pava, O. V. *J. Chem. Soc., Perkin Trans. 1* **1987**, 1607 [50, 225, 226].
[476] Bégué, J.-P.; Cerceau, C.; Dagbeavou, A.; Mathé, L. *J. Chem. Soc., Perkin Trans. 1* **1992**, 3141 [50, 162].
[477] Chicamatsu, H.; Taniguchi, M.; Kanemitsu, T. *Bull. Chem. Soc. Jpn.* **1986**, *59*, 2663 [50, 180].

[478]   Seebach, D.; Züger, M. F.; Giovannini, F.; Sonnleitner, B.; Fiechter, A. *Angew. Chem., Int. Ed. Engl.* **1984**, *23*, 151 [50, 225, 226].
[479]   Raddatz, P.; Radunz, H.-E.; Schneider, G.; Schwartz, H. *Angew. Chem., Int. Ed. Engl.* **1988**, *27*, 426 [50].
[480]   Braun, H.; Schmidtchen, F. P.; Schneider, A.; Simon H. *Tetrahedron* **1991**, *47*, 3329 [50, 132].
[481]   Konishi, J.; Ohta, H.; Tsuchihashi, G.-i. *Chem. Lett.* **1985**, 1111 [50, 178, 179].
[482]   Willstätter, R.; Bruce, J. *Chem. Ber.* **1907**, *40*, 3979, 4456 [53].
[483]   Zelinsky, N. D.; Kazanski, B. A.; Plate, A. F. *Chem. Ber.* **1933**, *66*, 1415 [53].
[484]   Kieboom, A. P. G.; van Benschop, H. J.; van Bekkum, H. *Recl. Trav. Chim. Pays-Bas* **1976**, *95*, 231 [53].
[485]   Kursanov, D. N.; Parnes, Z. N.; Loim, N. M. *Synthesis* **1974**, 633 [53, 55, 156].
[486]   Jardine, I.; McQuillin, F. J. *J. Chem. Soc. C* **1966**, 458 [53].
[487]   Berkowitz, L. M.; Rylander, P. N. *J. Org. Chem.* **1959**, *24*, 708 [53].
[488]   Brown, C. A. *Chem. Commun.* **1969**, 952 [54].
[489]   Siegel, S.; Smith, G. V. *J. Am. Chem. Soc.* **1960**, *82*, 6082, 6087 [54].
[490]   Tyman, J. H. P.; Wilkins, S. W. *Tetrahedron Lett.* **1973**, 1773 [55].
[491]   Brown, H. C.; Murray, K. J. *J. Am. Chem. Soc.* **1959**, *81*, 4108 [55].
[492]   Carey, F. A.; Tremper, H. S. *J. Org. Chem.* **1971**, *36*, 758 [55, 103].
[493]   Whitesides, G. M.; Ehmann, W. J. *J. Org. Chem.* **1970**, *35*, 3565 [56, 59, 61].
[494]   Corey, E. J.; Cantrall, E. W. *J. Am. Chem. Soc.* **1959**, *81*, 1745 [56].
[495]   van Tamelen, E. E.; Timmons, R. J. *J. Am. Chem. Soc.* **1962**, *84*, 1067 [56].
[496]   Ohno, M.; Okamoto, M. *Tetrahedron Lett.* **1964**, 2423 [56].
[497]   Newhall, W. F. *J. Org. Chem.* **1958**, *23*, 1274 [56].
[498]   Brown, H. C.; Brown, C. A. *J. Am. Chem. Soc.* **1963**, *85*, 1005 [56, 58, 59].
[499]   Hubert, A. J. *J. Chem. Soc. C* **1967**, 2149 [56].
[500]   Midgley, T.; Henne, A. L. *J. Am. Chem. Soc.* **1929**, *51*, 1293 [56].
[501]   Cope, A. C.; Hochstein, F. A. *J. Am. Chem. Soc.* **1950**, *72*, 2515 [57].
[502]   Ziegler, K.; Jakob, L.; Wollthan, H.; Wenz, A. *Justus Liebigs Ann. Chem.* **1934**, *511*, 64 [57].
[503]   Ziegler, K.; Häffner, F.; Grimm, H. *Justus Liebigs Ann. Chem.* **1937**, *528*, 101 [57].
[504]   Ziegler, K.; Wilms, H. *Justus Liebigs Ann. Chem.* **1950**, *567*, 1, 36 [57].
[505]   Bance, S.; Barber, H. J.; Woolmann, A. M. *J. Chem. Soc.* **1943**, 1 [58, 64, 242].
[506]   Lumb, P. B.; Smith, J. C. *J. Chem. Soc.* **1952**, 5032 [58].
[507]   Howton, D. R.; Davis, R. H. *J. Org. Chem.* **1951**, *16*, 1405 [58, 193, 194].
[508]   Campbell, K. N.; O'Connor, M. J. *J. Am. Chem. Soc.* **1939**, *61*, 2897 [58, 61, 64, 65].
[509]   Campbell, K. N.; Eby, L. T. *J. Am. Chem. Soc.* **1941**, *63*, 216, 2683 [58, 60].
[510]   Brown, C. A.; Ahuja, V. K. *Chem. Commun.* **1973**, 553 [58, 59].
[511]   Bellas, T. E.; Brownlee, R. G.; Silverstein, R. M. *Tetrahedron* **1969**, *25*, 5149 [58].
[512]   Savoia, D.; Tagliviani, E.; Trombini, C.; Umani-Ronchi, A. *J. Org. Chem.* **1981**, *46*, 5340, 5344 [58].
[513]   Newman, M. S.; Waltcher, I.; Ginsberg, H. F. *J. Org. Chem.* **1952**, *17*, 962 [58, 105].
[514]   Hennion, G. F.; Schroeder, W. A.; Lu, R. P.; Scanlon, W. B. *J. Org. Chem.* **1956**, *21*, 1142 [58, 105].
[515]   Harper, S. H.; Smith, R. J. D. *J. Chem. Soc.* **1955**, 1512 [58, 105].
[516]   Blomquist, A. T.; Burge, R. E., Jr.; Sucsy, A. C. *J. Am. Chem. Soc.* **1952**, *74*, 3636 [58].
[517]   Isler, O.; Huber, W.; Ronco, A.; Kofler, M. *Helv. Chim. Acta* **1947**, *30*, 1911 [58, 60, 105, 106].
[518]   Cram, D. J.; Allinger, N. L. *J. Am. Chem. Soc.* **1956**, *78*, 2518 [58].

[519] Tedeschi, R. J.; Clark, G., Jr. *J. Org. Chem.* **1962**, *27*, 4323 [58, 105].
[520] Brunet, J.-J.; Caubère, P. *J. Org. Chem.* **1984**, *49*, 4058 [58].
[521] Magoon, E. F.; Slaugh, L. H. *Tetrahedron* **1967**, *23*, 4509 [58, 59].
[522] Slaugh, L. H. *Tetrahedron* **1966**, *22*, 1741 [58].
[523] Zweifel, G.; Steele, R. B. *J. Am. Chem. Soc.* **1967**, *89*, 5085 [58, 59].
[524] Wilke, G.; Müller, H. *Chem. Ber.* **1956**, *89*, 444 [58, 59].
[525] Cope, A. C.; Berchtold, G. A.; Peterson, P. E.; Sharman, S. H. *J. Am. Chem. Soc.* **1960**, *82*, 6370 [58].
[526] Ashby, E. C.; Lin, J. J. *Tetrahedron Lett.* **1977**, 4481 [59].
[527] Chum, P. W.; Wilson, S. E. *Tetrahedron Lett.* **1976**, 15 [59].
[528] Benkeser, R. A.; Tinchner, C. A. *J. Org. Chem.* **1968**, *33*, 2727 [60, 64, 65].
[529] Dear, R. E. A.; Pattison, F. L. M. *J. Org. Chem.* **1963**, *85*, 622 [60].
[530] Svoboda, M.; Závada, J.; Sicher, J. *Collect. Czech. Chem. Commun.* **1965**, *30*, 413 [60].
[531] Blomquist, A. T.; Liu, L. H.; Bohrer, J. C. *J. Am. Chem. Soc.* **1952**, *74*, 3643 [60].
[532] Devaprabhakara, D.; Gardner, P. B. *J. Am. Chem. Soc.* **1963**, *85*, 648 [60].
[533] Moore, W. R.; Ward, H. R. *J. Am. Chem. Soc.* **1963**, *85*, 86 [60].
[534] Chanley, J. D.; Sobotka, H. *J. Am. Chem. Soc.* **1949**, *71*, 4140 [60, 105].
[535] Oppolzer, W.; Fehr, C.; Warneke, J. *Helv. Chim. Acta* **1977**, *60*, 48 [60, 105, 106].
[536] Dobson, N. A.; Raphael, R. A. *J. Chem. Soc.* **1955**, 3558 [60].
[537] Hershberg, E. B.; Oliveto, E. P.; Gerold, C.; Johnson, L. *J. Am. Chem. Soc.* **1951**, *73*, 5073 [61, 168].
[538] Weissberger, A. *J. Chem. Soc.* **1935**, 855 [61, 64, 65].
[539] Adams, R.; Marshall, I. R. *J. Am. Chem. Soc.* **1928**, *50*, 1970 [61, 62].
[540] Nishimura, S. *Bull. Chem. Soc. Jpn.* **1959**, *32*, 1155 [61, 109].
[541] Stocker, J. H. *J. Org. Chem.* **1962**, *27*, 2288 [62].
[542] Zelinsky, N. D.; Margolis, E. I. *Chem. Ber.* **1932**, *65*, 1613 [62].
[543] Hückel, W.; Wörffel, U. *Chem. Ber.* **1955**, *88*, 338 [63].
[544] Slaugh, L. H.; Raley, J. H. *J. Org. Chem.* **1967**, *32*, 369 [63].
[545] Campbell, K. N.; McDermott, J. P. *J. Am. Chem. Soc.* **1945**, *67*, 282 [63].
[546] Paquette, L. A.; Barrett, J. N. *Org. Synth., Collect. Vol.* **1973**, *5*, 467 [63].
[547] Birch, A. J. *J. Chem. Soc.* **1944**, 430 [63, 109, 113, 195, 309].
[548] Hückel, W.; Bretschneider, H. *Justus Liebigs Ann. Chem.* **1939**, *540*, 157 [63, 65, 67].
[549] Erman, W. T.; Flautt, T. J. *J. Org. Chem.* **1962**, *27*, 1526 [64].
[550] Hückel, W.; Schwen, R. *Chem. Ber.* **1956**, *89*, 150 [64, 67].
[551] Goodman, I. *J. Chem. Soc.* **1951**, 2209 [64, 66].
[552] Benkeser, R. A.; Arnold, C., Jr.; Lambert, R. F.; Thomas, O.H. *J. Am. Chem. Soc.* **1955**, *77*, 6042 [64].
[553] Adkins, H.; Reid, W. A. *J. Am. Chem. Soc.* **1941**, *63*, 741 [65, 66].
[554] Willstätter, R.; Seitz, F. *Chem. Ber.* **1923**, *56*, 1388; **1924**, *57*, 683 [65, 66].
[555] Stuhl, L. S.; Rakowski DuBois, M.; Hirsekorn, F. J.; Bleeke, J. R.; Stevens, A. E.; Muetterties, E. L. *J. Am. Chem. Soc.* **1978**, *100*, 2405 [65, 66, 67].
[556] Cook, E. S.; Hill, A. J. *J. Am. Chem. Soc.* **1940**, *62*, 1995 [65, 67].
[557] Birch, A. J.; Murray, A. R.; Smith, H. *J. Chem. Soc.* **1951**, 1945 [65].
[558] Hückel, W.; Schlee, H. *Chem. Ber.* **1955**, *88*, 346 [65, 67].
[559] Benkeser, R. A.; Kaiser, E. M. *J. Org. Chem.* **1964**, *29*, 955 [66].
[560] Kaiser, E. M.; Benkeser, R. A. *Org. Synth., Collect. Vol.* **1988**, *6*, 852 [66, 314].
[561] Hückel, W.; Wörffel, U. *Chem. Ber.* **1956**, *89*, 2098 [66, 67].
[562] Bass, K. C. *Org. Synth., Collect. Vol.* **1973**, *5*, 398 [66, 67].
[563] Crossley, N. S.; Henbest, H. B. *J. Chem. Soc.* **1960**, 4413 [66, 67].

[564]   Durland, J. R.; Adkins, H. *J. Am. Chem. Soc.* **1937**, *59*, 135; **1938**, *60*, 1501 [66, 68].

[565]   Burger, A.; Mosettig, E. *J. Am. Chem. Soc.* **1935**, *57*, 2731; **1936**, *58*, 1857 [66].

[566]   Phillips, D. D. *Org. Synth., Collect. Vol.* **1963**, *4*, 313 [66, 68].

[567]   Bamberger, E.; Lodter, W. *Chem. Ber.* **1887**, *20*, 3073 [66, 68].

[568]   Tarbell, D. S.; Weaver, C. *J. Am. Chem. Soc.* **1941**, *63*, 2939 [66, 68].

[569]   Entel, J.; Ruof, C. H.; Howard, H. C. *J. Am. Chem. Soc.* **1951**, *73*, 4152 [66].

[570]   Papa, D.; Schwenk, T.; Ginsberg, H. F. *J. Org. Chem.* **1951**, *16*, 253 [66].

[571]   Borowitz, I. J.; Gonis, G.; Kelsey, R.; Rapp, R.; Williams, G. J. *J. Org. Chem.* **1966**, *31*, 3032 [68].

[572]   Campaigne, E.; Diedrich, J. L. *J. Am. Chem. Soc.* **1951**, *73*, 5240 [68].

[573]   Greenfield, H.; Metlin, S.; Orchin, M.; Wender, I. *J. Org. Chem.* **1958**, *23*, 1054 [68].

[574]   Craig, L. C.; Hixon, R. M. *J. Am. Chem. Soc.* **1930**, *52*, 804; **1931**, *53*, 188 [68, 69].

[575]   Signaigo, F. K.; Adkins, H. *J. Am. Chem. Soc.* **1936**, *58*, 709 [68, 69, 156, 157, 213, 214].

[576]   Cantor, P. A.; VanderWerf, C. A. *J. Am. Chem. Soc.* **1958**, *80*, 970 [69].

[577]   Adkins, H.; Coonradt, H. L. *J. Am. Chem. Soc.* **1941**, *63*, 1563 [69, 72, 73, 74, 75].

[578]   Overberger, C. G.; Palmer, L. C.; Marks, B. S.; Byrd, N. R. *J. Am. Chem. Soc.* **1955**, *77*, 4100 [69].

[579]   Šorm, F. *Collect. Czech. Chem. Commun.* **1947**, *12*, 248 [69, 157].

[580]   Lukeš, R.; Trojánek, J. *Collect. Czech. Chem. Commun.* **1953**, *18*, 648 [69].

[581]   Freifelder, M. *J. Org. Chem.* **1963**, *28*, 602 [69].

[582]   Leonard, N. J.; Barthel, E., Jr. *J. Am. Chem. Soc.* **1949**, *71*, 3098 [69, 235].

[583]   Freifelder, M.; Wright, H. B. *J. Med. Chem.* **1964**, *7*, 664 [69].

[584]   Adkins, H.; Kuick, L. F.; Farlow, M.; Wojcik, B. *J. Am. Chem. Soc.* **1934**, *56*, 2425 [69, 70].

[585]   Freifelder, M.; Stone, G. R. *J. Org. Chem.* **1961**, *26*, 3805 [70].

[586]   Freifelder, M. *Adv. Catal.* **1963**, *14*, 203 [70, 74].

[587]   Lansbury, P. T. *J. Am. Chem. Soc.* **1961**, *83*, 429 [70].

[588]   Lansbury, P. T.; Peterson, J. O. *J. Am. Chem. Soc.* **1963**, *85*, 2236 [70, 150, 153].

[589]   Ferles, M.; Pliml, J. *Adv. Heterocycl. Chem.* **1970**, *12*, 43 [70, 71].

[590]   Marvel, C. S.; Lazier, W. A. *Org. Synth., Collect. Vol.* **1932**, *1*, 99 [70].

[591]   Ferles, M.; Tesařová, A. *Collect. Czech. Chem. Commun.* **1967**, *32*, 1631 [70, 142].

[592]   Ladenburg, A. *Chem. Ber.* **1889**, *25*, 2768 [70, 71].

[593]   Papa, D.; Schwenk, E.; Klingsberg, E. *J. Am. Chem. Soc.* **1951**, *73*, 253 [70].

[594]   Ladenburg, A. *Justus Liebigs Ann. Chem.* **1888**, *247*, 1, 51 [70].

[595]   Daeniker, H. U.; Grob, C. A. *Org. Synth., Collect. Vol.* **1973**, *5*, 989 [70].

[596]   Holík, M.; Ferles, M. *Collect. Czech. Chem. Commun.* **1967**, *32*, 3067 [71].

[597]   Ferles, M. *Collect. Czech. Chem. Commun.* **1959**, *24*, 2221 [71].

[598]   Ferles, M.; Attia, A.; Šilhánková, A. *Collect. Czech. Chem. Commun.* **1973**, *38*, 615 [71, 130].

[599]   Ferles, M. *Collect. Czech. Chem. Commun.* **1958**, *23*, 479 [71].

[600]   Ferles, M.; Holík, M. *Collect. Czech. Chem. Commun.* **1966**, *31*, 2416 [71, 235].

[601]   Profft, E.; Linke, H. W. *Chem. Ber.* **1960**, *93*, 2591 [71, 72].

[602]   Lukeš, R. *Collect. Czech. Chem. Commun.* **1947**, *12*, 71 [71].

[603]   Ochiai, E.; Tsuda, K.; Ikuma, S. *Chem. Ber.* **1936**, *69*, 2238 [72].

[604] Späth, E.; Kuffner, F. *Chem. Ber.* **1935**, *68*, 494 [72].
[605] Wibaut, J. P.; Oosterhuis, A. G. *Recl. Trav. Chim. Pays-Bas* **1933**, *52*, 941 [72].
[606] Smith, A.; Utley, J. H. P. *Chem. Commun.* **1965**, 427 [72, 129].
[607] Gribble, G. W.; Lord, P. D.; Skotnicki, J.; Dietz, S. E.; Eaton, J. T.; Johnson, J. L. *J. Am. Chem. Soc.* **1974**, *96*, 7812 [72, 236].
[608] Berger, J. G. *Synthesis* **1974**, 508 [72].
[609] Kikugawa, Y.; Saito, K.; Yamada, S. *Synthesis* **1978**, 447 [72, 73].
[610] O'Brien, S.; Smith, D. C. C. *J. Chem. Soc.* **1960**, 4609 [72].
[611] Remers, W. A.; Gibs, G. J.; Pidacks, C.; Weiss, M. J. *J. Org. Chem.* **1971**, *36*, 279 [72].
[612] Dolby, L. J.; Gribble, G. W. *J. Heterocycl. Chem.* **1966**, *3*, 124 [72].
[613] Young, D. V.; Snyder, H. R. *J. Am. Chem. Soc.* **1961**, *83*, 3160 [72].
[614] King, F. E.; Barltrop, J. A.; Walley, R. J. *J. Chem. Soc.* **1945**, 277 [72, 185, 186].
[615] Boekelheide, V.; Chu-Tsin, L. *J. Am. Chem. Soc.* **1952**, *74*, 4920 [72].
[616] Lowe, O. G.; King L. C. *J. Org. Chem.* **1959**, *24*, 1200 [72, 73].
[617] Galbraith, A.; Small, T.; Barnes, R. A.; Boekelheide, V. *J. Am. Chem. Soc.* **1961**, *83*, 453 [72, 73].
[618] Hückel, W.; Hagedorn, L. *Chem. Ber.* **1957**, *90*, 752 [72, 73].
[619] Hückel, W.; Stepf, F. *Justus Liebigs Ann. Chem.* **1927**, *453*, 163 [72, 73].
[620] Gribble, G. W.; Heald, P. W. *Synthesis* **1975**, 650 [72, 73, 237].
[621] Vierhapper, F. W.; Eliel, E. L. *J. Org. Chem.* **1975**, *40*, 2729 [73].
[622] Kocián, O.; Ferles, M. *Collect. Czech. Chem. Commun.* **1978**, *43*, 1413 [74].
[623] Birch, A. J.; Nasipuri, D. *Tetrahedron* **1959**, *6*, 148 [74].
[624] Witkop, B. *J. Am. Chem. Soc.* **1948**, *70*, 2617 [74].
[625] Boekelheide, V.; Gall, W. G. *J. Am. Chem. Soc.* **1954**, *76*, 1832 [74].
[626] Bohlmann, F. *Chem. Ber.* **1952**, *85*, 390 [74, 75, 76, 77, 78].
[627] Masamune, T.; Ohno, M.; Koshi, M.; Ohushi, S.; Iwadare T. *J. Org. Chem.* **1964**, *29*, 1419 [74, 75].
[628] Vierhapper, F. W.; Eliel, E. L. *J. Org. Chem.* **1975**, *40*, 2729 [74].
[629] Thoms, H.; Schnupp, J. *Justus Liebigs Ann. Chem.* **1923**, *434*, 296 [76].
[630] Butula, I. *Croat. Chem. Acta* **1973**, *45*, 313 [76].
[631] Bauer, H. *J. Org. Chem.* **1961**, *26*, 1649 [76].
[632] Hartmann, M.; Pannizon, L. *Helv. Chim. Acta* **1938**, *21*, 1692 [76].
[633] Evans, R. C.; Wiselogle, F. Y. *J. Am. Chem. Soc.* **1945**, *67*, 60 [76].
[634] Smith, V. H.; Christensen, B. E. *J. Org. Chem.* **1955**, *20*, 829 [76].
[635] Behun, J. D.; Levine, R. *J. Org. Chem.* **1961**, *26*, 3379 [76].
[636] Armand, J.; Chekir, K.; Pinson, J. *Can. J. Chem.* **1974**, *52*, 3971 [76].
[637] Schatz, F.; Wagner-Jauregg, T. *Helv. Chim. Acta* **1968**, *51*, 1919 [76].
[638] Neber, P. W.; Knoller, G.; Herbst, K.; Trissler, A. *Justus Liebigs Ann. Chem.* **1929**, *471*, 113 [76].
[639] Lund, H. *Acta Chem. Scand.* **1967**, *21*, 2525 [76, 77].
[640] Marr, E. B.; Bogert, M. T. *J. Am. Chem. Soc.* **1935**, *57*, 729 [76, 77].
[641] Bugle, R. C.; Oosteryoung, R. A. *J. Org. Chem.* **1979**, *44*, 1719 [76, 77].
[642] Maffei, S.; Pietra, S. *Gazz. Chim. Ital.* **1958**, *88*, 556 [76, 77, 78].
[643] Cavagnol, J. C.; Wiselogle, F. Y. *J. Am. Chem. Soc.* **1947**, *69*, 795 [76, 77].
[644] Broadbent, H. S.; Allred, E. L.; Pendleton, L.; Whittle, C. W. *J. Am. Chem. Soc.* **1960**, *82*, 189 [76, 77].
[645] Christie, W.; Rohde, W.; Schultz, H. P. *J. Org. Chem.* **1956**, *21*, 243 [76, 77].
[646] Lund, H.; Jensen, E. T. *Acta Chem. Scand.* **1970**, *24*, 1867; **1971**, *25*, 2727 [76, 78].
[647] Elslager, E. F.; Worth, D. F.; Haley, N. F.; Perricone, S. S. *J. Heterocycl. Chem.* **1968**, *5*, 609 [76, 78].

[648] Eckhard, I. F.; Fielden, R.; Summers, L. A. *Austr. J. Chem.* **1975**, *28*, 1149 [78].
[649] Karrer, P.; Waser, P. *Helv. Chim. Acta* **1949**, *32*, 409 [78, 79].
[650] Butula, I. *Justus Liebigs Ann. Chem.* **1969**, *729*, 73 [78].
[651] Lacher, J. R.; Kianpour, A.; Park, J. D. *J. Phys. Chem.* **1956**, *60*, 1454 [81].
[652] Ashby, E. C.; Lin, J. J.; Goel, A. B. *J. Org. Chem.* **1978**, *43*, 183 [81, 82].
[653] Baltzly, R.; Phillips, A. P. *J. Am. Chem. Soc.* **1946**, *68*, 261 [81, 87].
[654] Krishnamurthy, S.; Brown, H. C. *J. Org. Chem.* **1980**, *45*, 849; **1982**, *47*, 276 [81, 82].
[655] Masamune, S.; Rossy, P. A.; Bates, G. S. *J. Am. Chem. Soc.* **1973**, *95*, 6452 [81, 82].
[656] Hutchins, R. O.; Hoke, D.; Keogh, J.; Koharski, B. *Tetrahedron Lett.* **1969**, 3495 [81, 82, 125].
[657] Bell, H. M.; Vanderslice, C. W.; Spehar, A. *J. Org. Chem.* **1969**, *34*, 3923 [81, 82, 88].
[658] Grady, G. L.; Kuivila, H. G. *J. Org. Chem.* **1969**, *34*, 2014 [81, 83].
[659] Kuivila, H. G.; Menapace, L. W. *J. Org. Chem.* **1963**, *28*, 2165 [81, 82, 84, 308].
[660] Greene, F. D.; Lowry, N. N. *J. Org. Chem.* **1967**, *32*, 882 [81, 82].
[661] Kuivila, H. G. *Synthesis* **1970**, 499 [81].
[662] Kim, S.; Ko, J. S. *Synth. Commun.* **1985**, *15*, 603 [81].
[663] Kim, S.; Hong, C. Y.; Yang, S. *Angew. Chem., Int. Ed. Engl.* **1983**, *22*, 562 [81].
[664] Trevoy, L. W.; Brown, W. G. *J. Am. Chem. Soc.* **1949**, *71*, 1675 [82, 113, 114].
[665] Hofmann, K.; Orochena, S. F.; Sax, S. M.; Jeffrey, G. A. *J. Am. Chem. Soc.* **1959**, *81*, 992 [82].
[666] Gaoni, Y. *J. Org. Chem.* **1981**, *46*, 4502 [83].
[667] Sydnes, L.; Skattebol, L. *Tetrahedron Lett.* **1974**, 3703 [83].
[668] Fry, A. J.; Moore, R. H. *J. Org. Chem.* **1968**, *33*, 1283 [83].
[669] Winstein, S.; Sonnenberg, J. *J. Am. Chem. Soc.* **1961**, *83*, 3235, 3240 [83].
[670] Nagao, M.; Sato, N.; Akashi, T.; Yoshida, T. *J. Am. Chem. Soc.* **1966**, *88*, 3447 [83].
[671] Hartman, W. W.; Dreger, E. E. *Org. Synth., Collect. Vol.* **1932**, *1*, 357 [83].
[672] von Auwers, K.; Wissenbach, H. *Chem. Ber.* **1923**, *56*, 715, 730, 738 [83, 198, 221].
[673] Whalley, W. B. *J. Chem. Soc.* **1951**, 3229 [83, 171].
[674] Fuller, G.; Tatlow, J. C. *J. Chem. Soc.* **1961**, 3198 [83, 86].
[675] Schlosser, M.; Heinz, G.; Le Van Chan *Chem. Ber.* **1971**, *104*, 1921 [83].
[676] Haszeldine, R. N.; Osborne, J. E. *J. Chem. Soc.* **1956**, 61 [84].
[677] Hudlický, M.; Lejhancová, I. *Collect. Czech. Chem. Commun.* **1965**, *30*, 2491 [82, 84].
[678] Paleta, O.; Kvíčala, J.; Günther, J.; Dědek, V. *Bull. Soc. Chim. Fr.* **1986**, 920 [84].
[679] Bruck, P. *Tetrahedron Lett.* **1962**, 449 [84].
[680] Strunk, R. J.; DiGiacomo, P. M.; Aso, K.; Kuivila, H. G. *J. Am. Chem. Soc.* **1970**, *92*, 2849 [84].
[681] Newman, M. S.; Cohen, G. S.; Cunico, R. F.; Dauernheim, L. W. *J. Org. Chem.* **1973**, *38*, 2760 [84, 85].
[682] Conroy, H. *J. Am. Chem. Soc.* **1955**, *77*, 5960 [84].
[683] Pri-Bar, I.; Buchman, O. *J. Org. Chem.* **1986**, *51*, 734 [84].
[684] Knunyants, I. L.; Krasuskaya, M. P.; Mysov, E. I. *Izv. Akad. Nauk SSSR, Ser. Khim.* **1960**, 1412; *Chem. Abstr.* **1961**, *55*, 349 [85].

[685] Hudlický, M. *Chemistry of Organic Fluorine* Compounds; Ellis Horwood: Chichester, England, 1976; 1992, p 174 [85].

[686] Mettille, F. J.; Burton, D. J. *Fluorine Chemistry Reviews;* Tarrant, P., Ed.; Marcel Dekker: New York, 1967; Vol. 1, p 315 [85].

[687] Roth, W.; Schleyer, P. R. Z. *Naturforsch.* **1983**, *38B*, 1697 [85, 87].

[688] Wilcox, C. F., Jr.; Zajacek, J. G. *J. Org. Chem.* **1964**, *29*, 2209 [85, 86].

[689] Jefford, C. W.; Gunsher, J.; Hill, D. T.; Brun, P.; Le Gras, J.; Waegell, B. *Org. Synth., Collect. Vol.* **1988**, *6*, 142 [85].

[690] Hatch, L. F.; McDoland, D. W. *J. Am. Chem. Soc.* **1952**, *74*, 2911, 3328 [85].

[691] Haszeldine, R. N. *J. Chem. Soc.* **1953**, 922 [86].

[692] Gassman, P. G.; Pape, P. G. *J. Org. Chem.* **1964**, *29*, 160 [86].

[693] Warner, R. M.; Leitsch, L. C. *J. Labelled Compd.* **1965**, *1*, 42 [86, 217].

[694]. Burton, D. J.; Johnson, R. L. *J. Am. Chem. Soc.* **1964**, *86*, 5361; *Tetrahedron Lett.* **1956**, 2681 [86, 87].

[695] Dmowski, W. *J. Fluorine Chem.* **1985**, *29*, 273 [86].

[696] Taniguchi, M.; Takeyama, Y.; Fugami, K.; Oshima, K.; Utimoto, K. *Bull. Chem. Soc. Jpn.* **1991**, *64*, 2593 [87, 88, 89].

[697] Kämmerer, H.; Horner, L.; Beck, H. *Chem. Ber.* **1958**, *91*, 1376 [87, 88, 89, 109, 197, 298].

[698] Eliel, E. L. *J. Am. Chem. Soc.* **1949**, *71*, 3970 [87].

[699] Matsumura, S.; Tokura, N. *Tetrahedron Lett.* **1969**, 363 [87].

[700] Shapiro, S. L.; Overberger, C. G. *J. Am. Chem. Soc.* **1954**, *76*, 97 [88].

[701] Parham, W. E.; Wright, C. D. *J. Org. Chem.* **1957**, *22*, 1473 [88].

[702] Bell, H. M.; Brown, H. C. *J. Am. Chem. Soc.* **1966**, *88*, 1473 [88].

[703] Brown, H. C.; Krishnamurthy, S. *J. Org. Chem.* **1969**, *34*, 3918 [88].

[704] Brooke, G. M.; Burdon, J.; Tatlow, J. C. *J. Chem. Soc.* **1962**, 3253 [88].

[705] Bažant, V.; Čapka, M.; Černý, M.; Chvalovský, V.; Kochloefl, K.; Kraus, M.; Málek, J. *Tetrahedron Lett.* **1968**, 3303 [88, 95, 137, 138, 153, 194, 204, 206, 219, 231, 232, 233, 234].

[706] Rothman, L. A.; Becker, E. I. *J. Org. Chem.* **1959**, *24*, 294; **1960**, *25*, 2203 [88].

[707] Brunet, J.-J.; Taillefer, M. *J. Organomet. Chem.* **1988**, *348*, C5 [88].

[708] Node, M.; Nishide, K.; Ohta, K.; Fujita, E. *Tetrahedron Lett.* **1982**, *23*, 689 [88, 89].

[709] Wade, R. S.; Castro, C. E. *Org. Synth., Collect. Vol.* **1988**, *6*, 821 [88, 89].

[710] Beckwith, A.; Goh, S. H. *Chem. Commun.* **1983**, 907 [88, 89].

[711] Node, M.; Nishide, K.; Ohta, K.; Fujita, E. *Tetrahedron Lett.* **1982**, *23*, 689 [88].

[712] Jošt, K.; Rudinger, J.; Šorm, F. *Collect. Czech. Chem. Commun.* **1963**, *28*, 1706 [89].

[713] Erlenmeyer, H.; Grubenmann, W. *Helv. Chim. Acta* **1947**, *30*, 297 [89, 197].

[714] Gronowitz, S.; Raznikiewicz, T. *Org. Synth., Collect. Vol.* **1973**, *5*, 149 [89].

[715] Govindachari, T. R.; Nagarajan, K.; Rajappa, S. *J. Chem. Soc.* **1957**, 2725 [89].

[716] Levitz, M.; Bogert, M. T. *J. Org. Chem.* **1945**, *10*, 341 [89].

[717] Neumann, F. W.; Sommer, N. B.; Kaslow, C. E.; Shriner, R. L. *Org. Synth., Collect. Vol.* **1955**, *3*, 519 [89].

[718] Lutz, R. E.; Ashburn, G.; Rowlett, R. J., Jr. *J. Am. Chem. Soc.* **1946**, *68*, 1322 [89].

[719] Wibaut, J. P. *Recl. Trav. Chim. Pays-Bas* **1944**, *63*, 141 [89].

[720] Fox, B. A; Threlfall, T. L. *Org. Synth., Collect. Vol.* **1973**, *5*, 346 [89].

[721] Chambers, R. D.; Musgrave, W. K. R.; Drakesmith, F. G. British Patent 1134651, (1968); *Chem. Abstr.* **1969**, *70*, 57661 [89, 90].

[722] Banks, R. E.; Haszeldine, R. N.; Lathram, J. V.; Young, I. M. *Chem. Ind.* (*London*) **1964**, 835; *J. Chem. Soc.* **1965**, 594 [89, 90].

[723] Marshall, J. R.; Walker, J. *J. Chem. Soc.* **1951**, *1004*, 1013 [89].

[724] Brown, D. J.; Waring, P. *Austr. J. Chem.* **1973**, *26*, 443 [89].

[725] Sakamoto, T.; Kondo, Y.; Yamanaka, H. *Synthesis* **1984**, 252 [89, 90].

[726] Stephenson, E. F. M. *Org. Synth., Collect. Vol.* **1955**, *3*, 475 [89].

[727] Senkus, M. *Ind. Eng. Chem.* **1948**, *40*, 506 [91, 110].

[728] Nielsen, A. T. *J. Org. Chem.* **1962**, *27*, 1998 [91].

[729] Grundmann, C. *Angew. Chem.* **1950**, *62*, 558 [92].

[730] Artemev, A. A.; Genkina, A. B.; Malimonova, A. B.; Trofilkina, V. P.; Isaenkova, M. A. *Zh. Vses. Khim. Ova. im. D. I. Mendeleeva* **1965**, *10*, 588; *Chem. Abstr.* **1966**, *64*, 1975 [92].

[731] Feuer, H.; Bartlett, R. S.; Vincent, B. F., Jr.; Anderson, R. S. *J. Org. Chem.* **1965**, *30*, 2880 [92].

[732] McMurry, J. E.; Melton, J. *J. Org. Chem.* **1973**, *38*, 4367 [92].

[733] Barltrop, J. A.; Nicholson, J. S. *J. Chem. Soc.* **1951**, 2524 [92, 112].

[734] Aeberli, P.; Houlihan, W. J. *J. Org. Chem.* **1967**, *32*, 3211 [93].

[735] Guttieri, M. J.; Maier, W. J. *J. Org. Chem.* **1984**, *49*, 2875 [93, 98, 129, 233, 239, 240].

[736] Meyers, A. I.; Sircar, J. C. *J. Org. Chem.* **1967**, *32*, 4134 [93].

[737] Sinhababu, A. K.; Borchardt, R. T. *Tetrahedron Lett.* **1983**, *24*, 227 [93].

[738] Gilsdorf, R. T.; Nord, F. F. *J. Am. Chem. Soc.* **1952**, *74*, 1837 [93, 94].

[739] Varma, R. S.; Kabalka, G. W. *Synth. Commun.* **1984**, *14*, 1093 [93].

[740] Ranu, B. C.; Chakraborty, R. *Tetrahedron* **1992**, *48*, 5317 [93, 94].

[741] Aizpurna, J. M.; Oiarbide, M.; Palomo, C. *Tetrahedron Lett.* **1987**, *28*, 5365 [93].

[742] Brossi, A.; Van Burik, J.; Teitel, S. *Helv. Chim. Acta* **1968**, *51*, 1965 [93].

[743] Hass, H. B.; Susie, A. G.; Heider, R. L. *J. Org. Chem.* **1950**, *15*, 8 [93, 94].

[744] Varma, R. S.; Varma, M.; Kabalka, G. W. *Tetrahedron Lett.* **1985**, *26*, 6013 [93, 94].

[745] Burger, A.; Stein, M. L.; Clements, J. B. *J. Org. Chem.* **1957**, *22*, 143 [93, 94].

[746] Mourad, M. S.; Varma, R. S.; Kabalka, G. W. *J. Org. Chem.* **1985**, *50*, 133; *Synth. Commun.* **1984**, *14*, 1099 [93, 94].

[747] Butterick, J. R.; Unrau, A. M. *Chem. Commun.* **1974**, 307 [93].

[748] Heinzelman, R. V. *Org. Synth., Collect. Vol.* **1963**, *4*, 573 [93].

[749] Varma, R. S.; Varma, M.; Kabalka, G. W. *Synth. Commun.* **1985**, *15*, 1325 [93, 94].

[750] Lindemann, A. *Helv. Chim. Acta* **1949**, *32*, 69 [93].

[751] Brand, K.; Steiner, J. *Chem. Ber.* **1922**, *55*, 875, 886 [95, 96, 97, 133, 134].

[752] He, Y.; Zhao, H.; Pan, X.; Wang, S. *Synth. Commun.* **1989**, *19*, 3047 [95, 96, 134].

[753] Harman, R. E. *Org. Synth., Collect. Vol.* **1963**, *4*, 148 [95].

[754] Kamm, O. *Org. Synth., Collect. Vol.* **1932**, *1*, 445 [95, 96].

[755] Corbett, J. F. *Chem. Commun.* **1968**, 1257 [95, 96].

[756] Bigelow, H. E.; Robinson, D. B. *Org. Synth., Collect. Vol.* **1955**, *3*, 103 [95, 96].

[757] Nystrom, R. F.; Brown, W. G. *J. Am. Chem. Soc.* **1948**, *70*, 3738 [95, 96, 143, 180, 181, 215, 240, 241].

[758] James, B. D. *Chem. Ind. (London)* **1971**, 227 [95, 139, 216, 232].

[759] Meier, R.; Böhler, F. *Chem. Ber.* **1956**, *89*, 2301 [95, 96].

[760] Adams, R.; Cohen, F. L. *Org. Synth., Collect. Vol.* **1932**, *1*, 240 [95].

[761] Mendenhall, G. D.; Smith, P. A. S. *Org. Synth., Collect. Vol.* **1973**, *5*, 829 [95].

[762] Severin, T.; Schmitz, R. *Chem. Ber.* **1962**, *95*, 1417 [96].

[763] Alaimo, R. J.; Storrin, R. J. In *Catalysis of Organic Reactions;* Moser, W. R., Ed.; Marcel Dekker: New York, 1951; p 473 [96, 98].

[764] West, R. W. *J. Chem. Soc.* **1925**, *127*, 494 [96, 97].

[765] Dankova, T. F.; Bokova, T. N.; Preobrazhenskii, N. A. *Zh. Obshch. Khim.* **1951**, *21*, 787; *Chem. Abstr.* **1951**, *45*, 9517 [96, 99, 172].

[766] Kuhn, W. E. *Org. Synth., Collect. Vol.* **1943**, *2*, 448 [96].

[767] Kock, E. *Chem. Ber.* **1887**, *20*, 1567 [96, 97].

[768] Hurst, W. G.; Thorpe; J. F. *J. Chem. Soc.* **1915**, 934 [96].

[769] Ayling, E. E.; Gorvin, J. H.; Hinkel, L. E. *J. Chem. Soc.* **1942**, 755 [96, 97, 98].

[770] Entwistle, I. D.; Johnstone, R. A. W.; Povall, T. J. *J. Chem. Soc., Perkin Trans. 1* **1975**, 1300 [97].

[771] Hudlický, M.; Bell, H. M. *J. Fluorine Chem.* **1974**, *4*, 19 [97].

[772] Sachs, F.; Sichel, E. *Chem. Ber.* **1904**, *37*, 1861 [97, 98, 144, 314].

[773] Parkes, G. D.; Farthing, A. C. *J. Chem. Soc.* **1948**, 1275 [97].

[774] Brunner, W. H.; Halasz, A. U.S. Patent 3,088,978, 1963; *Chem. Abstr.* **1964**, *60*, 2826a [97, 98].

[775] Akita, Y.; Inoue, A.; Ishida, K.; Terui, K.; Ohta, A. *Synth. Commun.* **1986**, *16*, 1067 [98].

[776] Weil, T.; Prijs, B.; Erlenmeyer, H. *Helv. Chim. Acta* **1953**, *36*, 142 [99].

[777] Hendrickson, J. B. *J. Am. Chem. Soc.* **1961**, *83*, 1251 [100, 198].

[778] Suzuki, N.; Azuma, T.; Watanabe, K.; Nomoto, T.; Izawa, Y.; Tomioka, H. *J. Chem. Soc., Perkin Trans. 1* **1987**, 1951 [100].

[779] Kornblum, N.; Iffland, D. C. *J. Am. Chem. Soc.* **1949**, *71*, 2137 [100, 318].

[780] Newman, M. S.; Hung, W. M. *J. Org. Chem.* **1974**, *39*, 1317 [45, 100].

[781] Yasui, S.; Nakamura, K.; Ohno, A. *Tetrahedron Lett.* **1983**, *24*, 3331 [100].

[782] Newman, M. S.; Hung, W. M. *J. Org. Chem.* **1974**, *39*, 1317 [100].

[783] Reychler, A. *Chem. Ber.* **1887**, *20*, 2463 [100].

[784] Fischer, E. *Justus Liebigs Ann. Chem.* **1878**, *190*, 67, 78 [100].

[785] VanderWerf, C. A.; Heisler, R. Y.; McEwen, W. E. *J. Am. Chem. Soc.* **1954**, *76*, 1231 [100].

[786] Cama, L. D.; Leanza, W. J.; Beattie, T. R.; Christensen, B. G. *J. Am. Chem. Soc.* **1972**, *94*, 1408 [100, 211].

[787] Boyer, J. H. *J. Am. Chem. Soc.* **1951**, *73*, 5865 [100, 173].

[788] Rolla, F. *J. Org. Chem.* **1982**, *47*, 4327 [100, 101].

[789] Stanovnik, B.; Tišler, M.; Polanc, S.; Gračner, M. *Synthesis* **1978**, 65 [100].

[790] Bayley, H.; Standring, D. N.; Knowles, J. R. *Tetrahedron Lett.* **1978**, 3633 [101].

[791] Hassner, A.; Matthews, G. J.; Fowler, F. W. *J. Am. Chem. Soc.* **1969**, *91*, 5046 [101].

[792] Landa, S.; Mostecký, J. *Collect. Czech. Chem. Commun.* **1955**, *20*, 430; **1956**, *21*, 1177 [103].

[793] Brewster, J. H.; Osman, S. F.; Bayer, H. O.; Hopps, H. B. *J. Org. Chem.* **1964**, *29*, 121 [103].

[794] Krafft, M. E.; Crooks, W. J., III; Zorc, B.; Milczanowski, S. E. *J. Org. Chem.* **1988**, *53*, 3158 [103, 109, 110].

[795] Vowinkel, E.; Buthe, I. *Chem. Ber.* **1974**, *107*, 1353 [103].

[796] Nozaki, K.; Oshima, K.; Utimoto, K. *Tetrahedron Lett.* **1988**, *29*, 6125 [104].

[797] Iacono, S.; Rasmussen, J. R. *Org. Synth.* **1985**, *64*, 57 [104].

[798] Barrett, A. G. M.; Prokopiou, P. A.; Barton, D. H. R. *J. Chem. Soc., Perkin Trans. 1* **1981**, 1510 [104].

[799] Ueno, Y.; Tanaka, C.; Okawara, M. *Chem. Lett.* **1983**, 795 [104].

[800] Connor, R.; Adkins, H. *J. Am. Chem. Soc.* **1932**, *54*, 4678 [104, 108, 110, 217, 223].

[801]   Corey, E. J.; Carey, F. A.; Winter, R. A. E. *J. Am. Chem. Soc.* **1965**, *87*, 934 [104].

[802]   Ohloff, G.; Farnow, H.; Schade, G. *Chem. Ber.* **1956**, *89*, 1549 [104, 105].

[803]   Takaya, H.; Ohta, T.; Sayo, N.; Kumobayashi, H.; Akutagawa, S.; Inoue, S.-i.; Kasahara, I.; Noyori, R. *J. Am. Chem. Soc.* **1987**, *109*, 1596 [105].

[804]   Takaya, H.; Ohta, T.; Inoue, S.-i.; Tokunaga, M.; Kitamura, M.; Noyori, R.; *Org. Synth.* **1993**, *72*, 74 [105].

[805]   Hochstein, F. A.; Brown, W. G. *J. Am. Chem. Soc.* **1948**, *70*, 3484 [105, 108, 143].

[806]   Brewster, J. H.; Bayer, H. O. *J. Org. Chem.* **1964**, *29*, 116 [105, 106].

[807]   Birch, A. J. *J. Chem. Soc.* **1945**, 809 [105, 106].

[808]   Bohlmann, F.; Staffeldt, J.; Skuballa, W. *Chem. Ber.* **1976**, *109*, 1586 [105].

[809]   Corey, E. J.; Achiwa, K. *J. Org. Chem.* **1969**, *34*, 3667 [105].

[810]   Jones, T. K.; Denmark, S. E. *Org. Synth.* **1985**, *64*, 182 [105].

[811]   Vowinkel, E.; Wolf, C. *Chem. Ber.* **1974**, *107*, 907 [106, 107].

[812]   Vowinkel, E.; Baese, H. *Chem. Ber.* **1974**, *107*, 1213 [106, 107].

[813]   Musliner, W. J.; Gates, J. W., Jr. *J. Am. Chem. Soc.* **1966**, *88*, 4271 [106].

[814]   Cacchi, S.; Ciattini, P. G.; Morera, E.; Ortar, G. *Tetrahedron Lett.* **1986**, *27*, 5541 [106, 107, 126, 300].

[815]   Peterson, G. A.; Kunng, F.-A.; McCallum, J. S.; Wolff, W. D. *Tetrahedron Lett.* **1987**, *28*, 1381 [106, 107, 126].

[816]   Barth, L.; Goldschmiedt, G. *Chem. Ber.* **1879**, *12*, 1244 [106].

[817]   Severin, T.; Ipach, I. *Synthesis* **1973**, 796 [106, 130].

[818]   Mitsui, S.; Imaizumi, S.; Esashi, Y. *Bull. Chem. Soc. Jpn.* **1970**, *43*, 2143 [107].

[819]   Folkers, K.; Adkins, H. *J. Am. Chem. Soc.* **1932**, *54*, 1145 [107, 213, 214].

[820]   Gribble, G. W.; Leese, R. M.; Evans, B. *Synthesis* **1977**, 172 [107].

[821]   Brewster, J. H.; Bayer, H. O. *J. Org. Chem.* **1964**, *29*, 105, 110 [107, 108].

[822]   Breuer, E. *Tetrahedron Lett.* **1967**, 1849 [107].

[823]   Ferles, M.; Lebl, M.; Šilhánková, A.; Štern, P.; Wimmer, Z. *Collect. Czech. Chem. Commun.* **1975**, *40*, 1571 [107].

[824]   Marvel, C. S.; Hager, F. D.; Caudle, E. C. *Org. Synth., Collect. Vol.* **1932**, *1*, 224 [107].

[825]   Musser, D. M.; Adkins, H. *J. Am. Chem. Soc.* **1938**, *60*, 664 [109].

[826]   Adkins, H.; Krsek, G. *J. Am. Chem. Soc.* **1948**, *70*, 412 [109].

[827]   Stork, G. *J. Am. Chem. Soc.* **1947**, *69*, 576 [109, 112].

[828]   Gutsche, C. D.; Peters, H. H. *Org. Synth., Collect. Vol.* **1963**, *4*, 887 [109].

[829]   Gilman, N. W.; Sternbach, L. H. *Chem. Commun.* **1971**, 465 [110, 130].

[830]   Wiesner, K. University of New Brunswick, Fredericton, Canada, private communication, 1980 [110].

[831]   Lundt, I.; Pedersen, C. *Synthesis* **1986**, 1052 [110, 220, 221].

[832]   Woodburn, H. M.; Stuntz, C. F. *J. Am. Chem. Soc.* **1950**, *72*, 1361 [110].

[833]   Bailey, W. J.; Marktscheffel, F. *J. Org. Chem.* **1960**, *25*, 1797 [111].

[834]   Andrus, D. W.; Johnson, J. R. *Org. Synth., Collect. Vol.* **1955**, *3*, 794 [111].

[835]   Birch, A. J.; Subba Rao, G. S. R. *Austr. J. Chem.* **1970**, *23*, 1641 [111, 112].

[836]   Hutchins, R. O.; Learn, K. *J. Org. Chem.* **1982**, *47*, 4380 [111].

[837]   Hallsworth, A. S.; Henbest, H. B.; Wrigley, T. I. *J. Chem. Soc.* **1957**, 1969 [111, 112].

[838]   Hartung, W. H.; Simonoff, R. *Org. React. (N.Y.)* **1953**, *7*, 263 [111, 119, 130].

[839]   Reimann, E. *Justus Liebigs Ann. Chem.* **1971**, *750*, 109, 126 [111].

[840]   Olah, G. A.; Prakash, G. K. S.; Narang, S. C. *Synthesis* **1978**, 825 [112].

[841]   Reist, E. J.; Bartuska, V. J.; Goodman, L. *J. Org. Chem.* **1964**, *29*, 3725 [112].

[842]   Shriner, R. L.; Ruby, P. R. *Org. Synth., Collect. Vol.* **1963**, *4*, 798 [112].

[843]   Doyle, T. W. *Can. J. Chem.* **1977**, *55*, 2714 [112].

[844] Thoms, H.; Siebeling, W. *Chem. Ber.* **1911**, *44*, 2134 [112].
[845] Kariv-Miller, E.; Swenson, K. E.; Zemach, D. *J. Org. Chem.* **1983**, *48*, 4210 [112].
[846] Wilds, A. L.; Nelson, N. A. *J. Am. Chem. Soc.* **1953**, *75*, 5360, 5366 [112].
[847] Soffer, M. D.; Bellis, M. P.; Gellerson, H. E.; Stewart, R. A. *Org. Synth., Collect. Vol.* **1963**, *4*, 903 [113].
[848] Colon, I. *J. Org. Chem.* **1982**, *47*, 2622 [113].
[849] Fieser, L. F. *J. Am. Chem. Soc.* **1953**, *75*, 4395 [113, 175].
[850] Cornforth, J. W.; Cornforth, R. H.; Matthew, K. K. *J. Chem. Soc.* **1959**, 112 [113].
[851] Kupchan, S. M.; Maruyama, M. *J. Org. Chem.* **1971**, *36*, 1187 [113].
[852] Schobert, R. *Angew. Chem., Int. Ed. Engl.* **1988**, *27*, 855 [113].
[853] Kochi, J. K.; Singleton, D. M.; Andrews, L. J. *Tetrahedron* **1968**, *24*, 3503 [113].
[854] Fujisawa, T.; Sugimoto, K.; Ohta, H. *Chem. Lett.* **1974**, 883 [113].
[855] McQuillin, F. J.; Ord, W. O. *J. Chem. Soc.* **1959**, 3169 [113, 114].
[856] Eliel, E. L.; Delmonte, D. W. *J. Am. Chem. Soc.* **1956**, *78*, 3226 [114].
[857] Finan, J. M.; Kishi, Y. *Tetrahedron Lett.* **1982**, *23*, 2719 [114].
[858] Eliel, E. L.; Delmonte, D. W. *J. Am. Chem. Soc.* **1956**, *78*, 3226 [114].
[859] Eliel, E. L.; Rerick, M. N. *J. Am. Chem. Soc.* **1960**, *82*, 1362 [113, 114].
[860] Elsenbaumer, R. L.; Mosher, H. S.; Morrison, J. D.; Tomaszewski, J. E. *J. Org. Chem.* **1981**, *46*, 4034 [114].
[861] Chong, J. M. *Tetrahedron Lett.* **1992**, *33*, 33 [114, 115].
[862] Hutchins, R. O.; Taffer, I. M.; Burgoyne, W. *J. Org. Chem.* **1981**, *46*, 5214 [114].
[863] Brown, H. C.; Yoon, N. M. *Chem. Commun.* **1968**, 1549 [114].
[864] Brown, H. C.; Yoon, N. M. *J. Am. Chem. Soc.* **1968**, *90*, 2686 [114].
[865] Brown, H. C.; Ikegami, S.; Kawakami, J. H. *J. Org. Chem.* **1970**, *35*, 3243 [114].
[866] Brown, M.; Piszkiewicz, L. W. *J. Org. Chem.* **1967**, *32*, 2013 [114, 165].
[867] Zaidlewicz, M.; Uzarewicz, A.; Sarnowski, R. *Synthesis* **1979**, 62 [114, 115].
[868] Genus, J. F.; Peters, D. D.; Bryson, T. A. *Synlett* **1993**, 759 [114, 115].
[869] Cole, W.; Julian, P. L. *J. Org. Chem.* **1954**, *19*, 131 [114, 175].
[870] Criegee, R.; Zogel, H. *Chem. Ber.* **1951**, *84*, 215 [115].
[871] Paget, H. *J. Chem. Soc.* **1938**, 829 [115, 116].
[872] Richter, F.; Presting, W. *Chem. Ber.* **1931**, *64*, 878 [115, 116].
[873] Wallach, O. *Justus Liebigs Ann. Chem.* **1912**, *392*, 49, 60 [115, 116].
[874] Bruynde, P. *Justus Liebigs Ann. Chem.* **1945**, *20*, 551 [115].
[875] Windaus, A.; Linsert, O. *Justus Liebigs Ann. Chem.* **1928**, *465*, 148 [115].
[876] Pappas, J. J.; Keaveney, W. P.; Berger, M.; Rush, R. V. *J. Org. Chem.* **1968**, *33*, 787 [115].
[877] Kametani, T.; Ogasawara, K. *Chem. Ind. (London)* **1968**, 1772 [116].
[878] Burfield, D. R. *J. Org. Chem.* **1982**, *47*, 3821 [116].
[879] Pryde, E. H.; Anders, D. E.; Teeter, H. M.; Cowan, J. C. *J. Org. Chem.* **1960**, *25*, 618 [116].
[880] Fischer, F. G.; Löwenberg, K. *Chem. Ber.* **1933**, *66*, 665 [116, 117].
[881] Helferich, B.; Dommer, W. *Chem. Ber.* **1920**, *53*, 2004, 2009 [116, 117].
[882] Flippin, L. A.; Gallagher, D. W.; Jalali-Araghi, K. *J. Org. Chem.* **1989**, *54*, 1430 [117].
[883] Hoffmann, F. W.; Ess, R. J.; Simmons, T. C.; Hanzel, R. S. *J. Am. Chem. Soc.* **1956**, *78*, 6414 [119, 120, 319].
[884] Truce, W. E.; Perry, F. M. *J. Org. Chem.* **1965**, *30*, 1316 [119, 120, 121, 123].
[885] Mukaiyama, T.; Narasaka, K.; Maekawa, K.; Furusato, M. *Bull. Chem. Soc. Jpn.* **1971**, *44*, 2285; *Chem. Lett.* **1973**, 291 [119].

[886]   Welch, S. C.; Loh, J.-P. *J. Org. Chem.* **1981**, *46*, 4072 [119].
[887]   Mukaiyama, T.; Hayashi, M.; Narasaka, K. *Chem. Lett.* **1973**, 291 [119].
[888]   Brink, M. *Synthesis* **1975**, 807 [119].
[889]   Nayak, U. G.; Brown, R. K. *Can. J. Chem.* **1966**, *44*, 591 [119].
[890]   Birch, A. J.; Walker, K. A. M. *Tetrahedron Lett.* **1967**, 1935 [119].
[891]   Brown, E. D.; Iqbal, S. M.; Owen, L. N. *J. Chem. Soc. C* **1966**, 415 [119, 120].
[892]   Strating, J.; Backer, H. J. *Recl. Trav. Chim. Pays-Bas* **1950**, *69*, 638 [121, 123, 124, 125, 126].
[893]   Arnold, R. C.; Lien, A. P.; Alm, R. M. *J. Am. Chem. Soc.* **1950**, *72*, 731 [121].
[894]   D'Amico, J. J. *J. Org. Chem.* **1961**, *26*, 3436 [121].
[895]   Riegel, B.; Lappin, G. R.; Adelson, B. H.; Jackson, R. I.; Albisetti, C. P.; Dodson, R. M.; Baker, R. H. *J. Am. Chem. Soc.* **1946**, *68*, 1264 [121].
[896]   Ogura, K.; Yamashita, M.; Tsuchihashi, G. *Synthesis* **1975**, 385 [121, 122].
[897]   Drabowicz, J.; Oae, S. *Synthesis* **1977**, 404 [122].
[898]   Theobald, P. G.; Okamura, W. H. *Tetrahedron Lett.* **1987**, *28*, 6565 [122].
[899]   Truce, W. E.; Simms, J. A. *J. Am. Chem. Soc.* **1956**, *78*, 2756 [123].
[900]   Bordwell, F. G.; McKellin, W. H. *J. Am. Chem. Soc.* **1951**, *73*, 2251 [123].
[901]   Fehnel, E. A.; Carmack, M. *J. Am. Chem. Soc.* **1948**, *70*, 1813 [123, 166].
[902]   Gardner, J. N.; Kaiser, S.; Krubiner, A.; Lucas, H. *Can. J. Chem.* **1973**, *51*, 1419 [123].
[903]   Yamamoto, I.; Sakai, T.; Yamamoto, S.; Ohta, K.; Matsuzaki, K. *Synthesis* **1985**, 676 [123].
[904]   Cronyn, M. W.; Zavarin, E. *J. Org. Chem.* **1954**, *19*, 139 [123].
[905]   Dabby, R. E.; Kenyon, J.; Mason, R. F. *J. Chem. Soc.* **1952**, 4881 [123].
[906]   Kuo, Y.-C.; Aoyama, T.; Shioiri, T. *Chem. Pharm. Bull.* **1983**, *31*, 883 [123].
[907]   Trost, B. M.; Arndt, H. C.; Strege, P. E.; Verhoeven, T. R. *Tetrahedron Lett.* **1976**, 3477 [123].
[908]   Wertheim, E. *Org. Synth., Collect. Vol.* **1943**, *2*, 471 [124].
[909]   Field, L.; Grunwald, F. A. *J. Org. Chem.* **1951**, *16*, 946 [124, 125].
[910]   Ullmann, F.; Pasdermadjian, G. *Chem. Ber.* **1901**, *34*, 1151 [124].
[911]   Whitmore, F. C.; Hamilton, F. H. *Org. Synth., Collect. Vol.* **1932**, *1*, 479 [124, 125].
[912]   Adams, R.; Marvel, C. S. *Org. Synth., Collect. Vol.* **1932**, *1*, 504 [125].
[913]   Sheppard, W. A. *Org. Synth., Collect. Vol.* **1973**, *5*, 843 [125].
[914]   Kenner, G. W.; Murray, M. A. *J. Chem. Soc.* **1949**, *S1*, 178 [125, 126].
[915]   Krishnamurthy, S.; Brown, H. C. *J. Org. Chem.* **1976**, *41*, 3064 [125].
[916]   Kočovský, P.; Černý, V. *Collect. Czech. Chem. Commun.* **1979**, *44*, 246 [125, 312].
[917]   Cacchi, S.; Morera, E.; Ortar, G. *Tetrahedron Lett.* **1984**, *25*, 4821; *Org. Synth.* **1989**, *68*, 138 [126].
[918]   Closson, W. D.; Wriede, P.; Bank, S. *J. Am. Chem. Soc.* **1966**, *88*, 1581 [126].
[919]   Reber, F.; Reichstein, T. *Helv. Chim. Acta* **1945**, *28*, 1164, 1170 [126].
[920]   Fischer, E. *Chem. Ber.* **1915**, *48*, 93 [126].
[921]   Klamann, D.; Hofbauer, G. *Chem. Ber.* **1953**, *86*, 1246 [126, 127].
[922]   Ji, S.; Gortler, L. B.; Waring, A.; Battisti, A.; Bank, S.; Closson, W. D.; Wriede, P. *J. Am. Chem. Soc.* **1967**, *89*, 5311 [126, 127, 310].
[923]   Campbell, K. N.; Sommers, A. H.; Campbell, B. K. *Org. Synth., Collect. Vol.* **1955**, *3*, 148 [129].
[924]   Sabatier, P.; Mailhe, A. *C. R. Hebd. Seances Acad. Sci* **1907**, *144*, 784 [129].
[925]   Doldouras, G. A.; Kollonitsch, J. *J. Am. Chem. Soc.* **1978**, *100*, 341 [129].
[926]   King, T. J. *J. Chem. Soc.* **1951**, 898 [129].
[927]   Coulter, J. M.; Lewis, J. W.; Lynch, P. P. *Tetrahedron* **1968**, *24*, 4489 [129].

[928] Schmitt, J.; Panouse, J. J.; Hallot, A.; Pluchet, H.; Comoy, P.; Cornu, P. J. *Bull. Soc. Chim. Fr.* **1963**, 816 [129].
[929] Borch, R. F.; Bernstein, M. D.; Durst, H. D. *J. Am. Chem. Soc.* **1971**, *93*, 2897 [129].
[930] Kim, S.; Chang, Ho Oh; Jae, Suk Ko; Kyo, Han Anh; Yong, Jin Kim *J. Org. Chem.* **1985**, *50*, 1927 [129, 149, 204, 205].
[931] Benkeser, R. A.; Lambert, R. F.; Ryan, P. W.; Stoffey, D. G. *J. Am. Chem. Soc.* **1958**, *80*, 6573 [129].
[932] Birch, A. J. *J. Chem. Soc.* **1946**, 593 [130].
[933] Waser, E. B. H.; Mollering, H. *Org. Synth., Collect. Vol.* **1932**, *1*, 499 [130].
[934] Birkofer, L. *Chem. Ber.* **1942**, *75*, 429 [130].
[935] Dahn, H.; Zoller, P. *Helv. Chim. Acta* **1952**, *35*, 1348 [130].
[936] Dahn, H.; Solms, U.; Zoller, P. *Helv. Chim. Acta* **1952**, *35*, 2117 [130].
[937] Marchand, B. *Chem. Ber.* **1962**, *95*, 577 [130].
[938] Emmert, B. *Chem. Ber.* **1909**, *42*, 1507 [130].
[939] Brasen, W. R.; Hauser, C. R. *Org. Synth., Collect. Vol.* **1963**, *4*, 508 [130].
[940] Sugasawa, S.; Ushioda, S.; Fujisawa, T. *Tetrahedron* **1959**, *5*, 48 [130].
[941] Martin, E. L. *Org. Synth., Collect. Vol.* **1943**, *2*, 501 [130].
[942] Boyer, J. H.; Bunks, R. S. *Org. Synth., Collect. Vol.* **1973**, *5*, 1067 [130].
[943] Metayer, M. *Bull. Soc. Chim. Fr.* **1950**, 1048 [130].
[944] Schultz, E. M.; Bicking, J. B. *J. Am. Chem. Soc.* **1953**, *75*, 1128 [131].
[945] Clemo, G. R.; Raper, R.; Vipond, H. J. *J. Chem. Soc.* **1949**, 2095 [131].
[946] Shih, D. H.; Ratcliffe, R. W. *J. Med. Chem.* **1981**, *24*, 639 [131, 211].
[947] Anantharamaiah, G. M.; Sivanandaiah, K. M. *J. Chem. Soc., Perkin Trans. 1* **1977**, 490 [131].
[948] Haas, H. *J. Chem. Ber.* **1961**, *94*, 2442 [131].
[949] Hanna, C.; Schueler, F. W. *J. Am. Chem. Soc.* **1952**, *74*, 3693 [131].
[950] Schueler, F. W.; Hanna, C. *J. Am. Chem. Soc.* **1951**, *73*, 4996 [131].
[951] Hatt, H. H. *Org. Synth., Collect. Vol.* **1943**, *2*, 211 [131].
[952] Backer, H. J. *Recl. Trav. Chim. Pays-Bas* **1912**, *31*, 142 [131].
[953] Ingersoll, A. W.; Bircher, L. J.; Brubaker, M. M. *Org. Synth., Collect. Vol.* **1932**, *1*, 485 [131].
[954] Taylor, E. C.; Crovetti, A. J.; Boyer, N. E. *J. Am. Chem. Soc.* **1957**, *79*, 3549 [131, 132, 133].
[955] Berson, J. A.; Cohen, T. *J. Org. Chem.* **1955**, *20*, 1461 [131, 132].
[956] Boyer, S. K.; Bach, J.; McKenna, J.; Jagelmann, E., Jr. *J. Org. Chem.* **1985**, *50*, 3408 [131].
[957] Balicki, R. *Synthesis* **1989**, 645 [131].
[958] Brown, H. C.; Subba Rao, B. C. *J. Am. Chem. Soc.* **1956**, *78*, 2582 [131, 194, 197, 198, 199, 216, 222, 241, 305].
[959] Jankovský, M.; Ferles, M. *Collect. Czech. Chem. Commun.* **1970**, *35*, 2802 [131].
[960] Malinowski, M. *Synthesis* **1987**, 732 [132].
[961] Daniher, F. A.; Hackley, B. E., Jr. *J. Org. Chem.* **1966**, *31*, 4267 [132, 316].
[962] Olah, G. A.; Arvanaghi, M.; Vankar, Y. D. *Synthesis* **1980**, 660 [132].
[963] Yoshimura, T.; Asada, K.; Oae, S. *Bull. Chem. Soc. Jpn.* **1982**, *55*, 3000 [132].
[964] Tokitoh, N.; Okazaki, R. *Chem. Lett.* **1985**, 1517 [132].
[965] Grundmann, C.; Frommeld, H.-D. *J. Org. Chem.* **1965**, *30*, 2077 [132].
[966] Katritzky, A. R.; Monro, A. M. *J. Chem. Soc.* **1958**, 1263 [132].
[967] Rayburn, C. H.; Harlan, W. R.; Hanmer, H. R. *J. Am. Chem. Soc.* **1950**, *72*, 1721 [133].
[968] Skita, A. *Chem. Ber.* **1912**, *45*, 3312 [133, 134].

[969]   Guither, W. D.; Clark, D. G.; Castle, R. N. *J. Heterocycl. Chem.* **1965**, 2, 67 [133].

[970]   Mellor, J. M.; Smith, N. M. *J. Chem. Soc., Perkin Trans. 1* **1984**, 2927 [133].

[971]   Jardine, I.; McQuillin, F. J. *Chem. Commun.* **1920**, 626 [133, 134].

[972]   Alberti, A.; Bedogni, N.; Benaglia, M.; Leardini, R.; Nanni, D.; Pedulli, G. F.; Tundo, A.; Zanardi, G. *J. Org. Chem.* **1992**, 57, 607 [133, 134].

[973]   van Tamelen, E. E.; Dewey, R. S.; Timmons, R. J. *J. Am. Chem. Soc.* **1961**, 83, 3725 [133].

[974]   Cartwright, R. A.; Tatlow, J. C. *J. Chem. Soc.* **1953**, 1994 [133].

[975]   Ruggli, P.; Hölzle, K. *Helv. Chim. Acta* **1943**, 26, 1190 [133].

[976]   Thiele, J. *Justus Liebigs Ann. Chem.* **1892**, 270, 1, 44 [134].

[977]   Schmitt, Y.; Langlois, M.; Perrin, C.; Callet, G. *Bull. Soc. Chim. Fr.* **1969**, 2004 [134].

[978]   Oae, S.; Tsujimoto, N.; Nakanishi, A. *Bull. Chem. Soc. Jpn.* **1973**, 46, 535 [134].

[979]   Avalos, M.; Babiano, R.; García-Verdugo, C.; Jiménes, Y. L.; Palacios, J. C. *Tetrahedron Lett.* **1990**, 31, 2467 [134].

[980]   Engel, R.; Chakraborty, S. *Synth. Commun.* **1988**, 18, 665 [134, 135].

[981]   Brown, H. C.; Kulkarni, S. U. *J. Org. Chem.* **1977**, 42, 4169 [137].

[982]   Carothers, W. H.; Adams, R. *J. Am. Chem. Soc.* **1934**, 46, 1675 [137, 140].

[983]   Nystrom, R. F.; Chaikin, S. W.; Brown, W. G. *J. Am. Chem. Soc.* **1949**, 71, 3245 [137, 138, 139, 140, 149, 150, 153, 215].

[984]   Chaikin, S. W.; Brown, W. G. *J. Am. Chem. Soc.* **1949**, 71, 122 [137, 138, 139, 140, 143, 149, 150, 153, 166, 204, 205].

[985]   Krishnamurthy, S. *J. Org. Chem.* **1981**, 46, 4628 [137, 138, 176].

[986]   Shibata, I.; Yoshida, T.; Baba, A.; Matsuda, H. *Chem. Lett.* **1989**, 619 [137, 140, 141, 143, 144, 145].

[987]   Fung, N. Y. M.; de Mayo, P.; Schauble, J. H.; Weedon, A. C. *J. Org. Chem.* **1978**, 43, 3977 [137, 138, 140, 141, 149, 176].

[988]   Fujisawa, T.; Sugimoto, K.; Ohta, H. *J. Org. Chem.* **1976**, 41, 1667 [137, 138, 140].

[989]   Clarke, H. T.; Dreger, E. E. *Org. Synth., Collect. Vol.* **1932**, 1, 304 [137, 138].

[990]   Posner, G. H.; Runquist, A. W.; Chiapdelaine, M. J. *J. Org. Chem.* **1977**, 42, 1202 [137, 138, 140, 141, 143, 149, 176, 319].

[991]   Clemmensen, E. *Chem. Ber.* **1913**, 46, 1837; **1914**, 47, 51, 681 [137, 142, 151, 156, 163, 175, 311].

[992]   Herr, C. H.; Whitmore, F. C.; Schiessler, R. W. *J. Am. Chem. Soc.* **1945**, 67, 2061 [137, 142, 151, 157].

[993]   Brannock, K. C. *J. Am. Chem. Soc.* **1959**, 81, 3379 [138, 139].

[994]   Hennis, H. E.; Trapp, W. B. *J. Org. Chem.* **1961**, 26, 4678 [138].

[995]   Goetz, R. W.; Orchin, M. *J. Am. Chem. Soc.* **1963**, 85, 2782 [138, 139].

[996]   Rylander, P. N.; Steele, D. R. *Tetrahedron Lett.* **1969**, 1579 [139, 143].

[997]   Krishnamurthy, S.; Brown, H. C. *J. Org. Chem.* **1977**, 42, 1197 [139, 143, 166, 167].

[998]   Gemal, A. L.; Luche, J. L. *Tetrahedron Lett.* **1981**, 22, 4077 [139, 141, 176, 177].

[999]   Young, W. G.; Hartung, W. H.; Crossley, F. S. *J. Am. Chem. Soc.* **1936**, 58, 100 [139, 140, 143].

[1000] Skita, A. *Chem. Ber.* **1915**, 48, 1486 [140, 142, 165].

[1001] Adams, R.; McKenzie, S., Jr.; Loewe, S. *J. Am. Chem. Soc.* **1948**, 70, 664 [140].

[1002] Cook, P. L. *J. Org. Chem.* **1962**, 27, 3873 [140, 149, 150, 153, 165, 167].

[1003] Brown, B. R.; White, A. M. S. *J. Chem. Soc.* **1957**, 3755 [140, 141, 142, 153, 156].

[*1004*] Raber, D. J.; Guido, W. C. *J. Org. Chem.* **1976**, *41*, 690 [140, 141, 143, 149, 150, 153, 204, 205, 216].

[*1005*] Sarkar, D. C.; Das, A. R.; Ranu, B. C. *J. Org. Chem.* **1990**, *55*, 5799 [141, 143, 149, 158, 159, 307].

[*1006*] Fry, J. L.; Orfanopoulos, M.; Adlington, M. G.; Dittman, W. P., Jr.; Silverman, S. B. *J. Org. Chem.* **1976**, *43*, 374 [140, 141, 142, 151].

[*1007*] Mulholland, T. P. C.; Ward, G. *J. Chem. Soc.* **1954**, 4676 [141].

[*1008*] Burnette, L. W.; Johns, I. B.; Holdren, R. F.; Hixon, R. M. *Ind. Eng. Chem.* **1948**, *40*, 502 [141].

[*1009*] West, C. T.; Donnelly, S. J.; Kooistra, D. A.; Doyle, M. P. *J. Org. Chem.* **1973**, *38*, 2675 [141, 156, 199, 200, 201].

[*1010*] Hall, S. S.; Bartels, A. P.; Engman, A. M. *J. Org. Chem.* **1972**, *37*, 760 [140, 142].

[*1011*] Schwarz, R.; Hering, H. *Org. Synth., Collect. Vol.* **1963**, *4*, 203 [142].

[*1012*] Lock, G.; Stach, K. *Chem. Ber.* **1943**, *76*, 1252 [140, 142, 144, 157].

[*1013*] Tuley, W. F.; Adams, R. *J. Am. Chem. Soc.* **1925**, *47*, 3061 [143].

[*1014*] Jorgenson, M. J. *Tetrahedron Lett.* **1962**, 559 [143].

[*1015*] Hutchins, R. O.; Kandasamy, D. *J. Org. Chem.* **1975**, *40*, 2530 [143, 166, 167, 168].

[*1016*] Ranu, B. C.; Das A. R. *J. Org. Chem.* **1991**, *56*, 4796 [143, 166, 167].

[*1017*] Mincione, E. *J. Org. Chem.* **1978**, *43*, 1829 [143, 166].

[*1018*] Likhosherstov, M. V.; Arsenyuk, A. A.; Zeberg, E. F.; Karitskaya, I. V. *Zh. Obshch. Khim.* **1950**, *20*, 27; *Chem. Abstr.* **1950**, *44*, 7823 [143].

[*1019*] Heacock, R. A.; Hutzinger, O. *Can. J. Chem.* **1964**, *42*, 514 [144].

[*1020*] Hutchins, R. O.; Natale, N. R. *J. Org. Chem.* **1978**, *43*, 2299 [144, 148, 151, 167, 168, 187].

[*1021*] Kabalka, G. W.; Summers, S. T. *J. Org. Chem.* **1981**, *46*, 1217 [144, 148, 151, 156, 157, 167, 168].

[*1022*] Virgilio, J. A.; Heilweil, E. *Org. Prep. Proced. Int.* **1982**, *14*, 9 [144].

[*1023*] Nystrom, R. B. *J. Am. Chem. Soc.* **1955**, *77*, 2544 [144, 145, 205, 240, 304].

[*1024*] Kamitori, Y.; Hojo, M.; Masuda, R.; Yamamoto, M. *Chem. Lett.* **1985**, 253 [144, 145].

[*1025*] Bader, H.; Downer, J. D.; Driver, P. *J. Chem. Soc.* **1950**, 2775 [144, 145].

[*1026*] Woodward, R. B. *Org. Synth., Collect. Vol.* **1955**, *3*, 453 [144, 145].

[*1027*] Buck, J. S.; Ide, W. S. *Org. Synth., Collect. Vol.* **1943**, *2*, 130 [144].

[*1028*] Remers, W. A.; Roth, R. H.; Weiss, M. J. *J. Org. Chem.* **1965**, *30*, 4381 [144].

[*1029*] Eliel, E. L.; Badding, V. G.; Rerick, M. N. *J. Am. Chem. Soc.* **1962**, *84*, 2371 [145, 182].

[*1030*] Eliel, E. L.; Nowak, B. E.; Daignault, R. A.; Badding, V. G. *J. Org. Chem.* **1965**, *30*, 2441 [145, 146].

[*1031*] Eliel, E. L.; Nowak, B. E.; Daignault, R. A. *J. Org. Chem.* **1965**, *30*, 2448 [145, 146].

[*1032*] Eliel, E. L.; Doyle, T. W.; Daignault, R. A.; Newman, B. C. *J. Am. Chem. Soc.* **1966**, *88*, 1828 [145, 147, 182, 183].

[*1033*] Wolfrom, M. L.; Karabinos, J. V. *J. Am. Chem. Soc.* **1944**, *66*, 909 [145, 146, 183].

[*1034*] Pettit, G. R.; van Tamelen, E. E. *Org. React. (N. Y.)* **1962**, *12*, 356 [145, 146, 183].

[*1035*] Gutierrez, C. G.; Stringham, R. A.; Nitasaka, T.; Glasscock, K. G. *J. Org. Chem.* **1980**, *45*, 3393 [146, 182, 183].

[*1036*] Degani, I.; Fochi, R. *J. Chem. Soc., Perkin Trans. 1* **1976**, 1886; **1978**, 1133 [146, 147, 151].

[*1037*] Hill, A. J.; Nason, E. H. *J. Am. Chem. Soc.* **1924**, *46*, 2241 [146].

[*1038*] Eliel, E. L.; Daignault, R. A. *J. Org. Chem.* **1965**, *30*, 2450 [146, 147].

[1039]   Campbell, K. N.; Sommers, A. H.; Campbell, B. K. *J. Am. Chem. Soc.* **1944**, *66*, 82 [146].

[1040]   Winans, C. F.; Adkins, H. *J. Am. Chem. Soc.* **1932**, *54*, 306 [146, 147, 187, 188, 190].

[1041]   Allen, C. F. H.; van Allan, J. *Org. Synth., Collect. Vol.* **1955**, *3*, 827 [146].

[1042]   Bumgardner, C. L.; Lawton, E. L.; Carver, J. G. *J. Org. Chem.* **1972**, *37*, 407 [146, 147].

[1043]   Fischer, O. *Justus Liebigs Ann. Chem.* **1887**, *241*, 328 [146].

[1044]   Contento, M.; Savoia, D.; Trombini, C.; Umani-Ronchi, A. *Synthesis* **1979**, 30 [146, 166, 167].

[1045]   Karrer, P.; Schick, E. *Helv. Chim. Acta* **1943**, *26*, 800 [147].

[1046]   Smith, D. R.; Maienthal, M.; Tipton, J. *J. Org. Chem.* **1952**, *17*, 294 [147, 185, 186].

[1047]   Lycan, W. H.; Puntambeker, S. V.; Marvel, C. S. *Org. Synth., Collect. Vol.* **1943**, *2*, 318 [147].

[1048]   Caglioti, L. *Tetrahedron* **1966**, *22*, 487 [148, 157, 163, 164, 187, 238].

[1049]   Caglioti, L.; Magi, M. *Tetrahedron* **1963**, *19*, 1127 [148, 157].

[1050]   Hutchins, R. O.; Maryanoff, B. E.; Milewski, C. A. *J. Am. Chem. Soc.* **1971**, *93*, 1793 [148, 151, 187].

[1051]   Breitner, E.; Roginski, E.; Rylander, P. N. *J. Org. Chem.* **1959**, *24*, 1855 [149, 152, 165].

[1052]   Heusler, K.; Wieland, P.; Meystre, C. H. *Org. Synth., Collect. Vol.* **1973**, *5*, 692 [149].

[1053]   Eliel, E. L.; Nasipuri, D. *J. Org. Chem.* **1965**, *30*, 3809 [149, 158, 159, 160, 161, 162, 163].

[1054]   Santaniello, E.; Ponti, F.; Manzocchi, A. *Synthesis* **1978**, 891 [149, 150, 153].

[1055]   Thoms, H.; Mannich, C. *Chem. Ber.* **1903**, *36*, 2544 [149, 150].

[1056]   Whitmore, F. C.; Ottenbacher, T. *Org. Synth., Collect. Vol.* **1943**, *2*, 317 [149].

[1057]   Truett, W. L.; Moulton, W. N. *J. Am. Chem. Soc.* **1951**, *73*, 5913 [149, 150, 153].

[1058]   Yamashita, S. *J. Organomet. Chem.* **1968**, *11*, 377 [149, 150, 153, 154].

[1059]   Brown, H. C.; Mandal, A. K. *J. Org. Chem.* **1977**, *42*, 2996 [149, 150, 154].

[1060]   Brown, H. C.; Cho, B. T.; Park, W. S. *J. Org. Chem.* **1988**, *53*, 1231 [150, 154, 162, 224].

[1061]   Newman, M. S.; Walborsky, H. M. *J. Am. Chem. Soc.* **1950**, *72*, 4296 [151, 225].

[1062]   Tafel, J. *Chem. Ber.* **1909**, *42*, 3146 [150, 151].

[1063]   Schreibmann, P. *Tetrahedron Lett.* **1970**, 4271 [151, 164].

[1064]   Eisch, J. J.; Liu, Z.-R.; Boleslawski, M. P. *J. Org. Chem.* **1992**, *57*, 2143 [151, 156].

[1065]   Freifelder, M. *J. Org. Chem.* **1964**, *29*, 2895 [152, 155, 156, 157].

[1066]   Tyman, J. H. P. *J. Appl. Chem.* **1970**, *20*, 179 [153].

[1067]   Cram, D. J.; Abd Elhafer, F. A. *J. Am. Chem. Soc.* **1952**, *74*, 5828 [153, 155].

[1068]   Cutler, R. A.; Stenger, R. J.; Suter, C. M. *J. Am. Chem. Soc.* **1952**, *74*, 5475 [153].

[1069]   Ferles, M.; Attia, A. *Collect. Czech. Chem. Commun.* **1973**, *38*, 611 [154].

[1070]   Seebach, D.; Daum, H. *Chem. Ber.* **1974**, *107*, 1748 [154].

[1071]   Brown, H. C.; Pai, G. G. *J. Org. Chem.* **1982**, *47*, 1606 [154].

[1072]   Chong, J. M.; Mar, E. K. *J. Org. Chem.* **1991**, *56*, 893 [154].

[1073]   Evans, D. A.; Nelson, S. G.; Gagné, M. R.; Muci, A. R. *J. Am. Chem. Soc.* **1993**, *115*, 9800 [154].

[1074]   Kabuto, K.; Imuta, M.; Kempner, E. S.; Ziffer, H. *J. Org. Chem.* **1978**, *43*, 2357 [154, 163].

[1075]   Imuta, M.; Ziffer, H. *J. Org. Chem.* **1978**, *43*, 3530 [155, 321].

[1076] Prelog, V. *Pure Appl. Chem.* **1964**, *9*, 119 [155].
[1077] Howe, R.; Moore, R. H. *J. Med. Chem.* **1971**, *14*, 287 [155].
[1078] Brown, H. C.; Varma, V. *J. Am. Chem. Soc.* **1966**, *88*, 2871 [155, 158, 159, 161, 162].
[1079] Zagumenny, A. *Chem. Ber.* **1881**, *14*, 1402 [155, 156].
[1080] Gomberg, M.; Bachmann, W. E. *J. Am. Chem. Soc.* **1927**, *49*, 236 [155, 156].
[1081] Suzuki, H.; Masuda, R.; Kubota, H.; Osuka, A. *Chem. Lett.* **1983**, 909 [155, 156].
[1082] Lau, C. K.; Dufresne, C.; Bélanger, P.; Piétré, S.; Scheigetz, J. *J. Org. Chem.* **1986**, *51*, 3038 [155, 156].
[1083] Plattner, P. A.; Fürst, A.; Keller, W. *Helv. Chim. Acta* **1949**, *32*, 2464 [156].
[1084] Gribble, G. W.; Kelly, W. J.; Emery, S. E. *Synthesis*, **1978**, 763 [156].
[1085] Klages, A.; Allendorff, P. *Chem. Ber.* **1898**, *31*, 998 [156].
[1086] Read, R. R.; Wood, J., Jr. *Org. Synth., Collect. Vol.* **1955**, *3*, 444 [156].
[1087] Whalley, W. B. *J. Chem. Soc.* **1951**, 665 [156].
[1088] Cram, D. J.; Sahyun, M. R. V.; Knox, G. R. *J. Am. Chem. Soc.* **1962**, *84*, 1734 [157, 163, 187].
[1089] Andrews, G. C. *Tetrahedron Lett.* **1980**, *21*, 697 [157].
[1090] Hardegger, E.; Plattner, P. A.; Blank, F. *Helv. Chim. Acta* **1944**, *27*, 793 [157, 200].
[1091] Helg, R.; Schinz, H.; Eliel, E. L. *Helv. Chim. Acta* **1952**, *35*, 2406 [157, 223].
[1092] Eliel, E. L.; Ro, R. S. *J. Am. Chem. Soc.* **1957**, *79*, 5992 [157].
[1093] Eliel, E. L.; Senda, Y. *Tetrahedron* **1970**, *26*, 2411 [158, 159, 160, 161].
[1094] Eliel, E. L.; Rerick, M. N. *J. Am. Chem. Soc.* **1960**, *82*, 1367 [158].
[1095] Brown, H. C.; Deck, H. R. *J. Am. Chem. Soc.* **1965**, *87*, 5620 [158, 160, 161, 162].
[1096] Eliel, E. L.; Martin, R. J. L.; Nasipuri, D. *Org. Synth., Collect. Vol.* **1973**, *5*, 175 [158].
[1097] Richer, J. C. *J. Org. Chem.* **1965**, *30*, 324 [158, 159, 160, 161].
[1098] Kovacs, G.; Galambos, G.; Juvancz, Z. *Synthesis* **1977**, 171 [158, 160].
[1099] Eliel, E. L.; Doyle, T. W.; Hutchins, R. O. *Org. Synth., Collect. Vol.* **1988**, *6*, 215 [158, 160].
[1100] Haubenstock, H.; Eliel, E. L. *J. Am. Chem. Soc.* **1962**, *84*, 2363 [158, 159, 160, 161].
[1101] Haubenstock, H.; Eliel, E. L. *J. Am. Chem. Soc.* **1962**, *84*, 2368 [159, 160, 161].
[1102] Huffman, J. W.; Charles, J. T. *J. Am. Chem. Soc.* **1968**, *90*, 6486 [159].
[1103] Krishnamurthy, S.; Brown, H. C. *J. Am. Chem. Soc.* **1976**, *98*, 3383 [161, 162].
[1104] Coulombeau, A.; Rassat, A. *Chem. Commun.* **1968**, 1587 [161, 162].
[1105] Jackman, L. M.; Macbeth, A. K.; Mills, J. A. *J. Chem. Soc.* **1949**, 2641 [162].
[1106] Browne, P. A.; Kirk, D. N. *J. Chem. Soc. C* **1969**, 1653 [162, 163].
[1107] Kawasaki, M.; Suzuki, Y.; Terashima, S. *Chem. Lett.* **1984**, 239 [162].
[1108] Throop, L.; Tökés, L. *J. Am. Chem. Soc.* **1967**, *89*, 4789 [164].
[1109] Toda, M.; Hirata, Y.; Yamamura, S. *Chem. Commun.* **1969**, 919 [164].
[1110] Leonard, N. J.; Figueras, J. *J. Am. Chem. Soc.* **1952**, *74*, 917 [164, 175].
[1111] Sondheimer, F.; Velasco, M.; Rosenkranz, G. *J. Am. Chem. Soc.* **1955**, *77*, 192 [165].
[1112] Shepherd, D. A.; Donia, R. A.; Campbell, J. A.; Johnson, B. A. *J. Am. Chem. Soc.* **1955**, *77*, 1212 [165].
[1113] Adams, R.; Kern, J. W.; Shriner, R. L. *Org. Synth., Collect. Vol.* **1932**, *1*, 101 [165, 167].
[1114] Augustine, R. L. *J. Org. Chem.* **1958**, *23*, 1853 [165, 166].
[1115] Djerassi, C.; Gutzwiller, J. *J. Am. Chem. Soc.* **1966**, *88*, 4537 [165].
[1116] Sakai, K.; Watanabe, K. *Bull. Chem. Soc. Jpn.* **1967**, *40*, 1548 [165].
[1117] Ashby, E. C.; Lin, J. J. *Tetrahedron Lett.* **1976**, 3865 [166, 167].

[1118]　Wolf, H. R.; Zink, M. P. *Helv. Chim. Acta* **1973**, *56*, 1062 [166].

[1119]　Ashby, E. C.; Lin, J. J. *Tetrahedron Lett.* **1975**, 4453 [166, 167].

[1120]　Semmelhack, M. F.; Stauffer, R. D. *J. Org. Chem.* **1975**, *40*, 3619 [166, 167].

[1121]　Collman, J. P.; Finke, R. G.; Matlock, P. L.; Wahren, R.; Komoto, R. G.; Brauman, J. I. *J. Am. Chem. Soc.* **1978**, *100*, 1978 [166].

[1122]　Fetizon, M.; Gore, J. *Tetrahedron Lett.* **1966**, 471 [166].

[1123]　Davis, B. R.; Woodgate, P. D. *J. Chem. Soc.* **1966**, 2006 [166].

[1124]　Kergomard, A.; Renaud, M. F.; Veschambre, H. *J. Org. Chem.* **1982**, *47*, 792 [166].

[1125]　Brown, H. C.; Hess, H. M. *J. Org. Chem.* **1969**, *34*, 2206 [166].

[1126]　Gannon, W. F.; House, H. O. *Org. Synth., Collect. Vol.* **1973**, *5*, 294 [166].

[1127]　Meek, J. S.; Lorenzi, F. J.; Cristol, S. J. *J. Am. Chem. Soc.* **1949**, *71*, 1830 [166, 167].

[1128]　Hochstein, F. A. *J. Am. Chem. Soc.* **1949**, *71*, 305 [166, 167].

[1129]　Wilson, K. E.; Seidner, R. T.; Masamune, S. *Chem. Commun.* **1970**, 213 [166].

[1130]　Kim, S.; Moon, Y. C.; Ahn, K. H. *J. Org. Chem.* **1982**, *47*, 3311 [166, 167].

[1131]　Luche, J.-L. *J. Am. Chem. Soc.* **1978**, *100*, 2226 [166].

[1132]　Gemal, A. L.; Luche, J.-L. *J. Am. Chem. Soc.* **1981**, *103*, 5454 [166].

[1133]　Hutchins, R. O.; Natale, N. R.; Taffer, I. M. *Chem. Commun.* **1978**, 1088 [166].

[1134]　Ravikumar, K. S.; Baskaran, S.; Chandrasekaran, S. *J. Org. Chem.* **1993**, *58*, 5981 [166].

[1135]　Macbeth, A. K.; Mills, J. A. *J. Chem. Soc.* **1949**, 2646 [166, 167].

[1136]　Mozingo, R.; Adkins, H. *J. Am. Chem. Soc.* **1938**, *60*, 669 [167].

[1137]　Hayes, N. F. *Synthesis* **1975**, 702 [167].

[1138]　Sibi, M. P.; Sorum, M. T.; Bender, J. A.; Gaboury, J. A. *Synth. Commun.* **1992**, *22*, 809 [167].

[1139]　Kabalka, G. W.; Yang, D. T. C.; Baker, J. D. *J. Org. Chem.* **1976**, *41*, 574 [167].

[1140]　Biemann, K.; Büchi, G.; Walker, B. H. *J. Am. Chem. Soc.* **1957**, *79*, 5558 [168].

[1141]　Fischer, R.; Lardelli, G.; Jerger, O. *Helv. Chim. Acta* **1951**, *34*, 1577 [168].

[1142]　Harries, C.; Eschenbach, G. *Chem. Ber.* **1896**, *29*, 380 [168].

[1143]　Midland, M. M.; McDowell, D. C.; Hatch, R. L.; Tramontano, A. *J. Am. Chem. Soc.* **1980**, *102*, 867 [168, 169].

[1144]　Midland, M. M.; Graham, R. S. *Org. Synth.* **1984**, *63*, 57 [168, 169].

[1145]　Noyori, R.; Tomino, I.; Yamada, M.; Nishizawa, M. *J. Am. Chem. Soc.* **1984**, *106*, 6717 [168, 169].

[1146]　Ramachandran, P. V.; Teodorovič, A. V.; Rangaishenvi, M. V.; Brown, H. C. *J. Org. Chem.* **1992**, *57*, 2379 [168].

[1147]　Cohen, N.; Lopresti, R. J.; Neukom, C.; Saucy, G. *J. Org. Chem.* **1980**, *45*, 582 [168].

[1148]　Locher, R.; Seebach, D. *Angew. Chem., Int. Ed. Engl.* **1981**, *20*, 569 [169].

[1149]　Schlenk, H.; Lamp, B. *J. Am. Chem. Soc.* **1951**, *73*, 5493 [169].

[1150]　Hudlický, M.; Lejhancová, I. *Collect. Czech. Chem. Commun.* **1966**, *31*, 1416 [169].

[1151]　Huffmann, J. W. *J. Org. Chem.* **1959**, *24*, 1759 [169].

[1152]　Danheiser, R. L.; Savariar, S.; Cha, D. D. *Org. Synth.* **1989**, *68*, 32 [169].

[1153]　Kamiya, N.; Tanmatu, H.; Ishii, Y. *Chem. Lett.* **1992**, 293 [170].

[1154]　Wharton, P. S.; Dunny, S.; Krebbs, L. S. *J. Org. Chem.* **1965**, *29*, 958 [170, 171].

[1155]　Hirao, T.; Masunaga, T.; Hayashi, K.; Ohshiro, Y.; Agawa, T. *Tetrahedron Lett.* **1983**, *24*, 399 [170, 171].

[1156]　Inhoffen, H. H.; Kolling, G.; Koch, G.; Nebel, I. *Chem. Ber.* **1951**, *84*, 361 [170].

[1157]　McGuckin, W. F.; Mason, H. L. *J. Am. Chem. Soc.* **1955**, *77*, 1822 [171].

[*1158*] Julian, P. L.; Cole, W.; Magnani, A.; Meyer, E. W. *J. Am. Chem. Soc.* **1945**, *67*, 1728 [171].
[*1159*] Tashiro, M.; Nakamura, H.; Nakayama, K. *Org. Prep. Proced. Int.* **1987**, *19*, 442 [171, 172].
[*1160*] Shechter, H.; Ley, D. L.; Zeldin, L. *J. Am. Chem. Soc.* **1952**, *74*, 3664 [172].
[*1161*] Levai, L.; Fodor, G.; Ritvay-Emandity, K.; Fuchs, O.; Hajos, A. *Chem. Ber.* **1960**, *93*, 387 [172].
[*1162*] Nakamura, K.; Inoue, Y.; Shibahara, J.; Oka, S.; Ohno, A. *Tetrahedron Lett.* **1988**, *29*, 4769 [172].
[*1163*] Simpson, J. C. E.; Atkinson, C. M.; Schofield, K.; Stephenson, O. *J. Chem. Soc.* **1945**, 646 [172].
[*1164*] Oelschläger, H. *Chem. Ber.* **1956**, *89*, 2025 [173].
[*1165*] McMurry, J. E.; Melton, J. *J. Am. Chem. Soc.* **1971**, *93*, 5309; *J. Org. Chem.* **1973**, *38*, 4367 [173].
[*1166*] Gruber, W.; Renner, H. *Monatsh. Chem.* **1950**, *81*, 751 [173].
[*1167*] Birkofer, L. *Chem. Ber.* **1947**, *80*, 83 [173, 222].
[*1168*] Wolfrom, M. L.; Brown, R. L. *J. Am. Chem. Soc.* **1943**, *65*, 1516 [173].
[*1169*] Mamoli, L. *Chem. Ber.* **1938**, *71*, 2696 [174].
[*1170*] Mamoli, L.; Schramm, G. *Chem. Ber.* **1938**, *71*, 2698 [174].
[*1171*] Simon, H.; Rambeck, B.; Hashimoto, H.; Gunther, H.; Nohynek, G.; Neumann, H. *Angew. Chem., Int. Ed. Engl.* **1974**, *13*, 609 [174].
[*1172*] Guetté, J. P.; Spassky, N.; Boucherot, D. *Bull. Soc. Chim. Fr.* **1972**, 4217 [174].
[*1173*] Blomquist, A. T.; Goldstein, A. *Org. Synth., Collect. Vol.* **1963**, *4*, 216 [174].
[*1174*] Prelog, V.; Schenker, K.; Günthard, H.H. *Helv. Chim. Acta* **1952**, *35*, 1598 [174, 175].
[*1175*] Cope, A. C.; Barthel, J. W.; Smith, R. D. *Org. Synth., Collect. Vol.* **1963**, *4*, 218 [174].
[*1176*] Smith, W. T., Jr. *J. Am. Chem. Soc.* **1951**, *73*, 1883 [174].
[*1177*] Reusch, W.; LeMahieu, R. *J. Am. Chem. Soc.* **1964**, *86*, 3068 [174, 177].
[*1178*] Ho, T.-L.; Wong, C. M. *Synthesis* **1975**, 161 [174, 316].
[*1179*] Rosenfeld, R. S. *J. Am. Chem. Soc.* **1957**, *79*, 5540 [175, 211].
[*1180*] Mandel, M.; Hudlicky, T. *J. Chem. Soc., Perkin Trans. 1* **1993**, 741 [175].
[*1181*] Wharton, P. S.; Bohlen, D. H. *J. Org. Chem.* **1961**, *26*, 3615 [175].
[*1182*] Fehnel, E. A. *J. Am. Chem. Soc.* **1949**, *71*, 1063 [175].
[*1183*] Barluenga, J.; Aguilar, E.; Fustero, S.; Olano, B.; Viado, A. L. *J. Org. Chem.* **1992**, *57*, 1219 [175, 176].
[*1184*] Seebach, D.; Prelog, V. *Angew. Chem.* **1982**, *94*, 696; *Int. Ed. Engl.* **1982**, *21*, 654 [176].
[*1185*] Andrews, G. C. *Tetrahedron Lett.* **1980**, *21*, 697 [176].
[*1186*] Luche, J.-L.; Gemal, A. L. *J. Am. Chem. Soc.* **1979**, *101*, 5848 [176, 177].
[*1187*] Gemal, A. L.; Luche, J.-L. *J. Org. Chem.* **1979**, *44*, 4187 [176].
[*1188*] von Pechmann, H.; Dahl, F. *Chem. Ber.* **1890**, *23*, 2421 [177].
[*1189*] Rubin, M. B.; Ben-Bassat, J. M. *Tetrahedron Lett.* **1971**, 3403 [177, 178].
[*1190*] Seibert, W. *Chem. Ber.* **1947**, *80*, 494 [177, 178, 187].
[*1191*] Cusack, N. J.; Davis, B. R. *Chem. Ind. (London)* **1964**, 1426; *J. Org. Chem.* **1965**, *30*, 2062 [177, 178].
[*1192*] Wenkert, E.; Kariv, E. *Chem. Commun.* **1965**, 570 [177].
[*1193*] Bonnet, M.; Geneste, P.; Rodriguez, M. *J. Org. Chem.* **1980**, *45*, 40 [177].
[*1194*] Buchanan, J. G. S. C.; Davis, B. R. *J. Chem. Soc. C* **1967**, 1340 [177, 178, 179].
[*1195*] Fauve, A.; Veschambre, H. *J. Org. Chem.* **1988**, *53*, 5215 [180].
[*1196*] Rosenblatt, E. F. *J. Am. Chem. Soc.* **1940**, *62*, 1092 [180, 181].
[*1197*] Fatiadi, A. J.; Sager, W. F. *Org. Synth., Collect. Vol.* **1973**, *5*, 595 [180].

[1198] Blasczak, L. C.; McMurry, J. E. *J. Org. Chem.* **1974**, *39*, 258 [180, 181].
[1199] Boyland, E.; Manson, D. *J. Chem. Soc.* **1951**, 1837 [181].
[1200] Clar, E. *Chem. Ber.* **1939**, *72*, 1645 [181].
[1201] Howard, W. L.; Brown, J. H., Jr. *J. Org. Chem.* **1961**, *26*, 1026 [182].
[1202] Daignault, R. A.; Eliel, E. L. *Org. Synth., Collect. Vol.* **1973**, *5*, 303 [182].
[1203] Takano, S.; Abiyama, M.; Sato, S.; Ogasawara, K. *Chem. Lett.* **1983**, 1593 [182].
[1204] Borders, R. J.; Bryson, T. A. *Chem. Lett.* **1984**, 9 [182].
[1205] Eliel, E. L.; Krishnamurthy, S. *J. Org. Chem.* **1965**, *30*, 848 [182].
[1206] Eliel, E. L.; Pilato, L. A.; Badding, V. G. *J. Am. Chem. Soc.* **1962**, *84*, 2377 [182, 183].
[1207] Kikugawa, Y. *J. Chem. Soc., Perkin Trans. 1* **1984**, 609 [183].
[1208] Truce, W. E.; Roberts, F. E. *J. Org. Chem.* **1953**, *28*, 961 [183].
[1209] Georgian, V.; Harrison, R.; Gubich, N. *J. Am. Chem. Soc.* **1959**, *81*, 5834 [184].
[1210] Nishiwaki, Z.; Fujiyama, F. *Synthesis* **1972**, 569 [184].
[1211] Botta, M.; DeAngelis, F.; Gambacorta, A.; Labbiento, L.; Nicoletti, R. *J. Org. Chem.* **1985**, *50*, 1916 [184].
[1212] Gribble, G. W.; Leiby, R. W.; Sheehan, M. N. *Synthesis* **1977**, 856 [184, 185].
[1213] Borch, R. F.; Bernstein, M. D.; Durst, H. D. *J. Am. Chem. Soc.* **1971**, *93*, 2897 [184, 187].
[1214] Feuer, H.; Vincent, B. F., Jr.; Bartlett, R. S. *J. Org. Chem.* **1965**, *30*, 2877 [184].
[1215] Freifelder, M.; Smart, W. D.; Stone, G. R. *J. Org. Chem.* **1962**, *27*, 2209 [185].
[1216] Anziani, P.; Cornubert, R. *Bull. Soc. Chim. Fr.* **1948**, 857 [185, 186].
[1217] Rausser, R.; Finckenor, D.; Weber, L.; Hershberg, E. B.; Oliveto, E. P. *J. Org. Chem.* **1966**, *31*, 1346 [185].
[1218] Rausser, R.; Weber, L.; Hershberg, E. B.; Oliveto, E. P. *J. Org. Chem.* **1966**, *31*, 1342 [185].
[1219] Lloyd, D.; McDougall, R. H.; Wason, F. T. *J. Chem. Soc.* **1965**, 822 [185, 186].
[1220] Mukaiyama, T.; Yorozu, K.; Kato, K.; Yamada, T. *Chem. Lett.* **1992**, 181 [185].
[1221] Terentev, A. P.; Gusar, N. I. *Zh. Obshch. Khim.* **1965**, *35*, 125; *Chem. Abstr.* **1965**, *62*, 13068 [185, 186].
[1222] Prelog, V.; El-Neweihy, M. F.; Häflinger, O. *Helv. Chim. Acta* **1950**, *33*, 365 [185].
[1223] Curran, D. P.; Brill, J. F.; Rakiewicz, D. M. *J. Org. Chem.* **1984**, *49*, 1654 [185].
[1224] Ricart, G. *Bull. Soc. Chim. Fr.* **1974**, 615 [186].
[1225] Feuer, H.; Braunstein, D. M. *J. Org. Chem.* **1969**, *34*, 1817 [186].
[1226] Murphy, J. G. *J. Org. Chem.* **1961**, *26*, 3104 [186].
[1227] Goodwin, R. C.; Bailey, J. R. *J. Am. Chem. Soc.* **1925**, *47*, 167 [186].
[1228] Armand, J.; Boulares, L. *Bull. Soc. Chim. Fr.* **1975**, 366 [186].
[1229] Norton, T. R.; Benson, A. A.; Seibert, R. A.; Bergstrom, F. W. *J. Am. Chem. Soc.* **1946**, *68*, 1330 [187].
[1230] Emerson, W. S. *Org. React. (N. Y.)* **1948**, *4*, 174 [187].
[1231] Schellenberg, K. A. *J. Org. Chem.* **1963**, *28*, 3259 [187, 189].
[1232] Borch, R. F.; Durst, H. D. *J. Am. Chem. Soc.* **1969**, *91*, 3996 [187, 189, 190].
[1233] Winans, C. F. *J. Am. Chem. Soc.* **1939**, *61*, 3566 [187, 189].
[1234] Hancock, E. M.; Cope A. C. *Org. Synth., Collect. Vol.* **1955**, *3*, 501 [187, 189].
[1235] Mignonac, G. *C. R. Hebd. Seances Acad. Sci.* **1920**, *171*, 1148; **1921**, *172*, 223 [187, 189].
[1236] Balcom, D. M.; Noller, C. R. *Org. Synth., Collect. Vol.* **1963**, *4*, 103 [187].
[1237] Robinson, J. C., Jr.; Snyder, H. R. *Org. Synth., Collect. Vol.* **1955**, *3*, 717 [187, 190].

[1238] Barney, C. L.; Huber, E. W.; McCarthy, J. R. *Tetrahedron Lett.* **1990**, *31*, 5547 [187].

[1239] Borsch, R. F. *Org. Synth., Collect. Vol.* **1988**, *6*, 499 [187].

[1240] Ingersoll, A. W.; Brown, J. H.; Kim, C. K.; Beauchamp, W. D.; Jennings, G. *J. Am. Chem. Soc.* **1936**, *58*, 1808 [189].

[1241] Eschweiler, W. *Chem. Ber.* **1905**, *38*, 880 [189].

[1242] Muraki, M.; Mukaiyama, T. *Chem. Lett.* **1974**, 1447 [191, 194].

[1243] Bedenbaugh, A. O.; Bedenbaugh, J. H.; Bergin, W. A.; Adkins, K. D. *J. Am. Chem. Soc.* **1970**, *92*, 5774 [192].

[1244] Tafel, J.; Friedrichs, G. *Chem. Ber.* **1904**, *37*, 3187 [192].

[1245] Guyer, A.; Bieler, A.; Sommaruga, M. *Helv. Chim. Acta* **1955**, *38*, 976 [192, 194].

[1246] Ryashentseva, M. A.; Minachev, K. M.; Belanova, E. P. *Izv. Akad. Nauk SSSR, Ser. Khim.* **1974**, 1906; *Chem. Abstr.* **1974**, *81*, 135391 [192].

[1247] Powell, J.; James, N.; Smith, S. J. *Synthesis* **1986**, 338 [192, 232, 302].

[1248] Nystrom, R. F.; Brown, W. G. *J. Am. Chem. Soc.* **1947**, *69*, 2548 [192, 193, 194, 196, 204].

[1249] Černý, M.; Málek, J.; Čapka, M.; Chvalovský, V. *Collect. Czech. Chem. Commun.* **1969**, *34*, 1025 [192, 194, 204, 206, 215, 216].

[1250] Černý, M.; Málek, J. *Collect. Czech. Chem. Commun.* **1971**, *36*, 2394 [192, 194].

[1251] Yoon, N. M.; Pak, C. S.; Brown, H. C.; Krishnamurthy, S.; Stocky, T. P. *J. Org. Chem.* **1973**, *38*, 2786 [192, 194, 197, 198, 199, 200, 216, 227, 305].

[1252] Brown, H. C.; Subba Rao, B. C. *J. Am. Chem. Soc.* **1960**, *82*, 681 [192, 194, 197, 198, 199, 216, 241].

[1253] Kanth, J. V. B.; Periasamy, M. *J. Org. Chem.* **1991**, *56*, 5964 [192, 193, 194, 195, 196, 227].

[1254] Yoon, N. M.; Cho, B. T. *Tetrahedron Lett.* **1982**, *23*, 2475 [192, 193].

[1255] Kende, A. S.; Fludzinski, P. *Org. Synth.* **1985**, *64*, 104 [192, 193, 227].

[1256] Ranu, B. C.; Das, A. R. *J. Chem. Soc., Perkin Trans. 1* **1992**, 1501 [192].

[1257] Burton, H.; Ingold, C. K. *J. Chem. Soc.* **1929**, 2022 [193].

[1258] Farmer, E. H.; Hughes, L. A. *J. Chem. Soc.* **1934**, 1929 [193].

[1259] Castro, C. E.; Stephens, R. D.; Moje, S. *J. Am. Chem. Soc.* **1966**, *88*, 4964 [193, 196, 219].

[1260] Benedict, G. E.; Russell, R. R. *J. Am. Chem. Soc.* **1951**, *73*, 544 [193].

[1261] Brown, H. C.; Heim, P.; Yoon, N. M. *J. Am. Chem. Soc.* **1970**, *92*, 1637 [194].

[1262] Blackwood, R. H.; Hess, G. B.; Larrabee, C. E.; Pilgrim, E. J. *J. Am. Chem. Soc.* **1958**, *80*, 6244 [194].

[1263] Coleman, G. H.; Johnson, H. L. *Org. Synth., Collect. Vol.* **1955**, *3*, 60 [195].

[1264] Černý, M.; Málek, J. *Tetrahedron Lett.* **1969**, 1739 [195].

[1265] Černý, M.; Málek, J. *Collect. Czech. Chem. Commun.* **1970**, *35*, 1216 [195].

[1266] Benkeser, R. A.; Foley, K. M.; Gaul, J. M.; Li, G. S. *J. Am. Chem. Soc.* **1970**, *92*, 3232 [195, 197].

[1267] Burgstahler, A. W.; Bithos, Z. *J. Org. Synth., Collect. Vol.* **1973**, *5*, 591 [195].

[1268] Kuehne, M. E.; Lambert, B. F. *J. Am. Chem. Soc.* **1959**, *81*, 4278: *Org. Synth., Collect. Vol.* **1973**, *5*, 400 [195, 232, 233].

[1269] Johnson, W. S.; Gutsche, C. D.; Offenhauser, R. D. *J. Am. Chem. Soc.* **1946**, *68*, 1648 [195].

[1270] Eliel, E. L.; Hoover, T. E. *J. Org. Chem.* **1959**, *24*, 938 [195].

[1271] May, E. L.; Mosettig, E. *J. Am. Chem. Soc.* **1948**, *70*, 1077 [195].

[1272] Müller, A. *Org. Synth., Collect. Vol.* **1943**, *2*, 535 [196].

[1273] Paal, C.; Gerum, J. *Chem. Ber.* **1908**, *41*, 2273, 2278 [196, 217].

[1274] Dang, T. P.; Kagan, H. B. *Chem. Commun.* **1971**, 481 [196].

[1275] Page, G. A.; Tarbell, D. S. *Org. Synth., Collect. Vol.* **1963**, *4*, 136 [196].

[1276] Badger, G. M. *J. Chem. Soc.* **1948**, 999 [196].

[1277] Ingersoll, A. W. *Org. Synth., Collect. Vol.* **1932**, *1*, 311 [196].

[1278] Paal, C.; Hartmann, W. *Chem. Ber.* **1909**, *42*, 3930; **1918**, *51*, 640 [197].

[1279] Nystrom, R. F. *J. Am. Chem. Soc.* **1959**, *81*, 610 [197, 205, 215, 220, 221].

[1280] Yoon, N. M.; Brown, H. C. *J. Am. Chem. Soc.* **1968**, *90*, 2927 [197, 215, 220, 222, 232, 240].

[1281] Gaux, B.; Le Henaff, P. *Bull. Soc. Chim. Fr.* **1974**, 505 [197].

[1282] Rosenmund, K. W.; Zetzsche, F. *Chem. Ber.* **1918**, *51*, 578 [197].

[1283] Saburi, M.; Shao, L.; Sakurai, T.; Uchida, Y. *Tetrahedron Lett.* **1992**, *33*, 7877 [198].

[1284] Liebermann, C. *Chem. Ber.* **1895**, *28*, 134 [198].

[1285] Moppelt, C. E.; Sutherland, J. K. *J. Chem. Soc. C* **1968**, 3040 [198].

[1286] Mohrig, J. R.; Vreede, P. J.; Schultz, S. C.; Fierke, C. A. *J. Org. Chem.* **1981**, *46*, 4655 [199].

[1287] Allen, C. F. H.; MacKay, D. D. *Org. Synth., Collect. Vol.* **1943**, *2*, 580 [199].

[1288] McKennon, M. J.; Meyers, A. I. *J. Org. Chem.* **1993**, *58*, 3568 [199].

[1289] Rosenmund, K. W.; Zymalkowski, F.; Engels, P. *Chem. Ber.* **1951**, *84*, 711 [199, 200].

[1290] Tuynenburg, M. G.; van der Ven, B.; De Jonge, A. P. *Nature (London)* **1962**, *194*, 995 [200].

[1291] Steinkopf, W.; Wolfram, A. *Justus Liebigs Ann. Chem.* **1923**, *430*, 113 [200].

[1292] Ray, F. E.; Weisburger, E. K.; Weisburger, J. H. *J. Org. Chem.* **1948**, *13*, 655 [200].

[1293] Beyerman, H. C.; Boeke, P. *Recl. Trav. Chim. Pays-Bas* **1959**, *78*, 648 [200].

[1294] Martin, E. L. *Org. Synth., Collect. Vol.* **1943**, *2*, 499 [199, 201].

[1295] Kitamura, M.; Ohkuma, T.; Inoue, S.; Sayo, N.; Kumobayashi, H.; Akutagawa, S.; Ohta, T.; Takaya, H.; Noyori, R. *J. Am. Chem. Soc.* **1988**, *110*, 629 [200].

[1296] Durham, L. J.; McLeod, D. J.; Cason, J. *Org. Synth., Collect. Vol.* **1963**, *4*, 510 [201].

[1297] Weygand, C.; Meusel, W. *Chem. Ber.* **1943**, *76*, 503 [203, 204].

[1298] Rosenmund, K. W. *Chem. Ber.* **1918**, *51*, 585 [203, 204].

[1299] Barnes, R. P. *Org. Synth., Collect. Vol.* **1955**, *3*, 551 [203].

[1300] Brown, H. C.; Subba Rao, B. C. *J. Am. Chem. Soc.* **1958**, *80*, 5377 [203, 204, 301, 303].

[1301] Babler, J. H. *Synth. Commun.* **1982**, *12*, 839 [203].

[1302] Horner, L.; Röder, H. *Chem. Ber.* **1970**, *103*, 2984 [203].

[1303] Sroog, C. E.; Woodburn, H. M. *Org. Synth., Collect. Vol.* **1963**, *4*, 271 [204].

[1304] Santaniello, E.; Farachi, C.; Manzocchi, A. *Synthesis* **1979**, 912 [204, 205].

[1305] Soai, K.; Yokoyama, S.; Mochida, K. *Synthesis* **1987**, 647 [205].

[1306] Austin, P. R.; Bousquet, E. W.; Lazier, W. A. *J. Am. Chem. Soc.* **1937**, *59*, 864 [205, 206].

[1307] Morand, P.; Kayser M. *Chem. Commun.* **1976**, 314 [205, 206].

[1308] Morand P.; Salvator J.; Kayser, M. M. *Chem. Commun.* **1982**, 458 [205, 206].

[1309] Kayser, M. M.; Morand, P. *Can. J. Chem.* **1978**, *56*, 1524 [205, 206].

[1310] Bloomfield, J. J.; Lee, S. L. *J. Org. Chem.* **1967**, *32*, 3919 [205, 206].

[1311] Bailey, D.; Johnson, R. *J. Org. Chem.* **1970**, *35*, 3574 [205, 206].

[1312] Makhlouf, M. A.; Rickborn, B. *J. Org. Chem.* **1981**, *46*, 4810 [205, 206].

[1313] Wislicenus, J. *Chem. Ber.* **1884**, *17*, 2178 [206].

[1314] Zakharin, L. I.; Khorlina, I. M. *Tetrahedron Lett.* **1962**, 619 [208].

[1315] Winterfeld, E. *Synthesis* **1975**, 617 [208, 209].

[1316] Greenwald, R. B.; Evans, D. H. *J. Org. Chem.* **1976**, *41*, 1470 [208].

[1317] Weissman, P. M.; Brown, H. C. *J. Org. Chem.* **1966**, *31*, 283 [208].

[1318] Cha, J. S.; Kwon, S. S. *J. Org. Chem.* **1987**, *52*, 5486 [208, 303].

[1319] Muraki, M.; Mukaiyama, T. *Chem. Lett.* **1975**, 215 [208].
[1320] Arth, G. E. *J. Am. Chem. Soc.* **1953**, *75*, 2413 [209].
[1321] Wolfrom, M. L.; Wood, H. B. *J. Am. Chem. Soc.* **1951**, *73*, 2933 [209].
[1322] Wolfrom, M. L.; Anno, K. *J. Am. Chem. Soc.* **1952**, *74*, 5583 [209].
[1323] Sperber, N.; Zaugg, H. E.; Sandstrom, W. M. *J. Am. Chem. Soc.* **1947**, *69*, 915 [209].
[1324] Pettit, G. R.; Piatok, D. M. *J. Org. Chem.* **1962**, *27*, 2127 [209].
[1325] Baldwin, S. W.; Haut, S. A. *J. Org. Chem.* **1975**, *40*, 3885 [209].
[1326] Kraus, G. A.; Frazier, K. A.; Roth, B. D.; Taschner, M. J.; Neunschwander, K. *J. Org. Chem.* **1981**, *46*, 2417 [210].
[1327] Pettit, G. R.; Kasturi, T. R. *J. Org. Chem.* **1960**, *25*, 875; **1961**, *26*, 4557 [210].
[1328] Ager, D. J.; Sutherland, I. O. *Chem. Commun.* **1982**, 248 [210].
[1329] Zymalkowski, F.; Schuster, T.; Scherer, H. *Arch. Pharm. (Weinheim, Ger.)* **1969**, *302*, 272; *Chem. Abstr.* **1969**, *71*, 12277 [210].
[1330] Keinan, E.; Greenspoon, N. *Tetrahedron Lett.* **1982**, *23*, 241 [210].
[1331] Letsinger, R. L.; Jamison, J. D.; Hussey, A. S. *J. Org. Chem.* **1961**, *26*, 97 [210].
[1332] Beels, C. M. D.; Abu-Rabie, M. S.; Murray-Rust, P.; Murray-Rust, J. *Chem. Commun.* **1979**, 665 [210, 211].
[1333] Uhle, F. C.; McEwen, C. M., Jr.; Schröter, H.; Yuan, C.; Baker, B. W. *J. Am. Chem. Soc.* **1960**, *82*, 1200, 1206 [210, 211].
[1334] Bowman, R. E. *J. Chem. Soc.* **1950**, 325 [211].
[1335] Bergmann, M.; Zervas, L. *Chem. Ber.* **1932**, *65*, 1192, 1201 [211].
[1336] Wunsch, E.; Zwick, A. *Hoppe-Seyler's Z. Physiol. Chem.* **1963**, *333*, 108 [211].
[1337] Kovacs, J.; Rodin, R. L. *J. Org. Chem.* **1968**, *33*, 2418 [211].
[1338] Kuromizu, K.; Meienhofer, J. *J. Am. Chem. Soc.* **1974**, *96*, 4978 [211].
[1339] Carpenter, F. H.; Gish, D. T. *J. Am. Chem. Soc.* **1952**, *74*, 3818 [211].
[1340] Jackson, A. E.; Johnstone, R. A. W. *Synthesis* **1976**, 685 [211].
[1341] House, H. O.; Carlson, R. G. *J. Org. Chem.* **1964**, *29*, 74 [211, 212].
[1342] Fieser, L. F. *J. Am. Chem. Soc.* **1953**, *75*, 4377 [211].
[1343] Hansley, V. L. *J. Am. Chem. Soc.* **1935**, *57*, 2303 [212, 309].
[1344] Allinger, N. L. *Org. Synth., Collect. Vol.* **1963**, *4*, 840 [212].
[1345] Prelog, V.; Frenkiel, L.; Kobelt, M.; Barman, P. *Helv. Chim. Acta* **1947**, *30*, 1741 [212].
[1346] Blomquist, A. T.; Burge, R. E.; Sucsy, A. C. *J. Am. Chem. Soc.* **1952**, *74*, 3636 [212].
[1347] Bouveault, L.; Blanc, G. *Bull. Soc. Chim. Fr.* **1904**, *[3]*, *31*, 666, 1203; *C. R. Hebd. Seances Acad. Sci.* **1903**, *136*, 1676 [213].
[1348] Ford, S. G.; Marvel, C. S. *Org. Synth., Collect. Vol.* **1943**, *2*, 372 [213].
[1349] Manske, R. H. *Org. Synth., Collect. Vol.* **1943**, *2*, 154 [213].
[1350] Adkins, H.; Gillespie, R. H. *Org. Synth., Collect. Vol.* **1955**, *3*, 671 [213, 214].
[1351] Adkins, H. *Org. React. (N. Y.)* **1954**, *8*, 1 [213, 214, 219].
[1352] Adkins, H.; Folkers, K. *J. Am. Chem. Soc.* **1931**, *53*, 1095 [213, 217].
[1353] Lazier, W. A.; Hill, J. W.; Amend, W. Y. *Org. Synth., Collect. Vol.* **1943**, *2*, 325 [213].
[1354] Moffet, R. B. *Org. Synth., Collect. Vol.* **1963**, *4*, 834 [215].
[1355] Brown, H. C.; Narasimhan, S. *J. Org. Chem.* **1982**, *47*, 1604 [215, 216].
[1356] Brown, M. S.; Rapoport, H. *J. Org. Chem.* **1963**, *28*, 3261 [215, 216, 217, 219].
[1357] Santaniello, E.; Ferraboschi, P.; Sozani, P. *J. Org. Chem.* **1981**, *46*, 4584 [216].
[1358] Brown, H. C.; Narasimhan, S.; Yong Moon, Choi *J. Org. Chem.* **1982**, *47*, 4702 [216].
[1359] Brown, H. C.; Yong Moon, Choi; Narasimhan, S. *J. Org. Chem.* **1982**, *47*, 3153 [216, 232, 233].
[1360] Firestone, R. A. *Tetrahedron Lett.* **1967**, 2629 [216].

[*1361*]   Villani, F. J.; King, M. S.; Papa, D. *J. Org. Chem.* **1953**, *18*, 1578 [217].
[*1362*]   McClellan, W. R.; Connor, R. *J. Am. Chem. Soc.* **1941**, *63*, 484 [217].
[*1363*]   Marshall, J. A.; Carroll, R. D. *J. Org. Chem.* **1965**, *30*, 2748 [218, 242].
[*1364*]   Kadin, S. B. *J. Org. Chem.* **1966**, *31*, 620 [218, 242].
[*1365*]   Zurqiyah, A.; Castro, C. E. *Org. Synth., Collect. Vol.* **1973**, *5*, 993 [219].
[*1366*]   Snyder, E. I. *J. Org. Chem.* **1967**, *32*, 3531 [217, 219].
[*1367*]   Miller, A. E. G.; Biss, J. W.; Schwartzman, L. H. *J. Org. Chem.* **1959**, *24*, 627 [219, 239, 305].
[*1368*]   Schmidt, O. *Chem. Ber.* **1931**, *64*, 2051 [219].
[*1369*]   de Benneville, P. L.; Connor, R. *J. Am. Chem. Soc.* **1940**, *62*, 283 [219, 220].
[*1370*]   Dornow, A.; Bartsch, W. *Chem. Ber.* **1954**, *87*, 633 [219].
[*1371*]   Paul, R.; Joseph, N. *Bull. Soc. Chim. Fr.* **1952**, 550 [217, 219].
[*1372*]   Bouveault, L.; Blanc, G. *Bull. Soc. Chim. Fr.* **1904**, [*3*], *31*, 1209; *C. R. Hebd. Seances Acad. Sci.* **1903**, *137*, 328 [217, 220].
[*1373*]   Bates, E. B.; Jones, E. R. H.; Whiting, M. C. *J. Chem. Soc.* **1954**, 1854 [220].
[*1374*]   Richards, E. M.; Tebby, J. C.; Ward, R. S.; Williams, D. H. *J. Chem. Soc. C* **1969**, 1542 [220].
[*1375*]   Hardegger, E.; Montavon, R. M. *Helv. Chim. Acta* **1946**, *29*, 1203 [220].
[*1376*]   Paleta, O.; Ježek, R.; Dĕdek, V. *Collect. Czech. Chem. Commun.* **1983**, *48*, 766 [221].
[*1377*]   Weizmann, A. *J. Am. Chem. Soc.* **1949**, *71*, 4154 [221].
[*1378*]   Feuer, H.; Kucera, T. J. *J. Am. Chem. Soc.* **1955**, *77*, 5740 [222].
[*1379*]   von Pechman, H. *Chem. Ber.* **1895**, *28*, 1847 [222].
[*1380*]   Darapsky, A.; Prabhakar, M. *Chem. Ber.* **1912**, *45*, 1654, 2622 [222].
[*1381*]   Moore, A. T.; Rydon, H. N. *Org. Synth., Collect. Vol.* **1973**, *5*, 586 [222].
[*1382*]   Adkins, H.; Wojcik, B.; Covert, L. W. *J. Am. Chem. Soc.* **1942**, *55*, 1669 [223].
[*1383*]   Karrer, P.; Portmann, P.; Suter, M. *Helv. Chim. Acta* **1948**, *31*, 1617 [223].
[*1384*]   Nerdel, F.; Frank, D.; Barth, G. *Chem. Ber.* **1969**, *102*, 395 [223].
[*1385*]   Deol, B. S.; Ridley, D. D.; Simpson, G. W. *Austr. J. Chem.* **1976**, *29*, 2459 [224, 225, 236].
[*1386*]   Nakamura, K.; Kondo, S.-i.; Kawai, Y.; Ohno, A. *Tetrahedron Lett.* **1991**, *32*, 7075 [224, 321].
[*1387*]   Noyori, R.; Ohkuma, T.; Kitamura, M.; Takaya, H. *J. Am. Chem. Soc.* **1987**, *109*, 5856 [225, 226].
[*1388*]   Taniguchi, M.; Fujii, H.; Oshima, K.; Utimoto, K. *Tetrahedron* **1993**, *49*, 11169 [225].
[*1389*]   Nakamura, K.; Miyai, T.; Nozaki, K.; Ushio, K.; Oka, S.; Ohno, A. *Tetrahedron Lett.* **1986**, *27*, 3155 [225].
[*1390*]   Spiliotis, V.; Papahatjis, D.; Ragoussis, N. *Tetrahedron Lett.* **1990**, *31*, 1615 [225, 226].
[*1391*]   Seebach, D.; Eberle, M. *Synthesis* **1986**, 37 [225].
[*1392*]   Seebach, D.; Roggo, S.; Maetzke, T.; Braunschweiger, H.; Cercus, J.; Krieger, M. *Helv. Chim. Acta* **1987**, *70*, 1605 [225].
[*1393*]   Nutaitis, C. F.; Schultz, R. A.; Obaza, J.; Smith, F. X. *J. Org. Chem.* **1980**, *45*, 4606 [226].
[*1394*]   Howe, R.; McQuillin, F. J.; Temple, R. W. *J. Chem. Soc.* **1959**, 363 [226].
[*1395*]   Kraus, G. A.; Frazier, K. *J. Org. Chem.* **1980**, *45*, 4262 [226].
[*1396*]   Schmid, L.; Swoboda, W.; Wichtl, M. *Monatsh. Chem.* **1952**, *83*, 185 [227].
[*1397*]   Hartung, W. H.; Beaujon, J. H. R.; Cocolas, G. *Org. Synth., Collect. Vol.* **1973**, *5*, 376 [227].
[*1398*]   Putochin, N. J. *Chem. Ber.* **1923**, *56*, 2211 [227].
[*1399*]   Zambito, A. J.; Howe, E. E. *Org. Synth., Collect. Vol.* **1973**, *5*, 373 [227].
[*1400*]   Chang, Y.-T.; Hartung, W. H. *J. Am. Chem. Soc.* **1953**, *75*, 89 [227].
[*1401*]   Claus, C. J.; Morgenthau, J. L., Jr. *J. Am. Chem. Soc.* **1951**, *73*, 5005 [227].

[1402] Eliel, E. L.; Daignault, R. A. *J. Org. Chem.* **1964**, *29*, 1630 [228].
[1403] Bublitz, D. E. *J. Org. Chem.* **1967**, *32*, 1630 [228].
[1404] Wolfrom, M. L.; Karabinos, J. V. *J. Am. Chem. Soc.* **1946**, *68*, 724, 1455 [228, 299].
[1405] Liu, H.-J.; Bukownik, R. R.; Pednekar, P. R. *Synth. Commun.* **1981**, *11*, 599 [228].
[1406] Baddiley, J. *J. Chem. Soc.* **1950**, 3693 [228].
[1407] Weygand, F.; Eberhardt, G.; Linden, H.; Schäfer, F.; Eigen, I. *Angew. Chem.* **1953**, *65*, 525 [229].
[1408] Micovič, V. M.; Mihailovič, M. L. *J. Org. Chem.* **1953**, *18*, 1190 [229, 230, 231, 232].
[1409] Walborsky, H. M.; Baum, M.; Loncrini, D. F. *J. Am. Chem. Soc.* **1955**, *77*, 3637 [229].
[1410] Brown, H. C.; Tsukamoto, A. *J. Am. Chem. Soc.* **1961**, *83*, 2016, 4549 [229].
[1411] Málek, J.; Černý, M. *Synthesis* **1972**, 217 [230].
[1412] Ramegowda, N. S.; Modi, M. N.; Koul, A. K.; Bora, J. M.; Narang, C. K.; Mathur, N. K. *Tetrahedron* **1973**, *29*, 3985 [230].
[1413] Birch, A. J.; Cymerman-Craig, J.; Slayton, M. *Austr. J. Chem.* **1955**, *8*, 512 [230, 237].
[1414] Gilman, H.; Jones, R. G. *J. Am. Chem. Soc.* **1948**, *70*, 1281 [230].
[1415] Brown, H. C.; Kim, S. C. *Synthesis* **1977**, 635 [231].
[1416] Schön, I.; Szirtes, T.;Überhardt, T.; Csehi, A. *J. Org. Chem.* **1983**, *48*, 1916 [231].
[1417] Weygand, F.; Frauendorfer, E. *Chem. Ber.* **1970**, *103*, 2437 [231].
[1418] Wojcik, B.; Adkins, H. *J. Am. Chem. Soc.* **1934**, *56*, 2419 [231].
[1419] Sauer, J. C.; Adkins, H. *J. Am. Chem. Soc.* **1938**, *60*, 402 [231].
[1420] Wilson, C. V.; Stenberg, J. F. *Org. Synth., Collect. Vol.* **1963**, *4*, 564 [232].
[1421] Cope, A. C.; Ciganek, E. *Org. Synth., Collect. Vol.* **1963**, *4*, 339 [232].
[1422] Schindlbauer, H. *Monatsh. Chem.* **1969**, *100*, 1413 [232, 235, 236].
[1423] Borch, R. F. *Tetrahedron Lett.* **1968**, 61 [232, 233].
[1424] Umino, N.; Iwakuma, T.; Itoh, N. *Tetrahedron Lett.* **1976**, 763 [232].
[1425] Brown, H. C.; Heim, P. *J. Am. Chem. Soc.* **1964**, *86*, 3566; *J. Org. Chem.* **1973**, *38*, 912 [232].
[1426] Baille, T. B.; Tafel, J. *Chem. Ber.* **1899**, *32*, 68 [232, 234].
[1427] Galinovsky, F.; Stern, E. *Chem. Ber.* **1943**, *76*, 1034; **1944**, *77*, 132 [233].
[1428] Moffet, R. B. *Org. Synth., Collect. Vol.* **1963**, *4*, 354 [233].
[1429] Tafel, J.; Stern, M. *Chem. Ber.* **1900**, *33*, 2224 [233].
[1430] Kondo, Y.; Witkop, B. *J. Org. Chem.* **1968**, *33*, 206 [233, 234].
[1431] Merkel, W.; Mania, D.; Bormann, D. *Justus Liebigs Ann. Chem.* **1979**, 461 [234].
[1432] Lukeš, R.; Smetáčková, M. *Collect. Czech. Chem. Commun.* **1933**, *5*, 61 [234].
[1433] Lukeš, R.; Ferles, M. *Collect. Czech. Chem. Commun.* **1951**, *16*, 252 [234].
[1434] Butula, L.; Kolbah, D.; Butula, I. *Croat. Chem. Acta* **1972**, *44*, 481 [234].
[1435] Dunet, A.; Rollet, R.; Willemart, A. *Bull. Soc. Chim. Fr.* **1950**, 877 [234].
[1436] Herbert, R. M.; Shemin, D. *Org. Synth., Collect. Vol.* **1943**, *2*, 491 [234].
[1437] Shamma, M.; Rosenstock, P. D. *J. Org. Chem.* **1961**, *26*, 718 [235].
[1438] Holík, M.; Tesařová, A.; Ferles, M. *Collect. Czech. Chem. Commun.* **1967**, *32*, 1730 [235].
[1439] Ito, Y.; Yamaguchi, M. *Tetrahedron Lett.* **1983**, *24*, 5385 [236].
[1440] Fujita, M.; Hiyama, T. *J. Am. Chem. Soc.* **1985**, *107*, 8294 [236].
[1441] Lukeš, R.; Černý, M. *Collect. Czech. Chem. Commun.* **1959**, *24*, 1287 [236].
[1442] Kornet, M. J.; Poo, A. T.; Sip Ie, T. *J. Org. Chem.* **1968**, *33*, 3637 [236].

[1443]   Gribble, G. W.; Jasinski, J. M.; Pellicone, J. T.; Panetta, J. A. *Synthesis* **1978**, 766 [237].
[1444]   Brovet, D. *Ark. Kem.* **1948**, *20*, 70 [237, 238].
[1445]   Kornfeld, E. C. *J. Org. Chem.* **1951**, *16*, 131 [237].
[1446]   Cronyn, M. W.; Goodrich, J. E. *J. Am. Chem. Soc.* **1952**, *74*, 3936 [237].
[1447]   Sundberg, R. J.; Walters, C. P.; Bloom, J. D. *J. Org. Chem.* **1981**, *46*, 3730 [237].
[1448]   Karady, S.; Amato, J. S.; Weinstock, L. M.; Sletzinger, M. *Tetrahedron Lett.* **1978**, 403 [238].
[1449]   Staudinger, H. *Chem. Ber.* **1908**, *41*, 2217 [238].
[1450]   Williams, J. W.; Witten, C. H.; Krynitsky, J. A. *Org. Synth., Collect. Vol.* **1955**, *3*, 818 [238].
[1451]   von Braun, J.; Rudolph, W.; Kröper, H.; Pinkernelle, W. *Chem. Ber.* **1934**, *67*, 269, 1735 [237].
[1452]   Nicolaus, B. I. R.; Mariani, L.; Gallo, G.; Testa, E. *J. Org. Chem.* **1961**, *26*, 2253 [237].
[1453]   Newman, M. S.; Caflisch, E. G., Jr. *J. Am. Chem. Soc.* **1958**, *80*, 862 [238].
[1454]   Roberts, J. D. *J. Am. Chem. Soc.* **1951**, *73*, 2959 [238].
[1455]   Sprecher, M.; Feldkimel, M.; Wilchek, M. *J. Org. Chem.* **1961**, *26*, 3664 [238].
[1456]   Hesse, G.; Schroedel, R. *Justus Liebigs Ann. Chem.* **1957**, *607*, 24 [239].
[1457]   Brown, H. C.; Garg, C. P. *J. Am. Chem. Soc.* **1964**, *86*, 1085 [238, 239].
[1458]   Gardner, T. S.; Smith, F. A.; Wenis, E.; Lee, J. *J. Org. Chem.* **1951**, *16*, 1121 [239].
[1459]   Fuson, R. C.; Emmons, W. D.; Tull, R. *J. Org. Chem.* **1951**, *16*, 648 [239, 242].
[1460]   Backeberg, O. G.; Staskun, B. *J. Chem. Soc.* **1962**, 3961 [239].
[1461]   Staskun, B.; Backeberg, O. G. *J. Chem. Soc.* **1964**, 5880 [239].
[1462]   van Es, T.; Staskun, B. *J. Chem. Soc.* **1965**, 5775 [239].
[1463]   Freifelder, M. *J. Am. Chem. Soc.* **1960**, *82*, 2386 [240].
[1464]   Albert, A.; Magrath, D. *J. Chem. Soc.* **1944**, 678 [240].
[1465]   Schreifels, J. A.; Maybury, D. C.; Swartz, W. E., Jr. *J. Org. Chem.* **1981**, *46*, 1263 [240].
[1466]   Whitemore, F. C. *J. Am. Chem. Soc.* **1944**, *66*, 725 [240].
[1467]   Biggs, B. S.; Bishop, W. S. *Org. Synth., Collect. Vol.* **1955**, *3*, 229 [240].
[1468]   Greenfield, H. *Ind. Eng. Chem. Prod. Res. Dev.* **1967**, *6*, 142 [240].
[1469]   Suter, C. M.; Moffett, E. W. *J. Am. Chem. Soc.* **1934**, *56*, 487 [241].
[1470]   Nakamura, K.; Fujii, M.; Oka, S.; Ohno, A. *Chem. Lett.* **1985**, 523 [47, 93, 320].
[1471]   Schnider, O.; Hellerbach, J. *Helv. Chim. Acta* **1950**, *33*, 1437 [241].
[1472]   Bergmann, E. D.; Ikan, R. *J. Am. Chem. Soc.* **1956**, *78*, 1482 [241].
[1473]   Osborn, M. E.; Pegues, J. F.; Paquette, L. A. *J. Org. Chem.* **1980**, *45*, 167 [242].
[1474]   Profitt, A.; Watt, D. S.; Corey, E. J. *J. Org. Chem.* **1975**, *40*, 127 [242].
[1475]   Toda, F.; Iida, K. *Chem. Lett.* **1976**, 695 [242].
[1476]   Kornblum, N.; Fishbein, L. *J. Am. Chem. Soc.* **1955**, *77*, 6266 [242].
[1477]   Burger, A.; Hornbaker, E. D. *J. Am. Chem. Soc.* **1952**, *74*, 5514 [242].
[1478]   Fields, M.; Walz, D. E.; Rothchild, S. *J. Am. Chem. Soc.* **1951**, *73*, 1000 [242].
[1479]   Ferris, J. P.; Sanchez, R. A.; Mancuso, R. W. *Org. Synth., Collect. Vol.* **1973**, *5*, 32 [242].
[1480]   Zartman, W. H.; Adkins, H. *J. Am. Chem. Soc.* **1932**, *54*, 3398 [245].
[1481]   Becker, W. E.; Cox, S. E. *J. Am. Chem. Soc.* **1960**, *82*, 6264 [245].
[1482]   Bordwell, F. G.; Douglas, M. L. *J. Am. Chem. Soc.* **1966**, *88*, 993 [245].
[1483]   Reissert, A. *Chem. Ber.* **1913**, *46*, 1484 [234].
[1484]   Sigimura, H.; Yoshida, K. *J. Org. Chem.* **1993**, *58*, 4484 [224, 225].
[1485]   Takeshita, M.; Yoshida, S.; Kohno, Y. *Heterocycles* **1994**, *37*, 553 [49, 98].

[*1486*]   Wipf, B.; Kupfer, E.; Bertazzi, R.; Leuenberger, H. G. W. *Helv. Chim. Acta* **1983**, *66*, 485 [225].

[*1487*]   Singh, V. K. *Synthesis* **1992**, 605 [50].

[*1488*]   Fort, Y.; Vanderesse, R.; Caubere, P. *Tetrahedron Lett.* **1985**, *26*, 3111 [114].

[*1489*]   Hutchins, R. O.; Milewski, C. A.; Maryanoff, B. E. *Org. Synth., Collect. Vol.* **1988**, *6*, 376 [82, 125].

[*1490*]   Brown, H. C.; Park, W. S.; Cho, B. T.; Ramachandran, P. V. *J. Org. Chem.* **1987**, *52*, 5406 [50].

[*1491*]   Brown, H. C.; Narasimhan, S. *J. Org. Chem.* **1984**, *49*, 3891 [215].

[*1492*]   Baba, N.; Oda, J.; Inouye, Y. *Chem. Commun.* **1980**, 815 [224].

[*1493*]   Seki, M.; Baba, N.; Oda, J.; Inouye, Y. *J. Am. Chem. Soc.* **1981**, *103*, 4613 [224].

[*1494*]   Seki, M.; Baba, N.; Oda, J.; Inouye, Y. *J. Org. Chem.* **1983**, *48*, 1370 [224].

[*1495*]   Nakamura, K.; Ohno, A.; Oka, S. *Tetrahedron Lett.* **1983**, *24*, 3335 [210].

[*1496*]   Sakai, T.; Nakamura, T.; Fukuda, K.; Ameno, E.; Utaka, M.; Takeda, A. *Bull. Chem. Soc. Jpn.* **1986**, *59*, 3185 [49].

[*1497*]   Herzog, H. L.; Jevnik, M. A.; Perlman, P. L.; Nobile, A.; Hershberg, L. B. *J. Am. Chem. Soc.* **1953**, *75*, 266 [167].

[*1498*]   Dornow, A.; Gellrich, M. *Justus Liebigs Ann. Chem.* **1955**, *594*, 177 [93].

[*1499*]   Potin, D.; Dumas, F.; D'Angelo, J. *J. Am. Chem. Soc.* **1990**, *112*, 3483 [50, 218].

[*1500*]   Soderquist, J. A. *Aldrichimica Acta* **1991**, *24*, 15 [40].

[*1501*]   Mabic, S.; Castagnoli, N., Jr. *J. Org. Chem.* **1996**, *61*, 309 [235].

BIBLIOGRAPHY

REVIEWS IN *ORGANIC REACTIONS*

1. "The Clemmensen Reduction." Martin, E. L. *Org. React. (N. Y.)* **1942**, *1*, 155.
2. "The Meerwein–Ponndorf–Verley Reduction (Reduction with Aluminum Alkoxides)." Wilds, A. L. *Org. React. (N. Y.)* **1944**, *2*, 178.
3. "Replacement of the Aromatic Primary Amino Group by Hydrogen." Kornblum, N. *Org. React. (N. Y.)* **1944**, *2*, 262.
4. "The Preparation of Amines by Reductive Alkylation." Emerson, W. S. *Org. React. (N. Y.)* **1948**, *4*, 174.
5. "The Rosenmund Reduction." Mosettig, E.; Mozingo, R. *Org. React. (N. Y.)* **1948**, *4*, 362.
6. "The Wolff–Kishner Reduction." Todd, D. *Org. React. (N. Y.)* **1948**, *4*, 378.
7. "The Leuckart Reaction." Moore, M. L. *Org. React. (N. Y.)* **1949**, *5*, 301.
8. "Reductions by Lithium Aluminum Hydride." Brown, W. G. *Org. React. (N. Y.)* **1951**, *6*, 469.
9. "Hydrogenolysis of Benzyl Groups Attached to Oxygen, Nitrogen, or Sulfur." Hartung, W. H.; Simonoff, R. *Org. React. (N. Y.)* **1953**, *7*, 263.
10. "Catalytic Hydrogenation of Esters to Alcohols." Adkins, H. *Org. React. (N. Y.)* **1954**, *8*, 1.
11. "The Synthesis of Aldehydes from Carboxylic Acids." Mosettig, E. *Org. React. (N. Y.)* **1954**, *8*, 218.
12. "Desulfurization with Raney Nickel." Pettit, G. R.; van Tamelen, E. E. *Org. React. (N. Y.)* **1962**, *12*, 356.
13. "The Zinin Reduction of Nitroarenes." Porter, H. K. *Org. React. (N. Y.)* **1973**, *20*, 455.
14. "Clemmensen Reduction of Ketones in Anhydrous Organic Solvents." Vedejs, E. *Org. React. (N. Y.)* **1975**, *22*, 401.
15. "Reduction and Related Reactions of α,β-Unsaturated Compounds with Metals in Liquid Ammonia." Caine, D. *Org. React. (N. Y.)* **1976**, *23*, 1.
17. "Homogeneous Hydrogenation Catalysts in Organic Synthesis." Birch, A. J.; Williamson, D. H. *Org. React. (N. Y.)* **1976**, *24*, 1.
18. "Reductive Dehalogenation of Polyhaloketones with Low-Valent Metals and Related Reducing Agents." Noyori, R.; Hayakawa Y. *Org. React. (N. Y.)* **1983**, *29*, 4.
19. "Reductions by Metal Alkoxyaluminum Hydrides." Málek, J. *Org. React. (N. Y.)* **1985**, *34*, 1.
20. Rylander, P. N. *Hydrogenation Methods;* Academic: New York, 1985.
21. Rylander, P. N. In *Engelhard Catalysts and Precious Metals Chemicals Catalog;* Engelhard Corporation: Newark, NJ, 1985.
22. Štrouf, O.; Čásenský, B.; Kubánek, V. *Sodium dihydrido-bis(2-methoxyethoxy)-aluminate (SDMA);* Elsevier: New York, 1985.
23. Seyden-Penne, J. *Reductions by the Alumino- and Borohydrides in Organic Synthesis;* VCH Publishers: New York, 1991.
24. *Comprehensive Organic Synthesis;* Trost, B. M.; Fleming, I., Eds.; Pergamon: Oxford, England, 1991; Vol. 8.

MONOGRAPHS

1. Hudlický, M. *Reduction and Oxidation* (in Czech); Academia: Prague, Czechoslovakia, 1953.
2. Micovič, V. M.; Mihailovič, M. L. *Lithium Aluminum Hydride in Organic Chemistry;* Naučna Knjiga: Beograd, Yugoslavia, 1955.
3. Gaylord, N. G. *Reduction with Complex Metal Hydrides;* Wiley-Interscience: New York, 1956.
4. Rudinger, J.; Ferles, M. *Lithium Aluminum Hydride and Kindred Reagents in Organic Chemistry* (in Czech); Academia: Prague, Czechoslovakia, 1956.
5. Augustine, R. L. *Catalytic Hydrogenation;* Marcel Dekker: New York, 1965.
6. Zymalkovsky, F. *Katalytische Hydrierungen im Organisch-Chemischen Laboratorium* (in German); F. Enke: Stuttgart, Germany, 1965.
7. Rylander, P. N. *Catalytic Hydrogenation over Platinum Metals;* Academic: New York, 1967.
8. Augustine, R. L. *Reduction;* Marcel Dekker: New York, 1968.
9. Freifelder, M. *Practical Catalytic Hydrogenation;* Wiley-Interscience: New York, 1971.
10. House, H. O. *Modern Synthetic Reactions;* W.A. Benjamin: Menlo Park, CA, 1972.
11. James, B. R. *Homogeneous Hydrogenation;* John Wiley and Sons: New York, 1973.
12. Brown, H. C. *Organic Syntheses via Boranes;* Wiley-Interscience: New York, 1975.
13. Doyle, M. P.; West, C. T. Eds.; *Benchmark Papers in Organic Chemistry; Vol. VI, Stereoselective Reductions,* Dowden, Hutchison & Ross: Stroudsburg, PA, 1976.
14. McQuillin, F. J. *Homogeneous Hydrogenation in Organic Chemistry;* D. Reidel: Boston, MA, 1976.
15. Pizey, J. S. *Lithium Aluminium Hydride;* Ellis Horwood: Chichester, England, 1977.
16. Freifelder, M. *Catalytic Hydrogenations in Organic Synthesis. Procedures and Commentary;* Wiley-Interscience: New York, 1978.
17. Hájos, A. *Complex Hydrides and Related Reducing Agents in Organic Synthesis;* Elsevier: New York, 1979.
18. Rylander, P. N. *Catalytic Hydrogenation in Organic Syntheses;* Academic: New York, 1979.
19. Houben-Weyl *Methoden der Organischen Chemie: Reduktion* (in German); Müller, E.; Bayer, O., Eds.: G. Thieme: Stuttgart, Germany, 1980, 1981; Vols. IV/Ic and IV/Id.
20. *Comprehensive Organic Synthesis;* Fleming, I., Ed.; Pergamon: Oxford, England, 1991.

CHAPTERS IN *SYNTHETIC REAGENTS* MONOGRAPHS

Vol. 1: *Lithium Aluminium Hydride*; Pizey, J. S., Ed.; Ellis Horwood, Chichester, England, 1974; pp 101–294.

Vol. 2: *Raney Nickel;* Pizey, J. S., Ed.; Ellis Horwood, Chichester, England, 1974; pp 175–311.

Vol. 3: *Diborane;* Pizey, J. S., Ed.; Ellis Horwood, Chichester, England, 1977; pp 1–191.

SAFETY

Committee on Hazardous Substances in the Laboratory. *Prudent Practices for Handling Hazardous Chemicals in Laboratories;* National Academy: Washington, DC, 1981.

AUTHOR INDEX

The numbers following the author's names refer to the List of References (pp 323–362) where the text page on which that work is cited appears in square brackets after the reference.

SUBJECT INDEX

The index lists reagents, types of reactions, and types of compounds. Only those individual compounds shown in equations and procedures are listed. **Boldface** numbers refer to pages in Procedures (pp **291–322**).

Copy editing by Janet S. Dodd
Production by Paula M. Bérard
Jacket design by Amy O'Donnell

Typeset by Angie Miller, Blacksburg, VA
Printed and bound by Maple Press Company, York, PA